Biology
The Unity and Diversity of Life

David Kirk
Washington University

Ralph Taggart
Michigan State University

Cecie Starr
Del Mar, California

Wadsworth Publishing Company, Inc.
Belmont, California

The cover photograph shows bald cypress, Spanish moss, and a common egret on the wing, captured in time and space by James A. Kern at Fisheating Creek, Florida.

Biology Editor: Jack C. Carey

Production Editor: Catherine Aydelott

Designer: Ann Wilkinson

Illustrators: Darwen Hennings, Vally Hennings

Printed in the United States of America

2 3 4 5 6 7 8 9 10————82 81 80 79 78

Library of Congress Cataloging in Publication Data

Kirk, David, 1934–
  Biology.

  Bibliography: p.
  Includes index.
  1.Biology.   I.Taggart, Ralph, joint author.
  II.Starr, Cecie, joint author.   III.Title.
QH308.2.K57   574   77-17161
ISBN 0-534-00540-3

Preface

More and more frequently, biologists are becoming less and less reluctant to admit poetry into the classroom—the poetry of life, its beauty, its intensity. And that is not a bad thing. For some time, "life" has been spread out on the examination table, poked at, dissected into smaller and smaller bits until all that remained were electrons. Where in the electrons were the climax forests, the caribou thundering across the tundra, the cry and squirmings of the newborn child? One can readily commiserate with the molecular biologist Albert Szent-Györgyi, who looked up abruptly from his probings with the realization that somewhere along the line, "life" had slipped through his fingers. Yet today we have a profound legacy from those explorations of ever deeper levels of biological organization. We have at last penetrated the seeming paradox of life—its breathtaking complexity and its underlying simplicity. We now *know* where electrons fit in the climax forest, the caribou, the ongoing flow of life. It is a time of converging insights in biological thought; a time when investigations on many different levels resonate with one another; and a time for books to reflect it.

With this as our premise, the book we have written describes all events in the living world as part of a greater story. All of life is seen to share a common molecular beginning. Since the time of origin, life of any form has required a constant supply of energy and materials to maintain itself and reproduce. The amount and kinds of these environmental resources have changed over time, which has meant that living things have had to interact with one another in competition for them. Finally, diversity in form and behavior has arisen largely because of these interactions. For in group after group, variant forms have appeared that in some way were more adaptive to prevailing conditions. They tended to be the ones that survived and reproduced; they became increasingly represented in generations to follow. *Thus there has been unity in origins and in fundamental requirements for resources, and there has been an evolving panorama of diversity in the means by which these requirements have been met.*

Underlying the unity and diversity of life are certain principles of energy flow and evolution. Despite our unique human inventions that give us the illusion of being somehow beyond their reach, these principles ultimately govern all that we do, just as they govern all of life. This is the theme that threads through every chapter in the book.

With this theme to guide us, we have written a current, balanced survey of the mainstreams in the study of life. At the same time, we show how the biological perspective can add dimension to the way that students perceive the world and their place in it. We first present basic principles in enough depth to provide understanding of how the living world is supposed to work. Then, where appropriate, we are able to discuss such issues as overpopulation and pollution not as isolated problems *but as deviations from the working model.* It simply is not very instructive to rattle off the biological horror stories of our time as if they evolved in a vacuum. For billions of years, organisms have interacted with one another and with their environment; and it is through interaction that they achieved a state of dynamic stability. If there is one thing biology can give to this and future generations, it is an understanding that instability is an inevitable prospect whenever we ignore the fundamental meaning of those interactions.

In writing this book, we relied not only on our own judgment but on the teaching and research experience of more than sixty reviewers. Despite the broad range of interests and individual teaching approaches, more often than not we found strong agreement on how much coverage should be devoted to different topics, and on which applications might be most informative for students. Generally, examples relating to the human organism are presented not as the focus but as the counterpoint to chapters on basic biology. For instance, the discussions of how cells trap and use resources conclude with a look at what happens when not enough of the right kinds of resources are available for the millions of cells in each of *4 billion* human bodies. The seemingly remote world of the cell is related at once to the larger world in which starvation is rampant.

Of course, there probably never will be widespread agreement on what constitutes the "right" depth of coverage for photosynthesis, glycolysis and respiration, and human physiology. That is why we developed two tracks through this material. For those who desire minimal coverage, Chapter Three briefly presents photosynthesis, glycolysis, and respiration in an ecological context; they are shown to be the foundation for energy flow and materials cycling through the biosphere. Chapters Five and Six begin with an overview of the metabolic pathways of these strategies. Then, for instructors and students who would like more coverage, details of metabolism follow the overview. Similarly, the chapter on animal integration first describes the human nervous system and endocrine system. It continues with case studies that range from the integration involved in glucose metabolism to the stunning efficiency of the immune response. As formidable as such topics often appear to be, they are described in this book in a lively, accessible manner. Even so, *the second track through details is optional.*

As we were writing the manuscript, we were developing illustrations so that they could be reviewed together for accuracy. In selecting photographs, we used the biologist's eye to convey a feeling about life, and an awareness of its basic principles. We also maintained an abiding sympathy for students in terms of the way we illustrate difficult concepts. There is no reason (for instance) why photosynthesis or the human nervous system should be drawn as a schematic straight out of an electrical engineering handbook. And it is no accident that full-color illustrations fall where they belong. They are not clumped together as a single block, isolated from the concepts they are supposed to illustrate. Beyond this, *all* illustrations have been positioned with care in relation to the text, for our vision is most clear when we have concepts to see with. Without any idea of what the carbon cycle means, for instance, students might look at the photograph on pages 50–51 and see only the curiously heaped remains of a once-living thing. But if they are simultaneously reading about the concept of materials reuse, their eyes undoubtedly will move across the field, following the sweep of grass peppered with white flowers blooming—for these are living things, too, and vital links in the cycling of materials through the biosphere.

Even more basic study aids have been built into the text. Key words are printed in darker (boldface) type, and words of subordinate importance are printed in italics. In addition, italicized sentences, lists with boldface numbers, and key concepts are clearly displayed within the text. When students come across one of these devices, they can stop and think briefly about what they have just read. It is our signal to them that understanding the word or sentence is an important step in the accumulation of ideas. Throughout the text, strategically placed paragraphs provide orientation in terms of where students are in the formulation of concepts, and where they are headed with them; the first three paragraphs of Chapter Thirteen are examples. Key titles and subtitles within chapters act as signposts along conceptual roads; see, for instance, the Chapter Four subtitles briefly stating the function of cytoplasmic organelles. The glossary distills the main definitions that occur in the text. The index is comprehensive, simply because students may find a door to the text more quickly among finer divisions of topics. Finally, sections called "Perspectives" have been written to provide just that—perspective on the meaning of each subject, to bring in not only the past but extrapolations of possible futures for us and all of life.

In sum, we have written this book as a celebration of the world of life—of its lighter moments and its profound moments, of its fidelity to basic principles.

—D. Kirk, R. Taggart, C. Starr

Acknowledgments

We are indebted to our reviewers, who kept us attentive both to points of accuracy in their respective fields and to broader questions of pedagogical approach. We give special thanks to Laura Mays for her summary of recombinant DNA research, and to Herman Wiebe for his additions on plant physiology. The list on the following page is our way of thanking all those individuals who will know, in reading through the book, where they have left their imprint.

We are also indebted to many biologists and biologists-at-heart who contributed illustrations. Special thanks go to Roger Burnard of Ohio State University. His sense of the beauty inherent in life, so evident in his photographs, coincides with our own and his work has greatly enhanced the text. We thank Ronan O'Rahilly, M.D., Director of the Carnegie Embryological Laboratories, who kindly helped us prepare the illustrations on human embryonic development. We also thank Gary Grimes of Hofstra University. Early on we asked Dr. Grimes for some of his outstanding micrographs of ciliates. Our mutual interest in the potential of students soon became apparent, one thing led to another, and we ended up agreeing to give his graduate students in microscopy a chance to contribute to biology education by preparing original micrographs for the book. Even during the spring break, this enthusiastic group went to the laboratory voluntarily to pursue their assignment. You will see their work in these pages, along with the work of master microscopists.

We wish finally to thank some wonderful people at Wadsworth: Jim McDaniel for his continuing support; Cathy Aydelott for expertly routing this complex book through editorial production; Jean Graziano for her smooth liaison work; and Ann Wilkinson for giving so generously of her book design and production experience. The outcome is testimony to Ann's talent for making textbooks as beautiful as they are informative. We are grateful to Darwen and Vally Hennings for their sensitivity and skill in transforming our original sketches into final form, and to Howard Harrison and Dan Whiteman for their professionalism during the eleventh hour of book production. But most of all we wish to thank Jack Carey, biology editor at Wadsworth, for his unwavering encouragement and guidance. From the book's conception, through good times and bad times, through interpretation of myriad reviews, his enthusiasm, his subtle persuasion, and a few well-aimed baseball bats helped resolve issues at hand in ways that will, in the final analysis, best serve the student. His competence in book development is extraordinary in publishing. In no small way, this is as much his book as it is ours.

And for you, Lisa Starr and Denise Starr, who helped keep the reading level accessible, we accord all the respect due to young biologists in the making.

Reviewers

Donald Abbott, Stanford University
Kenneth Armitage, University of Kansas
George Ball, Gaston College
Walter Becker, Washington State University
Marjorie Behringer, University of North Dakota
James Bonner, California Institute of Technology
Richard Boohar, University of Nebraska
Ben Dolbeare, Lincoln Land Community College
Robert Grey, University of California, Davis
Frank Einhellig, University of South Dakota
Gordon Evans, Tufts University
Harold Eversmeyer, Murray State University
William Fennel, East Michigan University
Dorothy Frosch, Central State University
Berdell Funke, North Dakota State University
Donald Garren, Lake Land College
Robert Grey, University of California, Davis
Harlo Hadow, Coe College
Bruce Haggard, Hendrix College
Ted Hanes, California State University, Fullerton
Laszlo Hanzely, Northern Illinois University
Vernon Hendricks, Brevard Community College
James Henry, Illinois Central College
Joseph Hindman, Washington State University
J. Michael Jones, Emory and Henry College
Arnold Karpoff, University of Louisville
Norman Kerr, University of Minnesota
Charles Krebs, University of British Columbia
Frederick Landa, Virginia Commonwealth University
Frank Lang, Southern Oregon State College
Daniel Lee, State University of New York, Plattsburgh

William Mason, Auburn University
Laura Mays, Occidental College
Robert Macey, University of California, Berkeley
Francis McCarthy, California State College, Dominguez Hills
Jerry McClure, Miami University, Oxford, Ohio
Margaret McElhinney, Ball State University
G. Tyler Miller, Jr., St. Andrews Presbyterian College
John Minnich, University of Wisconsin
Michele Morek, Brescia College
Gerald Myers, South Dakota State University
Marian Reeve, Merritt College
Frank Salisbury, Utah State University
Robert Scagel, University of British Columbia
Ted Schwartz, University of California, San Diego
Valerie Seeley, Queensborough Community College
Arthur Shapiro, University of California, Davis
Frank Sivik, Broward Community College
Gordon Snyder, Schoolcraft College
David Stronck, Washington State University
M. Camilla Suddreth, Gaston College
Jack Thomas, Los Angeles Harbor College
Paul Thomson, Penn Valley Community College
Frank Toman, Western Kentucky University
Norman Tweed, Fort Steilacoom Community College
Warren Wagner, University of Michigan
George Washington, Jackson State University
Jon Weil, University of California, San Francisco
Jonathan Westfall, University of Georgia
Herman Wiebe, Utah State University
Stephen Wolfe, University of California, Davis
Richard Woodruff, West Chester State College

Contents in Brief

Table of Contents

Unit I

Principles Underlying the Diversity of Life

Chapter One

Figure 1.1 A tropical reef—monument to life in the restless seas.

The Diversity of Life

Imagine anyone saying "My interest is biology—the study of life." It sounds almost as preposterous as someone saying "My interest is astrophysics, the study of the origin and evolution of the entire universe." How could anyone really expect to learn all there is to know—must be known—in order to understand a subject so vast? Devoting a lifetime to it would be like dropping a rose petal into the Grand Canyon and waiting for the echo.

What has come about, of course, has been a sprouting of specialized interests. People are not just "biologists" now. They are specialists who must, of necessity, focus on one small aspect of the study of life—on one narrow topic within such already narrowed fields as biochemistry, cell biology, physiology, genetics, zoology, botany, behavior, and ecology. Even so, there is still so much to learn about in a given field that not much time can be spent sharing reflections, frustrations, and triumphs in exploring what Erwin Chargaff has called the lyrical intensity of life. There they sit—all those specialists above one part of the canyon or another, plucking petals to throw over the side.

Yet the amazing thing is that all those petals, dropping together, piling up over time, send back not only echoes but an occasional roar. Separate bits of understanding converge into insights that reverberate across the boundaries of specialization. The previously unexplained becomes explainable, parts fall into place, we sense the force of discovery. Thus today we hear the echoes of such concepts as "DNA" and "the genetic code" and "evolution" and "energy" ricocheting everywhere. No one can expect to be entirely untouched by their meaning—no one in science, no one in society. They are essential insights now, in the journey toward discovery for its own sake, and they are essential in the way we give direction to how we choose to live.

If we were to tell you, then, that in one course you can "learn biology," we would be liars. But if we were to help put a few of these insights, or basic principles, into your hands, and if you were to learn to use them—*not only as we have used them in this book but as a way of looking at the things you will come across in the world of your experience*—then you will have acquired something of enduring value.

We have chosen, as the framework for this book, a few basic principles that span all the specialized fields of biology because they span all of life in its many and varied expressions. *These are the principles of energy flow and evolution.*

They apply at every one of the levels of organization in nature, from molecules and cells to organisms and the entire biosphere. Some of these levels seem remote from everyday life. It is difficult, for instance, to picture how trillions of molecules make up the many billions of cells that form your body. More than this, it is difficult to imagine how so many separate parts could come together, let alone *stay* together for as long as you are alive. Even at the more tangible levels it is still difficult to accept some things; for example, to accept deep down that you have so much in common with earthworms in Missouri and flowers in Tibet—*with every one of the more than 10 million forms of life on earth!* But the links do exist, and they are explainable with principles of energy flow and evolutionary process.

In anticipation of our discussions about these principles, we will begin with the question of biological diversity, for it is an aspect of life that is the most readily apparent to all of us. No one need tell us, for instance, that fish live in water, humans live on land, and orange trees live in California, Florida, and Texas but not in Michigan or Maine. We know almost without thinking about it that life's diversity is linked somehow to diversity in environments. But let's go one step further, and compare the diversity that exists not between entirely different places but among organisms destined to live out their lives in the same place (Figure 1.2).

Life on a Coral Reef

Opening this chapter is a view of waves moving toward a tropical reef, lit momentarily by sunlight breaking through the clouds of an oncoming storm. As devoid of life as this place seems to be, it represents the earth's most ancient form of **ecosystem**—the interactions linking an entire community of organisms with one another and with their environment. It also represents one of the most diverse ecosystems of all.

Rising from the shallow sea floor is the reef's spine. Ages ago, tiny animals began anchoring themselves to rocks and sand beneath the warm, clear, agitated waters close to the edge of the land. Here they grew exuberantly, and reproduced, and died. But they left behind their skeletons as a foundation of limestone for more animals to build upon.

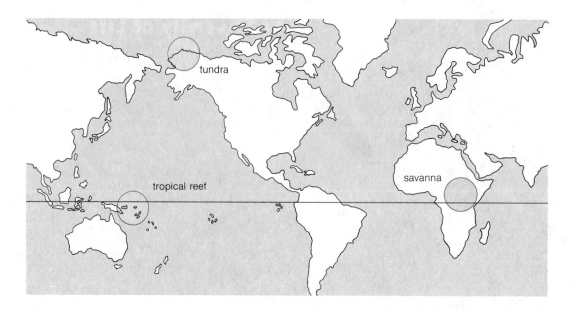

Figure 1.2 Where the photographic excursion in this chapter will take us.

tundra

tropical reef

savanna

As the residues and skeletons of generation after generation of organisms accumulated, the spine began growing vertically. But even as the reef was being constructed, relentless tides and currents were carving underwater caverns and ledges into it. It is on this sculptured monument to life that many diverse organisms have come to make their home.

The most spectacular are the master builders themselves: the corals—colorful, stonelike corals shaped like staghorns, domes, brains, flowers, mushrooms, cabbages, folded draperies, and fans. These corals are alive. They are meat-eating animals that send out tentacles from their hard chambers which ensnare organisms carried past on the turbulent tides.

Plants, too, thrive on the reef. On the ocean side, red algae encrust the coral foundation, from the bottom all the way to the top and across its jagged surface spine. In the shallow, calmer waters behind the reef, red algae give way to green. Then, in quiet inlets, the green forms give way to blue-green algae similar to the kind you see growing scumlike on stagnant ponds. The algae are no less than pastures upon which many small, transparent animals feed. These animals are food for still larger animals, such as brilliantly colored fish. In turn, the larger fish are food for toothy predators that lurk beneath dark ledges and in eroded hollows.

In other hollows are squat, predatory sea anemones, each with a fringelike crown of tentacles around its mouth. Each tentacle is studded with poisonous weapons for capturing prey (such as tiny fish). Yet cozying up to the deadly, waving tentacles *is* a certain fish! It is as edible as most others, but somehow it is protected from the weapons.

In fact, the two kinds of animals actually live together in a mutually rewarding way. The fish moves out to capture food, then it returns to its sedentary "housemate" to devour its meal. Scraps falling from its mouth land on the tentacles, which bend and shove each small bit into the anemone's mouth. By inadvertently sharing its meals in this way, the fish helps the anemone survive—just as the anemone helps the fish survive by giving it shelter from predators.

The patterns of predation are infinitely varied. Crawling over the surface of the reef are the sea stars, more commonly known as starfishes. When feeding, these amazing animals extend their stomach *outside* their body. The stomach is extended into coral chamber after coral chamber, where each hapless resident is digested before the stomach is pulled back out. In this case, however, predation is not so one-sided. It is true the corals are food for sea stars, nourishing them as they grow to maturity. But the time comes when the sea stars produce offspring—millions of immature forms (larvae) of sea stars. The larvae feed on microscopic algae, they grow—and they become food for the meat-eating corals! Now it is the corals' turn to feast, to grow, and to reproduce. In fact, they reproduce so much that they eventually repopulate the part of the reef that the parent sea stars had stripped clean. Just a few larvae escape. But in escaping, those few members of a new generation grow to become diner instead of dinner—and thereby initiate a new cycle of death and life.

Questions: Tens of thousands of individuals of hundreds of different kinds live in the underwater reef. Yet a mile out to sea, the waters are nearly vacant by comparison, a mere echo of the reef's explosive diversity. Why is such

Figure 1.3 (**a**) Underwater tropical reef. A few of the master builders: (**b**) pillar coral, (**c**) daisy coral, and (**d**) green tube coral.

Figure 1.4 Who eats whom on the reef. (**a**) Crown-of-thorns, a sea star that feasts on tiny corals. (**b**) Sea anemone, an animal with weapon-studded tentacles for ensnaring tiny animals floating past. (**c**) Sponges, with pores opened toward the oncoming food-laden currents. (**d**) Clownfish, curiously at home above the mouth of a sea anemone—a mouth through which other kinds of edible fish quickly disappear. (**e**) Green algae, plants that are food for various reef organisms. (**f**) Red algae, food for various animals (but not for this chambered nautilus, a shelled animal that swims expertly after shrimp and other prey). (**g**) A school of goatfish, which feed on small,

g

j

k

h

i

l

All photographs Douglas Faulkner

spineless animals on the sea floor. Goatfish are tasty to humans, also to large fish. (**h**) But some fish are not on the general menu. Here, a blue wrasse safely picks off and dines on parasites that prey on this large predatory fish. (**i**) Stone crab. Depending on the species, crabs eat plants, animals, and organic remains. The deadly moray (**l**) prefers meat. (**j**) Lion fish, with its fanned, poison-tipped spines warning away intruders. (**k**) Find the scorpion fish—a dangerous animal that lies camouflaged and motionless on the sea bottom, the better to surprise unsuspecting prey.

diversity concentrated in one small place? How have animals come to associate with or avoid one another—to hide in a predatory animal's tentacles, to eat some things but not others? How did predators become structurally attuned to their prey in such outlandishly effective ways as deployment of an inside-out stomach? With everything eating everything else, how does the community manage to remain relatively stable over time?

Before we try answering these questions, let's take a look at another kind of environment and see if the comparison yields any similarities or differences that might better help us in formulating some tentative answers.

Life in the Savanna

In the shadow of Kilimanjaro, a volcanic peak that rises high above the edge of the immense East African Rift Valley, grasslands sweep out to the northeast, first to Masai Amboseli and then on through Nairobi. This is the **savanna** —a region of warm grasslands punctuated with scattered stands of trees and shrubs. The natural vegetation of the African savanna supports more large ungulates (hoofed, plant-eating mammals) than any other environment in the world.

Roger K. Burnard

Figure 1.5 The Great Rift Valley, some 6,400 kilometers (4,000 miles) long. The sparsely wooded grasslands in this East African valley are home for a diverse array of animal life and, as you will read in later chapters, were the probable birthplace of the human species.

A dominant form is the Cape buffalo (Figure 1.6), a grazing animal that is unpredictable and most dangerous in its behavior. An adult male can weigh a ton. As you might well suspect, other animals do their utmost to avoid a whole Cape buffalo herd. But the savanna teems with many other, less formidable herds, including those of zebra and impala. These animals graze on grasses or short plants, usually at the edge of nearby woods. Here, too, is the tanklike rhinoceros, which dines on the leaves of low shrubs. And here, too, we find the giraffe, with legs and a neck so long that it can browse on the leaves of trees and taller shrubs, far above the reach of other ungulates.

a

b

c

Figure 1.6 A sampling of the ninety kinds of large plant-eating animals that live in the savanna—a clear example of diversity in a single environment. (**a**) A herd of Cape buffalo. Imagine yourself a predator this close to the herd and you get an idea of one of the benefits of group living. (**b**) Zebra mother and offspring. (**c**) Male and female impala on the alert, ready to take cover in the nearby woods. (**d**) The rhinoceros, another formidably decked-out plant eater. (**e**) The marvelous giraffe, browsing on vegetation high up.

d

e

All photographs Roger K. Burnard

a

b

c

Figure 1.7 Predators and scavengers of the savanna. (**a**) An adult lioness standing over a fresh kill. (**b**) A pride of cheetahs, large cats that hunt by day rather than by dusk. (**c**) Vultures waiting to clean up a carcass.

Buffalo, zebra, impala, rhinoceros, giraffe—these and *eighty-five other kinds of large plant-eating animals* live out their lives in the immense valley below Kilimanjaro. Somehow they manage to exist side by side in time, sharing available resources, moving with the seasons across the horizons— undulating westward and southward and back again as dry seasons follow rains, as scorched earth gives way to intense new life-sustaining growth.

All this beef on the hoof does not go unnoticed. Usually at dusk or afterward, usually near low-lying and dense vegetation that conceals her stealthy maneuverings, an adult lioness stalks the herds. Normally, only the weak and the careless will be taken during the intense heat of the afternoons, a time this large, furred cat considers more appropriate for dozing in some shaded place. Even a solitary Cape buffalo is not beyond the reach of this skilled huntress. But prey that are larger—and this includes adult

giraffe, elephant, rhinoceros, and hippopotamus—are rarely attacked. Neither are birds or small mammals such as hares given much attention.

A predator comparable in size to the lion and after some of the same prey is the cheetah (Figure 1.7). Unlike the lion, which hunts primarily at dusk, the cheetah hunts only during the day. And unlike the lion, which is no match for most of its prey in a prolonged chase (hence the reliance on stealth), the cheetah is successful precisely because of its speed and endurance. This large, long-legged cat is the fastest land animal on earth. It is capable of running down impala, gazelles, even hares in a flat-out race.

The large cats are not the only predators in this setting. Hyenas, too, are predators which kill in packs. But they serve still another function in the East African Rift Valley. Together with other scavengers such as vultures, jackals, and marabou storks, they devour dead animals and leftover

a

b

c

Figure 1.8 An interdependency of a most improbable sort, beginning with the plants that feed the elephants (**a**), the dung that leaves the elephants (**b**), the beetles that roll dung balls away and bury them (**c**), ending with the beetle larva (**d**) that hatches in the dung—and the remains of the dung itself, enriching the soil in which plants grow, eventually to feed the elephants (**e**).

d

e

carrion. In doing so, they help check potentially explosive increases in the populations of insects and rodents, which otherwise would flourish on a carrion-strewn plain.

So far the impression you may have of life in this place is one of an interdependency of intricate sorts. Sometimes it seems even of the most improbable sort.

Consider the mighty adult African elephant, standing almost two stories high at the shoulder and weighing in at $8\frac{1}{4}$ tons. This grazing animal eats quantities of plants, the remains of which leave the elephant as droppings of such considerable size as to be potential breeding grounds for the insect and rodent populations that are denied the carrion. Ah, but the plains of Africa remain free from such rampant infestations partly because of little dung beetles that rush to the scene almost simultaneously with the uplifting of an elephant tail. With singleminded purpose they carve out fragments of the dung into round balls, which they roll off and bury underground in burrows. In these balls they lay their eggs, thus providing a rich supply of nutrients that will sustain the developing offspring. The surface of the plains is tidied up, the beetle has its food source, and the remains of the dung are left to decay in the burrow—there to enrich the soil that nourishes the plants that sustain, among others, the elephants.

Questions: How can it be that ninety different kinds of grazing and browsing animals exist side by side in the time and space of one specific environment? Is there enough of the same kinds of food for all? If they are all grazing and feeding in the same environment, why are they so remark-ably diverse in appearance? What *mechanisms* brought about such diversity—and what mechanisms maintain it? If both cheetah and lion are large catlike predators, why is it that one runs down its dinner by day, and the other stalks its dinner by night? Why does the lion forego the abundant and surely tasty bird populations, which during parts of the year number perhaps *200 billion?* Of all potential sources of nutrients in the savanna, why does a certain kind of beetle rush forward to specialize only on the droppings of ele-phants (surely by any stretch of the imagination a most dangerous undertaking, requiring the fanciest of footwork)?

But these questions, while provocative, may be some-what unfair. The African savanna is, after all, the wellspring of more diversity—hence more complexity—than almost all other environments. So let's back up a bit and look at one of the environments that are more limited in the kinds of organisms they sustain, and see what kinds of patterns emerge.

Life in the Tundra

Tundra refers to a region where it's too cold for trees to grow but not cold enough to be perpetually frozen over with snow and ice. The Meade River (Figures 1.9 and 1.10) in the southerly arctic region of Alaska meanders through as wind-swept and desolate a stretch of tundra as you might ever expect to see. Temperatures here struggle up to an average

Figure 1.9 A sweep of seemingly uneventful tundra in the southerly arctic region of Alaska.

Roger K. Burnard

h

Figure 1.10 Same tundra, but during the short span of summer. Cottongrass (**a,b**) comes into bloom; (**c**) mosses and mushrooms appear. (**d**) Ptarmigan half-in and half-out of its winter dress, then, as the summer wears on, in its speckled brown summer feathers (**e**). Flowers of varied sorts appear (**f,g**). The caribou come to feed (**h**). One of the main predators of the tundra, the snowy owl (**i**), is linked with the main prey of the tundra—the lemming (**j**). Ah, and the mosquitoes (**k**).

i

j

k

All photographs Roger K. Burnard

Figure 1.11 Levels of biological organization in nature. In this book, you will be tracing the principles of energy flow and evolution as they apply at each one of these levels. For now, however, just become familiar with this framework, and refer to it every now and then as you read through the chapters to follow. At first the words may not mean much, but don't worry; they will take on meaning as you explore each level and the links between levels. In doing this, you will come to see that the principles holding the framework together apply to all the diverse forms of life on earth—ourselves included.

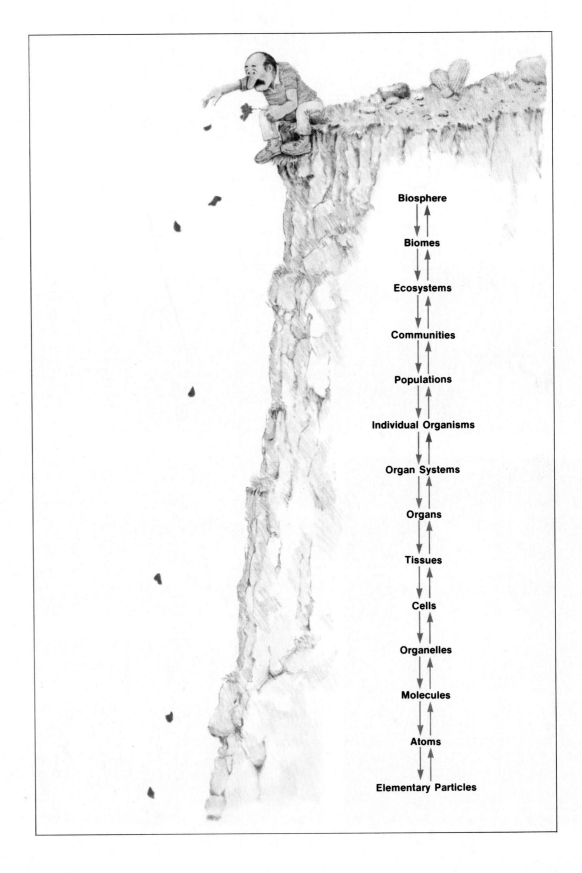

Biosphere

Biomes

Ecosystems

Communities

Populations

Individual Organisms

Organ Systems

Organs

Tissues

Cells

Organelles

Molecules

Atoms

Elementary Particles

5°C (41°F) in midsummer. They have been known to reach a balmy 21°C (70°F), only to drop to below freezing the next day. Temperatures plummet to −32°C (−26°F) in mid-winter. Generally, there is continuous light here for the three summer months. Then the skies gradually darken with the onset of the long arctic winter, when blizzards close in on the land.

Although the tundra is not completely covered with snow all year long, the brief summer thaw is not enough to warm much more than the surface soil. Just beneath the surface is the **permafrost,** a permanently frozen layer 610 meters (2,000 feet) thick in this region! Because of this impenetrable basement, drainage is poor, so the tundra becomes saturated with melt waters at the onset of summer. By late summer, the water evaporating from the boggy soil creates almost perpetual clouds until the first frosts clear the skies—before the next round of blizzards.

What sort of life can we expect to see in this harsh environment? As you might well imagine, the rich diversity of organisms that exists on the tropical reef and in the savanna cannot exist here. But as simple as the web of life in the tundra may be, it speaks eloquently of the same underlying forces in nature.

Most of the year, the low grasses are browned off. But during the short summer, when there is sunlight nearly twenty-four hours a day, there is intense new growth. Sedges and mosses thrive and the cotton grass bursts into bloom, then seed. Where the soil rises even slightly above the bog, shrubby, dwarfed willows and rushes flourish. Along the gravelly riverbanks they grow taller still, and are intermingled with a variety of colorful perennials (flowering plants that bloom year after year). The willows and cotton grass become covered with delicate filaments, much like the hair on a human arm.

It is during the resurgence of plant life that we become most aware of animal life in the tundra. In fact, it is impossible to be unaware of the flies and mosquitoes—thick, whirring clouds of them. Neither can we overlook the insects of other sorts, and spiders—all with dark coloration. Now the migratory birds such as the northern phalarope arrive in vast numbers. They stay long enough to nest and fatten up on the plants, the insects, and the spiders, and then they leave before the return of the long winter night.

Not at all anxious to leave is the ptarmigan, a permanent resident that changes its plumage from speckled brown during summer to white during winter. This bird feeds on willow and other stalks that poke through the drifts of snow and ice. Another animal that stays when the going gets rough is the lemming. What this tiny brown mammal lacks in size, it makes up for in sheer numbers. Feeding on the roots of sedges and grasses, lemmings live and reproduce throughout the year in their straw-lined underground burrows. Although they consume plants in great numbers, they also deposit droppings that fertilize the remaining plants and thereby stimulate more plant growth—at least for the short term. But every three to five years, the lemming population reaches the point that there is not enough food to go around, and these animals are forced to disperse in huge swarms to less crowded places.

Watching the movements with great interest are such lemming connoisseurs as the snowy owl and arctic fox. For them, the rise and fall of lemming populations through the seasons dictates how many of their own numbers will survive.

Few large animals walk the Meade River tundra. The most prominent are the caribou that move past during various times of the year. They feed on mosses and lichens during the summer and use their pointed hooves to scrape aside snow in their search for remnants of these plants during the winter. Sometimes wolves will stalk the herds here, weeding out—as on the African savanna—mostly the weak and the careless. With the coming of summer the mosquito, too, follows the herds. For then the caribou shed their thick winter coat and are vulnerable to this voracious insect. At times the air is so thick with mosquitoes that the animals walk about with eyes shut tight! Not only caribou but occasionally a human has been known to go quite mad under the relentless attack.

Questions: Why are plant life cycles completed so quickly along the Meade River, and by what mechanisms can they be completed so quickly? What other adaptations allow plants and animals to survive in this inhospitable place? For example, of what use are the "hairy" filaments on willows, the color changes in fur and feather, or the dark coloration of spiders and other creeping things? Given the more hospitable places on earth, why did the ancestral forms of these organisms come to live in the tundra at all? What built-in road maps and clocks do migratory birds use to find this place, rich as it is in plant life during the summer and relatively free of predators? Given how simple the web of life is in the tundra, why would environmentalists be so concerned about a single pipeline cutting across it, from the oil-rich North Slope of the continent to energy-hungry Americans to the south? Or is it because this web of life *is* so simple?

By the time you have completed this book, you may have insights into, if not answers for, these and all other questions raised in this chapter. For with these seemingly simple questions, you are about to relive the launching of countless petals and their cumulative echoes up and down the canyon called the **levels of biological organization in nature** (Figure 1.11). This "canyon" is no less than a framework that can be seen to be based on principles of evolution and energy flow. In this framework, the parts of life—beginning with certain large molecules—are organized into an increasingly greater, interconnected whole. You will be moving up through these levels as you read this book. As you do, you will see that what you learn about the expressions of diversity in the tropical reef, the savanna, and the tundra *applies to life in all times and in all places.*

Chapter Two

These huge reptiles, surrounded by the black lava, the leafless shrubs, and large cacti, seemed to my fancy like some antediluvian animals. The few dull-coloured birds cared no more for me than they did for the great tortoises. — Charles Darwin, on his first visit to the Galápagos Islands

The Evolutionary Road to Diversity

Tropical reef, savanna, tundra—how timeless life seems to be in these places! How reassuring it is to think that somewhere, beyond the clutter of cities, the land remains unchanged and life goes on in its eternal way. But the idea that life anywhere can remain permanently the same is no more than an assumption. With the camera we capture no more than scattered moments in time and space. Even thinking about all we have ever observed expands the picture only slightly. And adding to it all the recorded observations of all human generations past gives us no more than a glimpse of the immense sweep of life. Why, then, is the idea of permanence accepted almost without a second thought? Perhaps it is precisely because the idea *is* reassuring. We are, after all, very much aware of our own individual *impermanence*. Certainly we are the only organisms on earth that seek ways to deny it. That is one of the reasons why we as a species are not content simply to observe the nature of things; we are compelled to explain things in ways that make our lives a meaningful part of some greater picture that began before birth and extends beyond death.

Scientists, too, have responded to these same compelling tugs of history and hope. But in searching ever more deeply, ever more systematically through the greater picture, they have brought something into focus that no one had quite expected. Life, it seems, has been around for billions of years—which in itself is certainly heartening, and certainly a big picture. But it seems also that during those billions of years, life forms have undergone change after change of the most profound sort! To many individuals, a concept of change of this magnitude is not exactly reassuring. In fact, it's one of the reasons why some individuals tend to think of themselves as "nonscientists," as if the word somehow puts distance between themselves and "scientists." This separation is truly unfortunate, for science is an outgrowth of human abilities basic to all of us.

Human beings observe, they think about what they see, they search for explanations of what it means. Every human has the potential for **deductive reasoning**—starting with a general concept, then reasoning from one set of propositions to another in order to arrive at a specific conclusion. (For example, suppose you are mountain climbing and you suddenly lose your footing, which leaves you gripping a small ledge with your fingers. You can start with a general concept, then follow what seems like a reasonable set of propositions to its logical conclusion: "All people are heavier than air. I am a person, I must be heavier than air, and because there are 6,000 meters of air between me and the ground, if my fingers give out I shall be very sorry.") Every human also has the potential for **inductive reasoning**—putting together a tentative proposition based on specific observations, then testing the proposition with more observations. (If you observe a bird flying past while you are dangling from the ledge, you might suddenly reason from this observation that flapping your arms like the bird might keep you aloft. Within seconds of losing your grip, you will further observe that flapping is not the same thing as flying.) This example shows that deductive and inductive reasoning are interrelated processes. It also shows where such reasoning may lead when it is based on casual observation.

Sometimes propositions based on casual observations are not tested and they grow, over the years, into "facts." But such "facts" can be misleading, no matter how universally true they might seem to be. Will the sun rise tomorrow? Absolutely! How can we be sure? Tradition and our own observations tell us so! The sun rose today, yesterday, and all yesterdays past; therefore, it will rise tomorrow. But what of the Eskimo sledding across the snows during the long night of the arctic winter? What of the astronaut rapidly orbiting the earth or standing on the moon? Whether observations justify the belief that the sun will rise tomorrow depends on where one is standing—and on how one defines "tomorrow."

On the Scientific Method

In the broadest sense, the **scientific method** of approaching a question such as "Do life forms change over time?" would be (1) to consider all observations made so far and the range of conditions under which they have been made; (2) to work out an explanation that is consistent with all the observations; and (3) to predict what new observations will probably yield under new conditions. Thus the test of any scientific explanation, or theory, is basically quite simple: Does the explanation adequately account for all past observations—and as further observations are made, does the explanation hold up?

The scientific method is, in essence, a commitment to systematic observation. If all observations are to be accounted for, they must be made under well-defined conditions that may be reported to and repeated by any proficient person at any time. **Repeatability of observations** is a cornerstone of science. If a theory is valid, it shouldn't matter whether observations are made in Albuquerque or Kuala Lumpur or Anchorage; the theory should transcend cultural boundaries. That is why science needs a shared "language"—a common set of terms for measuring and describing all aspects of experience. Different workers must be able to use the same set of internally consistent terms if they are to test or challenge the accuracy, repeatability, and proper description of the observations—hence to test the theory itself.

Does such meticulous testing and demand for precise observation *always* lead to a neat package of truths, complete and perfect and demanding to be believed? Not at all. There are no absolute truths in science. Instead there is the **suspended judgment,** whereby a theory is tentatively said to be valid in that it is consistent with all observations at hand. You won't (or shouldn't) hear a scientist say, "There is no other explanation!" More likely you will hear, "Based on present knowledge, this is our best judgment at the moment." Even if a theory is one of the most basic of all—even if the weight of evidence is so impressive that it is regarded as a general principle—new observations may not fit with the rest and they may call for its replacement or modification. Far from being a disaster, such observations stimulate the development of even more general, more adequate, *yet always tentative* theories.

Obviously, individual scientists would rather propose right theories than wrong ones. But they must always ask: "Is my explanation consistent with all observations that have been made of what I hope to explain? *It is the external world, not internal conviction, that must form the testing ground for scientific beliefs.* Knowing this, scientists must keep reminding themselves to be objective. Of course, this doesn't mean all scientists are objective all of the time or even most of the time; no one can lay claim to that. It means only that scientists are expected as individuals to forsake pride and prejudice by testing their own beliefs even in ways that might prove them wrong. Even if an individual scientist doesn't, or won't, *others will*—for science proceeds as a community that is both cooperative and competitive. There is a sharing of ideas, with the understanding that it is equally as important to expose errors as it is to applaud insights.

It is true there are many different styles of scientific research and testing. Science is a creative process, and its insights can result from accident, from sudden intuition, or from methodical search. Some individuals adhere to standardized procedures; others may improvise as they go. Some tailor their work to reinforce an existing viewpoint; others deliberately take approaches that are likely to challenge it. But no matter what the individual approach, *the common element of all science lies in the process of testing existing knowledge, with the understanding that knowledge is an open system.*

Systematic observations, tentative propositions and explanations, relentless tests—in all these ways, scientific beliefs differ from systems of belief that are based on faith, force, authority, or simple consensus. It is not any "law" that forms the underpinnings on which science rests. Rather, it is the countless observations the "law" attempts to explain. A "law" may be invalidated by new evidence, but the observations remain, awaiting new and better tests of their meaning.

There are, in the history of science, a few individuals who challenged the longstanding beliefs held not only by the culture at large but by the scientific community within it. In biology, Charles Darwin and Alfred Wallace are among them. It will be useful to trace their story, for it will give us insight into one of the most powerful theories of our time—the principle of evolution by means of natural selection. More than this, it will show us the common denominator of their separate journeys was not so much specific training as it was an underlying attitude. They were willing to observe, to gather evidence, and to test the outcome—no matter how unsettling it might be—with the reasoning that is the hallmark of the human species *and the discipline that is the hallmark of science.*

Emergence of Evolutionary Thought

Part of being human is talking about observations; that is basically how each new generation learns to survive. But before observations can be talked about, names must be assigned to various aspects of the world and people must share an understanding of what the names mean. Since the time of Aristotle, the outstanding Greek naturalist who was one of the first to begin systematic inquiry into the natural world, there have been attempts to name and classify all living things. As long as "the world" mostly meant "Europe," any system of classification (or **taxonomy**) could consist merely of naming, drawing, and describing each known plant and animal, then grouping them together on the basis of which ones looked alike. Even at that time, it was obvious that many different organisms often share common features, and they came to be called members of the same "species." (The word originally meant little more than a taxonomic slot; it would later become the center of an intellectual storm.)

With the global explorations of the sixteenth century, however, naturalists were overwhelmed with descriptions of thousands upon thousands of plants and animals being discovered in Asia, Africa, the islands of the Pacific, and the New World. Some appeared to be remarkably similar to common European forms, but some were clearly unique to different lands. How could they possibly be cataloged? The naturalist Thomas Moufet, in attempting to sort through the bewildering array, simply gave up and recorded such gems as this description of grasshoppers and locusts: "Some are green, some black, some blue. Some fly with one pair of wings, others with more; those that have no wings they leap, those that cannot either fly or leap they walk." It was not exactly a time of profound insight.

Linnean System of Classification The impasse was finally broken in the mid-eighteenth century. Drawing on the work of earlier naturalists, Carl von Linné developed the **binomial system of nomenclature**—a way of classifying every organism by giving it two names. Both names are Latin or latinized words, because when von Linné devised the system, that was the language of scholars wherever Western civilization had taken root. (Scholar that he was, von Linné usually is known by his latinized name, Linnaeus.) The system is still used in every country of the world.

For instance, *Felis domesticus* is the scientific name for the common housecat. The first, capitalized name refers to the **genus** (plural, **genera**), a category into which distinct but in many ways similar species are grouped. (Other species of small, catlike animals in this genus include *Felis canadensis,* the lynx, and *Felis rufus,* the bobcat.) The second, uncapitalized name is the **species epithet.** By itself, the species epithet doesn't mean much, for it can also be the second name of a species found in an entirely different genus. (The common chicken, for instance, has the name *Gallus domesticus.* It takes on meaning only when used with the capitalized generic name or the first letter of it, as in *F. domesticus* and *G. domesticus.*)

Members of the same species share many common features that set them apart from other species in their genus. As you will read later, by far the most important of these features is an ability to interbreed; only members of the same species normally can do this. To the scholars of the eighteenth century, however, a definition based on interbreeding would not have made much sense, for they viewed all species as being permanent, unchanging groups. Who ever heard of members of a species *not* able to interbreed? Each kind of plant and animal was seen as having existed as a separate species since the time of creation and as having descended, *without change,* to the present. The "fixity" of all the world's diverse species was a "fact" of life; botanists and zoologists had only to discover, name, and catalog all the ones that existed.

Challenges to the Theory of Unchanging Life By the late eighteenth and early nineteenth centuries, however, investigations were beginning to turn up too many things that simply didn't fit with the concept of unchanging life. First, there was disquieting news from the infant science of

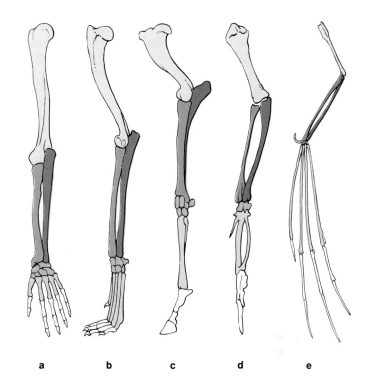

Figure 2.2 Homologous structures. Shown here, a forelimb from (**a**) a human, (**b**) a dog, (**c**) a horse, (**d**) a bird, and (**e**) a bat. The drawings are not to scale relative to one another. The homologous structures are shaded the same way from one animal to the next.

comparative anatomy (the comparison of the anatomical structures of different species). For instance, most mammals have two forelimbs and two hindlimbs. When anatomists dissected and compared forelimbs ranging from whale flippers to human arms to bat wings, they found them to be **homologous structures:** not only were they constructed from the same kinds of parts, they were constructed according to the same basic plan (Figure 2.2). Even though such parts had a different function in different species, without doubt they had an underlying similarity. But what did the similarity mean?

Equally puzzling were **vestigial structures:** body parts that have no apparent role in the functioning of an organism. For instance, like humans, snakes have a backbone. But unlike humans, they have no limbs. Now, if snakes were created as limbless creatures, why do some of them have what looks like remnants of a pelvic girdle—the set of bones to which hindlimbs are attached? Similarly, humans have a tiny appendix and what looks like the remnants of a tail. Many plant-eating animals have a large appendix, which functions in digestion. It's true that the human appendix is an offshoot of the intestinal tract. But it's too small to serve a digestive function and residues accumulate inside it, which makes it a target for infection. So what could be the point of having an appendix? And a tail! What could be the significance of *that?*

Even as comparative anatomy was turning up such unsettling news about animal body plans, biogeography was turning up some puzzling variations in the world distribution of plants and animals. For instance, marsupials (pouched mammals such as the kangaroo) are rare in most of the world, but they abound in Australia. There are no other kinds of Australian mammals except those brought in by humans. Various cactus species thrive in North and South American deserts, but they are nowhere to be seen in the deserts of Africa and Asia. But if all existing species had been created at the same time in the same place, as most scholars of the period believed, how could so many species be restricted to one part of the world or another? By the late eighteenth century, the zoologist George-Louis Leclerc de Buffon had an idea. Because species moving out from a single center would have come up against oceans or mountain barriers, perhaps there were several "centers of creation" for species throughout the world.

At the same time, Buffon's work in zoology led him to perceive relationships among different species. He even went so far to suggest that *perhaps species have become modified over time.* It would explain the similarities between some species; it would also explain the existence of vestigial structures. Was there any evidence to support his suggestion?

Buffon was keenly aware of layers of fossils in the earth, which to him suggested there may have been a progression of epochs in earth history. Fossils had been known about from antiquity, but they had been considered no more than mysteriously marked stones. Eventually they began to be accepted as the remains of once-living things. Beginning with the excavations by William Smith, an English canal builder, a pattern was becoming evident in the way these fossils were buried in the earth. In older, underlying rock layers, fossils of marine organisms were quite simple in structure. But in layers above them, they gradually became more complex until, in the uppermost (most recently deposited) layers, they closely resembled living marine organisms. The fossil record was a strong argument for a succession of different life forms. The time required for such changes and the reasons for change were not known — *but the very concept of change was at variance with the concept of the fixity of species.*

Many scientists now tried to explain the changing patterns of fossils within a traditional conceptual framework that did not allow for change. The nineteenth-century anatomist Georges Cuvier had spent twenty-five years comparing fossils with living organisms. He could not deny that there were changes in the fossil record, and that the changes were linked somehow to earth history. But his explanation for them came out as a theory of **catastrophism:**

There was only one time of creation, said Cuvier, which had populated the world with all species. Many had been destroyed in a global catastrophe. The few survivors repopulated the world. It wasn't that they were *new* species; biologists simply hadn't got around to discovering earlier fossils of them, which would date to the time of creation. Another catastrophe wiped out more species and led to repopulation by others, and so on until the most recent catastrophe, followed by the ascent of man. The trouble is, investigations never have turned up the kinds of fossils needed to support his theory. Rather they have turned up overwhelming fossil evidence against it (Chapter Fifteen).

Lamarck's Theory of Evolution One of Cuvier's contemporaries, Jean-Baptiste Lamarck, viewed the fossil record differently. Life, said Lamarck, had been created in the past in a simple state, and it gradually improved and changed into the level of organization we see today. The driving force for change was the "requirement" of organisms to improve themselves (the desire for improvement being centered somehow in nerve fibers). This desire directed a (vaguely defined) "fluida" to body parts in need of change (in a manner unspecified). For instance, Lamarck saw the ancestor of the modern giraffe as a short-necked animal. Pressed for the need to find food, this animal constantly stretched its neck to browse on leaves that were beyond the reach of other animals. Stretching directed some "fluida" to its neck, which made the neck permanently longer. The slightly stretched neck was bestowed upon offspring, who stretched their necks, also. Thus generations of animals desiring to reach higher and higher leaves led to the modern giraffe. Conversely, a vestigial structure was an organ no longer being exercised enough to get any "fluida" and therefore was withering away from disuse — with the withered form somehow passed on to offspring. The notion that changes acquired during an organism's life are brought about by environmental pressures and internal "desires," and that these changes can be transmitted to offspring, is the **Lamarckian theory of inheritance of acquired characteristics.**

Lamarck's contemporaries considered it a wretched piece of science. The problem was that Lamarck would make sweeping assertions but saw no need to back them up with actual observations. Why bother? Nature spoke directly through him, he loftily informed his colleagues. Many felt that, rather than advancing the case for changing life, Lamarck opened it up to ridicule that would spill over onto the more serious studies.

In retrospect, perhaps we can find kinder words for the man. His work in zoology was respected. And beneath his eccentricity, we find he had indeed put together the foundation for an evolutionary theory: *All species are interrelated, they gradually change over time, and the environment is a factor in that change.* It was his misfortune that he identified the crucial points but came up with an implausible explanation.

Figure 2.3 Charles Darwin, at about the time he accepted the position of ship's naturalist aboard H.M.S. *Beagle*. Darwin's mentor at Cambridge, the botanist Henslow, had this to say to him: "I have been asked to recommend a Naturalist as companion to Captain Fitz-Roy. I have stated that I consider you to be the best-qualified person I know. . . . I state this not in the supposition of your being a finished Naturalist, but as amply qualified for collecting and observing. . . . Don't put on any modest doubts about your qualifications, for I assure you I think you are the very man they are in search of."

The Theory of Evolution by Natural Selection

The Naturalist Inclinations of the Young Darwin   Charles Darwin was destined to develop an evolutionary theory that would have repercussions through the whole of Western civilization. Surely his early environment influenced that destiny. His grandfather was a well-known philosopher, physician, and naturalist, and one of the first to propose all organisms are related by descent. His father was a respected physician who married into the wealthy Wedgwood family, which meant the children had the means for indulging their interests. When Darwin was eight years old, he was an enthusiastic but haphazard collector of shells and minerals; at ten he was focusing on the names and habits of insects and birds. He cherished the time he spent alone observing the natural world. His pursuit of solitude led him soon enough to fishing and, later, hunting. His consuming interest at fifteen was not in school but in hunting for woodcock and partridge.

Perhaps it was boredom with school that from time to time led him to accompany his father during calls on patients; perhaps his father (who viewed his son's passion for hunting as a disgrace) interpreted this to be an interest in medicine. Whatever the reason, Darwin was packed off to study medicine at the University of Edinburgh. There he found the lectures "incredibly dull" but his trips with local fishermen (to collect shells) most interesting. In later life he admitted he felt no urge to study; after all, he was born well-to-do. But the clincher came when he attended his first session in the operating theater. Anesthetics were unknown in the early 1800s, and surgery was performed with the patient more or less under the influence of alcohol or opium. It was too much for Darwin, who could not stand the sight of human suffering. After one more session he ran from the operating theater and away from the university.

For a while he followed his own inclinations toward natural history. Then his father suggested a career in theology might be more to his liking than medicine (and more "respectable" than natural history), so Darwin packed for Cambridge. His grades were good enough to earn him a degree in theology. But most of his time he spent happily with companions of another sort: various faculty members with leanings toward natural history. It was the botanist John Henslow who perceived and respected Darwin's real ambitions; it was Henslow who guided him in such ways as arranging for him to take part in a training expedition led by an eminent geologist; it was Henslow who, at the pivotal moment when Darwin had to decide once and for all on a career, arranged that he be offered the position of ship's naturalist aboard H.M.S. *Beagle* (Figure 2.3).

The Voyage of the *Beagle*   The *Beagle* was about to sail to South America to complete earlier work on mapping the coastline. The prolonged stops at various islands, mountains, and rivers would present many opportunities to study the diversity of the natural world. Almost from the start of the voyage, the young man who had hated work suddenly began working enthusiastically—despite lack of adequate training, despite cramped quarters, and despite miserable bouts of seasickness. Throughout the journey to South America, he collected and examined marine life, and he read—particularly Henslow's parting gift, the first volume of Charles Lyell's *Principles of Geology*.

Amplifying the earlier ideas of the geologist James Hutton, Lyell argued that processes now molding the earth's

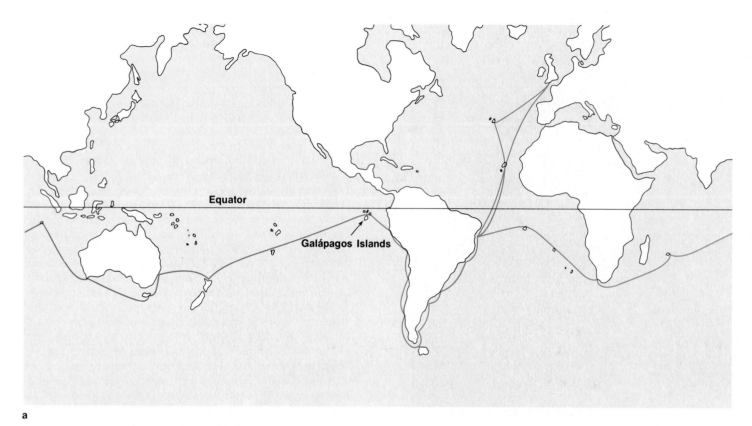

Equator

Galápagos Islands

a

The Bettmann Archive Inc.

b

Figure 2.4 (**a**) The five-year voyage of the *Beagle*, shown in (**b**) as it appeared in the Straits of Magellan. The Galápagos Islands, about 1,000 kilometers (600 miles) off the coast of Ecuador, support a number of unique plants and animals. The diversity of life on these isolated islands profoundly influenced Darwin's thinking.

surface—volcanic activity, the slow erosive action of wind and water, the gradual uplifting of mountain ranges—had also been at work in the past. This concept is called **uniformitarianism.** With it, Lyell convincingly extended earth history back in time, for known geologic processes would require not a few thousand but millions upon millions of years to reshape the landscape—*time enough, then, for species to evolve.*

It was on the Galápagos Islands (Figure 2.4) that Darwin began to see the implications of this concept. Even on that isolated cluster of volcanic islands, there was tremendous diversity of life. Particularly intriguing was the array of finch species. The birds of these species had strikingly different beaks, which seemed related to different foods in the environment. More than this, even though each species was confined to a single island or cluster of islands, *all* were similar to a single species of finch that Darwin had observed back on the South American mainland. Later it would occur to him that the finches might have descended from a single ancestral species, the mainland finch, which winds might have carried over to the Galápagos. If there was time enough for various descendants of the finch to change—and Lyell's concept strongly suggested that there was—then it would have profound implications for the whole question of species diversity!

Figure 2.5 A few examples of the more than 300 varieties of domesticated pigeons. Such forms are thought to have been derived, by selective breeding, from the wild rock dove (**g**).

Darwin and Wallace: The Theory Takes Form The *Beagle* returned to England in 1836, after nearly half a decade at sea. Darwin published his journal, *Voyage of the Beagle,* which was received enthusiastically. In the years to follow, his various publications established him as a respected figure in natural history. Despite the recognition, however, his consuming interest was the "species problem." *By what process do species evolve?* In his search for the answer, he patiently assembled his own data and then systematically analyzed the clues and difficulties they presented. *Whenever he developed an idea for explaining certain observations, he spent as much time trying to <u>disprove</u> it as he spent trying to support it.* The outcome would be impressively objective evidence for his concept of evolution.

Given a problem so complex, Darwin turned to a simpler question: Among domesticated animals, how do various **breeds** (varieties within a species) originate? Domesticated pigeons proved most interesting. He first determined that none of these flamboyant varieties was found in the wild, except for some that recently had escaped captivity. Then he noted that one species of pigeon, the wild rock dove, shares certain features with all these breeds (Figure 2.5). He concluded that the wild rock dove must have been their common ancestor.

But how did humans mold the rock dove into diverse breeds? The process, Darwin perceived, was one of *selection.* Rarely are any two animals of the same species exactly alike. In any generation, there is variation in such features as size, form, and color. Sometimes the variations are slight; now and then they are dramatic. In either case, plant or animal breeders choose the features they consider most desirable. For instance, instead of randomly breeding dairy cattle, farmers choose bulls for strength, form, and other traits, then they breed them with cows that produce quantities of high-quality milk. In this way, selective breeding encourages the perpetuation of some traits more than others.

Darwin reasoned that some sort of selection process must also be at work in the natural world. He gained insight into what that process must be when he read *Essay on the Principle of Population,* by Thomas Malthus. In his essay, Malthus had pointed out that populations tend to increase in ever greater numbers but food supplies do not. Thus, in the absence of controls, any population will outrun its resources. When that happens, its individuals must compete for available food. Suddenly the meaning of Darwin's own observations came into focus. There was indeed a struggle for existence in the natural world, in which the better adapted organisms had the competitive edge. It was this struggle that caused progressive adaptations, hence gradual changes in species. Darwin's view of how such adaptations are made is known as the **theory of natural selection.** The theory is expressed here in modern form:

1. Within a species, individual members vary in form and behavior, and some of the variant traits are inherited (passed from generation to generation).

2. The species as a whole produces more offspring than can survive to reproductive age. It's like a margin of safety that some *will* survive, *for the range of variations among all the offspring is the testing ground for selection.* Many are simply not adaptive and are weeded out.

3. The traits that *are* passed on are the ones that somehow improve chances for surviving and reproducing. Thus their bearers are likely to produce more offspring than other members of the species. As a result, the most adaptive traits show up among more individuals of the next generation.

4. But expressions of form and behavior mean nothing in themselves. They are adaptive or nonadaptive only in terms of prevailing environmental conditions. *Over time, interactions between heritable traits and the environment dictate the adaptive character of the species.*

5. Thus, in **natural selection,** heritable traits that are most adaptive in a given environment appear with increased frequency among individuals of a species, for their bearers contribute proportionally more offspring to the next generation.

Let's go one step further now, and see how natural selection might work to bring about not only gradual change within a species but the **evolution of new species:**

1. For one reason or another, a group of individuals sometimes gets separated from other individuals of the same species. When that happens, it is called a separate **population.**

2. The traits of individuals in the separate population are tested for responsiveness to a different physical or behavioral environment. Thus there is selection for *different* traits than there would have been if the group had remained with the original population.

3. When a population remains isolated from other populations of the same species, it gradually builds up its own pool of adaptive traits. The accumulation of traits that are adaptive to a different environment leads to a separate line of descent—a process called **divergence.**

4. After enough time, divergence becomes so great that **speciation** occurs: the accumulation of differences in form and behavior is so great that members of different populations can no longer interbreed even if they are brought together again. They have evolved into separate species.

American Museum of Natural History

Figure 2.6 Alfred Wallace. Although Darwin and Wallace had worked independently, they both arrived at the same concept of natural selection. Darwin tried to insist that Wallace be credited as originator of the theory, being the first to circulate a report of his work. Wallace refused; he would not ignore the decades of work Darwin had invested in accumulating supporting evidence.

Identifying the process by which the evolution of species may occur was a profound event in the history of biology. But now the theory had to be put to the test. For it is one thing to propose a theory; it is an entirely different thing to prove it is valid. *And how could convincing proof be given of the action of natural selection without knowing exactly what it acts upon—in other words, the physical nature of the heritable traits themselves?* The answer to that question would not be Darwin's to give. For it would not be until the rise of genetics, decades later, that the physical basis for inheritance would begin to be understood.

But Darwin continued his search, gathering notes and sifting his evidence for flaws in his reasoning. Then, in 1858, his careful search was interrupted: he received a paper from Wallace outlining the very theory he had been developing for more than two decades! Like Darwin, Wallace was a respected naturalist, with thirteen years of research in South

From the experiments of Dr. H. B. D. Kettlewell

Figure 2.7 An example of directional selection. (**a**) The light- and dark-colored forms of the peppered moth are resting on a lichen-covered tree trunk. (**b**) This is how they appear on a soot-covered tree trunk, which was darkened by industrial air pollution in certain parts of England.

America and the Malay Archipelago to his credit. Like Darwin, he had earlier been impressed with the writings of Malthus. But whereas Darwin had been working to document his ideas for more than twenty years, the concept of evolution by natural selection flashed into Wallace's mind and he wrote out his ideas in two days. This was the paper he sent off quickly to Darwin.

Despite the profound shock of seeing his theory presented by someone else, Darwin distributed Wallace's paper at once to his colleagues, suggesting that it be published. But Darwin's colleagues could not let him set aside the years that had gone into his own development of the theory. They prevailed on him to gather his notes into a paper that could be presented simultaneously with Wallace's. In 1858 both papers were presented to the prestigious Linnean Society, along with a letter Darwin had written several years earlier outlining the main points of this theory. One year later, Darwin published his book *On the Origin of Species by Means of Natural Selection*. Although the book was several hundred pages long and filled with detail, the scholarly Darwin wanted to call it an "abstract" of his theory! (His publisher talked him out of it.) The first half of his "big book" has only recently been published—in 1975.

Although many versions of the Darwin-Wallace story emphasize the controversy the book created in some quarters, Darwin's evidence was so overwhelming that the argument for evolving life was accepted almost at once by most naturalists and scholars from other disciplines. For as the naturalist Thomas Huxley commented, *the only rebuttal to the concept of evolution is a better explanation of the evidence—something which has yet to appear.*

Ironically, even though the idea of evolution had at last gained respectability, it would be almost seventy years before most of the scientific community would agree with Darwin and Wallace's unique insight—that natural selection is the means by which evolution occurs. In the meantime, their names would be associated mostly with the concept that life evolves—something others had proposed before them.

Examples of Natural Selection

In looking for examples of natural selection at work, it's important to keep in mind that selection doesn't *create* a new type of individual. Rather, there is selection for *existing* individuals having some combination of adaptive traits, and selection against those individuals having a less adaptive combination.

In England, for instance, there is a peppered moth (*Biston betularia*) that flies at night and rests during the day on lichen-covered tree trunks (Figure 2.7). Because of its

Figure 2.8 Adaptive radiation among Darwin's finches.

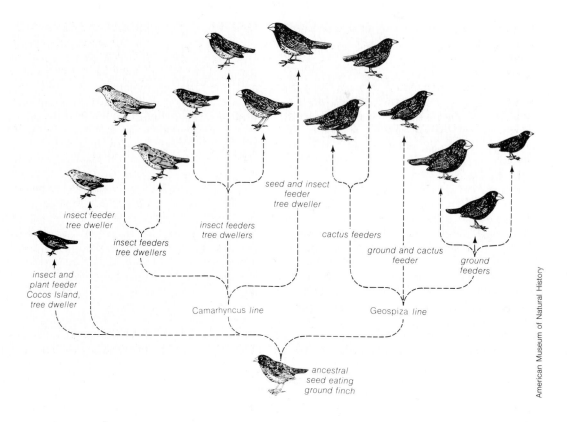

insect feeder
tree dweller

insect and
plant feeder
Cocos Island,
tree dweller

insect feeders
tree dwellers

insect feeders
tree dwellers

seed and insect
feeder
tree dweller

cactus feeders

ground and cactus
feeder

ground
feeders

Camarhyncus *line*

Geospiza *line*

ancestral
seed eating
ground finch

American Museum of Natural History

light color and speckled wing pattern, the moth is almost invisible against the trees. Sometimes hereditary factors in the moth population give rise to dark-colored (melanistic) individuals. But these variant forms don't leave many offspring. A dark moth on a light tree trunk is too conspicuous to bird predators, which act as a selective agent on the moth population. Because few of the dark moths survive to reproduce, fewer of their kind are represented in the next generation. But at one time, when the industrial revolution first swept through England, air pollution began to darken tree trunks and kill lichens in certain parts of the country. Because the moths' environment changed, the previously nonadaptive dark form was now the more adaptive: it blended with the soot-covered tree trunks! With the lighter moths now more visible to predators, the dark forms came to outnumber them 9 to 1. The increase in dark-colored moths in response to this environmental change was called **industrial melanism.** (With recent controls on air pollution, tree trunks are less sooty, lichens are becoming reestablished, and the dark forms are being selected against.)

The case of the peppered moth appears to be an example of **directional selection:** because of an increasing frequency of a certain heritable trait, the whole population tends to shift in a direction that corresponds to a specific direction of change in the environment.

Insects that develop resistance to insecticides provide another example of directional selection. With the first application of insecticide, most of the insects it is designed to kill are indeed killed. But some variant forms in the population may survive because they have a built-in resistance to the pesticide. If the traits contributing to the resistance can be passed on, then the next generation will contain more resistant individuals. Increasing numbers of resistant insects mean heavier applications of the insecticide—which selects for individuals having still more resistance. Gradually the nature of the population shifts (and new insecticides are needed).

By now, you may be thinking it takes a changing environment to trigger selection processes. But even when an environment remains much the same, there is selection for the *existing* species form—which has proved to be the most adaptive—and selection against extreme variant forms that do appear. In one experiment, red checkered butterflies were selectively bred in a laboratory until a dark form appeared in the population. Then the dark individuals were released into the kind of environment that favors the checkered form. (It was first ascertained that the area was free of any checkered individuals, which could interbreed with the dark ones and invalidate the experiment.) After several generations, the test population had largely reverted to the red checkered form. Such a process, which tends to select for and thereby maintain the form of a species already adapted to a given environment, is called **stabilizing selection.**

In another process, called **disruptive selection,** there is selection against the existing species form as a population expands in a nonuniform environment. Unlike directional selection (in which the character of the population shifts as a whole), disruptive selection tends to split up a population. Variant forms that are better adapted to some aspect of the environment become increasingly represented in different populations. When several populations of the same species become adapted for exploiting different aspects of the environment in specialized ways, they have undergone **adaptive radiation.** Darwin's finches, described earlier, are considered a classic case of adaptive radiation.

Some (but not all) evolutionary biologists suggest that adaptive radiation and speciation could arise without geographic isolation, as a consequence of disruptive selection. For example, assume an ancestral form of finch came over from South America and gradually populated the Galápagos Islands. As the finch populaton grew, competition for food may have become intense. Individuals having variant traits that allowed them to feed on as-yet un-exploited plants or insects would be favored, for they would not have to compete as intensely to survive. Each variant form would tend to cluster where its food was available, and the original population would eventually break up into specialized, partially isolated ones. Natural selection would continue to act on these specialized populations. If such were the case, enough differences might gradually have accumulated to prevent interbreeding; separate species would have evolved.

Phylogeny: A Classification Based on Evolutionary Relationships

The theory of evolution by natural selection has profoundly changed our perception of the nature of life. Today, each species is seen not as a static life form but as a current holder of one thread of life out of the past. In the evolutionary view, if we were to follow one of these threads back in time, we would sooner or later discover it once diverged from still another thread—from a less specialized ancestral form. Moving back further, we would find more and more separate threads of life diverged from common ancestors of simpler form, until at last there would be one strand. Billions of years ago it represented the common origin for all.

Currently, classification systems attempt to reflect the flow of life from this assumed common beginning to the increased specialization in form, function, and behavior we see today. The categories themselves are still traditional: they run from *species,* to *genus, family, order, class, phylum* (or *division*), and finally to *kingdom.* But a genus is now said to include only those species related by descent from a fairly recent common ancestor. A family includes all genera related by descent from a more remote common ancestor, and so on up to the highest (most generalized) levels of classification: phylum and kingdom. A scheme that takes into account the development of major lines of descent is known as a **phylogenetic system of classification.**

Constructing such systems is not easy. Throughout the past, environmental pressures have not been the same from group to group, environments themselves have changed, and the rates of evolution have varied from one group to the next. More than this, the evidence of interrelationships—the fossil record, comparative anatomy, genetics, biochemistry, reproductive biology, the behavior and ecology of living organisms, geology, geography—is not yet complete. Some information, such as parts of the fossil record, is lost forever because of geologic upheavals in the past. Even so, now that we have a better idea of where to look, and of what we are looking for, the gaps are filling in fast.

Regardless of its strengths, however, no classification system should be viewed as the final word in our understanding. As long as there are observations still to be made, different people will interpret relationships among organisms differently. Some group all organisms into two kingdoms (plants and animals); others group them into as many as twenty. In this book we use Robert Whittaker's five-kingdom model. (This scheme is sketched out in Figure 2.9 and in the Appendix.) Like the others, it helps summarize *current* knowledge about the evolution of life. It, too, is subject to change as new evidence turns up.

In the Whittaker model, the earliest forms of life have echoes in the kingdom **Monera.** The oldest known fossils suggest the first life forms were much like the simple one-celled bacteria and blue-green algae in this category. Most existing members of the kingdom **Protista** are also single-celled, and in some ways they are like monerans. But in other ways, some are like plants, others like animals, and still others like fungi! Whittaker sees them as living examples of the kinds of organisms that must have existed at a major evolutionary crossroad in the past. At that point, simple one-celled forms began evolving in ways that gave rise to the multicellular forms in the kingdoms **Plantae, Fungi,** and **Animalia.** Multicellular organisms are grouped into these three kingdoms on the basis of food-getting strategies: plants assemble their own food, fungi absorb food particles they have digested outside their body, and animals ingest other organisms as food.

But the separate "pools" of life in Figure 2.9 are not to be viewed as one flowing out of the other. *Representatives of all five kingdoms are alive today, side by side in time.* The branching routes simply suggest how they might have arrived at where they are now. This is the pattern of descent as many biologists now see it, as Darwin and Wallace might have anticipated it.

Figure 2.9 Simplified diagram of the five-kingdom model for classifying life forms. Only a few representative kinds of organisms are used to show the present scope of diversity. The general pattern suggests possible routes that may have led from the origin of life to this profusion of forms. Representatives of all kingdoms are still alive today.

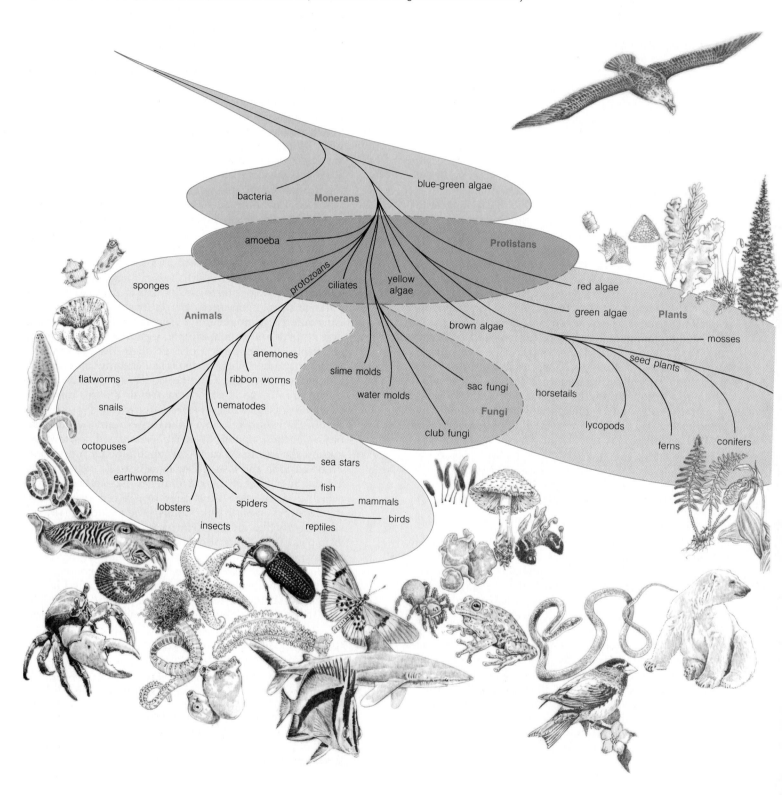

flowering
plants

Perspective

When Charles Darwin disembarked from the *Beagle* with his observations and thoughts on life's diversity, he set in motion a chain of events that made the study of life simpler and, at the same time, more complex. In one sense, an evolutionary view of diversity seems to leave us with far more problems than pre-evolutionary biologists faced. The species is no longer easy to define. The boundaries of taxonomic categories must be staked out most tentatively. Even the "natural" separation of plants and animals into two kingdoms is no longer convincing.

But the unfolding of more complex problems almost always follows any major advance in human understanding, for we happen to live in a complex world. *The important point is that there are greater insights as well.* All living things are now seen to be related by evolutionary threads extending back to the very dawn of life. Their many and diverse relationships are not random. They form a pattern of adaptive strategies that speaks of the reasons why life has been able to inhabit almost every accessible part of this planet. It is true that diversity is a puzzle. But we now know that the pieces fit, and that there is a solution.

The key, as we see it today, is natural selection, in which heritable traits that are most adaptive to a given environment occur with increased frequency among individuals of a species. It is a key to the future as well as the past. As environments continue to change, new selective pressures will appear. Selection will favor new variations in life forms, and those variant forms will in some large or small way influence the direction of evolution. We, the human species, have through the selection process become a major evolutionary force. Throughout this book, we will be looking at the kinds of pressure for change we are inflicting on the world of life, and at the consequences of those pressures at many levels. And perhaps, in that examination, we will begin to perceive not only what we have been, but what we and all of life might become.

Recommended Readings

Darwin, C. 1957. *Voyage of the Beagle.* New York: Dutton. In his own words, what Darwin saw and thought about during his first voyage around the world.

Moorehead, A. 1969. *Darwin and the Beagle.* New York: Harper & Row. Marvelously illustrated account of the voyage of the *Beagle*—the places Darwin visited, what he observed—and the home to which he returned.

Singer, C. 1962. *A History of Biology to About the Year 1900.* New York: Abelard-Schuman. Absorbing portrayals of the men and women who led the way in the development of the fields of biology.

Chapter Three

Figure 3.1 Sunlight on water—reflecting life's ongoing dependence on solar energy since its early evolution in the ancient seas.

Dynamics of the Living World

In viewing the diversity of life, sooner or later we come to the question of why it came about. The evidence is strong that natural selection is the process that brought it about, but *why* did it take place at all? We can gain insight into the question by observing organisms alive today. Although they vary in nearly every way we might imagine, they all have something in common: *If any organism is to retain what we call "life," energy must continually flow through it.* Even dormant life forms such as seeds survive because a trickling of energy sustains them. When they resume active growth, they stay alive only because they have built-in reserves to tide them over until they are able to take in more energy. Once energy stops flowing through them—once energy stops flowing through any organism—they become lifeless matter. The very fact that an organism is alive means it is heir to an unbroken stream of energy that has flowed through every generation since the beginning of life.

But this doesn't mean energy is always instantly available for the asking. First of all, *sources* of energy change from time to time. And if an individual does not have the means to adapt to an alternative energy source, that is the end of the individual. Second, *supplies* of energy normally are limited. When presented with an energy source, individuals of a population use it to grow and reproduce. Soon their increasingly numerous offspring compete intensely for it—with the most adaptive being the most likely to endure. Third, if an energy source happens to be another kind of organism, it is probably safe to say it does not exist simply to become dinner. Selection is constantly at work on prey, sifting through populations for less vulnerable forms. Of course, each new defense leads to selection for more efficient forms of predators, but that's another story. The point is, *directly or indirectly, the major impetus behind diversification has been competition for energy.*

This is not an abstract point that somehow applies to all other living things but has little to do with us. It has *everything* to do with us. We, the human population, are not now facing the world's first energy crisis just because we happen to be running out of food, or oil, or whatever. The whole history of life has been a story of successive energy crises, large and small. We will understand neither the nature of life nor the nature of our own energy problems until we come to understand the rules governing the flow of energy on which all the diverse strategies for living are based.

Two Principles of Energy

All events large and small, from the birth of stars to the death of a microorganism, are governed by two basic energy principles. The first may be expressed in this way:

The amount of energy in the universe is finite, and that amount never changes.

It's true energy exists in many forms, and it's possible to *convert* one form to another. Your body, for instance, can convert food into a form of energy that can power all your movements. But energy can't be created from nothing. We can create a "new" energy source only at the expense of a preexisting one. We can't get something for nothing.

The second principle is even more basic to an understanding of energy flow:

Left to itself, any system (the matter in a defined region), along with its surroundings (the rest of the universe), undergoes spontaneous energy conversions into less organized form. Each time that happens, some energy gets randomly dispersed in a form that is not readily available to do work.

If the first principle tells us we can't get something for nothing, this second principle tells us we can't expect to break even. For although the total amount of energy in the universe stays the same, the amount actually available in forms that can be used to do work is dwindling. The main reason it is dwindling is that all energy eventually tends to be converted to a form called heat—and heat energy tends to distribute itself so uniformly throughout a system that it cannot readily be converted back to another form. For instance, when you pour cold milk over hot cereal, the cereal cools down and the milk warms up. If you think of the milk and cereal as an isolated system, then once both are at the same temperature, nothing else will change; there is no temperature difference, so energy conversions proceed no further. The term **entropy** refers to how much energy in a system has been so dispersed (usually as evenly distributed heat) that it is no longer available to do work.

Entropy is constantly on the increase. As far as we can foresee, the universal rise in entropy will reach a maximum some billions of years hence, and nothing will ever change again. It happens, however, that life is one glorious pocket of resistance to this somewhat depressing flow toward oblivion. For the entropy of any local region can be lowered—*as long as that region is resupplied with usable energy being lost from someplace else.* Part of the energy streaming away from the sun is still in usable form as it reaches the earth. Plants intercept and convert some of this sunlight energy to chemical energy, which can be used at once to run life processes or stored away for later use. This stored energy sustains almost all other organisms, which feed directly or indirectly on plants. Thus, as long as the sun's energy flows into the interconnected web of life, the universal trend toward entropy can be postponed on earth.

Energy at Rest and On the Move

Energy exists in one of two basic forms: potential or kinetic. **Potential energy** is any form of energy "at rest." It is energy in a potentially usable form that is not, for the moment, being used. Each substance in the universe has a certain amount of potential energy. Because of its position in space, water behind a dam has gravitational potential energy; because of its composition, a piece of cake has chemical potential energy. But with any transformation, there is a drop in potential energy and a rise in **kinetic energy**—the energy of motion. Kinetic energy can be transferred from one region (or system) to another, and it can be changed from one form to another.

You can think of a rabbit, about to be pounced on by a coyote, as a store of chemical potential energy. As a twig snaps under the coyote's feet, the rabbit is startled and bounds away; some of its chemical potential energy becomes converted to kinetic energy. Well, the coyote is quicker than the rabbit, and that is the end of the rabbit. But it is *not* the end of its energy: part of it gradually gets stored away as chemical potential energy in the coyote.

We know, from the second energy principle, that a rise in entropy must have occurred all along the route. Because the coyote did not avoid the twig, it had to convert some of its own potential energy to mechanical energy for the chase. And as the coyote ran down the rabbit, some of its mechanical energy (as well as the rabbit's) was lost to the surrounding air as body heat. A silent coyote would have meant a more efficient energy conversion. The coyote would not have expended as much energy in securing dinner, and the dinner would have contained more potential energy on reaching its stomach. This example illustrates an important principle of energy use:

> **How efficiently energy is used in any given system depends on the precise route by which one form of energy is converted to another.**

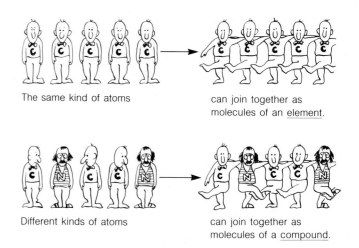

The same kind of atoms can join together as molecules of an <u>element</u>.

Different kinds of atoms can join together as molecules of a <u>compound</u>.

Figure 3.2 The difference between elements and compounds.

But this does not mean the most direct route is *always* the best route! Consider the controlled way small amounts of gasoline are exploded in a car engine. When gasoline is ignited, some of its chemical potential energy is converted to kinetic energy. Part of the kinetic energy is converted to mechanical energy of the pistons, the drive shaft, then the wheels. In one sense, the process is not too efficient: some energy is lost as heat with each conversion, and more is lost as moving parts rub together and become heated. The same amount of potential energy could be converted directly to kinetic energy by dropping a match into the gas tank and exploding all the gasoline at once. Of course, little would get converted to *useful* mechanical energy. Besides, there would be a steep rise in entropy, and a disastrously disorganized car. As you will read later, living cells also take a less direct (but safer) approach to using available energy.

When we talk about chemical conversions of the sort going on in the rabbit, the coyote, and the car, we are talking about two basic aspects of our world: matter and energy. "Matter" encompasses all substances—your body, the air, this book—and "energy" is what moves and changes them. Early in this century, an obscure patent office worker by the name of Albert Einstein proposed that matter and energy are fundamentally identical, hence interchangeable. This proposal seemed to violate common sense. But on August 6, 1945, a speck of matter hardly visible to the naked eye was completely converted to energy. The resulting explosion was so great that the Japanese city of Hiroshima was devastated—dramatic proof that seemingly remote scientific principles should never be dismissed as being too remote from human affairs.

The chemical conversions normally going on around us and within us do not destroy matter in this terrifying way, and the release of energy is far from complete. But it is useful to remember that *matter and its component parts are, in the essence, pent-up energy.* This concept is basic to our current view of how substances are put together and how they react with one another as energy flows through them.

Organization of Energy in Atoms

All the diverse substances in our everyday world are assembled from building blocks called **atoms.** Any combination of two or more atoms is known as a **molecule.** But sometimes different kinds of atoms come together, in which case they form molecules of a **compound.** If only the same kind of atoms come together, they form molecules of an **element** (Figure 3.2). Ninety-two kinds of atoms (elements) are known to occur in nature.

All atoms are made of similar kinds of parts. *But each kind of atom is unique because of its number of parts and because of the way those parts are arranged.* These two aspects of atomic structure dictate how atoms can combine to form molecules. They dictate what the properties of molecules will be, and how (if at all) different molecules will interact. Thus, to see why the molecules of life behave as they do, we must take a brief look at what goes on inside the atom.

Let's start with the way the parts of an atom are held together. When we want to join two blocks of wood together, we can use some epoxy glue. One of the "glues" holding atoms together and joining them to other atoms is of an entirely different sort: it's called **electric charge.** When something is electrically charged, it has energy that allows it to push away or to attract something else, without even touching it! This electric charge is of two kinds: positive (+) and negative (−). Two identical charges (+ + or − −) tend to repel each other. But two opposite charges (+ −) tend to attract each other.

One or more positively charged particles, called **protons,** reside in the center of every atom. Every atom except hydrogen also has in its center one or more **neutrons:** particles about the same size as protons but with no electric charge (Figure 3.3). Protons and neutrons make up the atom's **nucleus.** Orbiting the nucleus (in a region of space known, appropriately enough, as an orbital) are one or more negatively charged particles called **electrons.** Electrons zip about nearly as fast as the speed of light, so they are not the easiest things for the protons to hold onto. At any instant, they may be close to the nucleus and, at the next instant, farther away.

Hydrogen, which is found again and again in the molecules of living things, is the simplest atom of all. It has just one proton and one electron, so it's the easiest kind of atom to think about. When left alone, a hydrogen atom is said to be "at rest," or in its lowest energy level (even though its electron never stops its furious pace). But bombard it with heat or light, and its electron will get very

Common Elements in Living Things	Symbol	Atomic Number	Approximate Atomic Weight
Hydrogen	H	1	1
Carbon	C	6	12
Nitrogen	N	7	14
Oxygen	O	8	16
Sodium	Na	11	23
Magnesium	Mg	12	24
Phosphorus	P	15	31
Sulfur	S	16	32
Chlorine	Cl	17	35
Potassium	K	19	39
Calcium	Ca	20	40
Iron	Fe	26	55
Iodine	I	53	127

Figure 3.3 Atomic number and atomic weight.

No two elements have the same number of protons. Thus atoms of different elements can be ranked relative to one another: they can be assigned an *atomic number* equal to the number of protons in one atom of that element.

It's also possible to rank elements on the basis of weight. Electrons are practically weightless. But protons and neutrons are more massive, and they are both about the same size. As a first approximation, the *atomic weight* of an element is set by the total number of protons and neutrons in one atom of that element.

All atoms of the same element must, by definition, have the same atomic number. But different atoms of the same element may *vary* slightly in their number of neutrons—which means they will vary slightly in mass. These variant forms of an element are called *isotopes.* As you will see later, some isotopes have important uses in biological research.

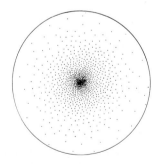

Figure 3.4 This is not an aerial view of ants in a sugar bowl but a model of how a hydrogen atom's solitary proton and solitary electron are arranged. The proton is somewhere in the center of all the dots. The spattering of dots indicates the likelihood of *finding* the electron at one of those dots in any given instant. An atom is too small, and an electron too fast, to ever see what's going on, but it's possible to calculate that this is the arrangement in all hydrogen atoms. (If we were to draw a line to show the circumference of this electron's orbital, we'd have the outline shown here. A real atom doesn't have a solid line around it.)

Figure 3.5 This is not a dying daisy but a model of how electron orbitals of an oxygen atom are arranged when it bonds with other atoms. The spherical orbital and two of the teardrops are filled, but the other two teardrops have only one electron each. Guess where other atoms bond to oxygen.

excited. The reason it does is that *an electron can absorb certain amounts of incoming light or heat energy.* When it does, the electron spends more time farther away from the proton's pull. Its added energy makes it swoop about in a new orbital farther from the nucleus. It is said to be in a higher energy level. Eventually it returns to its resting state, but it gives off the extra energy when it does. In many of the reactions of life, high-energy electrons give up some of their extra energy, which is then harnessed to power life processes.

Only certain energy levels are available to an electron. An electron may be found in one level, or the next level, or the next, but never in-between. It's something like standing on a ladder. Your potential energy increases as you climb, and it decreases as you descend. You can stand on any of the rungs, and you can move your feet from rung to rung. But you sure can't stand between rungs. It seems *all* electrons can only gain or lose energy like that: in steps of certain sizes. Living things respond to this property. For instance, electrons in certain atoms of a plant leaf can be excited by certain wavelengths (colors) of sunlight energy. The wavelengths have an amount of energy that corresponds exactly to specific "steps" on the energy ladder for those electrons.

We can see how energy organization affects atomic behavior by comparing a hydrogen and a helium atom. Both kinds are very small and far lighter than the molecules making up the oxygen-rich atmosphere. But they differ dramatically in their tendency to undergo chemical reactions, which was illustrated decades ago in a tragic way. Hydrogen is more abundant, hence cheaper, than helium. Thus the first lighter-than-air craft (blimps, zeppelins) were little more than hydrogen-filled "balloons" with a cabin slung beneath them. As long as the hydrogen gas was contained inside the balloon's rubber casing, it was safe enough. But during the 1930s, the zeppelin Hindenburg made a transatlantic flight from Germany—and exploded while landing in New Jersey. A small leak apparently developed in the casing, and hydrogen atoms began combining explosively with oxygen atoms in the surrounding air. Dozens of passengers died in the holocaust.

If the Hindenburg had been filled with helium (as blimps are today), the tragedy could have been avoided, for helium simply will not react with oxygen. It will not react with *any* substance under *any* normal conditions. Why is helium so different from hydrogen in this respect? The answer lies in the number and arrangement of its electrons. Each helium atom has two electrons zipping in the same orbital around two protons. In contrast, hydrogen and all other chemically reactive atoms have at least one orbital with only a single electron. Such atoms combine with one another, sharing electrons until all orbitals have at least a share of two electrons. From this comparison, we can identify some principles of energy organization in atoms:

1. Electrons occur in orbitals around a nucleus.

2. Each orbital may contain no more than two electrons.

3. Elements having an electron pair in each one of their orbitals tend to be **chemically nonreactive substances.**

4. Elements having orbitals with unpaired electrons tend to be **chemically reactive substances.**

Do these principles apply to large atoms having many protons and electrons? Yes they do. But large atoms require special orbitals so *all* the electrons can get as close as possible to the protons—yet stay as far away as possible from the electrons in all the different orbitals! (With their like charge, recall, electrons repel each other.) Thus we have orbitals shaped like spheres, teardrops, dumbbells, even flabby inner tubes. There is a limit to how many such orbitals can be squeezed into an energy level, of course. When that limit is reached, more orbitals are added at a higher energy level, farther away from the nucleus. For instance, an oxygen atom has eight protons and eight electrons. Two of its electrons are in a spherical orbital, just like that for a hydrogen atom "at rest." But the other six are distributed in four teardrop-shaped orbitals (Figure 3.5). Two of the teardrops are filled with two electrons each. The other two have only a single electron each—and it is precisely at these two places that oxygen shows a strong tendency to combine with other atoms.

It's important to know about the different shapes of atoms such as hydrogen and oxygen. They tell us why atoms bond as they do and give rise to specific shapes of molecules. In turn, the shapes of certain molecules give rise to specific shapes and structures of cells. And if molecules and structures are not assembled in predictable ways, cells don't function properly—or they don't function at all.

Bonds Between Atoms and Molecules

The number and kinds of molecules that can be formed by the interaction of just a few common elements is staggering. But the bonds themselves are fundamentally similar. *All bonds holding atoms together into molecules vary only in the degree to which electrons are shared.*

Covalent Bonds Let's return to the hydrogen atom, with its lone electron. Hydrogen atoms rarely exist by themselves. Most often two of them pair up as a molecule, which can be written as H_2. In a hydrogen molecule, the unfilled spherical orbital of one atom overlaps that of the other, so both electrons zip about both nuclei. When there is *equal sharing* of a pair of electrons, atoms are said to be joined by a **nonpolar covalent bond.** The single covalent bond of the hydrogen molecule may be written as H—H. (H—H means the same thing as H_2. It just tells us a little more about the molecule's structure. Such symbols, in which lines mean bonds, are "structural formulas.") Similarly, the oxygen we breathe exists as molecules of two oxygen atoms: O_2. But when two oxygen atoms join up, each of the two unfilled orbitals of one atom overlaps with an unfilled orbital of its partner. Because the molecule has two equally shared electron pairs (two covalent bonds), it may be written as O=O.

Does this mean *all* covalently bonded atoms share their electrons equally? Not at all. Consider what happens in a water molecule, or $\overset{O}{\underset{H\quad H}{\diagdown}}$. In each water molecule, all three atoms share electrons in such a way that their orbitals are filled up. But the shared electrons are attracted more toward the oxygen nucleus. On the average, the electrons spend more time at the oxygen "end" than they do at the hydrogen "end." So the oxygen atom tends to carry more negative charge than it does positive charge. The hydrogen atoms, which are deprived of a full share of electrons, tend to be positively charged. Because there is *unequal sharing* of electrons, the atoms are said to be joined by **polar covalent bonds.**

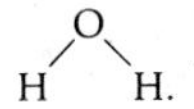

Ion Formation Polarity isn't an absolutely permanent feature of a water molecule. From time to time, with predictable frequency, electron sharing between the hydrogen and oxygen becomes even more unequal: an electron pair is drawn *entirely* into the domain of the oxygen atom! When that happens, one of the bonds is broken and the molecule temporarily separates into two charged parts: $\overset{O}{\underset{H\quad H}{\diagdown}}$ becomes OH$^-$ and H$^+$. When any atom or group of atoms draws extra electrons into its domain, it becomes a **negatively charged ion.** At the same time, the atom or group of atoms giving up its share of electrons becomes a **positively charged ion.**

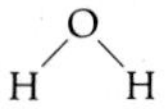

Ion formation in water is short-lived. Being oppositely charged, the two ions tend to re-form the broken covalent bond. At any one time, only two water molecules in a billion are separated into ions. Yet, these comparatively rare ions affect the stability of living systems. Many substances in cells take on different charge—hence different structure and reactivity—when ion concentrations change. To some extent, cells have ways of resisting such changes. For in-

pH Scale		A Few Familiar Substances
14	↑	Sodium hydroxide
13	increasing strength of bases	
12		Ammonia
11		Milk of magnesia
10		Phosphate detergents
9		Baking soda
8		
7	*neutral*	Human blood, sweat, tears
		Oysters, salmon, tuna, shrimp
6		Milk, butter, corn
5		Bread
		Bananas
4	increasing strength of acids	Beer
3		Sauerkraut, dill pickles, wine
		Peaches, apples, strawberries, oranges
2		Lemons, plums
1		Limes
		Stomach juices
0	↓	Hydrochloric acid

Figure 3.6 Ions and pH values of substances. In a glassful of pure water, only two molecules in a billion are ionized. Even so, this amounts to 15,000,000,000,000,000 H$^+$ ions and an equal number of OH$^-$ ions!

In pure water, the number of H$^+$ ions is always the same as the number of OH$^-$ ions. But many other substances, when dissolved in water, cause the number of H$^+$ ions to increase and the number of OH$^-$ ions to decrease. These substances are *acids.* Still other substances, when dissolved in water, cause the number of OH$^-$ ions to exceed the number of H$^+$ ions. They are *bases.*

The *pH scale* is a simple way of describing the relative abundance of both kinds of ions in a solution. When the pH is low (less than 7), it has an excess of H$^+$ ions; the solution is acidic. When the pH is high (above 7), it has an excess of OH$^-$ ions; the solution is basic. A pH of 7 means there are equal amounts of both kinds of ions; the solution is *neutral.*

stance, when something causes H$^+$ ions to flood in, special "buffer" molecules soak them up like a sponge. When OH$^-$ ions flood in, the buffers respond by doling out the extra H$^+$ ions they had been holding in reserve. (Of course, there are limits to how much the buffer molecules can counteract. In the American Southwest, cattle sometimes drink from pools of alkaline surface water, which have extreme concentrations of OH$^-$ ions. Because their cells can't respond to such extremes, the animals become poisoned.)

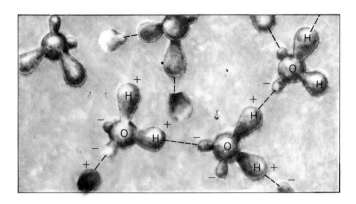

Figure 3.7 Hydrogen bonding between water molecules. Hydrogen bonds, shown in dashed lines, form between parts of opposite charge; thus two water molecules "share" a hydrogen atom between them at each hydrogen bond. Imagine billions upon billions upon billions of such molecules and hydrogen bonds, and you have an idea of what goes into a glass of water.

Hydrogen Bonds Covalent bonds and ion formation are important to life processes. So is the **hydrogen bond,** in which two molecules "share" a hydrogen atom. Because of hydrogen bonds, even molecules in a simple drop of pure water are not arranged randomly. Instead, the oppositely charged parts of neighboring molecules tend to face each other, even as the similarly charged parts tend to face away from each other (Figure 3.7). The hydrogen bonds between them are just strong enough to give water many of the properties of a much larger molecule. It freezes, and it boils at much higher temperatures than other substances made of molecules of similar size.

Hydrogen bonds give water another remarkable property. Say you stir some sugar into a cup of coffee. Without even being aware of it, you were witnessing water's astonishing capacity to dissolve other substances. But what does it mean, "to dissolve" something? If a molecule dissolves in water, it is a polar substance that undergoes direct chemical interaction with water molecules; it's **hydrophilic** ("water-loving"). If a molecule can't be dissolved in water, it's **hydrophobic** ("water-dreading"). Sugar happens to be hydrophilic. Within each sugar crystal, millions of molecules are stacked in a precise way. Projecting from the surface of each molecule is a series of oxygen–hydrogen parts, or —OH groups. Hydrogen bonds join adjacent sugar molecules at these groups. But drop a sugar crystal into coffee, and the crystal breaks apart at once. The sugar molecules are more attracted to the water molecules than they are to themselves. And the water molecules are more attracted to the sugar molecules than they are to each other. So the sugar goes into **solution**—it dissolves—which simply means its molecules become hydrogen-bonded with sur-rounding water molecules. Hydrogen bonds are not restricted to sugar and water. As you will soon see, they are essential to the structure of many substances of life.

The Versatile Carbon Atom

Of all the atoms found in living things, carbon is the most versatile. It combines with itself and with many other kinds of atoms to form more kinds of molecules than any other element. In fact, its importance to life is so great that for more than a century, nearly all carbon-containing molecules were known as organic ("organism-related") compounds to distinguish them from inorganic ("lifeless") compounds. The carbon atom is so accommodating because of its electron arrangement and its strong tendency to form covalent bonds. Two of its six electrons reside in the lowest energy level. But each of the other four electrons zips about alone in its own teardrop-shaped orbital—and each is in search of a partner. Thus other atoms can bond covalently to a carbon atom at all four of its "corners." With four hydrogen atoms, for instance, we have the simple compound methane (CH_4). It is called a **hydrocarbon** because it is made only of hydrogen and carbon atoms:

Notice that even this simple compound has a specific three-dimensional shape. The more complex a molecule becomes, the more complex is its shape—which means it can't be depicted very well on flat paper. That is why structural formulas are drawn as if the molecules they represent are squashed flat. Thus methane is also depicted as:

$$H-\underset{\underset{H}{|}}{\overset{\overset{H}{|}}{C}}-H$$

This four-cornered pattern of carbon bonding can produce an astonishing array of molecules, especially when carbon atoms bond covalently to other carbon atoms. They can be strung together like beads to form long chains, rings, and other shapes to which atoms of other elements can be attached. Imagine what a hydrocarbon chain would look like. Octane, a component of gasoline, is represented here:

$$H-\overset{\overset{H}{|}}{\underset{\underset{H}{|}}{C}}-\overset{\overset{H}{|}}{\underset{\underset{H}{|}}{C}}-\overset{\overset{H}{|}}{\underset{\underset{H}{|}}{C}}-\overset{\overset{H}{|}}{\underset{\underset{H}{|}}{C}}-\overset{\overset{H}{|}}{\underset{\underset{H}{|}}{C}}-\overset{\overset{H}{|}}{\underset{\underset{H}{|}}{C}}-\overset{\overset{H}{|}}{\underset{\underset{H}{|}}{C}}-\overset{\overset{H}{|}}{\underset{\underset{H}{|}}{C}}-H$$

Carbon atoms can be made to bond to one another to form a ring structure (which can be shown flat on the page or as a tilted, stripped-down version of the same thing):

Even when nothing is shown on a ring, it's understood that there is a carbon atom at every corner. Such ring structures are the framework for vitally important substances such as the sugar glucose. They also form the basis for such substances as pain-killing aspirin:

glucose   aspirin

As you've undoubtedly noticed by now, not only single atoms but groups of atoms such as the following can be attached to a basic carbon structure:

—OH   a hydroxyl group found in alcohol and sugar

an acid group, or —COOH, found in substances such as vinegar

an amino group, or —NH$_2$, found in the building blocks of proteins

These three common groupings are extremely important to life processes.

Molecules of Life

So far, the molecules we've described are fairly simple, for at most they contain no more than a few dozen atoms. How do we get from such molecules to the large molecules of life, which contain thousands—even millions—of atoms? Such "macromolecules" are not assembled atom by atom. First, smaller subunit molecules, called **monomers,** are formed. Then the monomers are assembled into long chains, or **polymers.** Generally, monomers are linked together through **dehydration synthesis.** In this process, one mon-

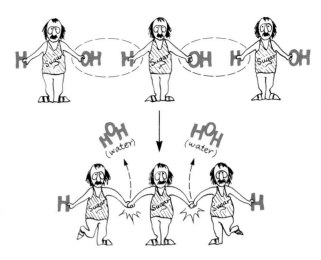

a *Dehydration synthesis of large molecule from monomer subunits*

b *Hydrolysis of large molecule into monomer subunits*

Figure 3.8 (**a**) In dehydration synthesis, many monomers (only three are shown here) are bonded covalently to form a polymer. A hydrogen atom must be stripped from one monomer and an —OH group stripped from another for bonding to occur. (**b**) Hydrolysis is basically the reverse of dehydration synthesis: the covalent bond is broken, a hydrogen atom is attached to one monomer, and an —OH group is attached to its previous partner. Although sugar molecules are used in this example, the processes apply to other large molecules of life, too. (These cartoon molecules obviously are simplified versions of the real thing—but then, so are structural formulas.)

omer is stripped of an H atom and another is stripped of an —OH group, then the two monomers become covalently bonded. The H and —OH released in the process are combined to form a new water molecule (Figure 3.8a).

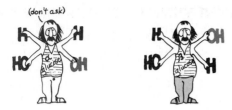

(don't ask)

a *Representations of the two forms of glucose*

b *Representation of how starch is formed*

c *Representation of how cellulose is formed*

Figure 3.9 Formation of two different carbohydrates from simple sugar monomers. (**a**) The sugar glucose comes in two forms, which differ only in the position of one —OH group. (**b**) When α-glucose monomers are linked together one after another, they form starch. (**c**) When β-glucose monomers are linked, the result is a different substance: cellulose, which has different properties.

Of course, life depends on more than an ability to build up macromolecules; it depends also on an ability to tear them down! How else could a cell dispose of large, worn-out molecules? How else could it break down large food molecules to extract their stored energy? Thus we have another process, called **hydrolysis** (which simply means "splitting by water"). It's basically the reverse of dehydration synthesis, for bonds holding monomers together into macromolecules are broken, and an H atom and an —OH group derived from water become attached to the ends of the broken fragments (Figure 3.8b).

Cells continuously recycle the atoms and molecules of life through dehydration synthesis and hydrolysis. At all times, these processes are controlled and regulated. Later chapters will cover some basic aspects of this control. For now, let's move to a survey of the kinds of macromolecules built up and torn down by the processes just outlined.

Carbohydrates When we talked about carbon compounds, we depicted a sugar molecule called glucose. All sugars such as glucose fall into a general class of molecules called **carbohydrates,** a name meaning "carbon to which water has been added." The most common sugars have the formula $C_5H_{10}O_5$ (the five-carbon sugars such as ribose) or $C_6H_{12}O_6$ (the six-carbon sugars such as glucose). Some are only a single sugar monomer; they are known as **monosaccharides.** When two sugar monomers combine, we have a **disaccharide** such as sucrose, or common table sugar ($C_{12}H_{22}O_{11}$). Macromolecules made of many sugar monomers are complex carbohydrates, or **polysaccharides** such as cellulose or starch. In some cases, polysaccharides are made of a single kind of sugar; in others they are made of two or more different kinds.

The differences between carbohydrates often can be simply traced to subtle differences in the links between their sugar monomers. Both cellulose and starch are composed entirely of glucose, as Figure 3.9 suggests. Cellulose is a tough, fibrous structural material (the main component of paper and wood), and most animals find it impossible to digest. Yet starch is one of their most important foods!

Later chapters will explain how carbohydrates are used as structural support and food reserves. For now, it's enough to note that carbohydrates are one of the main sources of energy needed to build *all* other macromolecules of life.

Lipids In the octane molecule portrayed earlier, all the carbon and hydrogen atoms share electrons equally. That means all the C—C and C—H bonds are nonpolar, which means the molecule has no chemical basis for interacting with water. But many such molecules *can* interact somewhat with water if they have polar groupings attached to them. **Lipids** are such molecules: all are hydrophobic structures to which at least some hydrophilic atoms or atomic groupings are attached. Lipids include fats, steroids, and phospholipids.

For instance, when three hydroxyl groups (—OH) are attached to a certain hydrocarbon, we have glycerol:

$$
\begin{array}{ccccc}
 & H & H & H & \\
H & - C & - C & - C & - H \\
 & | & | & | & \\
 & OH & OH & OH &
\end{array}
$$

When an acid group (—COOH) is attached to a certain hydrocarbon, we have a **fatty acid,** which has a hydrophilic "head" and a hydrophobic "tail":

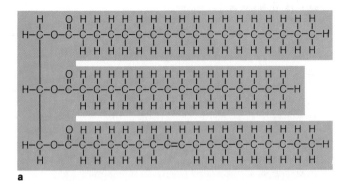

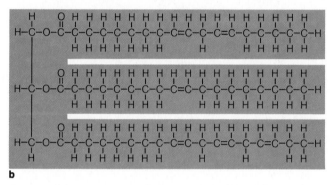

Figure 3.10 Two kinds of fats, formed by dehydration synthesis of glycerol and three fatty acid "tails." (a) Beef fat, which is essentially saturated. (b) Linseed oil, which is an unsaturated fat.

When a glycerol molecule is attached by dehydration synthesis to fatty acids, we have a molecule of **fat** (Figure 3.10). Butter and other animal fats are called *saturated* fats because some maximum possible number of hydrogen atoms is covalently bonded to the carbon backbone of their fatty acid component. Vegetable oils (such as corn and linseed oil) are *unsaturated* fats because some carbon atoms have a double covalent bond between them, which means the fatty acid component isn't fully saturated with hydrogen atoms.

Fats characteristically are insoluble in water. It is only at their hydrophilic attachment points that they can be broken apart by hydrolysis, which renders them soluble. That is why fats can be stockpiled as food storage materials in the watery environment of cells. What we call "obesity" is a sign that fat molecules are being stockpiled faster than the body can use them up. What we call "dieting" means living off fat reserves faster than replacements are taken in.

Steroids, another kind of lipid, are based on multiple ring structures

to which a variety of atoms can be attached. Cholesterol is a steroid compound. One group of steroids, produced in the adrenal glands, helps regulate sugar and salt levels in the blood. Another group, produced in the reproductive organs, helps regulate the functioning of those organs as well as sexual activity. Like fats, steroids are insoluble in water. That is why they can pile up in the blood vessels of people who are prone, by birth and/or by life-style, to accumulating lipids in the bloodstream (Chapter Twelve). When enough accumulate in a blood vessel to stop the flow of blood, a heart attack may be the result.

Other important lipid molecules are **phospholipids,** which are assembled by dehydration synthesis from glycerol, fatty acids, a phosphorus-containing compound, and usually a nitrogen-containing compound. Life probably never would have developed without lipids and phospholipids, for they are the basis of cell membranes.

Proteins Next to water, proteins are the most abundant molecules of a living organism. Some proteins act as structural building blocks; others act as enzymes, as you will see. A **protein** is built by dehydration synthesis from amino acid monomers. **Amino acids** are substances containing an amino group ($-NH_2$) and an acid group ($-COOH$). All proteins are made of some combination of twenty common amino acids, and all are assembled according to the same plan. Special covalent bonds (*peptide* bonds) link the acid group of one amino acid to the amine group of the next (Figure 3.11). The atoms linked in this way form a chain known as **peptide backbone.** The order of atoms in the backbone is somewhat monotonous:

$$N-C-C-N-C-C-N-C-C-N-C-C-N-$$

and so on.

Even though all proteins have the same kind of backbone, there is immense diversity in protein structure and function. The reason is that each kind of amino acid has a unique **R group:** a group of atoms that project from the protein backbone once the amino acid is locked in place. In theory, any amino acid may follow any other in the backbone, which means the number of possible R group sequences is extraordinary. We can calculate how many different dipeptides (molecules containing two linked amino acids) may be built from the twenty common amino acids. With twenty choices available for the first amino acid and twenty for the second, there are 20×20, or 400 possible dipeptide sequences. For a *small* protein with merely 100 amino acids, there are 20^{100} possible sequences. That number is billions of times larger than the number of stars in the known universe!

The specific sequence of amino acids in a protein determines what its structure will be. For it happens that the backbone we've been describing is not a stiff rod. Once

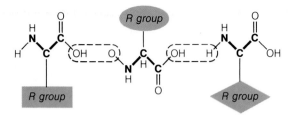

a

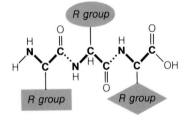

b

Figure 3.11 Assembling a protein. (**a**) Through dehydration synthesis, H atoms and —OH groups are removed from a series of amino acid monomers. The monomers differ from one another only in their R groups. (**b**) Covalent bonds form between a carbon atom of one monomer and a nitrogen atom of the next monomer in line. These so-called peptide bonds are shown here as beads. The rest of the protein backbone is shown in boldface.

Figure 3.12 Three-dimensional structure of the protein backbone in a hemoglobin molecule. Hemoglobin is a red pigment circulating in the blood of higher animals and carrying vital oxygen to and from tissues.

In different animals, there are slight differences in the amino acid sequence of hemoglobin. As a result, the backbone of the molecule bends in a slightly different way in each case, which gives rise to large differences in the way the molecule functions to pick up and release oxygen.

Such differences are adaptations to different oxygen levels in various kinds of environments, and to how much oxygen must be supplied to a given animal's tissues. The differences are vital. If a worm were forced to use insect hemoglobin, it would suffocate. Just as surely, a winged insect would plummet to the ground if it tried to fly with nothing but worm hemoglobin in its blood. (From R. Dickerson and I, Geis, *The Structure and Action of Proteins.* Menlo Park, Ca.: Benjamin. © 1969 Dickerson and Geis)

assembled, it can bend and twist randomly in the cellular environment. As it does, different R groups in different parts of the protein are brought together, and they attract or repel one another (depending on what kind of charged groups they encounter). The protein assumes precisely the shape in which the most attractive interactions possible can occur.

We can summarize the emergence of protein structure with three simple terms. **Primary structure** refers to the sequence of amino acids, strung one after the other in the backbone. **Secondary structure** refers to the way the backbone twists about its own axis as a result of hydrogen bonds, which tend to form between the C=O of one amino acid and the N—H of another some distance away (Figure 3.11). Often the backbone twists helically about its axis, much like a spiral staircase. If nothing else intervened, all proteins would be helical. (Indeed, helices are common to many proteins, such as those in skin and bone.) But most assume **tertiary structure:** they bend and fold asymmetrically in space, largely because of interactions among R groups. The final shape is no accident: it arises precisely because of the primary sequence of amino acids. In each case, the shape is exactly what is needed to interact chemically with some specific substance in the cellular environment, in a way that is important to cell functioning (Figure 3.12).

Nucleic Acids For all organisms, life depends on proteins made of a unique sequence of amino acids, of some specified length. One unit of heredity—a **gene**—dictates that sequence and length for each kind of protein. Recall that unlimited kinds of proteins are potentially available to the world of life. Hence there *could* be unlimited combinations of genes. The array of genes found in modern organisms is, by comparison, a mere drop in the bucket. But these are the genes that have *worked;* they have survived the tests of time and circumstance. Later you will read how these genes, and the proteins they specify, are the physical basis for evolution and life's diversity.

Each gene is a small stretch of nucleic acid molecule called **deoxyribonucleic acid,** or **DNA.** Like all nucleic acid molecules, DNA is built from subunits called **nucleotides,** which contain carbon, hydrogen, oxygen, and nitrogen. Only four kinds of nucleotides are needed to make DNA. The sequence in which they are assembled gives DNA, hence each gene, its unique character.

DNA is a double-stranded molecule that twists about its own axis. To build a given protein, these two strands must be temporarily unwound to expose the corresponding gene. Another class of nucleic acids "reads" the gene's chemical instructions and translates them into the actual protein. These are the **ribonucleic acids,** or **RNAs.** For now, it's enough to know that DNA is the storehouse of genetic instructions, and RNA is the means by which those instructions are translated into the proteins on which all forms of life are based.

Enzymes: Speeding Up Interactions Between Molecules

We began this chapter with a discussion of energy. But here we are, writing symbols of atoms and molecules. What's the connection? If all the atoms in the universe were frozen once and for all into particular shapes, there wouldn't be any connection. *And there would be no life.* But molecules move about. They collide with one another. And when they do, they often undergo **chemical reaction:** the bonds holding each molecule together are broken and parts become rearranged into new molecules. With chemical reactions, molecules and cell structures are built up or torn down, energy is stored away, and energy is released to drive life processes. Some are *exergonic* reactions: because some energy is lost during the rearrangement, the products end up with less energy than the starting substances (the reactants). Others are *endergonic:* because extra energy is fed into the reaction in some way, the products end up with more energy than the reactants. In living organisms, exergonic and endergonic reactions typically occur not in isolation but along interconnected routes called **metabolic pathways.** In these metabolic pathways, products from one reaction serve as starting substances for the next in the series; and energy released during one reaction is harnessed by the next reaction in line.

But left to themselves, the starting substances that are normally found in a living organism tend to rearrange themselves slowly. For life to continue, reactions that would normally take years or decades to occur on their own must be made to occur in a fraction of a second. Now, for *any* reaction to occur, molecules not only must collide, they must collide with a certain minimum energy. This minimum collision energy is the **activation energy** for the reaction. It's like striking a match to start wood burning. The heat energy makes wood molecules jostle faster, and they start colliding with oxygen molecules in the air. When

they collide with enough energy to react, bonds are broken, new bonds are formed, and light and heat energy are released. This released energy increases the energy collisions among other molecules, which leads to more energy changes. In this way, the wood's potential energy continues to be released. But without that initial "push" from the match, the average kinetic energy of the molecules is not enough for wood to begin reacting with the oxygen in the surrounding air.

Of course, if you have ever burned your fingers on a match, you know that high heat is not a good way of triggering reactions in living systems. Instead, metabolic reactions are made to occur at lower temperature through the action of a special class of protein molecules, which are known as **enzymes:**

An enzyme speeds up (catalyzes) the rate of a reaction because it lowers the activation energy for that reaction.

Each kind of enzyme interacts with a specific set of reactants by forming temporary bonds with them (Figure 3.13). In this state, the existing bonds of the reactants are weakened. Thus they combine more quickly with each other to form a molecule of product, because not as much activation energy is needed to break the bonds. Then they are released, and the enzyme—which has not in any way been altered by the reaction—is ready to unite another set of reactants. How fast the process can take place is staggering. Some enzymes can bind reactants, activate them, then release the products *10 million times a second!*

So far, we have been talking as if reactions proceed only from reactants to products. But many products can also revert to reactants, given enough activation energy. Often, activation energy is likened to a small hill over which reactants must be pushed before they can coast down the other side (Figure 3.14). But this analogy raises an intriguing question: In reverse reactions, what happens if the products contain *less* energy than the reactants? How do they climb the steeper side of the energy hill? The answer is this: When reactant molecules are present in high concentrations (or when they are heated, or under high pressure), they are more likely to move about and collide with enough energy to react, which they do. As time goes on, not as many reactant molecules are left, which means the reaction rate drops. But all the while, the concentration of *product* has been increasing. Thus, even though it takes a larger dose of activation energy to get the reverse reaction going, the higher concentration means product molecules will start colliding more often. And some fraction of them will have enough energy so that they will be pushed back over the energy hill.

As long as both reactant molecules and product molecules are present, most chemical reactions proceed in two directions. And eventually an **equilibrium point** is reached:

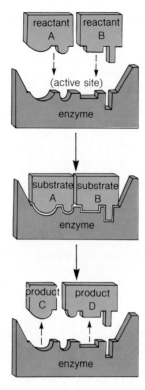

Figure 3.13 Model of enzyme action. Like all proteins, enzymes are folded into three-dimensional shapes. For enzymes, the folding creates a "groove" called the *active site*. The pattern of R groups projecting from the groove is an exact, three-dimensional complement of chemical groups present on the specific substrates an enzyme is meant to bind. This complementary "fit" is the reason why an enzyme can bind and react with specific substrates.

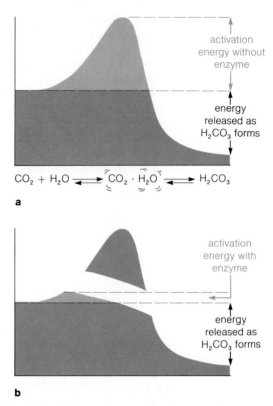

Figure 3.14 Energy hill diagram for the reaction between water and carbon dioxide. (**a**) Activation energy is like a small hill that reactants must be pushed over to get a reaction going. (**b**) An enzyme works to dig a tunnel through the energy hill (lower the required activation energy) in both directions.

the reaction runs as fast in the reverse direction as it does in the forward direction. Unless conditions change, there is no further *net* change in the concentrations of either reactants or products.

To illustrate the points made so far about chemical reactions, we can give a simple example of conditions under which equilibrium is reached, how equilibrium can be disturbed, and what effect enzymes have on achieving equilibrium. This example first describes how carbonic acid forms and breaks down in a bottle of cola, then how it forms and breaks down in your blood.

When soft drinks are manufactured in a bottling plant, carbon dioxide is pumped under high pressure into each bottle before it is sealed. At that time, some of the carbon dioxide molecules dissolve in the liquid, where they begin to combine with water molecules to form carbonic acid:

$$CO_2 + H_2O \longrightarrow H_2CO_3$$

$$\text{carbon} \quad \text{water} \qquad \text{carbonic}$$
$$\text{dioxide} \qquad\qquad\quad \text{acid}$$

After the bottle is sealed, this reaction continues. But as the concentration of product molecules increases, the reverse reaction occurs:

$$CO_2 + H_2O \longleftarrow H_2CO_3$$

Eventually the equilibrium point is reached and there is no further net change in the concentrations of carbon dioxide, water, or carbonic acid.

But when you pry off the bottle cap, you reduce the pressure above the liquid in the bottle. The carbon dioxide gas remaining above the liquid rushes outside. Some of the carbon dioxide dissolved in the water fizzles out, too, so that the rate of carbonic acid breakdown now exceeds the rate of carbonic acid formation. As a result, the reaction runs faster in the reverse direction. It continues to do so until a new equilibrium is reached. That happens when the rate of carbon dioxide leaving the liquid in the bottle is the same as the rate at which carbon dioxide wanders in from the surrounding air. It takes many, many hours, and by that time your drink is very flat.

The same kind of reactions go on in your bloodstream. When your body cells consume food molecules (carbohydrates, fats), carbon dioxide is given off as a by-product. But if carbon dioxide accumulates, it poisons cells, so it must be quickly removed. As blood passes by cells, carbon dioxide moves into the bloodstream, where the carbon dioxide concentration is lower. There, most of it combines with water molecules to form carbonic acid. When the blood reaches the lungs, the reaction is pulled in the reverse direction, for the blood gives up carbon dioxide to air in the lungs. Because of the lower carbon dioxide concentration, carbonic acid breaks down and reverts to reactants.

The rate at which carbonic acid forms and breaks down of its own accord is so sluggish that cells would be poisoned long before the reactions were completed. (It takes many hours, recall, for a carbonated drink to go flat.) Your blood is pumped so fast through your body that it has only an instant to pick up all the carbon dioxide it can in any given place. Seconds later it is in the lungs, where it has only an instant to unload its cargo. Fortunately, carbonic acid formation and breakdown is accelerated by *carbonic anhydrase,* an enzyme in your blood. In its action, carbonic anhydrase resembles the thousands of other enzymes that speed up thousands of other reactions:

Enzymes speed reactions to equilibrium. They affect only the rate at which a reaction approaches equilibrium; they do not change the proportions of products and reactants that will be present once equilibrium is reached.

Thus, if a bottle of cola were to contain carbonic anhydrase, it would go flat with amazing speed once you opened it—but it would not get any flatter than it otherwise would.

Before leaving the subject of enzymes, we should make one more point. *Not only do enzymes speed up the reactions that do occur, they help determine which ones can occur.* Each metabolic reaction requires a unique enzyme. Imagine two substances, A and B. They can react in one way to form products C and D, or in another way to form products E and F. If a cell contains only the enzyme that speeds up formation of C and D, the other products can't be assembled. But if the cell contains enzymes for both reactions, it can regulate how much of each set of products is formed—by regulating how much of either enzyme it constructs! As you will see later, a cell controls its entire chemistry by controlling the types and amounts of enzymes it deploys.

Energy Flow and the Cycling of Materials Through the Living World

Matter and energy, again, are the two basic aspects of our world. It is impossible to consider one without the other, for the ways in which organisms trap energy, store it in molecules, then break down those molecules to release energy are the most central events in the living world. These events involve systems of producing and using food at the level of cells, organisms, species, and populations. The energy-trapping process is known as photosynthesis; the energy-extraction processes are known as glycolysis and cellular respiration. Although later chapters will tell you more about them, the following overview of how these processes are linked together will give you a sense of how important they are for the stability of all forms of life on this planet.

We now assume that when life originated, the environment was rich in molecules based on carbon, hydrogen, oxygen, and nitrogen (Chapter Fifteen). The first life forms probably extracted energy from carbon-rich sugar molecules through a pathway similar to one living cells can still use. In this pathway, called **glycolysis,** energy is released as a sugar molecule such as glucose is broken down into small molecules such as lactic acid:

$$C_6H_{12}O_2 \longrightarrow 2(C_3H_6O_3) + \text{energy}$$
$$\text{lactic acid}$$

In itself, glycolysis is a one-way meal ticket. As energy-rich molecules are used up, small molecules lower in energy accumulate. The inefficiency of this pathway undoubtedly led to the world's first energy crisis, for as more and more organisms appeared, increasing demands must have been made on the stockpiles of energy-rich molecules.

Competition for dwindling energy apparently led to simple but effective adaptations in existing cell structures: in some organisms, the machinery for breaking down sugar molecules was modified and extended: it could be used for building up sugar molecules from carbon dioxide and water. How were the reactions made to run in reverse? They were driven by sunlight harnessed as a source of added energy! Such is the nature of **photosynthesis:** trapping and converting sunlight energy to a form that can be used to build large energy-rich molecules from small energy-poor ones:

$$6CO_2 + 6H_2O \xrightarrow{\text{sunlight}} C_6H_{12}O_6 + 6O_2$$

But the pathways of glycolysis and photosynthesis were still not completely efficient at extracting energy from food molecules. Well over 90 percent of the chemical

potential energy locked up in a sugar molecule was still locked up in the disassembled parts after glycolysis. More than this, lactic acid apparently was beginning to pile up in the environment, and oxygen that was locked in water molecules prior to photosynthesis was now accumulating in the atmosphere. In high enough concentrations, lactic acid and oxygen poison the very cells that produce them. Thus, with the accumulating waste products, early life forms were polluting the place in which they had to live. How could the by-products of life be disposed of? The energy crisis was there—yet so was the solution. Oxygen, one of the pollutants, was the key.

In a process called **cellular respiration,** the waste products could be stripped of their hydrogen atoms, which could be combined with oxygen to yield carbon dioxide and water:

$$2(C_3H_6O_3) + 6O_2 \rightarrow 6CO_2 + 6H_2O$$

In this process, the remaining potential energy stored in each sugar molecule could be released. With cellular respiration, life became permanent and self-sustaining. Its final products are precisely the materials needed to build sugar molecules in photosynthesis. Thus the cycling of carbon, hydrogen, and oxygen through the energy pathways of living organisms came full circle. Many similar cycles have come to exist for other essential elements, such as nitrogen and phosphorus. But it is through the so-called "carbon cycle" (Figure 3.15) that organisms are locked into the greatest interdependence in their search for energy. This legacy of the early community of life represents something we as a species have yet to acknowledge:

The only route to a stable system of life is through efficient use of energy and the complete recycling of material resources.

Figure 3.15 The carbon cycle, in which energy-rich molecules of carbon, hydrogen, and oxygen flow through all organisms on earth. In this recycling of matter through time, each birth is affirmation of our ongoing capacity for organization, each death a renewal.

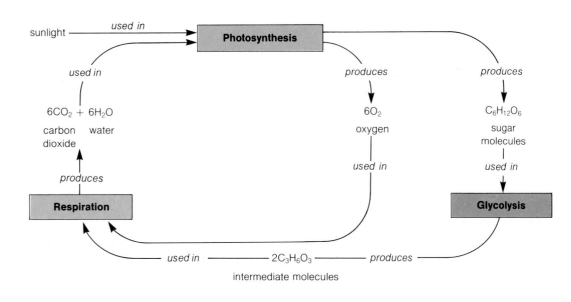

sunlight ——— *used in* ———→ **Photosynthesis**

used in

$6CO_2 + 6H_2O$
carbon   water
dioxide

produces ↑

Respiration

produces → $6O_2$
oxygen
used in

produces → $C_6H_{12}O_6$
sugar
molecules
used in ↓

Glycolysis

used in ——— $2C_3H_6O_3$ ——— *produces*
intermediate molecules

Perspective

Energy in space and time—that is the essence of the universe, the stars, the earth, all microorganisms, plants, and animals. Each living thing is organized energy. The shapes of living things originate in units of energy—protons and electrons—that come together into atoms. The number of electrons gives different kinds of atoms their unique character. The arrangement of electrons dictates the positions of atoms relative to one another, which dictates the unique forms of molecules, cells, and organisms.

And yet, perhaps one of the most difficult connections you are asked to perceive is the link between yourself—a living, intelligent being—and such remote-sounding things as the electrons of oxygen, hydrogen, nitrogen, and carbon. Is this really the stuff of humanity? Think back, for a moment, on the discussion of a simple water molecule. A pair of hydrogen atoms competing with an oxygen atom for a fair share of the electrons joining them doesn't exactly seem close to our daily lives. But from this simple inequality, the polarity of the water molecule arises. And that is a beginning for the organization of lifeless matter which leads, ultimately, to the organization of matter in *all* living things.

Imagine billions of water molecules jostling one another in space, all with that same asymmetry of charge. Now hydrogen atoms align themselves with the oxygen atom of neighboring molecules. In this way, hydrogen bonds form between molecules. This is the bond that spans the boundary between lifeless molecules and the molecules that make life possible.

For now you can imagine new kinds of molecules interspersed through the watery environment. Many will be nonpolar and will resist interaction with water; others will be polar and will respond by dissolving in it. And certain molecules, such as phospholipids, are both hydrophobic and hydrophilic. In swirling, agitated water (a primordial sea, perhaps, pounding against some ancient shore?) they form a two-layered membrane around a water droplet. As Chapter Four will indicate, such lipid bilayers are the basis for all cell membranes, hence all cells. The cell has, from the beginning, been the fundamental *living* unit.

With membrane organization, one small part of the environment can be isolated. With isolation, chemical reactions can be contained and controlled. The essence of life *is* chemical control. It comes through manipulating the energy pent up in each kind of molecule of life—carbohydrates, lipids, proteins, nucleic acids—and in each kind of atom needed to construct them and break them apart. It is not some mysterious force that explains this control. Instead it is a class of protein molecules—enzymes—that puts these molecules into action in precise, controlled ways. It is not some mysterious force that tells the enzymes when and what to build, and when and what to tear down. Instead it is

a chemical responsiveness to the kinds of molecules present in the environment, and to changes in that environment; it is a responsiveness built into the thousands or millions of atoms making up the molecules of life. And it is not some mysterious force that creates the enzymes themselves. DNA, the slender double strand of heredity, has the chemical structure—*the chemical message*—that allows molecule to faithfully reproduce molecule, one generation after the next. Those DNA strands lead to countless molecules in the trillions of cells in your body. A specific sequence of information in DNA stamped each cell of your body with individuality the instant you were conceived.

So yes, oxygen, hydrogen, nitrogen, and carbon represent the stuff of you, and us, and all of life. But it takes more than molecules to complete the picture. It is because of the way this stuff is organized and maintained by a constant flow of energy that we are alive. It takes outside energy from such sources as the sun to trigger the formation of new energy-rich molecules. Once molecules are assembled into organisms, it takes outside energy derived from food and water and air to sustain their organization. Individual plants, animals, and microorganisms are part of an interconnected web of energy and materials use in the environment. Should energy fail to flow inward at any one of these levels, matter in that level will join the universal trend toward increasing disorder.

For energy flows through time in only one direction—from forms rich in potential energy through forms having progressively less usable stores of it. Only as long as sunlight energy flows into the web of life—and only as long as there are molecules to recombine, rearrange, and recycle with the aid of that energy—does life have the potential to continue in all its rich diversity.

Life is, in short, no more *and no less* than a marvelously complex system of prolonging order. Sustained by energy transfusions, it continues because of its capacity for self-reproduction—the handing down of instructions and the means for organizing energy and materials to maintain life, generation after generation. Even with the death of individual plants, animals, and microorganisms, life is prolonged. For their molecules are released to be recycled once more, providing raw materials for new generations. In this flow of matter and energy through time, each birth is affirmation of our ongoing capacity for organization, each death a renewal.

Recommended Readings.

Roller, A. 1974. *Discovering the Basis of Life.* New York: McGraw-Hill. A fine introduction to the molecular basis of life. Paperback.

Watson, J. D. 1976. *Molecular Biology of the Gene.* Third edition. Menlo Park, California: Benjamin. This book is *the* classic on the molecular basis of life processes.

Unit II

The Shared Heritage of Life

Chapter Four

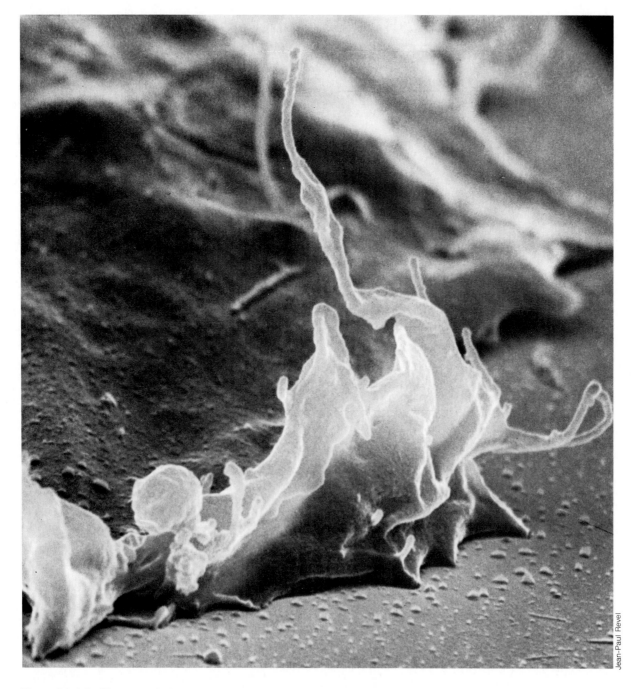

Figure 4.1 A ''cell''—a vaguely dreary name more suggestive of carton boxes than of things seething with life. Shown here, the dynamic ruffling edge of a hamster cell on the move.

Cell Structures and Their Function

Early in the sixteenth century, Galileo arranged two glass lenses in a cylinder. With this instrument he happened to look at an insect, and thereby came to describe the stunning geometric pattern of its tiny compound eyes. Thus Galileo, who was not a biologist, was the first to record a biological observation made through a microscope. The systematic study of the cellular basis of life was about to begin. First in Italy, then in France and England, biologists began to explore architectural details of a world whose existence had not even been suspected.

At mid-century Robert Hooke, "Curator of Instruments" for the Royal Society of England, was at the forefront of these studies. When Hooke first turned one of his microscopes to a thinly sliced piece of cork, he observed tiny, empty compartments that reminded him of prison chambers; he called them "cells." (They were actually dead cells, which is what tree bark is made of, although Hooke didn't think of them as being dead because he didn't know cells could be alive.) He also noted that cells in other kinds of plant materials contained "juices." He did not speculate on what the juice-filled structures might represent.

Given their simple instruments, it's amazing these pioneers saw all that they did. Antony van Leeuwenhoek, a shopkeeper, had the keenest vision of all and/or the greatest skill in glassworking. He devoted his spare time to constructing lenses and to observing everything he could politely get hold of, including sperm cells. He even observed a single bacterium—an organism so small it would not be seen again until the nineteenth century! But this was primarily an age of exploration, not interpretation. Once the limits of those simple instruments had been reached, biologists gave up interest in cell structure without ever having thought to look for explanations of what it was they had seen.

Then, in the 1820s, improvements in lens design tempted biologists back into the cell. And now it became clear that small structures were suspended in those cellular "juices." The botanist Robert Brown, for instance, reported the presence of a spherelike structure inside every plant cell he examined; he called it a "nucleus." By 1839, the zoologist Theodor Schwann had confirmed the presence of cells in animal tissues. About this time, he began working with Matthias Schleiden, a botanist who had concluded that cells are present in all plant tissues, and that the nucleus is somehow paramount in the reproduction of all cells. It was Schwann who distilled the meaning of these new observations into what came to be known as the first two principles of the **cell theory:**

All organisms are composed of one or more cells.

The cell is a basic unit of organization in all living things.

Another decade passed before Rudolf Virchow finished exhaustive studies of human cells and completed the basic theory with this third principle:

All cells arise from preexisting cells.

Not only was a cell viewed as the smallest living unit, the continuity of life was now seen to be arising directly from the division and growth of single cells. *Within each tiny cell, events were going on that had implications of the most profound sort for all levels of biological organization!*

Identifying these events has been difficult, for cells are incredibly small. For instance, if you were to line up 2,000 of your blood cells, you would have a string only about as long as a thumbnail. Some cells are less than 100 nanometers across. (A nanometer is a billionth of a meter.) The finest detail the modern light microscope can bring into focus is about 200 nanometers across, so you can see why early microscopists had to give up when they did. Only in the past few decades have biochemical studies and electron microscopy enlarged the picture of cell structure and function (Figures 4.2 and 4.3).

A Generalized Picture of the Cell

Today we know cells are not the simple bags of juices that early investigators thought them to be. Each living cell is a complex island of order in a universe tending toward disorder. Each taps energy from its environment, and *uses* it to acquire, break down, and build up certain molecules. In this way, the cell maintains the structures on which life depends. How does the cell carry out all these interrelated tasks? In the most fundamental sense, cell functioning depends on three things:

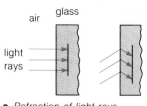

air glass

light rays

a *Refraction of light rays (The angle of entry and the molecular structure of the glass determine how much they will bend)*

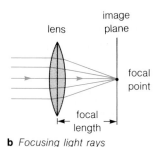

lens

image plane

focal point

focal length

b *Focusing light rays*

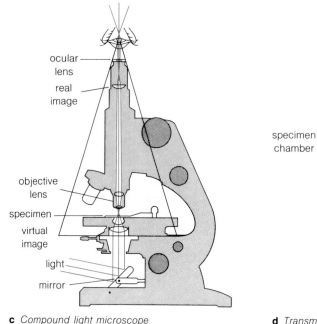

ocular lens

real image

objective lens

specimen

virtual image

light

mirror

c *Compound light microscope*

electron gun

condensing lens

specimen chamber

objective lens

projector lens

viewing window

fluorescent screen or photographic film

d *Transmission electron microscope*

Figure 4.2 The microscope: gateway to the cell.

Light Microscopes Light microscopy is based on the bending (refraction) of light rays. Light rays pass straight through the center of a simple convex lens (**b**). The farther away they are from the center, the more they bend. A lens must be curved so all light rays coming from the object being viewed will be channeled to a single place behind the lens. A two-lens system, the *compound light microscope,* is shown in (**c**).

One problem with compound light microscopes is spherical aberration: when you bring tiny objects close to the objective lens, they can blur. Another problem is that light comes in different wavelengths (colors). For instance, red is a long wavelength, blue is a short one. Blue and red light rays are bent differently as they pass through the same piece of glass, so they don't end up at the same place. If you sharply focus the red part of an image, the blue part may be out of focus, and vice versa. The distortion (chromatic aberration) causes the color haloes you sometimes see around images of small objects. Only in better microscopes does careful lens design overcome these problems.

If you wish to observe a *living* cell, it must be small or thin enough for light to pass through. (Some microorganisms and single cells are small enough; complex tissues are not. And when thin sections are made of cells in such tissues, the cells die.) Also, structures inside a cell can be seen only if they differ in color and density from their surroundings—but most are almost colorless and optically uniform in density. Specimens can be stained (exposed to dyes that react with some cell structures but not others), but staining usually alters the structures and kills the cells. Finally, dead cells begin to break down at once, so they must be pickled or preserved before staining. Most

observations have been made of dead, pickled, and stained cells. With the *phase contrast microscope* (in which small differences in the way different cell structures refract light are converted into larger variations in brightness), live cells *can* be observed as they actively move about. But the need for transparent specimens is even more critical.

No matter how good a lens system may be, when the magnification exceeds 2,000× (when the image diameter is 2,000 times as large as the object's diameter), cell structures will appear larger but not clearer. It's like what happens when you use a magnifying glass to enlarge a newspaper photograph. When you hold the magnifying glass too close, you see only black dots. There is no way to see a detail as small as or smaller than a dot; the dot would cover it up. In microscopy, something like dot size intervenes to limit *resolution* (the property dictating whether small objects close together can be seen as separate things). That limiting factor is the physical size of wavelengths of visible light. Red wavelengths are about 750 nanometers and violet wavelengths about 400 nanometers; all other colors fall in-between. If an object is smaller than about one-half the wavelength, light rays passing by it will overlap so much that the object won't be visible. The best light microscope can resolve detail only to about 200 nanometers.

Transmission Electron Microscope The vibrations of electrons are much smaller than the smallest visible light wavelengths. How fast an electron vibrates depends on how much energy it has. The more energy, the shorter the wavelength. It takes very little energy to excite an electron to wavelengths of about 0.005 nanometer—about 100,000 times shorter than those of visible light!

Ordinary lenses can't be used to focus such accelerated streams of electrons, because glass scatters electrons. But each electron carries an electric charge, so a magnetic field can be used to divert it along

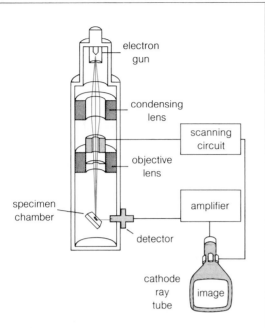

e *Scanning electron microscope*

certain paths. Accelerated electrons can be sent through an electro-magnetic field designed in a way to channel them to a focal point. The *transmission electron microscope* (**d**) uses such "magnetic lenses" of this sort.

The transmission electron microscope has greatly increased our knowledge of cell structure. But it, too, has limitations. Actual resolution of existing models ranges between 0.2 and 1.0 nanometer because there is an electrical equivalent to spherical aberration. Besides that, electrons must travel in a vacuum (otherwise they would be randomly scattered by molecules making up the air). Cells can't live in a vacuum, so living cells can't be observed at this higher magnification. Besides, specimens must be sliced extremely thin so that electron scattering will correspond to the density of different structures. (The more dense the structure, the greater the scattering and the darker the area in the final image formed.) Specimen fixation is crucial. Fine cell structures are the first to fall apart when the cell dies, and artifacts (structures that don't exist in a real cell) may result. Because most cell materials are relatively transparent to electrons, they must be stained with heavy metal "dyes," which can create more artifacts.

Scanning Electron Microscope With a *scanning electron microscope* (**e**), a narrow beam of electrons is played back and forth across a specimen's surface, which has been coated with a thin metal layer. The metal reflects or scatters electrons something like the way an object's surface reflects light. The reflection pattern is detected with equipment similar to a television camera, and an image is formed in much the same way that an image forms on a television screen. Scanning electron microscopy is restricted to specimen surfaces, and it does not approach the high resolution of the transmission instruments. But its images have a fantastic three-dimensional quality (Figure 4.3).

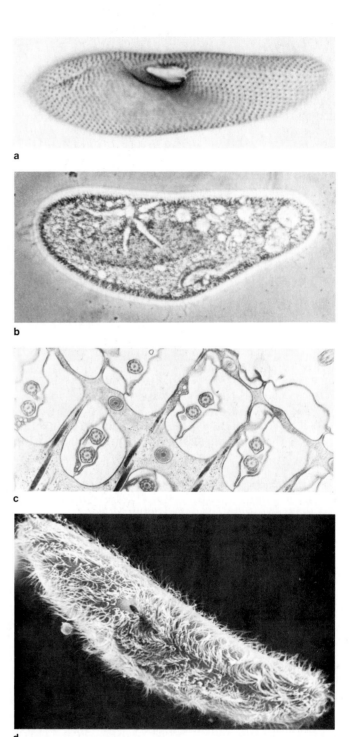

Figure 4.3 Comparison of image-forming abilities of microscopes. The specimen in all cases is *Paramecium.* (**a**) Conventional light microscope (bright-field). 750×. (**b**) Phase contrast microscope. 400×. (**c**) Transmission electron microscope (glancing section through the ventral surface). 17,800×. (**d**) Scanning electron microscope. 550×. (All photographs Gary W. Grimes and Steven W. L'Hernault)

First, *a selective barrier must be maintained between the cell's interior and the exterior environment.* The boundary must isolate the cell's organized internal parts from the surroundings. More than this, it must have the chemical means for capturing certain materials from the surroundings and moving them into the cell. It must also keep out other materials at all times or admit some only in required amounts. As conditions change inside or outside the cell, the boundary must be chemically responsive in precise ways to the change.

Second, *tens of thousands of precise chemical reactions must be promoted and regulated.* Because most reactions are linked to others in metabolic pathways, proper materials must be assembled and channeled to other reaction sites. But many cellular reactions are competitive or incompatible with one another. For instance, what one reaction builds up, another reaction is specifically designed to tear down. So the interior of the cell must have different reaction sites. Some sites might be loosely connected with others, some might be linked in a rigid way, and still others might be relatively isolated. Whatever the degree of isolation, the reactions must be capable of being called into and out of action in a controlled, integrated way.

Third, *the cell must have a set of coded blueprints and instructions for building every cell structure and operating every aspect of cell function.* Each cell must have the set of DNA molecules characteristic of the species to which that cell belongs. The chemical messages contained in DNA enable the cell to maintain itself as a living unit (Chapters Three and Seven). Obviously, the DNA must be in a form that can be passed on from generation to generation, so that each *new* cell that is formed comes equipped with the same vital chemical messages.

The cell's selective boundary later between the external and internal environment is the zone of the **plasma membrane** (along with any external cell-built structures such as cell walls). The boundary layer encloses the zone of the **cytoplasm,** a semifluid substance in which various cell structures are embedded. (Early investigators dubbed these structures *organelles,* meaning "little organs.") Suspended in the cytoplasm is a zone of hereditary control where DNA molecules are more or less concentrated. In all organisms but monerans, this zone is the **nucleus.** Pathways of chemical communication weave through and link these three zones together.

Figure 4.4 depicts how these three zones might be arranged in typical cells. It is important to keep in mind, however, that calling them "typical" is like calling a squid or a watermelon a "typical" animal or plant. Although they typify the basic plan, cells and their contents come in all manner of sizes, shapes, and elaboration. Variety among them is as remarkable as the variety that exists among whole organisms. Even so, these generalized drawings may help you keep the intact, *living* cell in focus as you now look at its functional parts.

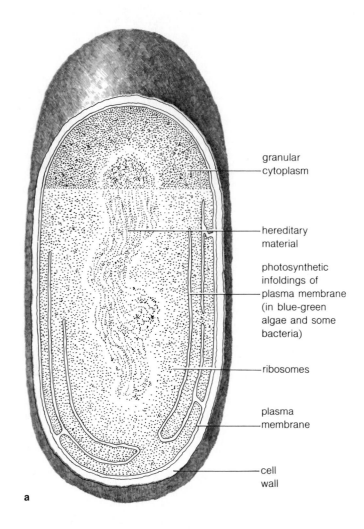

a

Figure 4.4 Generalized plans for cellular organization. In looking at these sketches, keep in mind that they are extremely generalized. Immense variation exists among single cells, just as it does among multicellular organisms.

(**a**) Bacteria and blue-green algae, as typified by this sketch, do not have a separate, membrane-enclosed nucleus. In fact, they don't have any membrane-enclosed organelles in which specialized reactions may take place. Molecules are relatively free to move about in the entire enclosed space. And the region in which hereditary material is found is variable.

(**b**) Plant cells and (**c**) animal cells do have semi-isolated, membrane-enclosed organelles. In the evolutionary sense, the most significant of these organelles is the nucleus. It represents isolation of the cell's hereditary material (DNA) from the clutter of other cell systems. Isolation means independent control, and independent control means the amount and complexity of DNA—hence the size and complexity of the organism—can be increased. The DNA can be parceled out during cell division without getting tangled or otherwise rendered useless by other cell parts and processes.

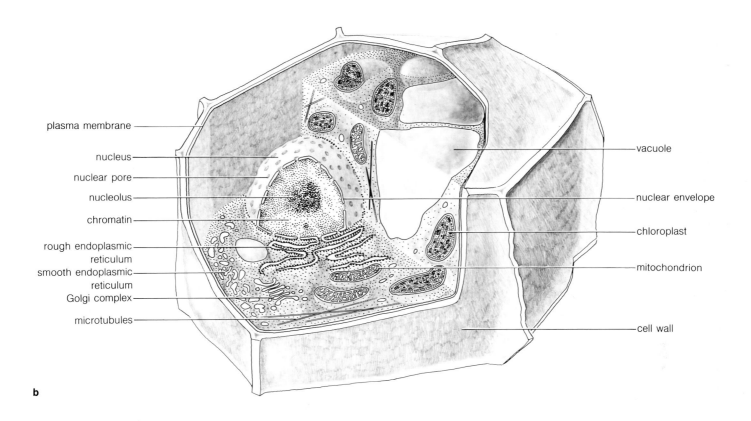

plasma membrane

nucleus

nuclear pore

nucleolus

chromatin

rough endoplasmic
reticulum

smooth endoplasmic
reticulum

Golgi complex

microtubules

vacuole

nuclear envelope

chloroplast

mitochondrion

cell wall

b

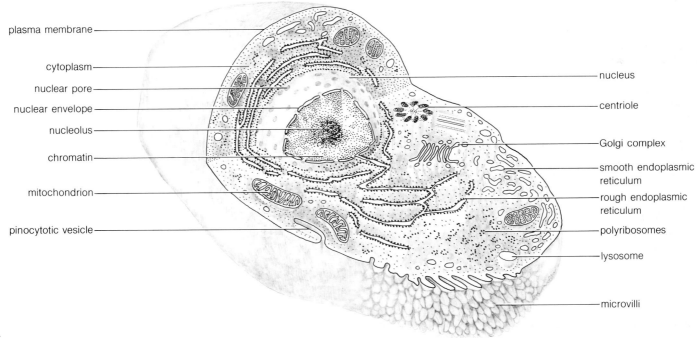

plasma membrane

cytoplasm

nuclear pore

nuclear envelope

nucleolus

chromatin

mitochondrion

pinocytotic vesicle

nucleus

centriole

Golgi complex

smooth endoplasmic
reticulum

rough endoplasmic
reticulum

polyribosomes

lysosome

microvilli

c

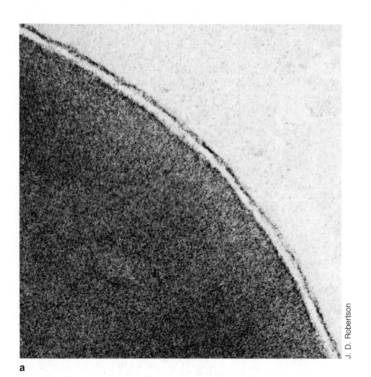

a

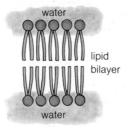

water

lipid
bilayer

water

b

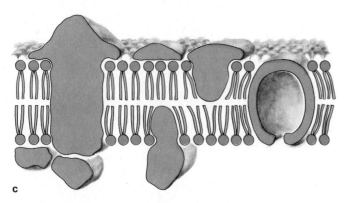

c

Figure 4.5 Cell membranes. (**a**) This is what a red blood cell membrane looks like at high magnification and in thin section. The three-layered (dark-light-dark) appearance is typical of all cell membranes. (**b**) Cell membrane structure. In an intact membrane surrounded by water, phospholipids form a bilayer with nonpolar ends pointed inward, tail to tail. (**c**) The current fluid-mosaic model of membrane structure.

Cell Walls Around the Plasma Membrane

Among bacteria, blue-green algae, many protistans, fungi, and plants, the thin plasma membrane is surrounded by a highly organized **cell wall:** a more or less continuous layer that not only confers rigidity and shape but also helps the plasma membrane hold in the cell's contents (Figure 4.4). Such walls are not impenetrable. They are usually porous, which means water and other materials can be exchanged across them.

Most cell walls are rich in carbohydrates. Among bacteria, carbohydrate chains completely surround the cell like a sack. Cell walls in plants, some fungi, and some protistans are made of cellulose fibers. During development, the cell lays down these small fibers as a primary cell wall. After the major growth phase, many types of plant cells also deposit an inner, secondary cell wall of cellulose. Cellulose walls are highly porous, but often they contain substances that alter permeability. For example, in the walls of water-conducting cells of some plants are deposits of cutin, which restricts water loss.

Animal cells don't produce walls, although in some cases they secrete materials to the outside. The animal way of life demands not rigid, external support but great flexibility and movement. To the extent that some animal cells assume a well-defined shape, it is usually because of a supporting network of fibers and filaments inside the plasma membrane.

Cell Membranes: Diversity on a Universal Theme

The plasma membrane, the boundary layer between the cell's interior and the outside world, holds in the cell's contents in much the same way that a plastic bag holds water. But unlike a plastic bag, the membrane actively transports some substances into and out of the interior even while it keeps others from traveling across. What is it about membrane organization that permits this selectivity? Like *all* membranes in a cell, the plasma membrane is composed mostly of proteins and lipids—especially phospholipids (Chapter Three). If the phospholipids are extracted from a membrane and added to a container of water, they assemble into a film on the water's surface. In this film the hydrophilic "heads" of the molecules are dissolved in the water, but the hydrophobic "tails" stick out of it. Because water normally presses in on both sides of a cell membrane, we can assume the lipid molecules take on a tail-to-tail bilayer organization, like that shown in Figure 4.5.

In our current view, the cell membrane is said to have a **fluid mosaic structure:** the lipid bilayer forms a sort of fluid sea in which diverse proteins float like icebergs. Studies of many membranes suggest that the lipid bilayer functions as the barrier itself, and that the proteins function in communication and metabolic activity. For example, some of the proteins bury their hydrophobic parts within the lipid bilayer and leave the hydrophilic parts projecting out; they may be sites where vital substances inside or outside the cell must become anchored to the membrane. Other proteins having a hydrophobic interior penetrate through the entire bilayer; they may act like a communication channel across the membrane. (Some may be direct channels for different kinds of molecules or ions. Others may be something like hollow balls, transporting water-soluble substances across the membrane in the manner of a revolving door.) Finally, some hydrophilic proteins merely associate with the highly polar "heads" of the phospholipid molecules on one side of the membrane or the other; they may serve to stabilize the lipid bilayer "sea."

Given that the plasma membrane is not a passive barrier but a dynamic part of cell functioning, imagine what would happen if it were to fold inward (or outward) even as it enclosed the cytoplasm of a simple cell. It most assuredly would increase the cell's membrane surface area, thereby providing more surface for more metabolic reactions! Now imagine what would happen if internal membrane material became "pinched off," so to speak, and wrapped around certain kinds of molecules and structures. Then it could form tiny, isolated chambers for special metabolic reactions. Such specialization would mean more efficiency for the cell: some reactions could be devoted exclusively to securing food energy, some to reproducing parts, or destroying used-up parts, and so forth. Thus, in one evolutionary view, there could have been selection long ago for variant cells having slight infoldings in their plasma membrane. Over time, cell lines would evolve that would have internal membranes developed to an amazing degree. Whatever its origins, internal membrane specialization is now a distinct characteristic of many cell types. The distinction is so significant that every cell—hence every organism—has been classified as belonging in one of two major divisions: the prokaryotes and eukaryotes.

The word **prokaryotic** means "before the nucleus." It refers to the fact that all prokaryotic cells (bacteria and blue-green algae) have no membrane-enclosed nucleus. All their life processes occur within a single, general chamber (the cell) bounded by a single membrane (the plasma membrane). In some cases, the plasma membrane may be infolded extensively into the cytoplasm. In photosynthetic prokaryotes, such infolding increases the membrane surface, which increases the possible reaction sites for photosynthesis. But these internal infoldings are not the same thing as separate organelles.

All other cells—those of protistans, fungi, plants, and animals—have a series of membrane-enclosed organelles, the most important of which is the nucleus. Hence they are called **eukaryotic,** which means "true nucleus." Specialized reactions are carried out in these compartments. Most of the remainder of this chapter is concerned with describing the structure and function of membranous organelles found in the diverse eukaryotic cells.

Organelles of the Cytoplasm

At any given moment, a cell is actively engaged in a multitude of tasks: energy metabolism, construction, storage, secretion of various substances, waste control, water regulation, structural support, and movement. In eukaryotic cells, these tasks are divided among various organelles in the cytoplasm. The structure and function of some of the more prominent organelles will be introduced here. Other chapters will take up the reactions going on within and between these organelles.

Energy Metabolism: Chloroplasts The **chloroplast,** found in the cells of plants and certain plantlike microorganisms, houses the photosynthetic machinery. In land plants, these organelles appear green, but in other photosynthetic organisms they appear green, yellow-green, red, or golden brown. Their color depends on the kinds of light-absorbing pigments embedded in their membranes. The pigment **chlorophyll,** for instance, appears green because it absorbs all wavelengths in the visible spectrum *except* those corresponding to green light, which it transmits. Chlorophyll is found in all chloroplasts. But sometimes it is masked by abundant accessory pigments.

A double membrane surrounds each chloroplast (Figure 4.6). Inside is a relatively formless substance (the stroma). Suspended in the stroma are membranous stacks (the grana) that look like piles of coins from the side. The membrane-bound disks in each stack are studded with particles that house enzymes and light-trapping pigments. The particles seem to be the sites where sunlight energy is trapped into energy-rich compounds. These compounds are transported to the stroma, where they are used as an energy source for building sugar molecules from carbon dioxide and water.

Energy Metabolism: Mitochondria Energy stored in organic molecules during photosynthesis must be released again and converted to a usable form if it is to power life processes. This energy conversion begins in the cytoplasm. But in all eukaryotic cells, it is *completed* in the tiny, membrane-enclosed organelle called the **mitochondrion** (plural, mitochondria). In these organelles, more than 90 percent of the pent-up energy in a food molecule is released. And the products of mitochondrial activity—molecules rich in a form of energy that can be put to use at once—flow to

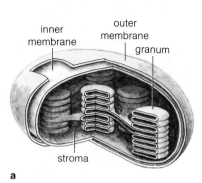

inner membrane   outer membrane   granum

stroma

a

Figure 4.6 (**a**) Generalized diagram of the membrane system of a typical chloroplast. (**b**) Electron micrograph of a chloroplast from a corn leaf cell, sliced lengthwise. 20,200×.

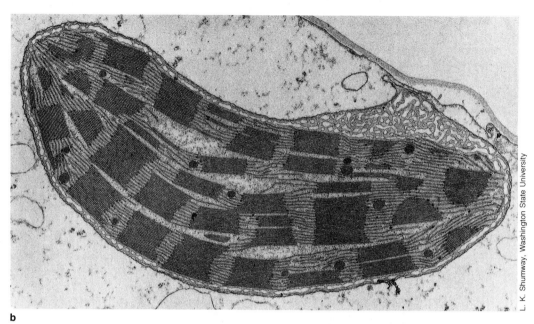

b

L. K. Shumway, Washington State University

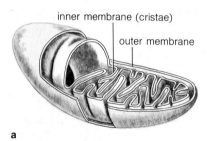

inner membrane (cristae)

outer membrane

a

Figure 4.7 (**a**) Generalized diagram of a mitochondrial membrane system. (**b**) Electron micrograph of a mitochondrion. 49,600×.

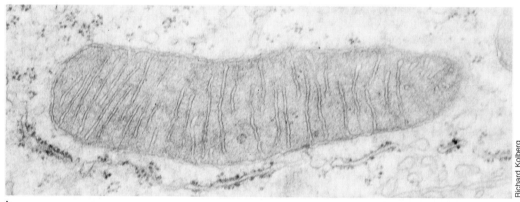

b

Richard Kolberg

the cytoplasm, the nucleus, and the plasma membrane. This energy conversion process is treated in Chapter Six.

Like chloroplasts, mitochondria have double membranes. The inner mitochondrial membrane is deeply infolded into a series of structures called **cristae** (Figure 4.7). In these membranes, the main energy conversions occur.

Because of its central role in energy metabolism, the mitochondrion has been called the powerhouse of the eukaryotic cell. Typically, the kinds of cells that demand high-energy output (for instance, muscle cells of a beating heart) have many more mitochondria than less active cells. They also have increased folding of the cristae, which increases reaction surfaces.

Interconversion of Substances: Microbodies Seldom in the real world is the supply of substances matched exactly to cellular needs. A cell must be able to convert excess amounts of one substance to a different kind that is needed but not available. Such conversions take place in many parts of the cell. But one type of organelle specializes in the task. In **microbodies** (Figure 4.8), fats can be converted to carbohydrates. Here, too, excess amounts of some amino acids can be converted for use as cellular fuel in place of sugars. Some of these conversion reactions produce hydrogen peroxide, which can poison cells. But microbodies contain enzymes keyed to hydrogen peroxide breakdown as well as enzymes keyed to its production—all in the same organelle! Thus microbodies are sometimes called "peroxisomes." As fast as hydrogen peroxide forms, these enzymes destroy it before it can destroy the cell.

Synthesis of Materials: Ribosomes Much of the energy produced in chloroplasts and mitochondria goes into the construction of new cell materials. One such task is the assembly of the all-important proteins. In all cells (prokaryotic and eukaryotic), the organelle responsible for building proteins is the **ribosome.** Ribosomes are not particular about the kinds of proteins they build. They simply build all the different kinds that the DNA tells them to (Chapter Eight). Intact ribosomes are composed of two subunits, as shown in Figure 4.9. These two-part construction sites are scattered individually through the cytoplasm. When in use, they are always found in groups known as **polyribosomes,** which are sometimes attached to cell membranes.

Synthesis of Materials: Endoplasmic Reticulum Many cells must produce at least two classes of proteins: some for use inside the cell and some to be exported for use on the outside. Proteins needed inside the cell are typically assembled on polyribosomes that float freely in the cytoplasm. But proteins destined for shipment elsewhere are typically assembled on polyribosomes that are attached to a membrane system called the **rough endoplasmic reticulum.** This membrane system varies in appearance. It may look like a series of tubes, vesicles, or flattened sacs. When it is highly developed, rough endoplasmic reticulum has sacs that are so large and flat they look like stacked sheets (Figure 4.10). Regardless of appearance, the functional plan is always the same: the membranes enclose and isolate some space from the cytoplasm. The polyribosomes are on the cytoplasmic side of the membranes. As they assemble proteins, the polyribosomes constantly send the newly formed molecules through special channels into the isolated space. In this way, proteins can be kept out of the way of the cell's metabolic machinery until they are needed.

In addition to rough endoplasmic reticulum, many cells contain a set of similar-looking vesicles. But in this case the membrane is not studded with polyribosomes; hence it is called **smooth endoplasmic reticulum.** The smooth version has different functions. Among other things, the smooth membranes contain the enzymes needed for assembling fats and other lipids. As with rough endoplasmic reticulum, the products are made on the cytoplasmic side of the membrane but are stockpiled in the space the membrane encloses.

Often the endoplasmic reticulum seems to form a continuous system of channels weaving through the cell. These channels evidently are material transport routes from one part of the cytoplasm to another. In such cases, the endoplasmic reticulum is a built-in communications and

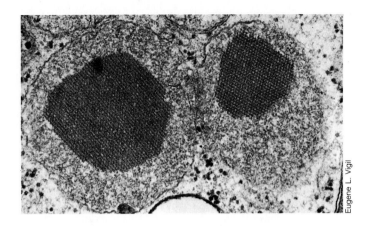

Figure 4.8 Microbodies from a castor bean plant cell, in thin section.

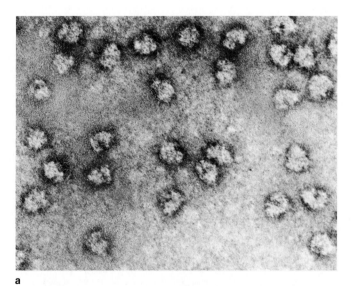

a

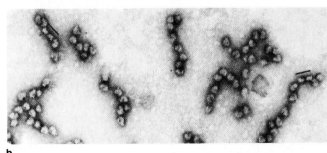

b

Figure 4.9 (**a**) Ribosomes and (**b**) polyribosomes. When the ribosomes are actively engaged in making protein, they are arranged in strings or whorls, called polyribosomes. (Nonomuri, Blobel, and Sabatini, *J. Molecular Biology,* 60:303—320, 1971)

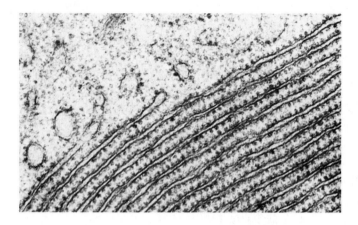

Figure 4.10 Rough endoplasmic reticulum, showing how the membrane surface facing the cytoplasm is studded with ribosomes. (Bloom and Fawcett, *A Textbook of Histology*, Saunders)

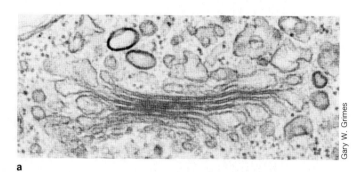

a

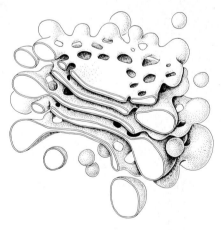

b

Figure 4.11 (**a**) Golgi complex, as it looks when sliced lengthwise. 53,300×. (**b**) Sketch of a Golgi complex. Notice the swollen sacs at the edges of the stack, and the free vesicles just beyond them. There may be a two-way flow of vesicles at the edge. Vesicles bearing products to be modified fuse at the edges. Once the contents are modified, vesicles pinch off and move to other cell regions.

guided transport system. Materials built in one region of the cell can be moved along to another region for further modification. In many cells, the system connects with both the outer membrane of the nucleus and the plasma membrane. Perhaps it is a channel for routing materials into and out of the control center of the cell.

Synthesis and Secretion of Materials: The Golgi Complex In addition to the two kinds of endoplasmic reticulum, a third membrane system is involved in the cell's program for assembling substances, packaging them up, and transporting them. It is the **Golgi complex** (so named for its discoverer). As Figure 4.11 shows, the Golgi complex differs from the other two kinds of membrane systems in its appearance. Its functions are also somewhat different. It is in the Golgi complex that most large carbohydrate molecules are built. Also, many proteins must have one or more carbohydrate chains attached to them before they can be shipped out of the cell, and such attachments are made in the Golgi complex. Vesicles containing proteins in need of such changes travel from the rough endoplasmic reticulum through the cytoplasm, until they reach and fuse with the Golgi membranes. Once their contents have been appropriately modified, they move on to the plasma membrane, where they release their contents to the cell's exterior. It is thought that the Golgi complex may figure in the assembly of new cell membranes.

Storage: Vesicles, Plastids, and Vacuoles What happens to all the materials being built in the cytoplasm? If they are not intended for export, special storage organelles concentrate them in places out of the way of metabolic activity. In animal cells, products are stored in small, membrane-bound sacs called **vesicles.** In plant cells, **plastids** are such storehouses (Figure 4.12). Although the chloroplast described earlier is known mostly as the organelle of photosynthesis, it also functions as a storage plastid. In some cells it contains starch grains that have been built up from sugar monomers. Leucoplasts are colorless plastids that also store starch grains. (A potato is little more than tightly packed leucoplasts.) Chromoplasts store various plant pigments such as those giving flowers and carrots their characteristic colors. Other plastids contain oils and similar high-energy products as food reserves.

Many plant cells contain a prominent **vacuole:** a large, membrane-enclosed chamber taking up much of the interior (Figure 4.12). Some vacuoles may be storage areas for metabolic by-products the plant cell can't use. Many of these by-products are in solution; others are insoluble and look like crystals suspended in the fluid. As you will read later, animal cells contain structures that are also called vacuoles. But in animal cells they are active in material transport, not storage.

Dismantling and Disposal: Lysosomes As part of their normal functioning, cells break down a variety of structures and molecules. Malfunctioning or worn-out organelles, particles brought into the cell, and so forth are dismembered, which frees their molecules for other purposes. In animals, the organelle that does this is the **lysosome,** a membranous container for a host of enzymes. These so-called hydrolytic enzymes are able to dismantle virtually all the molecules composing a cell. Obviously the enzymes could not be allowed to float about in the cytoplasm; they would destroy everything in sight. Instead they are isolated in small vesicles, the lysosomes.

The disposal of worn-out organelles such as the ones shown in Figure 4.13 proceeds in the following manner. First a membrane is wrapped around the organelle to form a sac. Then several lysosomes fuse with the sac and their enzymes pour into it. The structures of the worn-out organelle are hydrolyzed to ever smaller molecules, which eventually become small enough to be transported across the membrane. The cell can reuse these small molecules in further building programs. Particles such as bacteria may be broken down in the same way.

Perhaps you are wondering why the enzymes don't dismantle the lysosome membranes along with everything else. Although the picture is far from clear, the cell apparently expends considerable energy to maintain the membranes, which counters the constant enzyme attacks. Once a cell dies, it falls apart rapidly. Without energy expenditure, lysosome membranes simply can't be maintained. And once the membranes disintegrate, the enzymes flood into the cytoplasm, there to destroy indiscriminately.

Lysosomes have been found only in animal cells. But recent evidence suggests the central vacuole in plant cells plays much the same role. Structures resembling lysosomes have also been found in fungi and protistans, but whether they should actually be called lysosomes is still not clear.

Shape and Movement Within the Cell: Microtubules and Microfilaments Very few cells are spherical, even when freed from the physical confines of multicellularity. And few cells are internally rigid. If you were to look through a microscope at any living cell, you undoubtedly would be impressed with its constant motion. Cellular contents stream about endlessly in regular patterns. Chloroplasts move in response to the apparent changes in the sun's position during the day. Vesicles form at cell boundaries, pinch off from the plasma membrane, and move toward specific regions inside. Even animal cells from an organ that usually stays put (say, the liver) move actively when they are separated from the tissue and placed in a culture dish. Until recently, the shape and movement of many diverse cells defied explanation. But studies over the past decade have begun to tell us something about them. Two structures —microtubules and microfilaments—have been implicated.

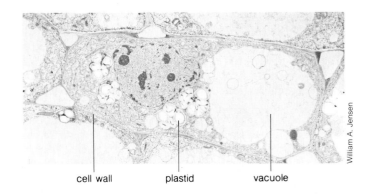

cell wall   plastid   vacuole

Figure 4.12 Plastids and a central vacuole in a cotton plant cell.

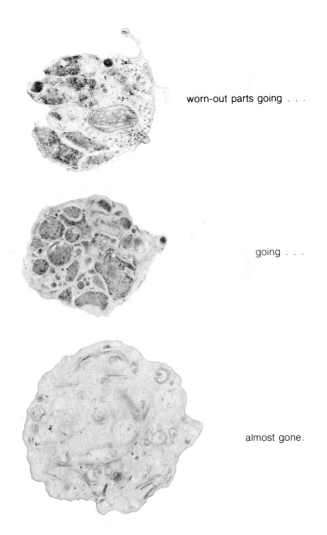

worn-out parts going . . .

going . . .

almost gone.

Figure 4.13 Digestion of organelles as seen in a lysosome. 12,000×. (Gary W. Grimes)

A **microtubule** is a small, hollow cylinder made of the protein tubulin. This cylinder can lengthen or shorten by the controlled addition and removal of tubulin subunits from one end. The list of events in which microtubules take part is impressive. First, they form the basic "skeleton" of many asymmetric cells and cell extensions. Sometimes this framework is just beneath the plasma membrane. Second, many slow cellular movements occur by the controlled assembly and disassembly of microtubules. Examples include the movement of chromosomes during cell division, and the elongation of some cell parts for trapping prey. Third, microtubules are involved in many forms of rapid movement such as the beating of cilia and flagella, which will be described shortly.

Just as important for cell shape and movement are the **microfilaments,** structural elements that are only about one-fifth the diameter of microtubules. Microfilaments are composed of **actin,** a major protein in your muscles, and invariably they are associated with molecules similar to **myosin,** the other major muscle protein. Myosin and actin serve one basic function: contraction.

If a cell having microfilaments attached to its plasma membrane is made to contract, the cell narrows at one end. If a cell having a ring of microfilaments around its middle is made to contract, the cell pinches in two (as it does during cell division). If a cell having microfilaments attached to its plasma membrane and to internal sacs is made to contract, the sacs are drawn toward the membrane surface. In a cell crawling across a culture dish, microfilaments running from the lower front surface to the upper midsurface are involved in contractions that move the cell body forward.

Recent evidence suggests that microtubules, microfilaments, and even finer fibers are highly organized in ways that anchor and rearrange internal parts. The cell biologist Keith Porter sees the patterns of radial and parallel filaments, of fibrous mesh and bundles as no less than "cytomusculature" by which organelles are constantly moved about during the dynamic life of a cell.

Movement Through the Environment: Flagella and Cilia
Two structures involved in rapid cell movement are **cilia** and **flagella.** Both follow the same basic structural plan, but cilia are shorter and usually more numerous where they do occur. Both are built of microtubules, which project from the cell body in a sheath that is an extension of the plasma membrane (Figure 4.14).

Complex flagella and cilia are further evidence of the basic unity of life. They appear among protistans, and on the gametes (sex cells) of many fungi, plants, and animals. Cilia project from the cells of numerous animal organs, where they help move fluids and particles. An extraordinary number of animal sensory organs (those of sight, hearing, touch, and so forth) incorporate modified cilia, as you will see in illustrations throughout this book.

a

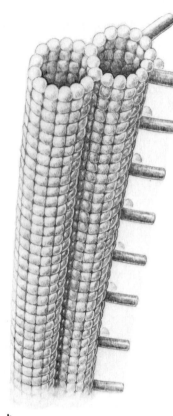
R. Barton
b

The basic organization of eukaryotic cilia and flagella is shown in Figure 4.14. Nine microtubule pairs are arranged radially about two central microtubules, a pattern called the **9 + 2 arrangement.** Recent evidence suggests that this microtubule system is generated by a structure called a **centriole.** After the cilium or flagellum has been formed, the centriole remains as the basal body for the motile structure (Figure 4.14).

How do cilia and flagella move? According to one view, numerous "arms" project from each paired microtubule (Figure 4.14). These arms are made of a protein that takes part in converting chemical energy to the mechanical energy of movement. The arms attached to one microtubule pair make contact with the next microtubule in line. As the protein molecules are energized, they bend, which exerts a downward push on the adjacent microtubule. The push causes microtubules to slide past each other, which causes the flagellum or cilium to bend. In cytoplasmic streaming, similar arms are thought to make contact with and propel other organelles through the cytoplasm.

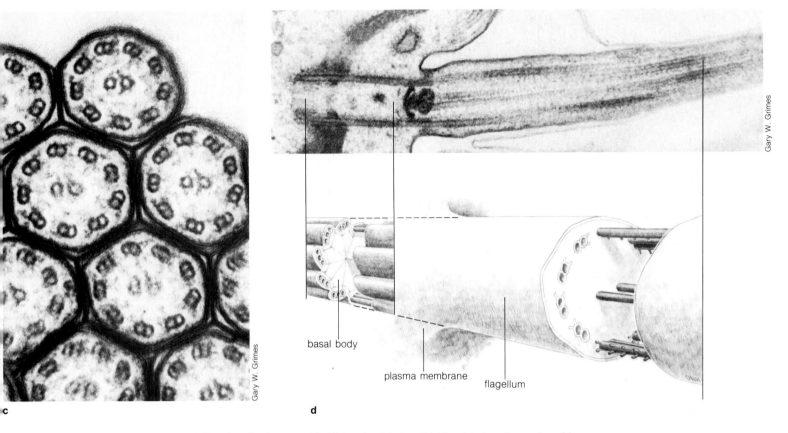

c

d

Gary W. Grimes

basal body

plasma membrane

flagellum

Figure 4.14 Cell movement (**a**) Protein subunits assembled into microtubules. (**b**) Microtubule pairs are found in eukaryotic flagella and cilia. The "arms" projecting from one side of each pair interact with the adjacent pair to bend the motile structure. (**c**) Cilium in cross-section. 133,500×. (**d**) Cilium in longitudinal section. 70,000×.

The Membrane-Enclosed Nucleus

All eukaryotic cells contain a large organelle that is isolated from the cytoplasm by a double membrane. This organelle is the nucleus (Figure 4.15); it is absolutely essential for cell survival and self-reproduction. If its nucleus is removed, a cell embarks on a course of declining metabolic activity that ends in death. How the nucleus controls reproduction as well as the everyday activity of the cell will be considered in later chapters. For now, we will simply introduce some aspects of nuclear structure.

The Porous Nature of the Nuclear Envelope The matrix of the nucleus is called the **nucleoplasm** to distinguish it from the cytoplasm. It is enclosed in a double membrane, called the **nuclear envelope.** The outer membrane (facing the cytoplasm) is richly peppered with ribosomes, and in some cases it is continuous with endoplasmic reticulum. Some

experiments suggest these ribosomes are engaged in building proteins to be used inside the nucleus.

At regular intervals, pores break through the nuclear envelope. Each pore, which is about 70 nanometers wide, is largely filled with a mass of material. The filling material has a diameter of about 120 nanometers (which means it extends beyond the pore itself), but there seems to be an opening through the center. Although little is known about the function of this organization, it seems to exert extremely precise control over movement of molecules into and out of the nucleus.

The Nucleolus One prominent feature of the nucleus is an internal mass that usually looks more dense than its surroundings in electron micrographs. This is the **nucleolus,** the site where ribosomes are assembled (Figure 4.15). During division, the nucleolus breaks apart and its components become dispersed throughout the nucleus. After nuclear division, the nucleolus forms once again.

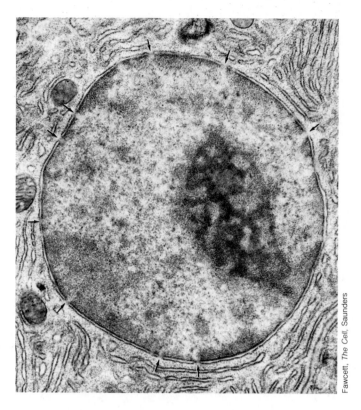

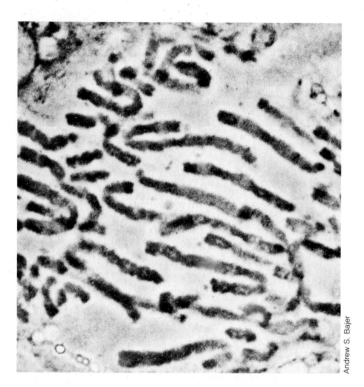

Figure 4.15 Electron micrograph of the nucleus from a bat's pancreatic cell. Arrows point to pores in the nuclear envelope. Dark region is the nucleolus.

Figure 4.16 Chromosomes. Chromosomes are large enough to be seen with a light microscope. These are chromosomes of an endosperm cell from the seed of a flowering plant.

Chromatin and Chromosomes Much of the nuclear interior is filled with **chromatin,** which is now known to be masses of DNA and protein fibers. The protein serves to strengthen the DNA molecules and to regulate what information is read from the DNA. During division, the chromatin fibers are assembled into discrete bodies called **chromosomes** (Figure 4.16). Later chapters will cover the role of chromatin and chromosomes in the cell division process. For now, it is enough to think of chromosomes as convenient ways to transfer copies of the cell's hereditary information to two distinct nuclei, which are formed prior to division. Chromatin is the dispersed form in which DNA is actually being used by the cell; visible chromosomes are the form in which chromatin is condensed and packed up for distribution during division.

Perspective

All living things are made of one or more cells; each cell is a basic unit of structure and function; and new cells can arise only from cells that already exist. Beginning with this cell theory, we have looked at the internal structures of this fundamental living unit. In doing so, we have seen that each cell is a complex island of organization designed to extract energy from its surroundings in order to sustain and reproduce itself. Each cell is alive. Whether we are talking about a single-celled organism floating in a shallow sea or a single cell in your own body, it is alive. At any instant, it is actively engaged in maintaining its position in space, in extracting energy, in building new molecules and storing others, and in getting rid of wastes produced during metabolism.

The single-celled alga floating about requires its plasma membrane to retain its individuality and to keep out the sea. It traps sunlight and various molecules from its environment, thus securing energy and raw materials. Some of these resources it uses at once to sustain itself; some it stores in reserve. That single cell also has the means to dispose of used-up cell parts as well as the by-products of its existence. Similarly, each cell in your body holds its place in relation to all others and thereby helps maintain an identifiable "you." At any moment, it is actively engaged in extracting oxygen and food molecules from the blood streaming through your body. Each cell uses some of these molecules immediately to carry out the daily tasks of living;

Table 4.1 Structures Typically Found in Prokaryotic and Eukaryotic Cells

Cell Structure	Primary Function	Prokaryotes	Eukaryotes			
		Monerans	Protistans	Fungi	Plants	Animals
Cell wall	Protection, support	✔	✔	✔	✔	
Plasma membrane	Selective boundary layer	✔	✔	✔	✔	✔
Photosynthetic pigments	Light-energy conversions	✔	✔		✔	
Membrane-bound chloroplasts	Light-energy conversions		✔		✔	
Other plastids	Storage		✔		✔	
Mitochondria	Energy metabolism		✔	✔	✔	✔
Ribosomes	Assembling molecules	✔	✔	✔	✔	✔
Endoplasmic reticulum	Assembling molecules		✔	✔	✔	✔
Golgi complex	Assembling, secreting molecules		✔	✔	✔	✔
Vacuoles	Variable (e.g., storage, digestion)		✔	✔	✔	✔
Lysosomes	Breaking down molecules, structures		?	?	?	✔
Microbodies	Interconversions of molecules		✔	✔	✔	✔
Microtubules	Structural support, internal movement		✔	✔	✔	✔
Complex flagella, cilia	Movement through environment		✔	✔	✔	✔
Centriole	Gives rise to microtubule system		✔	✔*	✔*	✔
DNA molecules	Information storage	✔	✔	✔	✔	✔
Membrane-bound nucleus	Genetic control		✔	✔	✔	✔
Chromosomes	Packaging DNA during division		✔	✔	✔	✔
Nucleolus	Assembling ribosomes		✔	✔	✔	✔

*Occurs in some groups but not in others.

others it converts to molecules that can be stashed away as a hedge against times when nutrients are in short supply. Like the photosynthetic alga, each one has the cellular machinery that allows it to be self-contained in its environment.

As you will be reading in Chapter Fifteen, a long history of chemical evolution apparently preceded the appearance of the first cell. But once the basic cell form had evolved—plasma membrane, cytoplasm, ribosomes, and DNA molecules capable of being duplicated—then natural selection could go to work on a unit of organization that could gradually become adapted to different environments. Where environments changed, so apparently did this living unit change through modifications on the basic plan; hence the differences between prokaryotic and eukaryotic cells. Plasma membrane folded inward and outward, thus providing more reaction surfaces. Folded membranes became specialized chambers for dividing up essential tasks among different cell parts. Once the hereditary material was isolated from the machinery of the cytoplasm, it could become more complex. But in all cases, modifications to the basic cellular plan have accrued as modifications on a basic theme: the extraction and use of energy and materials, in ways that will be discussed in chapters to follow.

Recommended Readings

Bloom, W. and D. W. Fawcett. 1975. *A Textbook of Histology.* Philadelphia: W. B. Saunders. Advanced reading, but excellent reference book for anyone going on in the life sciences.

Fawcett, D. W. 1966. *The Cell. An Atlas of Fine Structure.* Philadelphia: W. B. Saunders. One of the most outstanding collections of micrographs on cell fine structure ever assembled.

Kessel, R. and C. Shih. 1974. *Scanning Electron Microscopy in Biology: A Student's Atlas of Biological Organization.* New York: Springer-Verlag. Stunning micrographs provide insights into cell structures and products.

Loewy, A. and P. Siekevitz. 1963. *Cell Structure and Function.* New York: Holt, Rinehart & Winston. Valuable as a permanent reference for anyone going on in the life sciences.

Wolfe, S. 1972. *Biology of the Cell.* Belmont, California: Wadsworth. More advanced treatment, but a classic.

Chapter Five

Figure 5.1 A flowering plant in the sun and wind, one of many life forms whose cells trap the sunlight energy that directly or indirectly sustains all existing communities of life.

How Cells Trap Resources

Just before dawn in the Midwest the air is dry and motionless; the heat that has scorched the land for weeks still rises from the earth and hangs in the air of the new day. There are no clouds in sight; there is no promise of rain. For hundreds of miles in any direction you care to look, crops stretch out, withered or dead. All the sophisticated agricultural methods in the world can't save them now, for in the absence of one vital resource—water—life in each cell of each of those many thousands of plants has ceased.

In Los Angeles, a student reading the morning newspaper complains to no one in particular about the hike in food prices that the Midwest drought will mean. In Washington, D.C., economists busily calculate the crop failures in terms of decreased tonnage available for domestic consumption and for export, and what it means to the nation's balance of payments. In Africa, a child with bloated belly and spindly legs waits passively for death. Even if food from the vast agricultural plains of America were to reach her now, it would be too late. Deprived of vital food resources, cells of her body will never grow normally again.

This chapter is about the acquisition and use of materials and energy. It gets into cellular pathways that might at first seem to be far removed from the world of your interests. But knowledge of these pathways is a key to understanding what it means to acquire food, why energy must be expended to get it, why food supplies are not keeping up with our increasing demands, and what might be done to replenish the supplies. Thus you will be reading a chapter that combines two topics that often are treated separately: *how cells trap materials, and how they trap energy*. As you will soon perceive, the two are not separable—especially when it comes to addressing concerns of the sort we have just outlined. Directly or indirectly, these concerns will touch your life in the decades to follow.

Part 1
Acquisition of Materials

Regardless of how small or large, how simple or complex an organism may be, the act of taking in food ultimately boils down to getting essential kinds of molecules into every single cell. There are several ways in which this is done, but let's begin with the acquisition of whole bits of food.

Phagocytosis and Pinocytosis: Wholesale Gulping Many cells have the means to take in not only large molecules but entire organisms. Such is the process of **phagocytosis** ("cell eating"). The amoeba, a single-celled organism, is phagocytotic. When an amoeba encounters an appetizing particle, part of its fluid membrane and cytoplasm extend outward in the manner shown in Figure 5.2a. This extension completely surrounds the particle, thereby creating a vacuole which is moved into the amoeba's interior. There, lysosomes fuse with the vacuole. As they do, they release enzymes that break the particle into manageable bits, which readily diffuse across the vacuole membrane and into the cytoplasm. Phagocytosis is also the means by which white blood cells devour bacterial cells invading your body.

Another process for moving materials into the cell is **pinocytosis** ("cell drinking"). In pinocytosis, a depression forms where small particles in the surrounding environment adhere to the plasma membrane. Vacuoles seem to form as a response to this adhesion (Figure 5.2b). As the membrane dimples inward, it forms a vesicle that contains some of the extracellular fluid. The vesicle pinches off to become a small vacuole inside the cytoplasm. Once inside, the vacuole may dump its contents into the cytoplasm or fuse into large, membrane-enclosed storage vacuoles.

Passive Transport As dramatic as phagocytosis and pinocytosis may be, an even more vital aspect of acquiring materials is based on processes that are, by comparison, much more subtle. Like energy, matter tends to move in the direction of increasing disorder. This tendency applies to the random movement of molecules in any given system. For instance, when you drop a cube of sugar into a cup of water, at first the sugar molecules are found in one region of the cup, just as the water molecules are found everywhere except in the space the sugar cube occupies. For each substance, you have created a **concentration gradient:** its molecules are concentrated more in one region than in another. But gradually, both kinds of molecules undergo **diffusion:** each substance moves from its region of greater concentration to the region where it is less concentrated. Diffusion occurs along the gradient because of the energetic movements of individual molecules, which jostle about until they are uniformly dispersed. Thus, sugar molecules tend to become dispersed throughout the cup, just as water

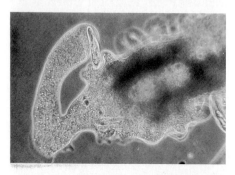

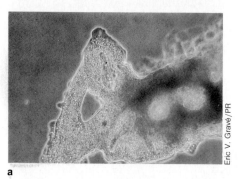

a

Eric V. Gravé/PR

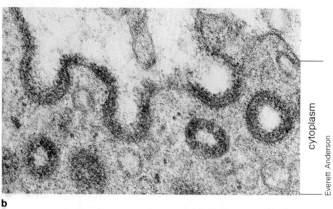

cytoplasm

Everett Anderson

b

Figure 5.2 (a) Phagocytosis. Here, a single-celled organism, *Amoeba*, is trapping another single-celled organism, *Paramecium*. (b) Pinocytosis. Notice how the plasma membrane dimples inward where nutrients (dark regions) accumulate, then pinches off to form separate vacuoles.

molecules tend to become dispersed even in the space the sugar cube once occupied.

Diffusion occurs in gases, liquids, even solids to some extent. The *rate* of diffusion depends on several things. Diffusion speeds up with increases in pressure. It speeds up with increases in temperature (heat energy causes molecules to move faster). Even the size of molecules affects diffusion rates (small molecules move faster than large ones at any temperature). Finally, the greater the difference in concentration between two regions, the faster the net movement will be.

What does all this mean for the living cell? Compared to its watery surroundings, a cell has much greater concentrations of some substances. If diffusion alone governed the movement of all substances, it wouldn't be long before much of the cell's insides diffused out! Fortunately, diffusion and the cell's plasma membrane act in concert to assure the passive movement of some substances into and out of the cell in ways that maintain the concentrations of substances that have to stay put. The membrane is *differentially permeable:* it permits movement down some of the concentration gradients that exist between a cell and its surroundings, and prohibits others. Thus certain molecules travel freely across the membrane, others are kept from diffusing across it quite so readily, and large molecules assembled within the cell are kept from crossing it at all.

The name **passive transport** applies to any movement across a plasma membrane that is simply a response to a concentration gradient; the cell does not directly spend energy for the movement to occur. Many substances, including water, can be transported passively into and out of a cell. But the movement of *water* molecules along a gradient that takes them across any differentially permeable membrane is given a special name: **osmosis.**

A simple experiment based on osmosis tells us something about how different kinds of cells have become adapted to different environments. Imagine you have just made a plastic bag permeable in a way that permits the passage of water molecules but blocks the passage of larger molecules. Suppose you fill the bag with water containing a 2-percent concentration of table sugar. (The water in such a system would be the **solvent,** and the dissolved material the **solute.**) Suppose you now put the bag in a container of distilled water. Distilled water is the most concentrated kind, because all dissolved materials have been removed from it; hence water molecules can get much closer to one another. This means water is less concentrated inside the bag (where sugar molecules take up space) than it is on the outside. In this case, diffusion does *not* lead to uniform distribution of substances throughout the system. Water molecules move in—but sugar molecules can't move out (they are too large to pass through the holes you made in the "membrane"). Soon the bag swells with water (Figure 5.3a). Because the bag can hold only so much, fluid pressure builds up and it bursts.

What happens when the water is more concentrated inside the bag than on the outside? Obviously the net water movement will be out of the bag, which will shrivel up (Figure 5.3b). Only if the water concentration is the same on both sides of the bag will there be no net movement of water in either direction (Figure 5.3c).

The same kinds of things happen to living cells. Imagine a red blood cell being immersed in distilled water. Because the cell contains many large organic molecules that can't cross its plasma membrane, internal pressure builds up as water moves in. This kind of cell simply can't function under such conditions. It is doomed to burst, because it has no mechanism for disposing of the excess water. If such a cell is immersed in a solution having a sugar concentration that's greater than the one inside, the cell is still doomed: it will lose water and shrivel up. Only when red blood cells are immersed in an environment where the concentration of dissolved materials is the same as the concentration inside the cells will there be no change in cell volume.

The thought of your red blood cells being so vulnerable to shriveling up and bursting might make you a little uneasy, until you remember that neither distilled water nor dissolved table sugar flows through your veins. As in other complex multicellular animals, your body fluids have the same solute concentrations as the cells they bathe. (In fact, that's why surgeons never use distilled water to bathe tissues during an operation. They use a "physiological saline solution." It's a fluid containing dissolved salts in essentially the same concentration as the blood.

For any cell of any organism to survive, it must be adapted to the solute concentrations in the surrounding environment. A plant cell survives in fresh-water environments, which contain very little dissolved materials compared to the cell's interior, because that kind of cell has built-in mechanisms for controlling the inward or outward flow of water. When incoming water causes pressure to build up in the cell, the rigid wall surrounding the plasma membrane keeps the cell from bursting. The increasing internal pressure gradually forces water out through the membrane. When the outward flow equals the rate of inward diffusion, the internal pressure is constant. The cell has achieved equilibrium. Some single-celled protistans have adapted to life in fresh water in a different way. They have their own water pumping system, called **contractile vacuoles.** As water accumulates inside, these cells expend energy to pump the excess into the vacuoles and make them contract—a movement that causes the water to squirt back outside. In salt-water environments, where dissolved substances are present in higher concentrations than in typical cells, various adaptations again make life possible. Seagulls, for example, take in considerable salt from the kinds of food they eat and from the sea water they drink. They are able to keep their body fluids in balance with their cells only by accumulating the solute in special salt glands, which later excrete crystals of sea salt from the body.

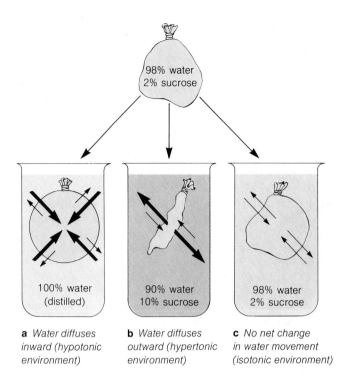

a Water diffuses inward (hypotonic environment)

b Water diffuses outward (hypertonic environment)

c No net change in water movement (isotonic environment)

Figure 5.3 Osmosis in different environments. This experiment shows why it's important for a cell to be "matched" to its environment. Without any special mechanisms for actively taking in or expelling water molecules, passive diffusion alone can't keep the cell from shriveling up or bursting if conditions in the environment change.

Even in environments where solute and water concentrations are constantly *changing,* it's possible to find organisms getting along quite well. An estuary (a place where a river flows into the sea) may be almost entirely fresh water during low tide. Under these conditions, the single-celled protistans living there use contractile vacuoles to rid themselves of the excess water flooding into their systems. During other periods, the estuary may contain water with solute concentrations equal to those inside the cells, so water regulation isn't needed. But during high tide, the estuary becomes much more salty and the protistans must again expend energy, this time to transport water into the cell. Whenever a cell must expend energy to move materials across its plasma membrane, we are talking about a different kind of transport mechanism—one that allows cells to compensate for variable conditions on the outside. It is called active transport.

Active Transport Many of the raw materials a cell needs—potassium ions, for instance—simply aren't all that abundant in the environment. Thus diffusion alone can't bring in all the potassium ions required. The cell has to employ some mechanism for *accumulating* potassium ions in concentrations that are higher inside the cell than they are on the outside. But many other raw materials—sodium ions, for instance—are overly abundant in such environments as salt water and the body fluids of animals. These ions diffuse freely across a plasma membrane, but they would kill any cell if they were to reach the same concentration inside as exists on the outside! So the cell must have some mechanism for *preventing* the accumulation of sodium.

Regardless of whether it's stockpiling some materials or getting rid of undesirable amounts of others, a cell has to do work to get them moving against the concentration gradient. Moving materials into or out of the cell against the concentration gradient is known as **active transport.** Unlike passive diffusion, which is usually slow and somewhat indiscriminate, active transport is rapid and highly selective.

A cell actively transports *many* substances in one direction or another, and it does so with remarkable specificity. It can do this because of special enzymes embedded in its membrane mosaic. Each of these enzymes has an active site that binds only those ions or molecules it is supposed to help transport. There is great variation in how these enzymes function, but in each case some energy-yielding reaction is coupled with the transport process.

Active transport can be maintained only as long as a cell continues to expend energy to drive these reactions. As soon as the flow of energy stops, active transport shuts down and materials can move across the plasma membrane only by simple diffusion. In this event, the cell's contents become less organized and more like the surroundings until eventually the cell dies.

A Shortage of Nitrogen in a Sea of Plenty It is one thing to say that many of the materials a cell must take in are not abundantly available. But added to this, *many materials that are available often are in a chemical form that cells cannot use directly.* In such cases, the community of life survives only because there is an efficient recycling of atoms through the cells of different kinds of organisms. The waste product of one species becomes the starting material for another. Such is the nature of the cycling of carbon, hydrogen, and oxygen, as outlined in Chapter Three. Let's now look to the manner in which nitrogen is recycled, for it is a cellular process that has far-reaching consequences for all of us (Figure 5.4).

The life of any cell depends on proteins, and protein structure depends on nitrogen. Nitrogen (N_2) makes up about 80 percent of the atmosphere, so it would seem to be abundant just about everywhere. Yet, for the cells of most organisms, nitrogen of the air is totally useless until certain kinds of single-celled organisms in the community convert it to a different chemical form. Thus, even though we breathe in abundant N_2 molecules, we can't use them for body building; we breathe them right back out again.

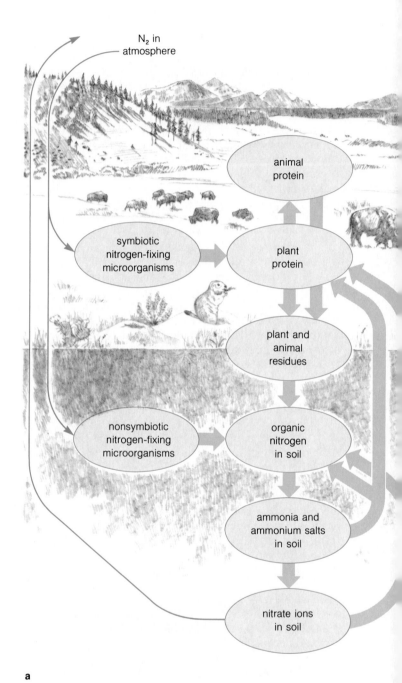

Figure 5.4 A brief look at the availability of nitrogen—one of many materials vital to cell functioning. (**a**) The nitrogen cycle in a stable natural ecosystem and (**b**) the modified nitrogen cycle in an agricultural system. These diagrams have been greatly simplified to show only the main processes described in the text. The important difference between them is that (**a**) is self-contained and self-perpetuating, but (**b**) is not.

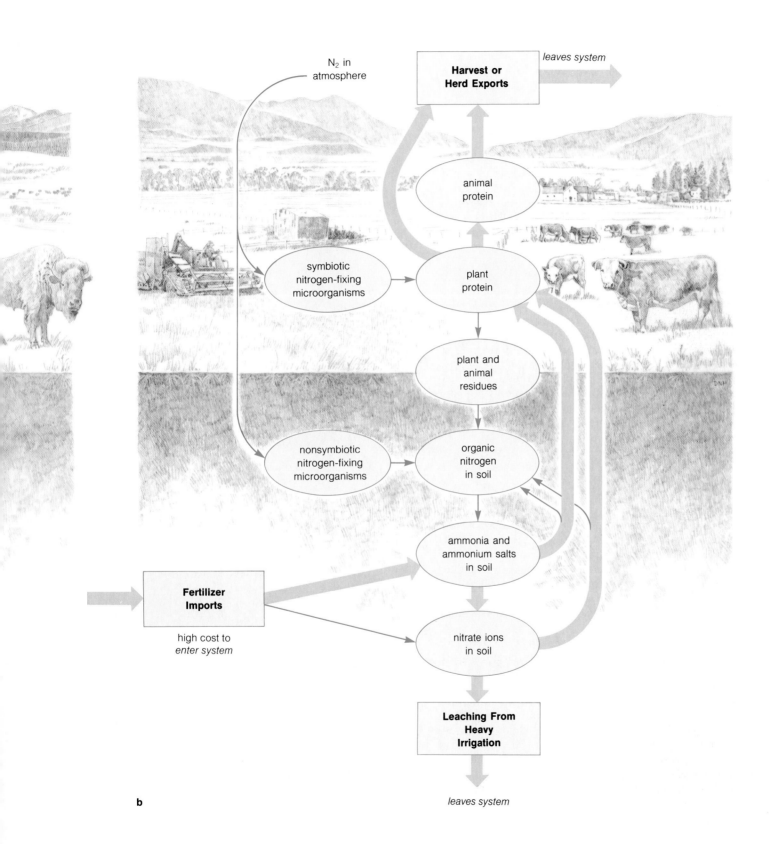

Millions upon millions of humans suffer from a shortage of nitrogen in this sea of plenty. The only form of nitrogen our cells can use comes from nitrogen-containing proteins found in the plants and animals we eat. From these proteins, our cells can assemble their own required proteins. But it happens that in many parts of the world, the quantity and quality of protein is not high enough to permit normal cell maintenance and growth. Thus there is rampant **protein starvation** at the cellular level, which translates into starvation among individuals and whole populations.

How can it be that nitrogen is recycled through natural communities, yet is not available to us? In natural communities (in places where human agriculture has not disturbed the land), nitrogen flows from the atmosphere to the soil, to microorganisms, to plants, to animals, then back to the soil. The overall movement is known as the **nitrogen cycle** (Figure 5.4a). In a process called **nitrogen fixation,** single cells of blue-green algae and a few kinds of bacteria take nitrogen gas from the air, convert it first to ammonia (NH_3), then use it to build amino acids and proteins. The bacteria of another important group are also able to fix nitrogen, but only when they live in association with the roots of certain plants such as the legumes (peas, beans, clover, and the like). The plants provide the bacteria with energy-rich sugar molecules, and the bacteria provide the plants with a rich supply of amino acids. (Species living together in a way that benefits one or both species are said to have entered a **symbiotic relationship.**) Free-living and symbiotic nitrogen-fixing microorganisms are essential in the cycling of nitrogen.

Plants themselves do not fix nitrogen, and most plants do not have symbiotic nitrogen fixers in their roots. Nitrates (NO_3^-) in the soil are almost their sole source of nitrogen. Where do nitrates come from? The soil teems with single-celled microorganisms that busily engage in **nitrification:** the conversion of ammonium ions (NH_4^+) to nitrite (NO_2^-) by one group of bacteria, then the conversion of the nitrite into nitrate by a different group. Some of these nitrates are recycled through the community of life. In a process called **denitrification,** some are also recycled to the atmosphere by bacteria that convert nitrogen-containing compounds to N_2. In all these ways, nitrogen is produced and recycled in small but dependable amounts.

The minute a plot of land is cultivated for agriculture, however, these delicate interrelationships are disrupted (Figure 5.4b). Plants are cut, burned, and/or uprooted. And animals are driven off—which effectively removes the nitrogen locked up in their proteins from the community. Initially the decaying matter remaining in the soil is a rich store of organic nitrogen, so nitrate production continues for the short term and crops can be grown. But at the end of each growing season, crops are harvested and carried off to nourish people and livestock elsewhere, so still more nitrogen leaves. Because of heavy tilling and irrigation of the soil, nitrates run off into streams, rivers, and lakes.

Unless the crops are of a kind that live together with nitrogen-fixing bacteria, the land soon becomes unable to support crops at all—unless new nitrogen is added. Enter fertilizer.

Today most fertilizer is in the form of ammonia or ammonium salts, because young plants of the most common crops (cereal grains) grow faster on ammonia ions than on nitrate. But adding ammonia to the soil bypasses the soil microorganisms, so to speak. Thus the planting of such crops disrupts the natural cycling of nitrogen even more. Most ammonia fertilizer is manufactured by a chemical process that takes nitrogen from the air and combines it with hydrogen to form ammonia. Enormous amounts of energy are needed to do this—not energy from the unending stream of sunlight but energy from tremendous amounts of oil. As long as we assumed the supply of oil was unending, there was little concern about the energetic cost of fertilizer production. *We simply overlooked the fact that in many cases, we are pouring more energy into the soil in the form of fertilizer than we are getting out of it in the form of food to sustain our cells!* Only recently has there been growing awareness of an energy principle introduced earlier in this book: you can't get something for nothing. Unlike natural systems, in which resources are recycled in accordance with this principle, our agricultural systems exist only because of constant, massive infusions of rapidly dwindling resources. We will return to our problem of acquiring food materials after we consider the related problem of acquiring energy. In this second part of the chapter, as in the first, we will begin with systems for acquisition at the cellular level and then move on to see what the limits and potentials of these systems hold for our own future.

Part 2
Acquisition of Energy

Photosynthesis: Where the Flow of Energy Begins As you read earlier, the flow of energy through all forms of life on earth begins with photosynthesis:

In photosynthesis, sunlight energy is trapped and converted to chemical energy, which is used to build food molecules from carbon dioxide and water.

Organisms capable of building their own food molecules, given an energy source and some simple inorganic molecules, are known as "self-feeders," or **autotrophs.** They include plants, blue-green algae, and some bacteria. Non-photosynthetic organisms are "other-feeders," or **heterotrophs:** being unable to build their own food molecules, they ingest autotrophs, each other, or organic waste products. They include animals, fungi, and many microorganisms.

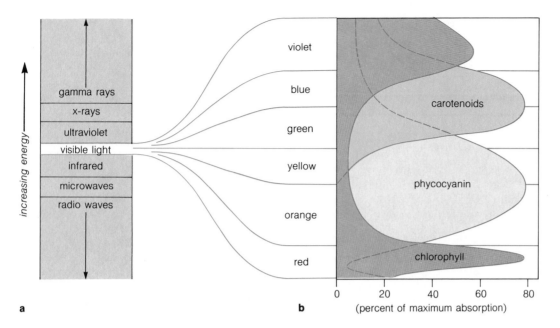

violet

blue

green

yellow

orange

red

gamma rays

x-rays

ultraviolet

visible light

infrared

microwaves

radio waves

increasing energy

carotenoids

phycocyanin

chlorophyll

0   20   40   60   80
(percent of maximum absorption)

a

b

Figure 5.5 (**a**) Electromagnetic spectrum and (**b**) generalized ranges of wavelength absorption for chlorophyll and for two accessory pigments. The peaks indicate which colors (wavelengths) of light are used most efficiently. (**c**) Light-trapping pigments are organized in photosynthetic membranes such as those seen in this chloroplast.

There are special systems for trapping light in photosynthetic autotrophs. Light travels in packets of energy called photons, and the sizes of these energy packets correspond to different wavelengths of light (Figure 5.5a). It happens that the different pigment molecules in photosynthetic membranes absorb photons of different wavelengths. For example, Figure 5.5b shows the range of absorption for the most important pigment molecule: chlorophyll. Clearly, chlorophyll is most efficient at using light from the red end of the visible spectrum, and it's also capable of using blue light, which is at the other end. But this doesn't mean that wavelengths in-between go unexploited. **Accessory pigments** such as the carotenoids and phycocyanin absorb intermediate wavelengths and pass on some of the energy they absorb to cholorophyll.

All these light-trapping pigments are not sprinkled randomly through photosynthetic membranes. They are highly organized into tiny functional units called **photosystems.** Each photosystem consists of several hundred chlorophyll molecules, various accessory pigments, and an array of enzymes. *The photosystem is the functional light-trapping unit in the initial excitation of chlorophyll.* And it is here that the first stage of photosynthesis—the light reactions—begins.

The Light Reactions Two distinct light-trapping systems are used to trigger the **light reactions,** which are concerned with harnessing photon energy and converting it to usable forms of energy inside the cell. They are named **photosystems I and II.** When light strikes either photosystem,

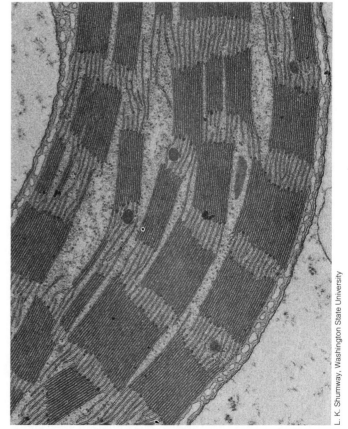

c

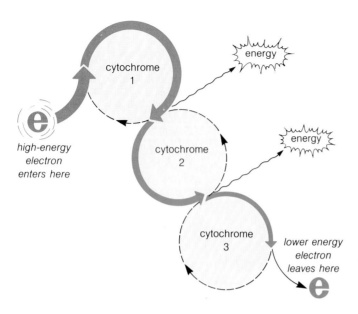

Figure 5.6 Electron transport chain. When a cytochrome molecule in the chain accepts an electron, it becomes reduced (solid line). When it gives up the electron to the next cytochrome, it becomes oxidized (dashed line) and thereby can pick up another electron entering the chain. Each time the electron is transferred, it gives up some of its extra energy.

pigment molecules gather up the photon energy. They transfer it to a special chlorophyll–protein molecule (called P700 in photosystem I, and P680 in the other). When enough energy is transferred to one of these special molecules, some of its electrons become so excited they actually pop out of the system in pairs. But they don't get very far at all. Nearby in the photosynthetic membrane is an acceptor molecule. It captures the high-energy electron pairs as fast as they become free and transfers them to an **electron transport chain** (Figure 5.6). Such chains are nothing more than a series of molecules capable of rapidly accepting and giving up electrons in a stepwise fashion.

Most electron transport chains contain cytochromes (iron–protein molecules). They work because the iron atoms in cytochromes can shift back and forth between being reduced and being oxidized. In living systems, a "reduced" atom or molecule is generally one that has taken on an extra electron from a hydrogen atom (sometimes the whole hydrogen atom comes along for the ride). It becomes "oxidized" when it gives up its extra electron (or hydrogen atom). Thus each electron transfer in the chain is really an **oxidation–reduction reaction.** By way of analogy, imagine people lined up in a bucket brigade that reaches from a stream to a campfire that's burning out of control. To keep

the buckets moving, each person must give up a bucket to the next person in line in order to free the hands to grab another bucket. The buckets must be taken away at the end of the line, of course, otherwise everybody in line would soon be left holding a bucket with no one to pass it to, and nothing would get done. Similarly, cytochromes in an electron transport chain are positioned next to one another. As the first in line accepts an electron, it becomes reduced. Then it gives up the electron to an oxidized cytochrome next in line, which thereby becomes reduced—even as the first cytochrome returns to the oxidized state, ready to accept another electron (Figure 5.6). The last cytochrome in the chain gives up its electron to an oxidized acceptor molecule, which carts off the electron and thereby keeps the chain clear for operation.

The point of these oxidation–reduction reactions is the generation of usable forms of energy. Because no energy transformation is perfect, some of the excess energy of an electron is lost each time it's transferred from molecule to molecule. And at certain transfer points, the amount of energy given off is enough to *phosphorylate* (tack a phosphate group onto) a molecule called ADP. The result is the formation of adenosine triphosphate, or **ATP**—a very special high-energy molecule. By moving electrons down an electron transport chain, a cell is able to build up a store of chemical potential energy in ATP molecules, the universal energy currency of all cells (Figure 5.7).

Before discussing the way electrons move through the whole set of light reactions, let's first consider a simpler, more ancient pathway based only on photosystem I and a single electron transport chain. Electrons excited by light energy in photosystem I are passed to an electron acceptor molecule, then through the transport chain. There, some of the energy they give up is used to generate ATP (Figure 5.8). Eventually they give off all their excess energy and return to the place they came from—photosystem I.

Because the electrons travel full circle each time, the process is said to be cyclic. Because they first gain energy from photons and then contribute that energy to phosphorylating the ADP, the entire pathway is known as **cyclic photophosphorylation.**

For every two electrons entering the cyclic photophosphorylation pathway, the energy yield is one ATP molecule.

Cyclic photophosphorylation probably represents the way living cells first employed light energy to produce usable cellular energy. But in existing plants, it is now used only in times of stress. Apparently, natural selection long ago went to work on the cyclic pathway, expanding it to include a second photosystem and a second transport chain. When the *two* photosystems function together, electrons don't flow in a cycle. Instead, *new* electrons obtained from water molecules at one end of the light reactions are

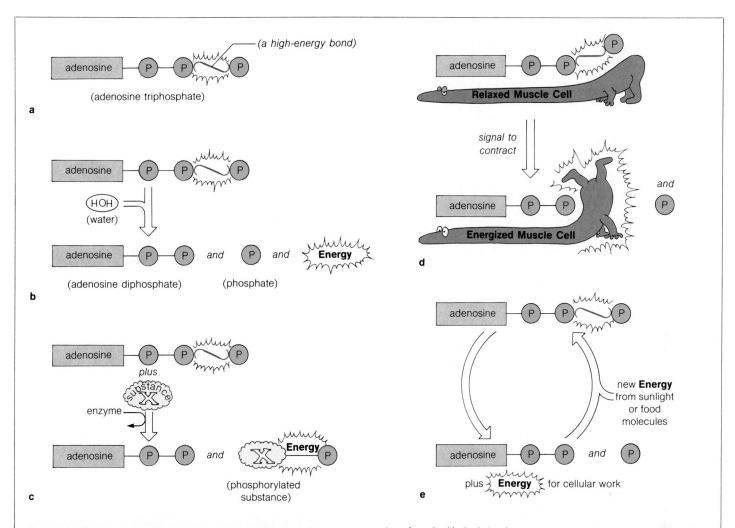

Figure 5.7 ATP: the main energy currency of the cell.

(**a**) Energy enters living cells mainly in two forms: food molecules (such as sugar and fat) or sunlight. Before it can be put to work, it must be converted to more usable forms. The most common form is a nucleotide molecule called **ATP** (adenosine triphosphate).

(**b**) ATP is rich in energy because of covalent bonds linking two of its three phosphate groups together. They are high-energy bonds, for if they are split apart (say, by hydrolysis), energy is released in a usable form. When ATP is simply hydrolyzed into **ADP** (adenosine diphosphate) and an inorganic phosphate group (P_i), the energy given off does little more than heat the surroundings. It would be like turning on an electric motor without connecting it to anything. The motor would whir, it would heat up the air, electricity would be dissipated— and nothing would get done. But if you took a pulley belt and connected the motor to, say, a sewing machine, that same energy could be harnessed to do something useful. *In living cells, enzymes make the connection between ATP breakdown and the energy-requiring reactions involved in the performance of useful work.*

(**c**) For instance, the phosphate group split off the ATP can be transferred, with the help of an enzyme, to some other molecule in the cell. That molecule acquires much of the potential energy lost by the ATP and becomes more reactive, so it can interact with other atoms or molecules to produce a substance the cell needs.

(**d**) The energy released by ATP breakdown may also be used to cause an enzyme to change its shape. Such changes can be used to perform mechanical work in the cell (to move something in a predictable way). When you raise your arm, it is the bending of many billions of enzymes that causes the motion. In other cases, the movement of an enzyme results in an unwanted molecule being thrust out of the cell; or it might result in some vital nutrient being brought in. The number and kind of reactions coupled to ATP breakdown are endless.

(**e**) Where does all the ATP needed for cellular reactions come from? If ATP were used up like nonreturnable bottles, cells (and organisms) would be in deep trouble. For instance, the mass of ATP you break down every day adds up to more than your total body weight! Fortunately, leftover ATP molecules aren't thrown away after use. They are recharged with energy and prepared for reuse by phosphorylation. Thus a single ADP molecule may be phosphorylated, broken down, and rephosphorylated thousands or even millions of times in a single day!

Figure 5.8 Cyclic photophosphorylation, which takes place in photosynthetic membranes and which yields one energy-rich ATP molecule. The process is described in the text.

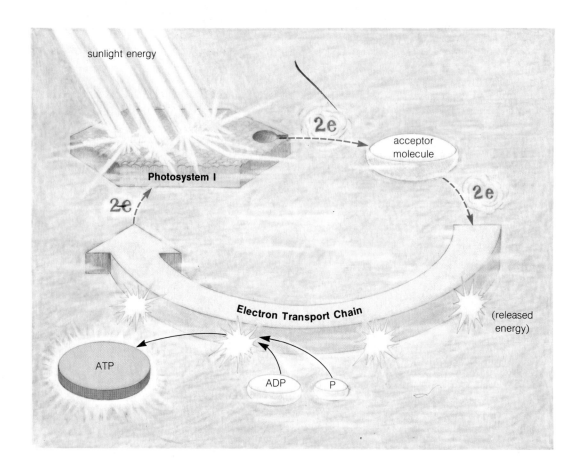

sunlight energy

2e

acceptor molecule

2e

Photosystem I

2e

Electron Transport Chain

(released energy)

ATP

ADP

P

ultimately converted to a form that can be used in producing sugars and other organic molecules. This pathway is known as **noncyclic photophosphorylation.** It is the heart of photosynthesis.

The first step of the noncyclic pathway begins when water is split into H$^+$ and OH$^-$ and their associated electrons (Figure 5.9). The two electrons derived from each water molecule flow into photosystem II, where they are excited and passed to the first electron transport chain. Here they give up much of their energy, and here two ATP molecules are formed. But when the electrons reach the end of this chain, they do not return to photosystem II. They now enter photosystem I. They are excited once again and passed to a second transport chain. But the flow of electrons through this chain does not yield more ATP. Instead, a molecule even more rich in usable energy than ATP is produced at the end of the chain. It is called **NADPH$_2$.** Apparently NADPH$_2$ is produced when an NADP molecule in the cellular environment is combined first with the two electrons zipping through the photosystems, then with the two H$^+$ ions released from each water molecule at the start of the noncyclic route.

For every two electrons entering the noncyclic photophosphorylation pathway, there is a <u>greater</u> energy yield of one NADPH$_2$ and two ATP molecules.

Even beyond the production of ATP and NADPH$_2$, the noncyclic pathway has had important consequences for the community of life. For noncyclic photophosphorylation produces oxygen as a by-product. The oxygen so produced has changed the entire nature of the earth's atmosphere. And it has made possible the extraordinary energy-yielding form of metabolism called cellular respiration, to be described in the next chapter.

The Dark Reactions Once the light reactions have produced ATP and NADPH$_2$, the photosynthetic cell has the means for converting carbon dioxide from its surroundings to carbohydrates and other molecules it needs to maintain itself, grow, and reproduce. (The conversion pathways are called the **dark reactions** because they don't depend directly

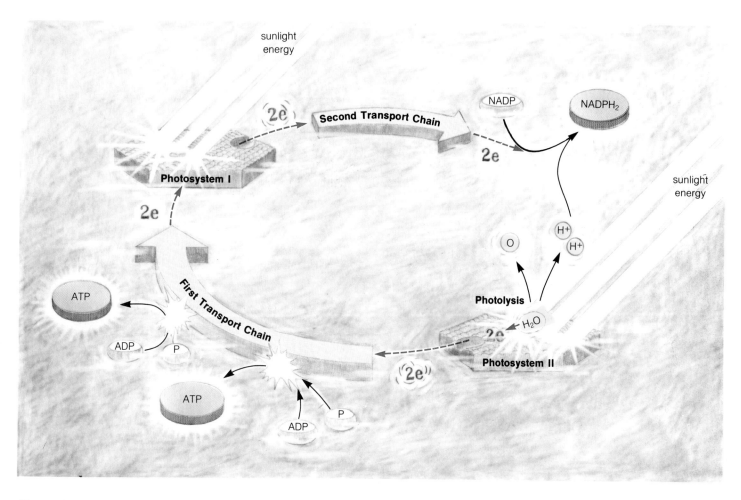

Figure 5.9 Noncyclic photophosphorylation, with its greater energy yield of one $NADPH_2$ molecule and two ATP molecules. The process is described in the text.

on sunlight. They *can*, in principle, proceed without it as long as ATP and $NADPH_2$ are available. But these molecules normally are produced only during daylight, so the "dark" reactions don't usually proceed for very long in the dark.)

To see how these conversions work, let's follow the reaction whereby six molecules of carbon dioxide (CO_2) are used to form one molecule of the sugar glucose ($C_6H_{12}O_6$). The reactions begin when an enzyme hooks up carbon dioxide to ribulose diphosphate (RuDP). The result is a highly unstable compound that immediately breaks apart into two molecules of a substance called phosphoglyceric acid, or PGA. This reaction sequence, in which carbon dioxide is removed from the air and combined with organic molecules, is called **carbon dioxide fixation.** For every six

CO_2 molecules fixed, twelve PGA molecules are produced (Figure 5.10).

The twelve PGA molecules now enter a reaction series by which they are first converted to phosphoglyceraldehyde, or PGAL. This conversion alone uses up no less than twelve ATP and twelve $NADPH_2$ molecules from the light reactions! The twelve PGAL molecules are then fused into different intermediate molecules, which are broken apart and rearranged in different ways. Two eventually are rearranged into a glucose molecule. In reactions that use up six more ATP molecules, the remainder are reassembled into RuDP. This entire reaction series is called the **Calvin– Benson cycle:** it yields enough RuDP molecules to replace the six used up in carbon dioxide fixation, as well as one glucose molecule.

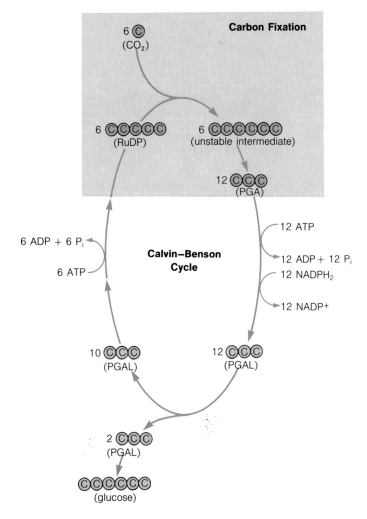

Carbon Fixation

6 Ⓒ
(CO₂)

6 ⒸⒸⒸⒸⒸ
(RuDP)

6 ⒸⒸⒸⒸⒸⒸ
(unstable intermediate)

12 ⒸⒸⒸ
(PGA)

6 ADP + 6 Pᵢ

6 ATP

Calvin–Benson Cycle

12 ATP

12 ADP + 12 Pᵢ

12 NADPH₂

12 NADP+

10 ⒸⒸⒸ
(PGAL)

12 ⒸⒸⒸ
(PGAL)

2 ⒸⒸⒸ
(PGAL)

ⒸⒸⒸⒸⒸⒸ
(glucose)

Figure 5.10 Summary of the dark reactions of photosynthesis. (Only the carbon atoms of the different molecules are depicted.)

Figure 5.11 relates the dark reactions to the overall process of photosynthesis. We may summarize them here in this way:

In the first stage of the dark reactions, carbon dioxide from the air is combined with RuDP, then the carbon atoms are locked into stable intermediate compounds.

In the second stage, chemical energy in NADPH₂ and ATP is used in the conversion of the carbon-containing intermediates to food molecules, and the RuDP used up at the start of the reactions is replaced.

Although the dark reactions require a total of eighteen ATP and twelve NADPH₂ molecules to produce each molecule of glucose, the process is amazingly efficient. For the ADP, inorganic phosphate, and NADP+ leftovers from the Calvin–Benson cycle are returned to the sites of the light reactions—where they can be reconverted to NADPH₂ and ATP!

What does an autotroph do with glucose produced in this way? It uses some as building blocks for its cellulose walls. As soon as night falls, the cell uses glucose as a food source until the following day. But usually an autotrophic cell produces more glucose during the day than it can possibly use or even hold. So the excess is converted into a different chemical form, which can be reserved for future needs. Some plants (sugar beets and sugar cane) store excess sugars as sucrose. But by far the most common storage form is starch. In simple plants such as algae, starch is assembled and stored within the chloroplast that produces the glucose. And certain land plants such as potatoes send the sugars produced in leaf cells down to special stem or root regions for storage as starch.

Chloroplasts are also capable of assembling lipids and amino acids, in addition to carbohydrates such as glucose. The PGA produced in carbon dioxide fixation can be used to make all the amino acids needed for building cell proteins and all the lipids needed for building membranes. Indeed, more than 90 percent of the carbon fixed by some green algae is used in the construction of proteins and lipids. Having a brief life cycle and enjoying plenty of water and sunlight, these algae put most of their photosynthetic activity into growth and reproduction, and very little into storage molecules.

A Modified Pathway for Photosynthesis The scheme we've been describing is the way most green plants acquire carbon, but certain modified pathways do exist. The most important is the C4 pathway. In the first step of C4 photosynthesis, ATP provides the energy needed to combine carbon dioxide and a three-carbon molecule into a compound having four carbon atoms (hence the name "C4"). The carbon atoms find their way into the Calvin–Benson cycle, where they are used in the normal way to produce carbohydrates and regenerate RuDP.

You might at first suspect the C4 pathway of being a waste of energy, because an extra ATP molecule is needed for carbon dioxide fixation. But in hot, dry climates, it has a selective advantage. First, there is plenty of sunlight to drive photophosphorylation, so abundant ATP can be produced at virtually no "cost" to the cell. Second, water is typically scarce in hot, dry regions—which has a direct effect on whether carbon dioxide fixation proceeds at all. Why is this so? The leaves of land plants have openings for controlling the amount of water vapor loss (Chapter Eleven). Hot, dry weather causes the openings to close and keep water in the

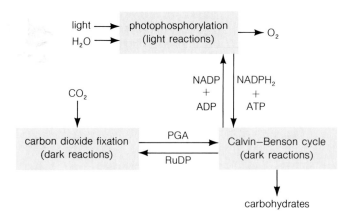

Figure 5.11 Summary of the reaction systems in photosynthesis. Like all reaction systems in a cell, the systems of photophosphorylation, carbon dioxide fixation, and the Calvin–Benson cycle can't operate in isolation. Each demands materials produced in another part of the system, and its products in turn are used by still other parts.

plant—which also happens to keep carbon dioxide (and other gases) out. Sufficient amounts of carbon dioxide must diffuse in through the openings if it is to be dissolved on the moist surface of interior cells and taken up in carbon dioxide fixation.

In plants without the C4 pathway, the enzyme that couples carbon dioxide to RuDP simply stops functioning when carbon dioxide levels fall below certain concentrations (about 300 parts per million). When the enzymes are inactive, carbohydrate assembly grinds to a halt. In C4 plants, however, the enzyme responsible for carbon dioxide fixation is active even when concentrations of the gas are quite low. This means C4 plants can continue to build food molecules even when leaf openings are nearly closed! The extra ATP is a small price to pay, given the presence of ample sunlight for generating more ATP.

About 100 different genera of plants use the C4 pathway. It exists in many tropical and desert grasses and in such plants as sugar cane and corn (which have tropical ancestors). In many cases, a single genus contains both C4 and non-C4 plants. Thus there are enough indications that we might be able to breed crop plants selectively to make the C4 pathway available to them. The advantage would be that crops may be raised in hot, relatively dry regions and during times of the year when rainfall is sparse. Water is a critical limiting factor on crop growth throughout much of the world, so even a topic as obscure-sounding as "the C4 pathway" may be one route toward increasing food production.

Photosynthesis and the Food Production Crisis Many biologists are studying mechanisms such as nitrogen fixation and the C4 pathway. Their work becomes especially urgent when we consider that the number and kinds of organisms that can be supported in any area depend on the amount of food being produced. It doesn't matter whether we are talking about the tropical reef, the savanna, the tundra, or our own agricultural systems: *every* ecosystem depends on the photosynthetic autotrophs that build the food foundation for all. This foundation is called the **net primary productivity:** it is the total dry weight of materials the autotrophs produce that is available to other life forms (after, of course, the autotrophs have used what they themselves need to grow and reproduce). Later chapters will cover some of the ways in which net primary productivity can be put to use in natural ecosystems. In the present context, we may simply introduce these two ideas:

1. The photosynthetic production of food by green plants on land and algae in the seas is the only significant food energy source available to natural ecosystems.

2. To remain stable, a natural ecosystem must operate on a balanced energy budget, year after year. (An ecosystem may be inefficient because of some kind of environmental constraint, and it may "underspend" its energy budget. But *no* system, no matter how efficient, can remain stable if it "overspends" the food energy base available to it.)

Assuming these two points are valid, does that mean we can get around our own food shortages simply by increasing primary production through modern agricultural practices? For instance, couldn't we export modern mechanized farming to parts of the world that still rely on primitive farming methods? Let's take a look. Table 5.1 compares the net primary production from a few natural ecosystems and from agricultural systems based on mechanized farming. The values for the natural ecosystems are not in any way maximum; they simply are frequently encountered. The values for agricultural systems include average as well as maximum yields. It is probably safe to say that the *average* productivity for any agricultural system will be lower than the productivity for the ecosystem that would exist naturally on the same site. Why, then, do the maximum figures for agriculture look so promising? The reason is that mechanized farming is sustained by heavy imports of fertilizers, herbicides, and pesticides, as well as by intensive cultivation and irrigation. Virtually all food products are shipped away from the site. Soil fertility plummets, so the application of more ammonia, phosphates, and other minerals is a one-way energy drain. Right now, fertilizer production is based on nonrenewable resources such as petroleum, natural gas, and mineral deposits. Further resources (usually petrochemicals) are used to manufacture pesticides and herbicides. Do we really need these last two

Table 5.1 Comparing the Net Primary Productivity of a Few Natural Ecosystems With Agricultural Systems

	Net Primary Productivity (grams/meter^2/year)	
	Average	Maximum*
Natural ecosystems		
Desert**	40	—
Shortgrass prairie***	60	—
Tallgrass prairie	446	—
Deciduous forest	1,560	—
Pine forest	3,180	—
Agricultural systems		
Wheat	344	1,250
Oats	359	926
Corn	412	790
Rice	497	1,440
Hay	420	940
Potatoes	385	845
Sugar beets	765	1,470
Sugar cane	1,725	6,700

*Represents intensive cultivation under best of conditions for soil and climate, using the most "advanced" agricultural practices (heavy irrigation; fertilizer, herbicide, and pesticide application; and so forth).
**12.5 centimeters (5 inches) annual rainfall.
***32.5 centimeters (13 inches) annual rainfall.

Table 5.2 A Few of the Energy Costs of Acquiring Food Energy With Modern Agriculture

Energy Requirement	Energy Source
Photosynthesis	Solar
Manual labor	Food energy*
Mechanization	Fossil fuels
Mining ores	
Shipping ores	
Refining metals	
Shipping metals	
Manufacturing farm machinery	
Shipping machinery	
Maintaining machinery (repeats all of above)	
Using machinery (plowing, seeding, cultivating, harvesting, storing)	
Fertilizers	
Mining or obtaining raw materials	
Shipping raw materials	
Manufacturing products	
Shipping products	
Application of products	
Processing food materials	
Packaging food materials	
Shipping food materials	
Distributing food materials	

*Derived directly or indirectly from photosynthetic autotrophs.

products? We do as long as we have field after field of the same kinds of crops and as long as we have no other means for stabilizing them. For our vast monocrops are also the natural food for many insects, and the fertilizers are energy sources for other plants (weeds). Thus we are locked into competition with insect and weed populations for agricultural resources that can fan their runaway growth—just as they have fanned our own population growth (Chapter Twenty-One).

The balance sheet for resources looks more alarming when we examine all the energy costs of mechanized farming (Table 5.2). And this table doesn't even list *all* the energy needed to produce a crop. *The amount of energy needed in modern mechanized farming at the very least equals and usually exceeds by several times the total energy value of the crop it produces!* We are, in modern societies, running in the red on the most basic energy conversion of all: food production.

Projections into the future are risky, but some guidelines are available. Most will be considered in Chapters Twenty-One and Twenty-Two. But in the context of this discussion, we can say this: *Agricultural systems must be redesigned to minimize external energy input, other than the solar energy needed to maintain photosynthesis among the autotrophic cells which, directly or indirectly, feed all of life.*

In this same light, *agricultural systems must eventually be redesigned to depend more and more on plant species that do not require massive inputs of material resources.* As you will be reading in the next chapter, the soybean (a legume) is such a crop. Because it is symbiotic with nitrogen-fixing bacteria, it does not require nitrogen-containing fertilizers. Another possibility is modification of food crops through the direct transfer of hereditary material from one kind of organism to another (Chapter Ten). Recently the nitrogen-fixing capacity of a symbiotic bacterium has been introduced into a eukaryotic cell with which it bears no obvious relationship in a natural community. And the cell line retains the nitrogen-fixing capacity for generations! Obviously it is a long way from a single cell to nitrogen-fixing crops, but the first step in that journey has been taken. In these and other ways, we can look to the cellular level of organization for potential solutions to many of the most urgent problems we are facing today.

Perspective

We embarked on our journey into the cell with a basic premise in mind—that for a cell to exist, a steady stream of energy must flow through it. We have also looked at the functional zones within a cell, where energy is trapped and used. In this chapter we put the cell in its environment. For as long as it remains alive, the cell maintains a stable, constant internal environment by precisely adjusting its activities to prevailing conditions in the external environment. Achievement of this state is called **homeostasis.** You will read more about the cell's homeostatic control mechanism in later chapters. Here, you have seen that achieving this kind of dynamic stability begins with the acquisition and use of (1) materials, and (2) energy.

Let's reflect on that first activity. Cells are adapted to the prevailing conditions of their environment. If they live in such places as a fresh-water lake, with its low solute concentrations, they have devices for pumping out excess water that diffuses naturally into their body. If they live in places such as the sea, with its high salt concentrations, they might use structural devices to pump water in and/or pump salt out. Some molecules are transported passively into and out of cells, others are actively sought, stockpiled, and discharged. All such manipulations of materials help cells achieve homeostasis.

But both the acquisition and use of resources require energy. In a natural ecosystem, energy acquired by the autotrophic cells percolates through the entire array of organisms; and materials such as nitrogen, once used, are returned to the site where they can be picked up and used again. Through reliance on limitless energy streaming from the sun, and through dependence on recycled materials, natural ecosystems are self-sustaining over long periods. In contrast, our agricultural systems are not self-sustaining. Expensive and rapidly diminishing energy sources are used to create fertilizers, pesticides, and herbicides, which are carted onto crop fields. Then materials are carted away in the form of harvests and herds. Such systems have strayed far from the biological model of trapping and using resources.

Of course, many other factors impinge on this issue, not the least of which is the fact that the human population is expanding at an astounding rate. But in trying to cope with the sheer numbers of people demanding to be fed, we must not simply assume modern technology will automatically lead to increased food production. For one thing, it's unrealistic to hope for a completely self-sustaining agricultural system. It would, after all, demand that all the food products be returned to the site, leaving nothing for us to eat. Even so, it is realistic to strive for a system that can be balanced with more modest energy inputs from renewable resources. Natural selection, operating for billions of years, has achieved such balances. Artificial selection, carefully

conceived, might possibly do the same. Recycling human and animal wastes in an agriculturally useful way, designing crop plants for disease and insect resistance, incorporating natural systems for nitrogen fixation in many crop plants, and perhaps engineering crops genetically for incorporating more efficient pathways such as C4 photosynthesis—all these things would go a long way toward increasing production while decreasing our reliance on nonrenewable energy resources.

Our present strategy is to meet *immediate* needs with the technology now available, for the immediate reality holds the specter of starvation for much of the world without that technology. But it is imperative that, in the long run, agricultural policies be brought in line with biological principles governing the stability of all living systems. Efforts in this area must be part of a fundamental redirection in our thinking about materials and energy use, human population growth, and all other factors that have led to the instability that leaves us vulnerable.

Recommended Readings

Kirk, D. et al. 1975. *Biology Today.* Second edition. New York: Random House. One of the most up-to-date summaries of what is known about photosynthesis can be found in chapter 15 of this introductory text.

Lehninger, A. 1971. *Bioenergetics.* Second edition. Menlo Park, California: Benjamin. A brief, well-written description of energy flow through living systems.

There are several good summaries of the world food production crisis. We recommend those listed at the conclusion of the next chapter, which expands on this subject.

Chapter Six

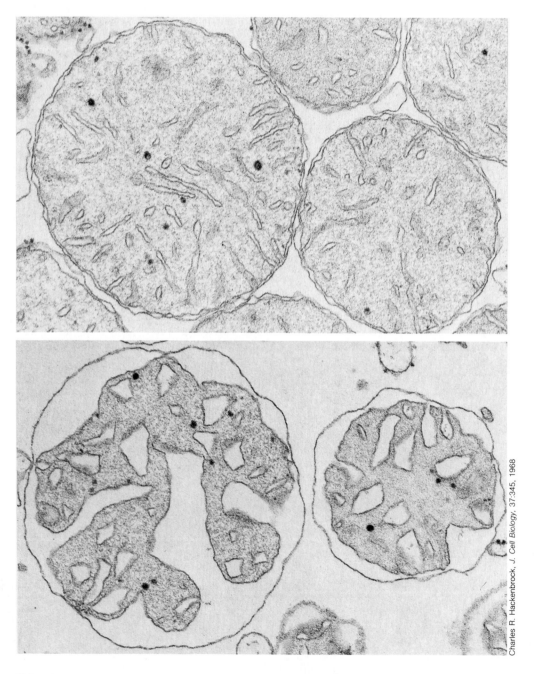

Charles R. Hackenbrock, *J. Cell Biology.* 37:345, 1968

Figure 6.1 Mitochondria at rest, and then engaged in the fine art of rapidly releasing energy stored up in molecules within the eukaryotic cell.

How Cells Use Resources

When you look at a single living cell beneath a microscope, you are watching a form seething with activity. In its movements it is seeking out and taking in small molecules—food, raw materials—suspended in the water droplet on the slide. To power those tiny movements of itself and parts of itself, the cell is tapping into internal reservoirs of molecules it had previously stored away, and it is dismantling them now for their chemical potential energy. Even as you observe it, the cell is using these molecules for building and maintaining itself: its membranes and extracellular structures and organelles, its pools of enzymes, its concentrations of chemical compounds, its information-storage system. It is alive; it is growing; it can reproduce itself. Multiply all this activity by *60,000 billion cells* and you have an inkling of the extent of activity going on in your own body even as you are sitting quietly, doing nothing more than observing that single cell! Thus, how cells use resources is not by any stretch of the imagination a story pieced together by biologists and of interest only to biologists. It is also the story of how *you* use resources—how you have the means to move, to think, to laugh, to do all the myriad things that go into being alive.

An Overview of Different Energy-Extraction Pathways

Depending on the type of organism we are talking about, there is more than one kind of resource a cell can use, and there is more than one way to use it. To keep things simple, however, let's focus on the manner in which energy extracted from a glucose molecule is converted to the energy of ATP, for it can be used as an example of energy-extraction pathways in general. The steps involved are outlined in Figure 6.2. You may find it useful to look at this illustration before we begin, and to keep this important point in mind:

Glucose breakdown and the associated energy transfers leading to ATP formation are central to the functioning of most prokaryotic and eukaryotic cells.

The initial break-up of a glucose molecule is known as **glycolysis** (*glyco-*, "sugar"; *-lysis*, "break apart"). It is not a haphazard event. Certain chemical compounds must first be gathered up from the surrounding environment. Special enzymes are organized to receive and modify parts of the glucose molecule as it is broken apart. The product of each enzyme-catalyzed reaction is then passed on to another enzyme in a predictable sequence of reactions. Energy extraction is, in short, systematic business.

Still, some cells are more systematic about it than others. In the case of the simpler prokaryotic cells, much of the energy locked up in glucose is still locked up in its disassembled parts when they are done with it. These simple cells live in places where there is no oxygen, so they must rely solely on *anaerobic* pathways (which can proceed in the absence of oxygen atoms). One of these anaerobic routes, glycolysis, is the springboard for glucose breakdown. In fact, glycolysis is the springboard for glucose breakdown in almost all prokaryotic and eukaryotic cells, regardless of whether oxygen is present or not. Briefly, a glucose molecule (which has six carbon atoms) is split in half (into two three-carbon fragments). The fragments are rearranged into two pyruvate molecules. (Pyruvate is the ionized form of pyruvic acid.) During this juggling act, the cell gains a total ot two ATP molecules.

In numerous kinds of cells living under anaerobic conditions, the pyruvate continues down one of two routes (Figure 6.2). In the first, a carbon atom is split away from each pyruvate molecule to form carbon dioxide. The two-carbon compounds remaining are converted to a substance called ethyl alcohol (ethanol). This pathway is known as **alcoholic fermentation.** We may summarize the reaction in the following way:

$$C_6H_{12}O_6 \longrightarrow 2C_2H_6O + 2CO_2$$
(ethanol)

$$\boxed{\text{energy yield} = 2\ \text{ATP}}$$

Yeasts (small, single-celled fungi) use this pathway. The carbon dioxide by-product of the reactions is the reason yeast dough rises; the alcohol produced is the reason beer, wine, and distilled spirits have the kick that they do. In terms of energy conversion efficiency, though, a yield of

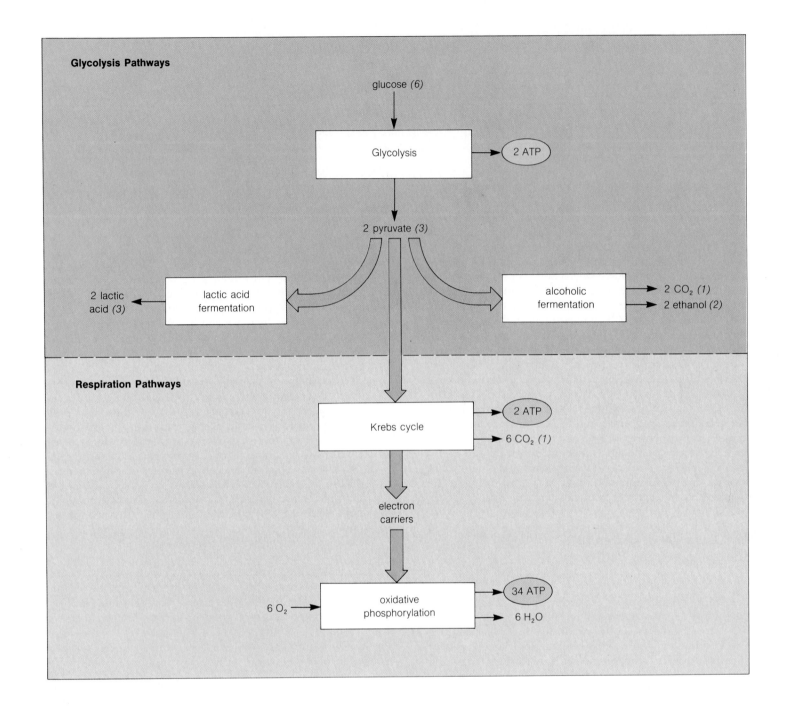

Figure 6.2 Overview of glycolysis and cellular respiration. Glucose is the starting material in this example. Boldface arrows indicate three routes by which it can be dismantled. The route taken depends on the organism and on environmental conditions. (Italic numbers in parentheses refer to how many carbon atoms there are in each molecule of the compound listed.)

A simple way to remember the *net* energy yield of these three routes is to add up the ATP molecules from the start (in all cases, glycolysis) to the finish. Only two ATP molecules are produced on the route ending in lactic acid fermentation; only two are produced on the route ending in alcoholic fermentation. Thirty-eight ATP molecules are produced on the route ending with the pathways of cellular respiration.

Figure 6.3 On alcoholic fermentation. In every country throughout the world, through all recorded history and probably before then, people have engaged in the fermentation of various natural materials such as grapes and other fruits, cactus, rice, and other grains.

For those students with an interest in history, we will attend here to the practical problem of why a bottle of grape juice being tended so carefully in the back of a closet turns into sour-tasting vinegar instead of something that will revolutionize the wine industry.

The uninitiated usually open the bottle at intervals to sniff their creation and to determine how things are progressing. Their inquisitiveness has two undesirable effects. It introduces oxygen into the bottle, and it permits microorganisms to enter. The combination has dire consequences, for certain ever-present bacteria will, upon encountering ethanol in the presence of oxygen, metabolize it into acetic acid—better known as vinegar. Conditions in the bottle must be kept anaerobic.

But this doesn't mean the trick is simply to seal up the bottle. Carbon dioxide building up inside might cause the bottle to explode its contents all over the closet. Simple devices have been designed, over the centuries, that let carbon dioxide bubble out even while keeping oxygen from getting in.

If you are experimenting with this kind of microbiology, you will also discover that yeasts do far more than merely convert sugar molecules in the grape juice into alcohol. They add a host of other metabolic by-products that impart to the wine certain subtle qualities, the absence of which is stunningly apparent with the first taste. If you use ordinary baker's yeast for your first experiment, you will soon understand why wineries jealously guard their yeast cultures as professional secrets.

The Bettmann Archive Inc.

only two ATP molecules from each glucose molecule traveling the glycolysis/alcoholic fermentation route is pretty dismal.

In the other anaerobic route, the pyruvate coming from glycolysis can be rearranged into lactic acid, which is a three-carbon compound. This pathway is known as **lactic acid fermentation:**

$$C_6H_{12}O_6 \longrightarrow 2C_3H_6O_3$$
(lactic acid)

energy yield = 2 ATP

Milk or cream turned sour is a sign of microorganisms having used pyruvate in lactic acid fermentation. (The sourness depends on how much lactic acid has accumulated.) With a yield of only two ATP molecules, however, the route proceeding from glycolysis through lactic acid fermentation isn't any more efficient than the other fermentation route described.

It is only when we get to cellular respiration that we cross a truly major threshold in energy extraction processes. **Cellular respiration** actually consists of two closely linked pathways: the Krebs cycle and the oxidative phosphorylation sequence. Many prokaryotic cells and almost all eukaryotic cells can follow these pathways, which begin with the pyruvate made available by glycolysis. In the **Krebs cycle,** pyruvate fragments are completely broken down to carbon dioxide. By itself, the cycle yields only two more ATP molecules for each glucose molecule. But as the pyruvate is broken down, its hydrogen atoms—with their high-energy electrons—are transferred to nucleotides (such as NAD) present in the cell. The nucleotides cart them off to an electron transport chain similar to the one used in photosynthesis. *And as the high-energy electrons move through the transport chain, they give off energy in increments, some of which are enough to phosphorylate ADP present in the cell.* This is the pathway of **oxidative phosphorylation.** It yields thirty-four ATP molecules for every glucose molecule entering the reaction sequence!

To keep this electron transport chain going, something must take away all the electrons and hydrogen ions being dumped at the end (Chapter Five). Oxygen serves this role of electron acceptor. Hence oxidative phosphorylation is said to be an *aerobic* process: it can't proceed unless oxygen atoms are present. More than this, when the electrons and hydrogen ions are coupled with oxygen, water is formed. Thus the aerobic pathway helps recycle materials (carbon dioxide and water) through the environment, in addition to increasing the total energy yield from the breakdown of each glucose molecule:

$$C_6H_{12}O_6 + 6O_2 \rightarrow 6CO_2 + 6H_2O$$

energy yield = 38 ATP

Taken as a whole, glycolysis, the Krebs cycle, and oxidative phosphorylation is the major route by which usable energy becomes available in most living things. This overall route is summarized in Figure 6.2.

By now, you might be asking yourself why a cell bothers to go through all this trouble. Wouldn't it be far easier to completely break down the whole glucose molecule all at once? But if you think back on what happens when a match is dropped into a gasoline tank (Chapter Three), you already have an inkling of why the most direct energy conversion route is *not* the best route for a cell. With the conversion of just one glucose molecule to carbon dioxide and water, tremendous energy is released. If a cell were simply to pop off the molecule in a single step, most of the energy would be released as heat. The sudden burst of heat would play havoc with the cell—which would literally cook to death! It is this potentially destructive release of heat energy that calls for a moderate approach to energy metabolism. It is the reason why a cell must resort to a step-by-step attack on energy-rich molecules (in the case of a glucose molecule going through the aerobic route just described, about 140 steps in all). Whatever the details of any given energy-extraction pathway, the point is to get the most energy possible without getting fried in the process.

And now, if you would like to (or are asked to) take a closer look at these energy-releasing pathways, you can browse through the following two sections, which are included to round out the picture just presented.

A Closer Look at Glycolysis

Why does an anaerobic cell even bother to convert pyruvate to lactic acid (or ethanol) and carbon dioxide? It gets no more energy once pyruvate has formed—so why doesn't it quit and leave well enough alone? For the answer, let's go back to where glycolysis begins.

Figure 6.4 lists the reaction steps in glycolysis. Because glucose is a relatively stable (unreactive) molecule, it is made reactive by hooking a phosphate group onto it—not once, but twice. The resulting sugar diphosphate is split into a phosphoglyceraldehyde (PGAL) molecule and a dihydroxyacetone phosphate (DHAP) molecule. These molecules are structural isomers (they have the same atoms but in a different arrangement), and the DHAP is readily converted to PGAL. So at this point we say we have two PGAL molecules.

But now another phosphate group is tacked on to each PGAL—and hydrogen atoms are released during the reaction. It takes molecules of the nucleotide NAD (nicotinamide adenine dinucleotide) to pick up the hydrogen ions and electrons. When they do, they become high-energy reduced dinucleotides: $NADH_2$. The problem is that there is only so much NAD in the cytoplasm of a cell. And without NAD, nothing would be around to accept the hydrogen atoms released during the breakdown of PGAL, glycolysis would rapidly grind to a halt, and the cell would be doomed from that point onward.

How does the cell maintain enough NAD? When the pyruvate from glycolysis travels the alcoholic fermentation route, acetaldehyde formed during the conversion of pyruvate acts as a hydrogen *acceptor*. It strips the $NADH_2$ of the two hydrogen atoms it acquired during PGAL breakdown, and it becomes converted to ethanol in the process. In the lactic acid fermentation route, the pyruvate itself acts as a hydrogen acceptor, stripping the $NADH_2$ of its two hydrogen atoms on the way to becoming lactic acid. Thus, even though the anaerobic cell gets no more usable energy when pyruvate is converted, the conversion does have this advantage:

The final steps of the fermentation pathways serve to regenerate NAD.

In short, the two fermentation pathways that have been discussed here represent two different adaptive strategies for recycling NAD.

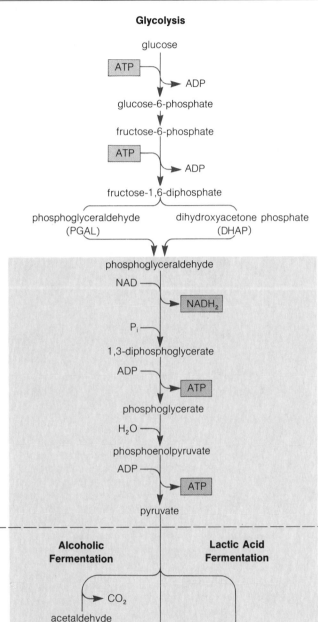

Glycolysis

glucose

ATP

ADP

glucose-6-phosphate

fructose-6-phosphate

ATP

ADP

fructose-1,6-diphosphate

phosphoglyceraldehyde (PGAL) dihydroxyacetone phosphate (DHAP)

phosphoglyceraldehyde

NAD

NADH₂

Pᵢ

1,3-diphosphoglycerate

ADP

ATP

phosphoglycerate

H₂O

phosphoenolpyruvate

ADP

ATP

pyruvate

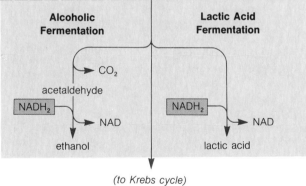

Alcoholic Fermentation   **Lactic Acid Fermentation**

CO₂

acetaldehyde

NADH₂

NAD

NADH₂

NAD

ethanol   lactic acid

(to Krebs cycle)

A stable 6-carbon sugar

picks up a phosphate group from an ATP molecule,

thereby becoming a reactive 6-carbon sugar phosphate

that is promptly modified into another 6-carbon sugar phosphate.

The modified form picks up a phosphate group from another ATP molecule

to become an even more reactive 6-carbon sugar diphosphate,

which is split into two 3-carbon sugar phosphates: PGAL and DHAP.

DHAP can be converted to PGAL, so beyond this point, each step occurs <u>twice</u> for each glucose molecule being torn down.

PGAL gives up hydrogen to an NAD coenzyme, which produces the energy carrier molecule NADH₂.

In the process, PGAL is converted to a 3-carbon acid and picks up another phosphate group,

which is promptly used to produce ATP from ADP.

The resulting molecule is immediately relieved of a water molecule,

which produces an acid known as PEP.

PEP is capable of donating its phosphate group to ADP, so more ATP is formed

and this leaves only pyruvate (the ionized form of pyruvic acid).

If the pyruvate now enters the alcoholic fermentation sequence, the whole chain of reactions leading from glucose to ethanol ends up with a net energy yield of two ATP molecules.

If the pyruvate enters the lactic acid fermentation sequence, the net energy yield is only two ATP molecules.

In both cases the NADH₂ is restored to NAD, which can take part again in the reaction with PGAL.

Figure 6.4 Pathways of glycolysis. All the steps in the shaded area must occur twice for every glucose molecule being dismantled along the glycolysis pathways.

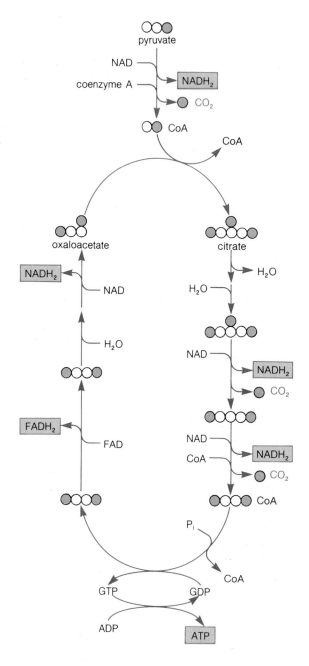

Step 1. As pyruvate enters the mitochondrion, a carbon atom is removed as CO_2. At the same time, two hydrogen atoms are removed and transferred to NAD, producing $NADH_2$. The result is the conversion of one of the two remaining carbon atoms to an acid group. The two-carbon compound is linked to a molecule called coenzyme A, or CoA.

Step 2. The Krebs cycle proper begins when the two-carbon acid is coupled with a four-carbon compound having two acid groups (oxaloacetate). This produces citrate, a six-carbon compound having three acid groups. CoA is released for reuse.

Step 3. After two intermediate reactions, an acid group is split off the six-carbon compound and is released as CO_2. At the same time, two hydrogen atoms are removed, forming a second molecule of $NADH_2$.

Step 4. Now the third and last carbon atom derived from the pyruvate is split off as another CO_2 molecule. Once again hydrogen atoms are given off to produce $NADH_2$. In the process, a new acid group is linked to coenzyme A. The rest of the cycle works to convert this new four-carbon molecule with two acid groups to the related compound oxaloacetate, with which the cycle begins once again.

Step 7. And another $NADH_2$ molecule is produced by the transfer of two more hydrogen atoms. The product of this last reaction is the same four-carbon compound that the cycle starts with. Thus, we are ready for another turn of the Krebs cycle.

Step 6. Two more hydrogen atoms are removed, but this time they are passed on to FAD, producing $FADH_2$.

Step 5. During the next interconversion, the energy released when the CoA is split off is used to produce an ATP molecule. (Actually the related compound GTP is formed by addition of a phosphate group to GDP, but GDP then transfers its phosphate to ADP.)

Figure 6.5 The Krebs cycle, the first stage of cellular respiration. In this sketch, each ○ signifies the presence of a carbon atom in the molecule, and each ⬤ signifies an acid group (—COO⁻) or a carbon dioxide molecule produced when an acid group is split off. Remember that for each glucose molecule being metabolized, two pyruvate molecules have been formed. So two turns of the Krebs cycle must occur for each molecule of glucose being oxidized.

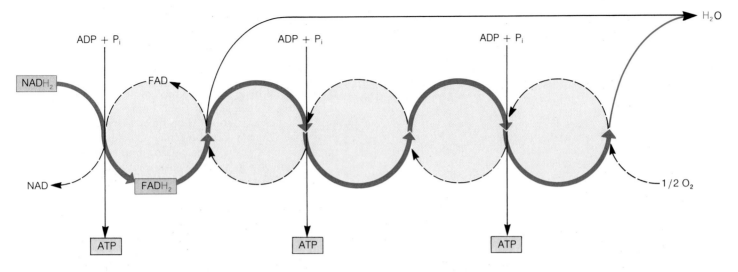

Figure 6.6 Simplified version of the energy flow through the oxidative phosphorylation sequence. $NADH_2$ from the Krebs cycle enters the sequence, transferring its hydrogen atoms (with their electrons) to FAD. The energy released in the reaction couples P_i to ADP, forming an ATP molecule. (The sequence also can start directly with $FADH_2$ from the Krebs cycle, but the cell loses its chance to form an extra ATP molecule when it does.) The electrons derived from the hydrogen atoms are shuttled through the chain to an oxygen acceptor, which carts them off. Oxygen must be available to keep the chain clear for operation.

A Closer Look at Cellular Respiration

When the products from glycolysis are broken down further in the Krebs cycle, they are completely oxidized. Two kinds of molecules serve as acceptors for the hydrogen ions and electrons being released. One is NAD; the other is FAD (flavin adenine dinucleotide, Figure 6.5). They are extremely important components of the two stages of cellular respiration:

In the Krebs cycle, NAD and FAD are reduced to $NADH_2$ and $FADH_2$, which are used in the energy transfers of oxidative phosphorylation.

In the oxidative phosphorylation sequence, $NADH_2$ and $FADH_2$ are stripped of hydrogen atoms and the high-energy electrons associated with them. The energy of these electrons is used in the formation of thirty-four ATP molecules.

Electrons from $NADH_2$ and $FADH_2$ move through the specified sequence because of the way the cytochrome molecules making up the chain are lined up in the membranes of a mitochondrion. The first in line has a slightly greater affinity for electrons than $FADH_2$; hence the elec-

trons are drawn to it. Then the first cytochrome releases its hold on each newly acquired electron because of the greater pull of the second cytochrome sitting next to it, and so on down to the most voracious acceptor of all: the oxygen near the end of the chain. Once an oxygen atom has picked up two electrons (thereby acquiring two negative charges), it combines with the two hydrogen ions separated from $NADH_2$ or $FADH_2$ at the beginning of the electron transport chain.

As Figure 6.6 shows, for every $NADH_2$ molecule coming from the Krebs cycle, the cell can produce three ATP molecules. ($FADH_2$ from the Krebs cycle can also start the sequence. But notice that it enters the sequence at a point past the first ATP production step. The cell gets a return of only two ATP molecules when $FADH_2$ is used instead.) The Krebs cycle is not the only source of $NADH_2$, however. Glycolysis (Figure 6.4) also leads to the production of two of these molecules, which can take part in oxidative phosphorylation under aerobic conditions. Under anaerobic conditions, though, cells call this source of hydrogen donors into play in order to produce lactic acid (or alcohol) and thereby regenerate NAD, which is essential for cell functioning.

Your own lungs and circulatory system are designed to provide every cell in your body with enough oxygen to carry out aerobic metabolism under normal conditions. But what

Table 6.1 Summary of the ATP Yield From the Aerobic Metabolism of One Glucose Molecule

Source of Energy	Energy Yield
Glycolysis pathway	2 ATP
Krebs cycle	2 ATP
Oxidative phosphorylation:	
8 NADH$_2$ from Krebs cycle	24 ATP
2 cytoplasmic NADH$_2$ from	
glycolysis pathway	6 ATP
2 FADH$_2$ from Krebs cycle	4 ATP
Total:	38 ATP

happens in muscle cells, for example, when you exercise so rapidly that oxygen uptake no longer corresponds to your pace? Consider the cellular events that occur when you decide to see how fast you can run in a mile. When you reach the point that you are no longer supplying individual cells with enough oxygen, you are causing the transport chains of oxidative phosphorylation to load up with the electrons that oxygen normally would be carrying away. NADH$_2$ molecules are no longer stripped of their electrons and hydrogen atoms, so there is a shortage of NAD. Without NAD, the Krebs cycle stops, and the pyruvate from glycolysis begins to pile up.

At this point, your muscle cells revert to the less efficient lactic acid pathway. In response to the NAD shortage, pyruvate must move down this pathway instead. So lactic acid builds up in muscle cells, which are forced to operate on the low ATP output from glycolysis. They can do this for a short time, assuming enough glucose is available. But the build-up of lactic acid, which is toxic in high concentrations, causes muscle cramps in your legs. If you heed the message and stop to rest, the lactic acid will soon be flushed out of your muscles by the bloodstream. Once that happens, your muscle cells revert to the efficient aerobic pathways.

Just how efficient is aerobic metabolism compared to the anaerobic routes? As Table 6.1 shows, it is about *eighteen times* more efficient than anaerobic glycolysis in converting the energy in glucose to ATP.

Extracting Energy From Substances Other Than Glucose

Cells, again, can use a wide variety of resources other than glucose in obtaining energy. For instance, carbohydrates (both simple sugars and complex carbohydrates such as starch) can be metabolized as an energy source because they

can be converted to glucose or some other compound, which can flow directly into the glycolysis pathway (Figure 6.7). There, they are broken down with the same enzymes and in the same manner as glucose itself. Similarly, fats can be broken down into substances that flow into either the glycolysis pathway or the Krebs cycle. But fats are much more reduced than sugars; they contain far more hydrogen atoms than oxygen atoms. This means more NADH$_2$ can be produced when a fat molecule is broken down. Thus, metabolizing 1 gram of fat yields about *twice* as much ATP as metabolizing 1 gram of carbohydrate. Fats are highly efficient storage molecules because they are such a concentrated energy source. (An overweight person finds an accumulation of body fats difficult to burn up for precisely this reason.)

Proteins also may be used as an energy source, once hydrolysis releases their component amino acids. The amino acids are converted to pyruvate or to one of the intermediate compounds of the Krebs cycle. The energy yield from protein breakdown is roughly the same as that from carbohydrate breakdown. Normally, though, a cell uses the amino acids not as an energy source but as raw materials for assembling its own proteins. It's just that sometimes protein intake exceeds cellular requirements, so the excess is sent down energy-extracting pathways. And sometimes protein will be sent down those pathways even if the cell is *not* getting enough protein to maintain itself! This happens among individuals suffering from starvation. Because they are not taking in enough carbohydrates and fats to meet their energy requirements, their cells disassemble proteins making up their own body. Draining away the very proteins needed to maintain a healthy body is a last-ditch measure to supply the energy needed to keep the body alive.

Pathways for Interconverting Material Resources

Although most heterotrophic cells take in a variety of carbohydrates, fats, amino acids, and nucleotides, rarely are these molecules in precisely the right chemical forms and proportions needed to build new cellular materials. Here again, the basic metabolic pathways we have just considered are brought into action. In addition to providing routes for breaking down molecules, they take part in the interconversion of one kind of molecule into other kinds that the cell requires.

Carbohydrates and lipids, for instance, are constantly needed as sources of energy and as raw materials for maintaining, enlarging, and reproducing cell structures. In the case of carbohydrates, the molecules needed are built up largely through modifications of glycolysis and through rearrangements similar to the Calvin–Benson cycle of

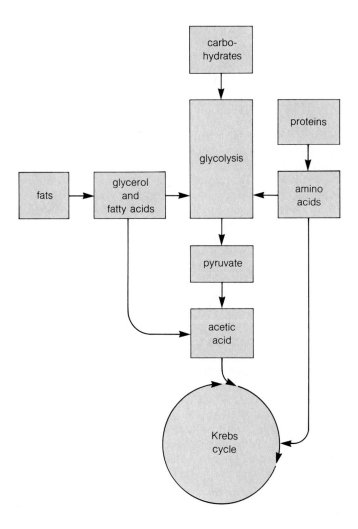

Figure 6.7 Where substances other than glucose flow into the energy-extraction pathways of a cell.

photosynthesis. Such rearrangements require more cellular energy (ATP) to drive the reactions. *This means it always takes more energy to build up a carbohydrate molecule than can be recovered when the same molecule is broken down.* Given an adequate supply of ATP, however, the intermediates of glycolysis and the Krebs cycle can serve as raw materials for all the kinds of sugars a cell might need, regardless of the type of carbohydrates taken in as food. Disaccharides can be built from appropriate monosaccharides, or the monosaccharides can be linked together into larger molecules such as starch and cellulose (Chapter Three), depending on the type of cell.

As another example, cells must have their own array of proteins as structural materials and enzymes. Autotrophic cells are able to synthesize all the twenty or so amino acids

they need for building proteins. For them, intermediates of the Calvin–Benson cycle, glycolysis, and the Krebs cycle are essential ingredients. Provided enough nitrogen is available, these intermediates can be rearranged and converted to various amino acids. Heterotrophic cells build amino acids in similar ways. But many heterotrophs, including ourselves, have lost the ability to build all the different kinds needed. In fact, we can make only twelve of the twenty amino acids required for assembling body proteins. The remainder are **essential amino acids;** we must get them from the food we eat for protein assembly to proceed. Regardless of whether they are synthesized by the body or derived from food, it takes at least eight times as much energy to build a protein as can be recovered when the protein is broken down. Given the importance of proteins to the life of a cell, it's a small price to pay.

Vitamins, Minerals, and Normal Body Functioning

The catch-all category **vitamins** refers to various organic molecules that an organism needs to help build essential compounds but can't produce for itself. Most plant cells can synthesize all these molecules, so plants generally have no vitamin requirements. But animal cells have, in general, lost the ability to make all the vitamins they need. Thus vitamins, like essential amino acids, must come from the diet.

Our own cells need at least thirteen different vitamins. For example, riboflavin (vitamin B_2) is needed for assembling FAD. Nicotinic acid (vitamin B_3, or niacin) is needed for assembling NAD and NADP. Ascorbic acid (vitamin C) is essential for building collagen (the main protein of skin, tendon, and bone). Its absence from the diet results in the deterioration of collagen-containing connective tissue, a disease known as **scurvy.** The US Food and Drug Administration has established recommended daily allowances (RDAs) for these and other vitamins. Food packaging laws now require that many food products be labeled with their specific contribution toward fulfilling these recommended daily allowances.

Despite the plethora of vitamin supplements that can be purchased in grocery stores as well as pharmacies, the most reasonable way of supplying your cells with the vitamins they need is to consume a well-balanced assortment of wholesome foods. This also assures you of adequate carbohydrates, fats, proteins, and minerals. In recent years there have been many claims that massive doses of certain vitamins are spectacularly beneficial. To date, however, there is no clear evidence that vitamin intake exceeding the recommended daily allowances leads to better health. In most cases, excessively large doses of vitamins are merely wasted. For individuals taking in large amounts of vitamin

C, for instance, direct chemical analysis shows that any amount above the recommended daily allowance ends up in the urine immediately after it is absorbed from the gut. The body simply will not hold more vitamin C than it needs for normal functioning. Abnormal intake of at least two other vitamins—A and D—not only is unnecessary, it actually causes serious disorders. *Both shortages and massive excess of foodstuffs can disturb the delicate feedback relationship that characterizes physiological health.*

In addition to vitamins, all cells need inorganic materials known as **minerals.** (Some minerals are called trace elements because they are needed only in extremely small amounts.) Plant cells, for example, need magnesium in order to make chlorophyll. Most cells require both calcium and magnesium in a host of enzyme-mediated reactions. All cells need phosphorus for phosphorylation; they need sodium and potassium for maintaining ion balances. These last two minerals are especially important in muscle and nerve cell functioning. All cells need iron for building cytochromes. Many animal cells need additional amounts of iron to produce hemoglobin, the oxygen-carrying protein in the blood.

Occasionally, diets must be supplemented when a mineral is missing from foods found in the environment. For instance, all vertebrates require iodine for synthesis of thyroid gland hormones. At one time, a disease known as **goiter** was common in certain regions (such as the Great Lakes region of North America) where water and soil are deficient in iodine. In this disease, the dietary deficiency causes hypothyroidism (a decrease in thyroid hormone production), which in turn triggers the action of a thyroid-stimulating hormone to compensate for its absence. The outcome is a nonfunctional enlargement of the thyroid gland in the throat. (Adding iodized salt to the diet of individuals living in these regions reduced the incidence of the disease.)

Resources and the Human Condition

From a biological perspective, the main limiting factor on our existence as a species is food. Said another way, where do we get the energy and materials needed to sustain 60,000 billion cells multiplied by 4 *billion* human bodies? It is beyond the scope of this book to address all the economic and social implications of this question, but we can outline the biological realities of what it takes to maintain a human individual in a given environment. Simply becoming aware of these realities will not in itself lead you to a simple answer, of course, for the problem is complex. However, this basic information may help you in dealing with difficult decisions that you most assuredly will be facing in the near future, both as an individual and as a member of society.

Calories: A Biological Energy Crisis The first reality is that most humans in the world are malnourished. Most people can't get enough energy (through fat and carbohydrates) or protein. Of 4 billion people, perhaps only 500 million have truly adequate diets. The unique problem facing these well-fed few (primarily those who live in North America, Western Europe, and the USSR) is **obesity:** energy intake exceeds body requirements and results in a weight gain. Another 1 billion people get enough energy but not enough protein. For the remaining $2\frac{1}{2}$ billion, the situation is desperate. Thus the human species as a whole is not able to meet its present food requirements (Figure 6.8).

How much energy a human individual needs each day varies with size, age, degree of activity, and physiological state. For us, as for all organisms, these energy needs are measured in kilocalories. (A kilocalorie is the same thing as 1,000 calories—the amount of energy needed to heat 1,000 grams of water by 1°C. Because energy can be converted from one form to another, the energy of most reactions of life can be expressed in kilocalories even if the energy is used for some purpose other than simply heating water.) About 2,700 kilocalories/day are needed for an adult male of "average" size who engages in normal activities; about 2,000 are needed for the "average" adult female. More energy is needed for more strenuous activities; somewhat less is needed for low activity levels. Physiological state is also important. A pregnant woman requires about 2,300 kilocalories/day, and a lactating (breastfeeding) woman about 2,500. The high metabolic activity of children in their midteens demands between 2,400 and 2,800 kilocalories/day. When caloric intake falls below these energy requirements, there is a loss in weight or, for growing children, a slackening in the growth rate. Emaciation is one result; mental and physical deterioration are others.

Caloric intake that is chronically below the minimum requirement marks the onset of starvation. The total number of deaths from starvation each year is impossible to estimate. (Governments, no matter how large or affluent, are reluctant to give out such information. More than this, individuals who do starve to death usually come from sectors of society that are least likely to be accounted for in the ebb and flow of events. And in their weakened condition, starving individuals are prime targets for death by disease.) Undoubtedly the annual figure reaches into the millions; certainly it is greater than the toll of our greatest wars. Yet starvation is not a cyclic social aberration like war. As you will see in later chapters, it is a steady, relentless, and increasingly prevalent commentary on the human condition and the environment that sustains it.

Protein: A Matter of Quality as Well as Quantity Protein deficiency complicates the problem of food resources (Figure 6.8). For us, the essential amino acids are phenylalanine (and/or tyrosine), isoleucine, leucine, lysine, threonine,

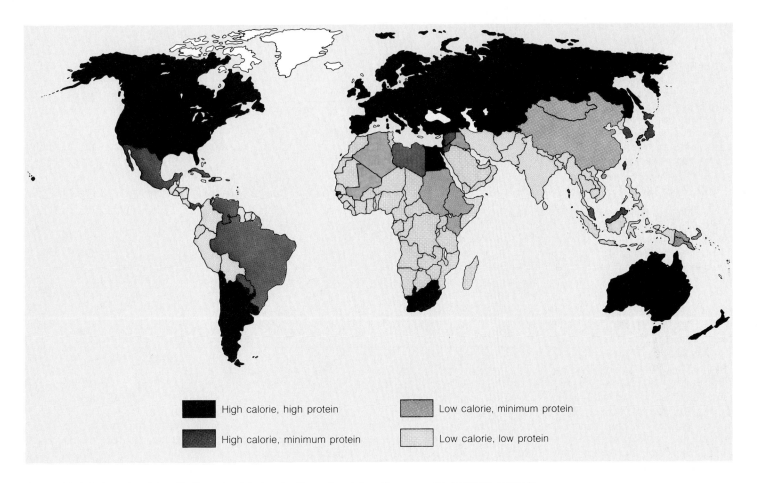

Figure 6.8 Caloric and protein utilization on a global scale. (From *Population, Resources, Environment: Issues in Human Ecology,* second edition, Paul Ehrlich and Anne Ehrlich. W. H. Freeman and Company. Copyright © 1972)

tryptophan, cysteine (and/or methionine), and valine. Our cells simply cannot make these materials; they must be provided by the diet. More than this, our cells can build the other "nonessential" amino acids *only* if the total amino acid intake is adequate. Although our cells can interconvert the "nonessential" ones, they can't make them from nothing. It takes an amino acid of one sort to make another amino acid; to build one, a cell has to tear down another.

Protein deficiency is serious at any age, but it is particularly distressing among young children, for it is during the early years that rapid brain growth and development must occur. Large amounts of protein are needed to sustain the growth phase. If not enough protein is supplied just before and just after birth, irreversible mental retardation occurs. But even mild protein starvation can retard growth and affect physical and mental performance.

The basic minimum daily requirement for protein probably ranges between 0.214 and 0.227 gram for every 454 grams (1 pound) of body weight. If we add another 30 percent as a safety factor against individual differences, we can take 0.28 gram as a realistic figure. This translates to about 43 grams (about 1½ ounces) of "pure" protein every day for an "average" adult male, about 35 grams for an "average" adult female.

Unfortunately, quantity alone is not enough. *Cells must receive the essential amino acids at the same time, and in just the right proportions, before they can assemble their own proteins.* Suppose the kind of protein you eat today has seven of the eight essential amino acids in adequate amounts—but has only half the required lysine. Even if you eat 43 grams of that source of protein, you don't meet your daily protein needs. You would have to eat twice as much again in order to get enough lysine.

To compare proteins from different sources, nutritionists use a measure called **net protein utilization** (or **NPU**). NPU values range from 100 (all essential amino acids

Figure 6.9 Protein content versus net protein utilization (NPU) for several commonly used foods. NPU values range from a low of 40 (kidney beans) to a high of 95 (eggs). Of the sources shown, eggs come closest to the ideal amino acid ratios. Meat, cheese, and fish have relatively high NPU values. Plant proteins such as grains and beans have lower NPU values, mostly because they are usually deficient to some extent in two essential amino acids—lysine and tryptophan. (The protein content of milk and eggs shown here is somewhat misleading because of their high water content.)

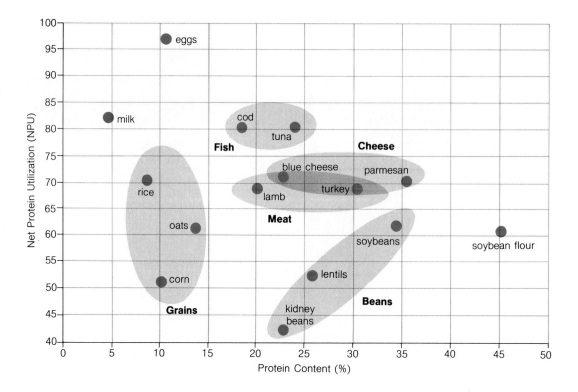

present in the ideal proportions) to 0 (one or more amino acids absent, which renders the protein useless when eaten alone). Balancing the diet with protein from different sources can make up for deficiencies in any one of them, however. Figure 6.9 is a plot of NPU values versus the actual protein content of some common foods.

The staple foods for much of the world are cereal grains. As you can see from Figure 6.9, cereal grains are low in protein content and in NPU values. In contrast, beans are high in protein, partly as a result of their symbiotic relation with nitrogen-fixing bacteria that increase the amount of nitrogen available to these plants (Chapter Five). Although the NPU values for beans are no higher than those of cereal grains, beans are deficient in *different* amino acids. So when beans are eaten *with* grain, the one food enhances the other. What one lacks, the other has—which raises the overall NPU value. Interestingly, long before the terms "protein" and "NPU values" emerged, the Algonquin Indians of North America thought up succotash—a dish that combines corn and beans, and thereby produces an amino acid balance of the sort nutritionists now recommend.

The High Cost of Protein Conversion It takes nearly 1,000 grams (about 2 pounds) of dried corn to meet the daily protein needs of an adult male (Table 6.2). Multiply this amount by a large population and it's easy to see even why

high-energy crops still can't satisfy protein needs unless there is enough variety in the diet. Of course, those of us in affluent countries don't think of "corn" when we hear the word "protein." We think of "meat." It's true that less than 257 grams (9 ounces) of meat are enough to satisfy daily needs. The problem with meat, however, is the large energy investment needed to produce it. Essentially, we use cattle as converters of "low-grade" plant protein into "high-grade" meat protein. That cattle can make the conversion at all is due largely to symbiotic microorganisms that live in the **rumen** (the first two of four separate chambers in the stomach of these animals). The microorganisms secrete enzymes that break down the tough cellulose fibers of plants and convert them to fatty acids. These fatty acids are used as an energy source for building (among other things) the amino acids that cattle are unable to produce for themselves. This supplementary source of amino acids, originating as it does from the symbiotic relationship within the rumen, enables cattle to build proteins at a rapid rate.

The cost of the conversion depends somewhat on external factors. At one time, cattle were raised on the open range (semiarid land unsuitable or only marginally suitable for producing crops). The resource being consumed (grasses) could not be used in other ways, so the conversion was profitable even though between 3,175 and 4,536 grams (7 and 10 pounds) of grasses were needed to produce only 454 grams (1 pound) of meat. In modern agriculture, cattle spend very little grazing time on the open range. Instead they are confined to feedlots, pens in which they are fed high-quality grains to bring them to market size. The conversion cost is about the same as before: about 10 to 1. But here we are talking about grains that have potential as a *direct* food source for humans. A tremendous part of the grain produced each year in the United States becomes feed for beef and dairy cattle, as well as for pigs, chickens, and turkeys. And most of our export grain goes to developed countries, where again it is used as livestock feed.

To some, the answer is simple. Humans can subsist most efficiently on plant proteins. *When domestic cattle are inserted between crop production of those plants and individual consumers, there is a 10-to-1 conversion loss in both calories and protein.* But can you honestly envision the entire nation giving up table meat? That simply will not happen, at least not in the foreseeable future. It will be more realistic to concentrate on working out alternatives to the ideal.

For example, consider that 454 grams (1 pound) of hamburger now costs about a dollar. The same quantity of turkey costs only 50 cents. The difference in cost stems from two things. Turkeys are more efficient than beef cattle in protein conversion (it takes less feed to produce the same amount of meat), and turkeys require less preparation to make them ready for market. More useful protein can be produced at lower cost, and the market price reflects it.

Consider also the potential of fish protein. Fish harvesting itself represents only 5 percent or so of world protein. And it is estimated that we are already harvesting more than half of what the world's oceans are capable of supporting. Therefore, massive efforts to step up harvesting would seriously deplete fish populations—and would not appreciably change the world protein picture as it did. Fish farming, however, may be an entirely different matter. For centuries, small-scale fishponds have been an integral part of Asian agriculture. Human and livestock wastes are emptied into the ponds, which contain controlled numbers of various fish species. The nutrients in the wastes fan the growth of microorganisms that are food for some fish; plants thriving in the enriched water are food for other fish that can grow to about 1,575 grams (35 pounds). There is a potential problem in that such ponds often nurture disease-causing microorganisms, but proper waste treatment methods could eliminate the problem. Thus efforts are under way to develop a cycle in which (1) grains are used to feed livestock, (2) the runoff from feedlots is used to stimulate plant growth in nearby ponds that incorporate adequate waste treatment facilities, (3) the plants are used to feed select varieties of fish, (4) pond water is used to fertilize the grain fields, and (5) the fish are harvested for food.

There are also some signs of a shift to more efficient use of plant proteins. Consider soybeans, with an NPU value of 60 and a protein content of 34 percent (Table 6.2). Adults can meet their daily recommended allowances with only 210 grams (7.3 ounces) of soybeans. And if soybeans are

Table 6.2 Comparison of the Efficiency of Some Single Protein Sources in Meeting Minimum Daily Requirements

Source	Protein Content (%)	Net Protein Utilization (NPU)	Amount Needed to Satisfy Minimum Daily Requirement	
			(Grams)	(Ounces)
Eggs	11	97	403	14.1
Milk	4	82	1,311	45.9***
Fish*	22	80	244	8.5
Cheese*	27	70	227	7.2
Meat*	25	68	253	8.8
Soybean flour	45	60	158**	5.5**
Soybeans	34	60	210**	7.3**
Kidney beans	23	40	468**	16.4**
Corn	10	50	860**	30.0**

*Average values.
**Dry weight values.
***Equivalent of 6 cups. The figure is somewhat misleading, for most of the volume of milk is water. Milk is actually a rich source of high-quality protein.

processed into flour, the NPU value is retained and the protein content jumps to 45 percent. This means a mere 158 grams (5.5 ounces) of dried flour per day is enough to meet the body's need for protein. Soybean production has increased greatly in recent years. Progress is also being made in efforts to breed plants for higher protein content or higher NPU values. For instance, corn and rice are low in lysine, which limits their potential NPU values. Varieties with higher lysine content are being developed by careful breeding and selection experiments. The problem is that these high-lysine varieties demand even more nitrogen-rich fertilizers than are used now. The correlated fertilizer cost tends to be more than small subsistence farmers can bear—yet people living at the subsistence level are the ones who need them the most. It also raises the agricultural energy deficit even more. Present efforts to incorporate nitrogen-fixing bacteria into the crop plant system itself could improve matters.

Other possible (although radical) solutions include raising bacteria on petroleum or sewage sludge and harvesting their edible by-products, or mass culturing protein-rich algae. But it is quite clear that such fundamentally different approaches—no matter how technologically or biologically feasible—will require monumental reeducation if people of affluent countries are to accept such proteins as alternatives to their T-bone steaks.

Other Variables As valuable as these and similar research efforts may be, in themselves they are not enough to make the world food crisis diminish much. The problem is awesomely complex, and its variables must be unraveled on many different levels besides that of improving NPU values and protein content. At each of these levels, simplistic programs have already been proposed. But it's important to recognize that carrying them out would churn up new problems in their wake. To see why this is so, let's briefly consider just four of these suggestions.

First suggestion: *Improve crop production on existing land.* In this view, we can (1) improve plant varieties for higher yields, and (2) export modern agricultural practices and equipment to developing countries. This thinking is the basis for the so-called **green revolution.** It has not been as easy to do as it sounds. High-yield crops require fertilizers, pesticides, and ample irrigation. The plain truth is that poor countries and their subsistence farmers do not have the economic base to make the investment needed to take widespread advantage of the new crop strains. Where such improvements have become available, the farmers have become dependent on industrialized producers of fertilizers and machinery. Of necessity, the cost of these items has been reflected in the market price of the food. Thus food becomes too expensive for the population in the area.

In the long term, even industrialized countries may have to reassess their own agricultural systems. One reason is that the available energy sources used to drive mechanized farming are dwindling and becoming more expensive for everyone. One of the few realistic alternatives is to build agricultural systems around devices for harnessing solar energy. Some devices, such as low-cost, solar-powered microcircuits capable of harnessing heat energy from the sun and converting it to direct currents of electrical energy, are on the drawing boards. Even so, achieving stability in any agricultural system—regardless of the energy source— means that some of the crops must be returned to the land, which will obviously decrease crop yields available for human consumption.

Second suggestion: *Open up new areas for agriculture.* Almost 3½ billion acres are now under cultivation. Some people have proposed that perhaps another 7 or 8 billion acres could be converted to agriculture. Ignoring for a moment the environmental cost of such expansion, the best land available for agriculture is already being used for agriculture. What remains as arable land (land capable of supporting cultivation) is less desirable. In some areas such as Asia, population and food problems are already severe even though more than 80 percent of the arable land is already intensively cultivated. Much of what remains is found in tropical regions of Africa and South America. However, for reasons that will become evident in Chapter Twenty-Two, land that supports a rich tropical forest simply will not support crops for more than a few years after it has been cleared. Desert areas are another possibility, assuming water (and fertilizers) can be brought in. But desert irrigation brings its own problems, such as salt build-up in the soil. (There isn't enough rain to wash away the salts that accumulate because of the rapid evaporation of

irrigation water in the hot climate.) Besides, where will the water come from? Water, too, is not an unlimited resource; witness the serious drought in the western United States during the winter of 1977. Desalinizing ocean water (reducing its salt content) is not yet economically feasible for large-scale efforts. As long as desalinization processes are based on energy from petroleum, they won't help us much in the short term.

The cost of opening up new lands must also be accounted for. Four hundred dollars per acre is not an unreasonable estimate, considering extra costs such as clearing, building roads, developing transport and storage systems, and so forth. And if each one of the acres so cleared will support only one person, it might well cost 28 billion dollars per year—every year—just to keep abreast of the current rate of population growth.

Third suggestion: *Equalize food distribution.* This is the easy one—just make sure everybody on the planet gets an equal share. It is probably the most impractical suggestion of all, simply because we are dealing with human beings. In any nation with an agricultural surplus, farmers must pay taxes on their land and buy seed, fertilizers, and machinery. They expect some return for their labors in producing above and beyond what they personally need. They want to be paid for their crops. If a government wishes to give the crops to other people, the government must directly or indirectly pay the farmers. In other words, it is the taxpayer who foots the bill, voluntarily or involuntarily. Each passing year sees a boost in energy costs and inflation, and it becomes increasingly difficult to sell a program of international assistance to the general public. Lacking such programs, farmers will sell the surplus to whoever can pay the highest prices (mainly, other industrialized nations).

One solution is to have developing countries with certain essential resources get a fair return for them on the world market (witness oil prices). But not all countries have such resources, and those that do are as reluctant to use them to help their poorer neighbors as are the industrialized nations. The oil-rich countries invest their profits in the developed world—even as they burn off natural gas in the Persian Gulf in amounts that could produce all the fertilizer the developing countries could use.

As long as each nation exists to satisfy the parochial interests of its own citizens, the world food situation cannot improve. It seems obvious to some that a world government with interests that embrace all of humanity is required. But any rapid transition to such a government just as obviously would be so cataclysmic that this third suggestion is superfluous. Slower social evolution may occur, but it almost certainly will be too slow to prevent decades of unimaginable misery.

Fourth suggestion: *Stabilize or reduce world population.* Ours is a finite world with finite resources. It can support a large number of individuals at a bare subsistence level; it can, as it does now, support a small number of individuals in first-class accommodations while confining the remainder to steerage. Alternatively, it could support a smaller number in dignity and modest comfort. But to achieve this alternative condition, we must not simply stop, we must back up in terms of population growth. We can dream of science pulling some new technological rabbit out of a hat. It is not a vain dream, for there are probably innumerable ingenious rabbits hidden down among the folds and shadows. *But any solution must be in accord with the supply of material resources and the principles of energy flow.* So it's a numbers game we're playing, and the manner in which we choose to play will affect our own lives to some extent. But more importantly, it will restrict the options of all those who follow. The question on the horizon is this: Who gets first-class or even second- and third-class tickets—and who gets left behind?

Perspective

We began this chapter by looking at the pathways that exist in a cell for breaking down food molecules—pathways that release the energy contained in those molecules in manageable amounts, that break them down into small enough units that can be reassembled in new cell structures. We looked at the food molecules needed to keep cells alive—at the kinds of food as well as the proportions in which they must be taken in on a daily basis.

We then asked, What are the consequences of materials and energy flow when we are talking about the life and death of a single multicellular organism, a human individual? We explained what happens when someone's cells are deprived of essential resources and, briefly, what we can and cannot do about conditions that are the basis for deprivation. Thus the last part of this chapter undoubtedly has had a sobering effect on you, if you are at all concerned about any of the issues presented, because the answers simply are not yet ours. We have passed the point, however, where we can claim ignorance of the questions. As a species, as nations, and as individuals we must resolve the conflict between the limitations on resources and the individuals who need them.

Our world contains vast resources—not resources that can be ripped from the ground and used for limited interests, but subtle sources of biological information on how to survive and lead productive lives on what is still available to us. Our species has dreamed of great accomplishments, but the luxury of dreams can be purchased only with a secure base for survival. Until now, our base has been a form of technological savagery—action without knowledge or regard for the consequence. With luck and with effort, future generations may look back on this as the first generation to seek the beginnings of a stable relationship in the total world of life.

Recommended Readings

Borgstrom, G. 1973. *Focal Points.* New York: Macmillan. Pages 172–201 contain an excellent discussion of the limitations of existing world agriculture policies.

Brown, L. and G. Finsterbush. 1972. *Man and His Environment: Food.* New York: Harper & Row. Excellent overview of the problem of food production.

Food: Readings From Scientific American. 1973. San Francisco: Freeman. Collection of important articles.

Kirk, D. et al. 1975. *Biology Today.* Second edition. New York: Random House. Chapter 14 contains an excellent summary of glycolysis and respiration.

Lappé, F. 1975. *Diet for a Small Planet.* Revised edition. New York: Ballantine Books. Excellent paperback with arguments (and recipes) for revising our eating habits.

Wade, N. 1973. "World Food Situation: Pessimism Comes Back Into Vogue." *Science,* vol. 181, pp. 634–638. Summary of the deteriorating world food situation.

Wilkes, H. and S. Wilkes. 1972. "The Green Revolution." *Environment,* 14:8, pp. 32–39. Covers the problems in exporting American agricultural technology to developing nations.

Chapter Seven

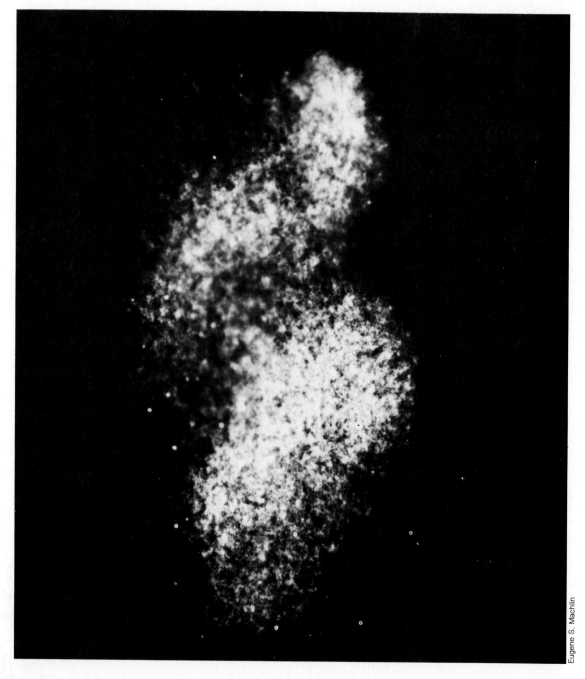

Eugene S. Machlin

Figure 7.1 DNA, hereditary molecule in all living things—source of life's unity, and the basic fabric into which subtle variations have been embroidered to give us life's diversity.

DNA: The Basis of Heredity

Trapping energy, releasing pent-up energy, moving materials about—it's almost impossible to believe that each tiny cell can deploy *thousands* of intricate reactions and control each one with a unique enzyme. Yet that is what a cell must do to stay alive in its environment. What could possibly be the source of all these reactions? What could be the source of the enzymes that hurry the chemical activities along, at the exact time and to the precise extent required? Is it some sort of inherent wisdom that permits each cell to sense what is needed and to hammer out appropriate enzymes on some cellular forge? If this were true, then each cell—and each multicellular organism—should be able to move freely from place to place, quickly churning out a new assortment of enzymes to meet each new challenge. But there is a direct and inescapable relationship between a given life form and the environment in which it thrives. Many life forms, for instance, abound in coastal waters. But when they are accidentally washed up on the beach—which is an entirely different kind of environment—they perish. Still others crowd the water's edge, but they perish when they are accidentally swept into the water. None of them has any means whatsoever of instantly creating new enzymes to meet new conditions.

Somehow, each cell must come into the world with a specific set of instructions for survival in a specific kind of place. Because we know cells arise only from preexisting cells, we can assume that the instructions must be inherited from the preceding generation. In each new generation, those instructions are tested by the environment. To the extent that they provide their bearer with the means for responding effectively to both unchanging conditions and any unanticipated changes in the environment, they have the potential to be passed on:

Hereditary instructions must ensure constancy in structure and function even while allowing room for subtle change.

But where in the cell do these hereditary instructions reside? In what form do they exist, and how do they work? A little more than a century ago, there began a most remarkable chain of events that culminated, in just the past three decades, in the answers to these questions. Because of these answers, we have come to appreciate even more the true meaning of the basic unity and the extraordinary diversity of life.

The Search for the Hereditary Molecule

It is one of the marvelous aspects of biology that seemingly unconnected lines of research often converge in unexpected ways to reveal a basic principle. So it happened that the events leading to the discovery of the physical basis of heredity began indirectly in 1868, when the German scientist Johann Friedrich Miescher isolated an acidic substance from the nucleus of certain cells. He had discovered what came to be known as **deoxyribonucleic acid,** or **DNA.** His source of this new substance was so unromantic as to belie the true elegance of his discovery, for Miescher had deliberately sought out cells that were almost entirely nuclear in composition, with very little cytoplasm. The cells he used came from the pus of open wounds and, later, the sperm of a fish.

At the time, few biologists appreciated the potential meaning of this nucleic acid, because only a few were beginning to suspect the nucleus as being the hereditary control center of the cell. In fact, more than seventy-five years would pass before DNA would be recognized as a substance of profound biological importance. And it would be through research in many different quarters that the impact of Miescher's discovery would eventually change the way we view the world of life.

The Puzzle of the "Transforming Principle" Another clue came in 1928, when a British medical officer named Fred Griffith was waging a solitary war against *Diplococcus pneumonia,* a bacterium that causes pneumonia. Griffith had isolated two distinct strains of this bacterium. He called one strain the "S" form, because it formed colonies having a smooth appearance. When he injected living S cells into laboratory mice, the mice promptly contracted pneumonia and died. Samples of blood removed from the mice were teeming with bacteria; the S form clearly was a pathogenic (disease-causing) strain. The other strain was called the "R" form because of the rough surface appearance of its colonies on culture plates. When Griffith injected live R cells into mice, nothing happened. (The reason later became evident through microscopic study. The killer S form shrouds its cell wall with a smooth protective capsule, which resists attack by the normal defenses of the host cell. The R form possesses no such capsule; hence the host's defense system destroys the R form before it can cause the disease.) But the real surprise came when Griffith injected mice with a mixture of live R cells and *dead* killer S cells. Incredibly, the combination led to pneumonia and death—and blood samples from the dead mice were teeming with *live* S cells!

Obviously dead cells can't be brought back to life. The dead killer cells must have transferred the capacity to form capsules to the vulnerable R cells. But how? Griffith ruled out the possibility that the live cells had become temporarily modified simply as a result of picking up capsule material from the dead ones. He was able to show they were permanently "transformed," as he put it, *for they now passed on the capsule-forming capacity to all their offspring.* The transformation had to involve a change in the hereditary system of the bacterium itself.

News of Griffith's puzzling results reached the American bacteriologist Oswald Avery and his associates. Soon they were able to cause a similar bacterial transformation in the test tube. When they exposed a culture of R cells to an extract of dead S cells, some R cells were permanently transformed into the lethal strain. For a decade they struggled to identify the chemical nature of this "transforming principle." Then, in 1944, they announced: "The evidence presented supports the belief that a nucleic acid of the deoxyribose type is the fundamental unit of the transforming principle of *Pneumococcus.*" The implication was clear. The capsule-forming ability was a heritable feature that could be passed on from one bacterium to another in a DNA molecule. DNA had to be the hereditary material!

Was the work of Avery's group acclaimed and the importance of Miescher's original work finally understood? Not quite. Their remarkable evidence was almost uniformly ignored by the scientific community, which was still clinging stubbornly to a notion (by now a generation old) that only proteins could serve as carriers for hereditary information. Proteins, after all, were diverse and complex; nucleic acids were thought to be much too simple to contain all the instructions needed to build and maintain any living organism. Nevertheless, this line of research did not go entirely unnoticed. The American biochemist Erwin Chargaff, for one, came under the influence of Avery's message. As Chargaff was to say later, "This discovery, almost abruptly, appeared to foreshadow a chemistry of heredity and, moreover, made probable the nucleic acid character of the gene."

Enter Bacteriophage Still more clues were about to converge. Earlier, in the middle decades of the nineteenth century, it seemed that medical advances following the discovery of bacteria as disease-causing agents would enable biologists to trace all infections back to such microorganisms. But later it became apparent that a host of diseases, including influenza, polio, and the common cold,

could *not* be explained in this way. As tiny as bacteria are, some "invisible" infecting particles had to be smaller still. These infecting agents came to be known as **viruses.** Their structure would not be revealed until the advent of electron microscopy.

Today we know viruses are particles of nucleic acid encased in protein. Some are shapeless blobs, others are rodlike, and still others have a distinct head, tail, and sheath (Figure 7.2). By themselves, viruses are not alive. In fact, many kinds can even be crystallized and stored in a bottle on a shelf. But within each virus is hereditary information that can be used to create more viruses; all that's needed is the "machinery" to put that information to use. Cells, inadvertently, provide it.

In the 1930s, Max Delbrück, Alfred Hershey, and Salvador Luria initiated research into the hereditary system of a special kind of virus, a **bacteriophage** that infects and destroys certain bacterial cells. *Escherichia coli,* which flourishes in your gut, is one such target victim.

Bacteriophages do not deliberately go about ambushing victims in a culture dish or a plant or an animal. They haven't any means whatsoever of moving about on their own. The only way they can infect a potential host is to accidentally bump into it. But once contact does occur, a prescribed chemical behavior is set in motion. The protein mosaic of the bacterial surface is no less than a chemical invitation to specific bacteriophages, which accept the invitation by means of complementary proteins in their tail. It is a highly specific fit: each kind of bacteriophage can bond *only* to one kind of protein complex on the cell surface, and the complex is different for each bacterial species.

Once a bacteriophage binds to a cell wall, its sheath contracts and the contents of its head are injected into the victim. Within sixty seconds a strange thing happens. The cell stops making all the things it normally would make and produces an entirely foreign set of enzymes. Subsequently, all the enzyme-mediated activities of the bacterium are devoted to building new bacteriophages! Within twenty minutes, the infected cell undergoes lysis. Its cell wall breaks down completely and its contents are released— which now include deadly bacteriophage particles. By subverting the synthetic machinery of the host, the bacteriophage has managed to reproduce itself.

For Alfred Hershey and his colleague Martha Chase, the intriguing question became this: What part of the bacteriophage was being injected into the host? Whatever it was, it had to be the chemical blueprint specifying "build bacteriophages." Knowing that a bacteriophage contains only DNA and protein, they narrowed their question to whether the injected substance was protein, or DNA, or a combination of the two. For the answer, they turned to a basic chemical difference between the two kinds of molecules.

Protein contains sulfur but no phosphorus; DNA contains phosphorus but no sulfur. It happens that both of

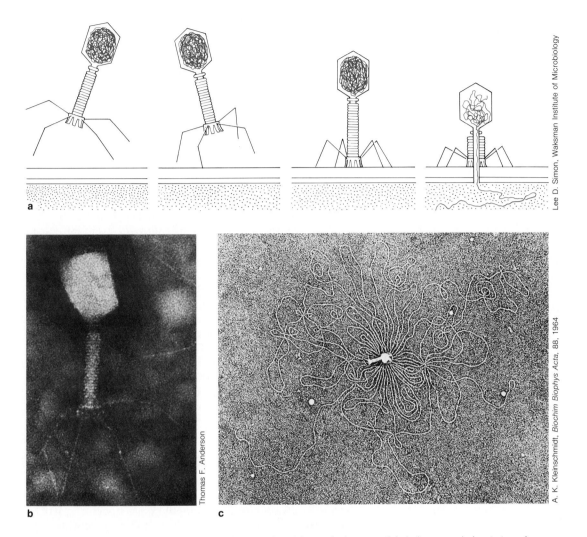

Figure 7.2 Bacteriophage. (a) This is what happens when a bacteriophage particle being moved about at random in the environment makes contact with an *E. coli* bacterium. Proteins in its tail fibers "recognize" proteins in the bacterial cell wall, the sheath contracts, and the contents of the head are injected into the cell. (b) Electron micrograph of a "T-even" bacteriophage. (c) In this now-famous electron micrograph, the contents of a bacteriophage's head appear to be a single tangled strand of a length that seems incredible, considering the small size of the place from which it was released!

these elements have radioactive isotopes: ^{35}S and ^{32}P. Suppose bacterial cells are grown on a culture medium containing both kinds of isotopes. The cells will take up radioactive sulfur and use it to help make amino acids, and they will take up radioactive phosphorus and use it to help make nucleotides. Suppose the cells have been infected with bacteriophage. Following lysis, the new bacteriophage particles will contain labeled protein and labeled DNA, for they can be built *only* from the radioactive materials drawn from their hosts. If these particles are used to infect unlabeled bacteria, it should be possible to determine whether it is the radioactive material in the protein or in the DNA that is being injected inside them.

Hershey and Chase performed this experiment. They found the radioactive sulfur remained outside the bacterial cell and the radioactive phosphorus ended up inside. More than this, they found the radioactive phosphorus was also present in the DNA of certain members of the next bacteriophage generation. Here was further proof that DNA, not protein, is the molecule of heredity. Later work would confirm this general conclusion: *In every living cell, DNA is the repository of hereditary information.*

DNA Structure: The Riddle of the Double Helix

Now the search was on to find out precisely how a DNA molecule is constructed, for in that construction had to be the secret of life's capacity for self-reproduction. It happens that only four different kinds of nucleotide bases are needed to make a DNA molecule. All four include the same five-carbon sugar, deoxyribose, which is a ring structure:

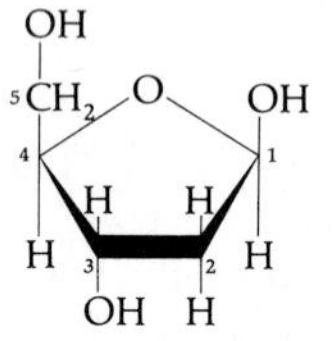

And they all have the same phosphate group:

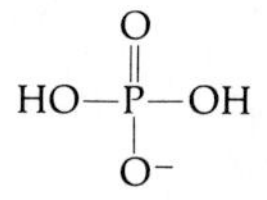

But each nucleotide has a different nitrogen-containing base. Two of these bases are single-ring *pyrimidines;* they are called **cytosine** and **thymine.** The other two are double-ring *purines;* they are called **adenine** and **guanine:**

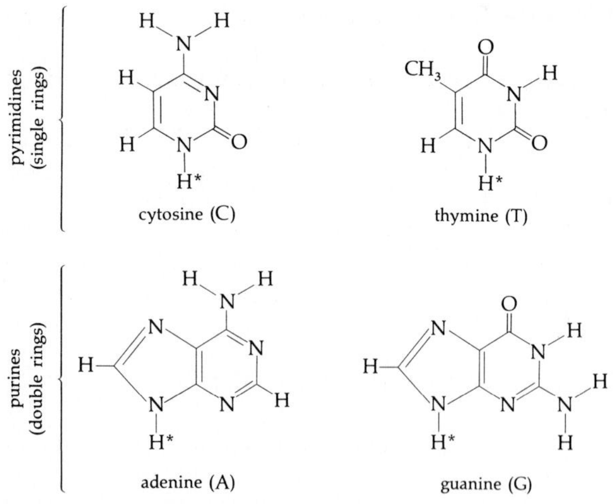

cytosine (C)   thymine (T)

adenine (A)   guanine (G)

The asterisk indicates where each base bonds to the sugar ring structure. If you were to use a flat formula to show how the phosphate group and one of these bases hook up with the sugar to form a nucleotide, you'd end up with something like this:

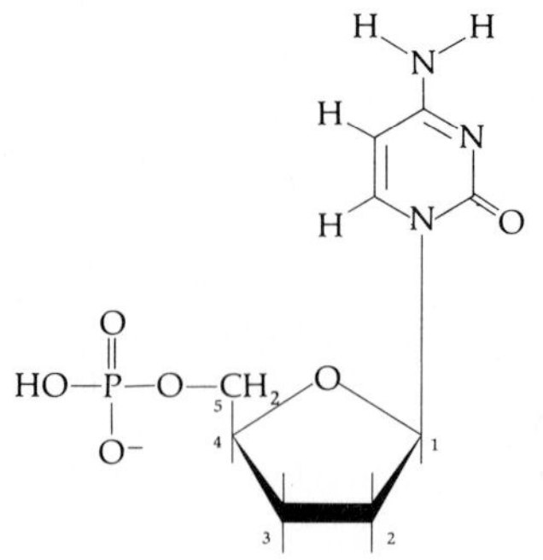

Notice how the phosphate group attaches to carbon 5 of the sugar ring structure, and how the nitrogen-containing base attaches to carbon 1. It's possible for nucleotide bases to be strung together into a long chain, with phosphate groups connecting the sugars and with bases sticking out to one side, as shown in Figure 7.3.

By the early 1950s, this much was known about the component parts of DNA. In addition, through the work of Chargaff and his colleagues, three portentous clues had been uncovered:

1. The relative amounts of the four bases vary greatly from one species to the next—yet the relative amounts are always the *same* among all members of a single species.

2. In the DNA of all species of living things, the amount of pyrimidine present is always equal to the amount of purine present.

3. Most significant of all, the amount of adenine present always equals the amount of thymine, and the amount of cytosine always equals the amount of guanine. Chargaff called this A = T and C = G relationship **nucleotide complementarity.**

While this work was going on, Maurice Wilkins and his associates were using x-ray diffraction methods to determine the physical structure of DNA. (The atoms in a crystal of any chemical substance can bend a narrow beam of x-rays, and an atomic structure having a regular pattern will bend the x-rays in a regular way. If a piece of film placed behind the crystal is exposed by the x-rays, a pattern of dots and streaks will show up on it. Each dot represents a beam diverted by a particular kind of repeating atomic group. It's possible to use the distances and angles between these dots to calculate the positions of atomic groupings relative to one another.) It was in Wilkins' laboratory that Rosalind Franklin identified some intriguing aspects of DNA structure. For one thing, the molecule had to be long and thin, with a constant 2-nanometer diameter along its entire length. For another, its structure had to be highly repetitive: her data showed some structural element being repeated every 0.34 nanometer and another being repeated every 3.4 nanome-

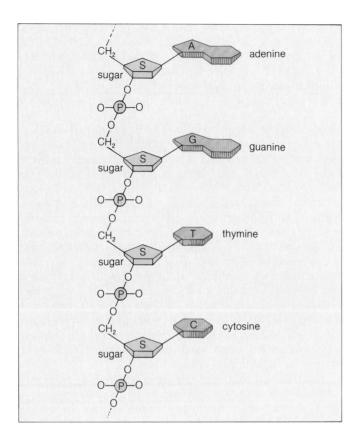

Figure 7.3 How nucleotides can be attached together into a long chain structure.

ters. Finally, the molecule had to be helical; it had the general shape of a circular stairway.

The promise of success was in the wind. It would be only a matter of time before someone would take the separate clues of Miescher, Chargaff, Wilkins, Franklin, and so many others and use them to reveal the answer. In 1953, in a one-page paper entitled "A Structure for Deoxyribose Nucleic Acid," James Watson and Francis Crick did just that.

Watson had recently completed graduate work at Indiana University; Crick was still a graduate student at Cambridge University when Watson joined up with him there. They became aware of Franklin's unpublished data, including her findings that strongly suggested the DNA molecule had to be helical. Undoubtedly it resembled the secondary structure of proteins, which Linus Pauling and his coworkers had only recently identified as a single helical coil held in place by weak hydrogen bonds. The important difference—and this was Watson and Crick's momentous insight—*was that the DNA helical coil had to be double, with two strands wound one around the other.*

The Pattern of Base Pairing Because purines were known to be larger structures than pyrimidines, Watson and Crick knew these nucleotides could not be arranged randomly in DNA. If they were, the molecule would bulge in purine-rich regions and narrow down in pyrimidine-rich areas—but Franklin's data clearly implied a uniform diameter for the whole molecule. Knowing that Chargaff had proved the amount of purine was always the same as the amount of pyrimidine, Watson and Crick guessed that purines must somehow be paired physically with pyrimidines in the structure. But *how* were they paired? Suddenly, after hours of arranging and rearranging paper cut-outs of the nucleotides, they stumbled onto the answer. In a certain orientation, certain groupings of adenine and thymine would form a pair of hydrogen bonds with each other—and in a very similar orientation, certain groupings of guanine and cytosine could form three hydrogen bonds with each other!

If two DNA chains were arranged in space so their nucleotide bases *faced* each other, then hydrogen bonds might easily bridge the gap between them, like rungs on a ladder (Figure 7.4). Watson and Crick began constructing scale models of how this "ladder" might look. And they found the only arrangements possible in their model of a DNA double helix were purine–pyrimidine pairs: A–T and G–C! Thus Chargaff's principle of nucleotide complementarity, discovered years earlier, came out in new dress as Watson and Crick's **principle of base pairing.**

In the ladder model, the only way purine–pyrimidine pairs could be aligned was to have the two strands running in opposite directions and to twist each strand into a helix, as Figure 7.4 shows. Because Watson and Crick built the model to scale with known atomic sizes, they could calculate that there was a base pair every 0.34 nanometer and a complete twist of the helix every 3.4 nanometers—and the helix diameter itself came out to exactly 2 nanometers! In all three respects, then, their model fit Franklin's data.

What about the remainder of Chargaff's data? Here, too, was agreement. With such base pairing, cytosine would always be present in proportions equal to guanine, and thymine in proportions equal to adenine. Yet any pair could follow any other in the DNA chain. For example, in one small stretch of DNA, the sequence might be:

```
T A T C T A
| | | | | |
A T A G A T

      or

G G G T G G
| | | | | |
C C C A C C
```

In this way there could be variation in the total amount of A–T relative to the total amount of G–C present in the DNA

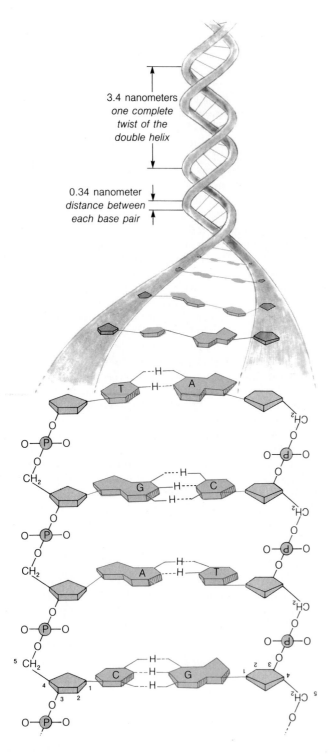

Figure 7.4 Representation of a DNA double helix. Notice how the two sugar–phosphate backbones run in opposite directions. This is the only arrangement in which one nucleotide base can become aligned with and bonded to its complementary base in the DNA molecule.

of different species. Thus the model of DNA reflected at once the properties of unity and diversity required of the hereditary material: *The structure of DNA invariably is built on A–T and G–C base pairs, but the number of possible sequences of these two kinds of base pairs is nearly infinite!*

Unwinding the Double Helix–The Secret of Self-Replication Once the configuration of the DNA molecule had been completely deduced, it immediately became apparent how such a molecule might duplicate itself. If the two hydrogen-bonded strands were able to unwind from each other, their bases would be exposed to the cellular environment. Their comparatively weak hydrogen bonds could be readily broken. Assuming the right enzymes were present, nucleotide subunits could be captured, hydrogen-bonded to the exposed bases on the two unwound strands, and then linked together as shown in Figure 7.5. The only base sequence that could be made complementary to one strand would be *an exact duplicate* of the base sequence present on the other, original strand. Direct experimental evidence has since proven conclusively that before every cell division, DNA strands are indeed replicated in this manner in all the myriad forms of life on earth.

In this way, chemical blueprints for life have been sent down from organism to organism in an unbroken line since the beginning of life. During replication, when the double helix is unzipped, the message of those blueprints is conserved in the arrangement of nucleotide bases sticking out from the two sugar-phosphate backbones, which remain intact. Because each of the two original strands is paired up with a new, complementary strand (in other words, because half of the "old" molecule is used again to form half of each "new" molecule), the process is called **semiconservative replication** (Figure 7.5).

We now know, through various experiments, that replication can begin *before* the whole DNA molecule unwinds completely. This is true for both the DNA of bacteria and viruses (which is circular) and eukaryotic DNA (which is linear). In prokaryotes, the DNA circle opens up at one particular spot, and replication proceeds from this "initiation point." In eukaryotes, the double helix opens slightly at many different points to expose many small units (replicons). In the DNA of some species, there may be thousands of these units, each with its own "initiation point" and "termination point."

When the signal comes to replicate the DNA in either prokaryotic or eukaryotic cells, the strands separate at the initiation point and a construction team of enzymes moves in. These **construction enzymes** attach to the exposed regions of each separated strand and begin making a new complementary strand on each of them. The double helix continues unzipping rapidly ahead of the enzymes, which slide down the strands until they reach the termination point (Figure 7.6).

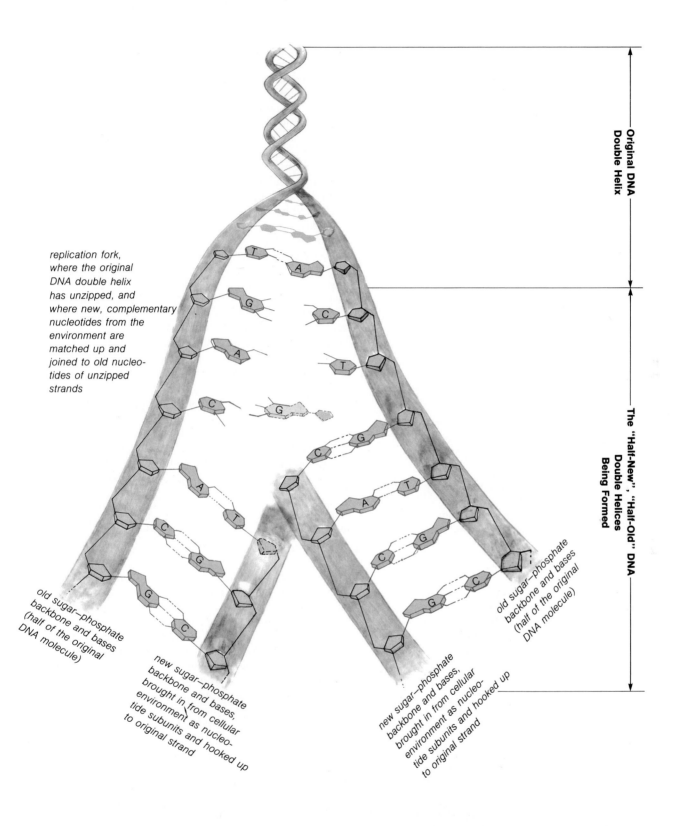

Original DNA Double Helix

The "Half-New", "Half-Old" DNA Double Helices Being Formed

replication fork, where the original DNA double helix has unzipped, and where new, complementary nucleotides from the environment are matched up and joined to old nucleotides of unzipped strands

old sugar–phosphate backbone and bases (half of the original DNA molecule)

new sugar–phosphate backbone and bases, brought in from cellular environment as nucleotide subunits and hooked up to original strand

new sugar–phosphate backbone and bases, brought in from cellular environment as nucleotide subunits and hooked up to original strand

old sugar–phosphate backbone and bases (half of the original DNA molecule)

Figure 7.5 DNA replication, as described in the text.

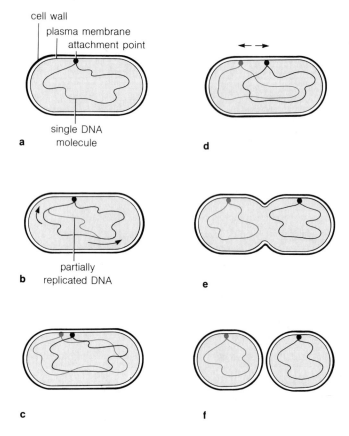

cell wall
plasma membrane
attachment point

single DNA
a   molecule

d

partially
b   replicated DNA

e

c

f

Figure 7.6 The prokaryotic plan for replicating DNA and parceling it out to the next generation of cells. (**a**) Generalized drawing of a bacterium before replication of its single DNA molecule, which is attached to the plasma membrane at a single point. (**b**) Replication begins at some point on the DNA; replication is thought to proceed in two directions away from the initiation site. (**c**) The replicated DNA is also attached closely nearby. (**d**) Membrane growth occurs between the two attachment points, which moves the two DNA molecules apart. (**e**) Once the DNA molecules are at opposite ends of the cell, the middle of the cell undergoes constriction. (**f**) With the growth of plasma membrane and cell wall material, the cytoplasm is divided in two.

Mutation: How the Molecule Assuring Hereditary Continuity Undergoes Change

If information is to be transmitted precisely from generation to generation, DNA obviously must be copied precisely. But it is not difficult to imagine that every so often, something might slip up in the fast-paced movement of the replication process. Also, external influences—cosmic or atomic radiation, ultraviolet radiation, various chemical compounds that find their way into the cell—can alter the chemical messages

in the strands. Fortunately, there is a category of enzymes that function solely to counter inevitable accidents. These **repair enzymes** monitor the general form of the DNA molecule, and if they recognize a distortion, they work to restore the molecule to its correct configuration.

The recognition process is chemical, but it is not all that discriminating. Repair enzymes recognize when and where the DNA molecule differs from the usual form, but they don't "memorize" the entire sequence and act to restore it to what it should be. For instance, a break in the strands can be joined back together—but if a short nucleotide sequence has been lost at the break point, the enzymes have no way of recognizing the loss and replacing the lost bases. They simply will join the broken ends back together, leaving a "deletion" of possibly crucial information. Similarly, if an improper pairing of bases occurs during replication (as it does at rare intervals), the double helix will be distorted at that point. If, for example, A on the original strand is accidentally paired with G on the new strand, the repair enzymes will have no way of determining whether the correct base pair should be A–T or C–G! The enzymes will clip out a region of one strand in the area of the distortion and restore a sequence that does adhere to the base-pairing rules. But there is a 50-percent chance that they will perpetuate the error instead of correcting it. Any such change that is introduced into a DNA molecule is known as a **mutation.**

Other chapters will discuss the effects of mutations on the functioning of cells, individuals, and ultimately populations and species. For now, it's enough to say this:

The overall precision of DNA replication and repair is the source of life's fundamental unity, but the rare "mistakes"—the mutations—account for its diversity.

In the evolutionary view, if DNA replication and repair had been absolutely perfect from the start, life might never have progressed beyond the first simple cells to appear on earth, some $3\frac{1}{2}$ billion years ago (Chapter Fifteen). Since then, most mutations were probably harmful, making their bearers defective in some vital process and marking them for extinction. But suppose some mutations led to enzymes that could operate more efficiently under different conditions. For example, perhaps they would be less sensitive to temperature changes. Such mutations might confer on their bearers the ability to survive in new or different environments even as others perished—to leave more offspring even as others left fewer. Such mutations would be perpetuated and would become increasingly represented in populations of organisms. And gradually the pattern of life on earth would change from simplicity to startling diversity. We who are alive today are, in this sense, little more than a collection of $3\frac{1}{2}$ billion years of environmentally selected errors!

Mitosis: Parceling Out the Replicated DNA of Eukaryotes

Replicating DNA in a form that can be passed on faithfully to a new generation of cells may not seem like such an amazing feat—until you stop to think about what is going on, and where it's taking place. Consider what must happen to the DNA inside *E. coli*, a relatively simple one-celled organism. Although a typical *E. coli* is only 2 micrometers long, its tiny body houses a circular DNA molecule that's about 1,220 micrometers long! For replication to proceed, the molecule cannot possibly be crammed haphazardly in the cell body; it must be twisted and folded in a very specific way. Recall that the double helix makes one complete turn every 3.4 nanometers. This means a DNA molecule 1,220 micrometers long makes nearly 360,000 turns. As replication proceeds, either the DNA must spiral past a construction enzyme or the enzyme must spiral down the DNA (with the new strands whipping around behind it) at an incredible speed. Through all this whirling, the DNA must be taken out of its complex folds and then folded up again once replication is complete. There are about 3,600,000 nucleotide pairs in one molecule of *E. coli* DNA—which means at least 7,200,000 nucleotides must be gathered up from the surroundings in a precise order to make two new DNA strands. All this unwinding, unfolding, untwisting, base pairing, and rewinding and refolding of the two new DNA molecules must be finished in the time between cell divisions—*as little as twenty minutes*. Once done, the two DNA molecules are parceled out in the manner shown in Figure 7.6.

Such is the fate of the relatively simple DNA of a tiny, relatively simple prokaryote. Imagine what it must take to duplicate and parcel out all the DNA of a eukaryotic organism as complex as yourself! It takes more than one DNA molecule; it takes a whole *set* of many different DNA molecules to house all the hereditary instructions needed for building and maintaining your body.

As you might well imagine, DNA replication for eukaryotes requires even more organization and control. The isolation of the nucleus from the cytoplasm of eukaryotic cells makes that organization and control possible. Earlier, you saw a cross-section of the nucleus of a nondividing eukaryotic cell (Figure 4.15). You may have concluded, from this figure, that the nucleus is grainy and far less structured than the organelles of the surrounding cytoplasm. But what you were looking at was actually a slice across a mass of fibers that had been cut during the sectioning process. This fibrous mass is **chromatin:** it contains the eukaryotic DNA and certain proteins associated specifically with DNA. (The function of these proteins, which are known as histones and nonhistones, is not clear. There is as much of these proteins in chromatin as there is DNA itself. Some think they are important in

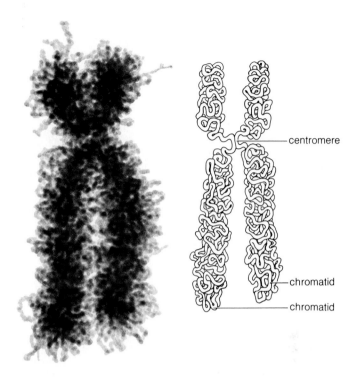

Figure 7.7 A human chromosome as it appears in the duplicated form just before nuclear division. The two chromatids, produced by prior DNA replication, remain attached at the centromere. (From E. J. DuPraw, *DNA and Chromosomes,* Holt, Rinehart and Winston, 1970)

maintaining the structure of the DNA-protein fibers. Some think they may even play a role in regulating DNA activity.)

When a eukaryotic cell prepares to reproduce its hereditary material, the chromatin becomes organized into distinct fibrous bodies called **chromosomes,** the number of which depends on the species. Figure 7.7 shows the general structure of a chromosome as it appears just before nuclear division, which will now be described:

1. The process by which the nucleus of a eukaryotic cell divides in two is called **mitosis.**

2. Prior to mitosis, the DNA found in the nucleus is replicated.

3. With the onset of mitosis, the replicated DNA and its associated proteins become progressively more coiled and more compact. Soon they become visible under the microscope as distinct chromosomes.

4. Each chromosome can be seen to consist of two identical structures called **chromatids.** One DNA molecule with its associated proteins is coiled up as one "chromatid," and its duplicate counterpart is coiled up as the other

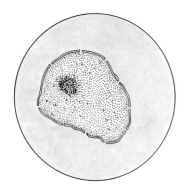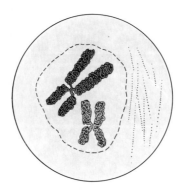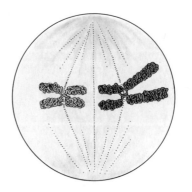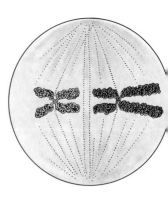

Interphase

Toward the end of interphase, chromatin (which has already been duplicated) condenses into visible chromosomes. The nucleolus beings to break down.

Prophase

Mitosis begins. Chromosomes become distinctly visible beneath the light microscope. They clearly are double in nature. Microtubules of the mitotic spindle begin forming. The nuclear membrane breaks down entirely.

Early Metaphase

The centromere of each chromosome becomes attached to microtubules of the forming mitotic spindle. Each chromosome is drawn to the cell's equator by the microtubules.

Late Metaphase

Now the spindle fiber apparatus is completed. The shorter (chromosomal) microtubules attach at the centromere to connect each chromatid to one pole. Other microtubules run from pole to pole.

—— **Mitosis** ——

Figure 7.8 Mitosis: the process by which the nuclear material of a eukaryotic cell is divided in two. Only two different kinds of chromosomes are shown in this representation of an animal cell. That is the minimum number of kinds of chromosomes in any eukaryotic cell. Imagine how complicated the picture becomes for those cell types that contain more than a hundred different kinds of chromosomes! (Later on, you may want to compare this process with that of meiosis, a reduction-division process illustrated on pages 142 and 143.)

"chromatid." The two are joined at the **centromere:** a localized, differentiated region of the chromosome.

5. During mitosis, the two sister chromatids of each chromosome are systematically lined up. The centromere of each chromosome breaks, and the sister chromatids are pulled apart into two equivalent parcels.

6. At the close of mitosis, new nuclear membrane forms around the two parcels, hence each new nucleus ends up with a complete set of hereditary instructions.

Figure 7.8 shows the main events in this process. Although these events flow smoothly, one into the other, it is useful to consider them as four sequential stages: prophase, metaphase, anaphase, and telophase.

When the chromatin begins to coil up into clearly visible chromosomal bodies, we have arrived at **prophase**—the first stage of mitosis. During prophase, the nucleolus disappears entirely and the nuclear membrane breaks down. At this stage, the chromosomes seem to become progressively more compact in the region where the nucleus

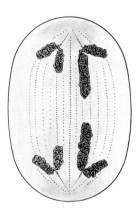

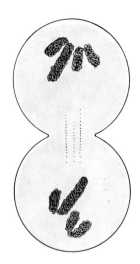

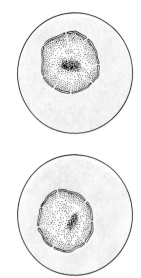

Anaphase

Each centromere splits, which allows the sister chromatids to separate. Microtubules of the spindle apparatus guide the two chromatids of each chromosome to opposite poles.

Telophase

Microfilaments begin to constrict the cell at the equatorial plane of this animal cell. (In plant cells, vesicles would begin to condense into a cell plate.) New nuclear membranes start forming and the chromosomes unwind. Mitosis is now complete.

Interphase

A distinct nucleus with a nucleolus has formed in each of the two daughter cells, which now embark on a new cycle of growth.

once was. During prophase, a series of microtubules (Chapter Four) begins forming in the region of the nucleus.

As **metaphase** begins, the microtubules increase in number and become organized into a bipolar structure called a **mitotic spindle.** The microtubules forming the mitotic spindle are the structures actually responsible for separating the sister chromatids from each other. Of course, if you were told at the outset that this is their function, you might ask why some of them (the *continuous* microtubules) extend from pole to pole right past the chromosomes,

without attaching to them at all! But as you will soon perceive, they have an important role to play in the separation of these chromosomes. The role of the other microtubules is more readily apparent. In early metaphase, two sets of *chromosomal* microtubules become attached to each chromosome in the region of the centromere. One set extends from one chromatid toward one pole, and one set extends from the other to the opposite pole.

The chromosomal microtubules are not attached directly to the chromatid but to the centromere. (The region in

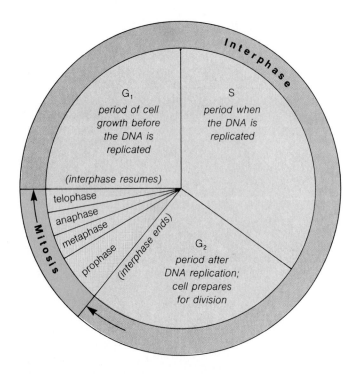

Figure 7.9 Where mitosis fits in the eukaryotic cell cycle. This drawing has been generalized; there is great variation in the length of different phases from one cell to the next. For instance, for some developing embryos, interphase may last for less than half an hour; yet human brain cells remain in interphase from birth to death.

which the centromere is located between two chromatids differs from one chromosome to the next, but it is always the same place for chromosomes of a given type.) As more and more chromosomal microtubules attach to each centromere, the chromosomes are subjected to a tug-of-war, being pulled first toward one pole and then toward the other. Finally, by the end of metaphase, the tug-of-war ends in a draw, with each chromosome lined up halfway between the two poles at the "equator" of the mitotic spindle.

As **anaphase** begins, these events give way to what seems to be a brief pause in activity. But when this "pause" ends, it suddenly becomes clear that each centromere region has been split. Each H-shaped chromosome now separates into two I-shaped members! Thus separated, the duplicate chromatids of each chromosome move apart, toward opposite poles of the spindle.

In most cells, the movement is the combined result of two different processes. The chromosomal microtubules shorten, which *pulls* each chromosome toward its destination. And the continuous microtubules elongate, which *pushes* the poles apart before division takes place. Once the two members arrive at opposite poles, anaphase is over.

During **telophase,** the chromosomes begin to unwind. A newly forming nuclear membrane surrounds each of the chromosomal clusters, and eventually two new nuclei emerge. Mitosis is complete. In the meantime, the cell itself usually has been dividing also (cytokinesis). In plant cells, vesicles typically gather in the region that formed the equator of the mitotic spindle. Upon fusing together, the vesicles form a **cell plate** that grows into a complete partition between the two newly forming cells. In an animal cell, a ring of microfilaments forms around the cell's midsection, just beneath the plasma membrane. The ring then contracts, which pinches the cell in two.

When does a eukaryotic cell actually undergo mitosis? This nuclear division process takes up only a small slice of a larger cycle of events known as the **cell cycle** (Figure 7.9). Depending on the type of cell and on environmental conditions, mitosis can last anywhere from a few minutes to an hour or so. The cell spends the rest of the time in **interphase.** Unlike mitosis, interphase is highly variable in duration. It may last for only a few minutes, but in some cell types it can last more than a century! It is during interphase that most of the activities of a cell take place; it is the time of construction, growth, and maintenance of the cell body. It is during interphase that eukaryotic DNA is duplicated once more, before the next round of mitosis begins.

Evolutionary Implications of Haploid and Diploid Cells

We have considered how mitosis serves to parcel out eukaryotic DNA in the form of chromosomes during nuclear division. But what is the nature of the chromosomes being parceled out? Sometimes a cell has only *one* of each type of chromosome characteristic of the species, in which case the cell is **haploid.** Because haploid cells have only a single set of chromosomes, they have only a single copy of each of the genes needed for cell functioning. In one sense, they are vulnerable to the effects of a mutation. They are like a student with a single set of class notes who loses an important page, or has a page damaged in some way. If the lost or damaged information is important, the results will be immediately apparent: confusion for the student. It can be far worse for the haploid cell. Because the only copy of an important gene may be lost or damaged, death inevitably will follow.

In most environments, it would be far more adaptive to have duplicate sets of chromosomes. And in fact, most eukaryotic cells do have *two* sets of each kind of chromosome characteristic of the species; they are **diploid.** One chromosome of each pair has been derived from one parent and its equivalent has been derived from the other parent (Chapter Ten). Thus, even if a gene in one chromosome of a diploid cell undergoes mutation, the equivalent gene on the other chromosome remains functional. For instance, if a mutation appears in a stretch of DNA that governs the construction of a key enzyme, the mutant DNA may give rise to an enzyme that functions less effectively under existing conditions. But the diploid cell may survive anyway because it has a corresponding nonmutated version in its second set of hereditary instructions. And that version may see it through.

But suppose the environment warms up for some reason. Suppose, further, that the mutant enzyme (defined by the mutant DNA) just happens to function better at higher temperature than the nonmutated version. Then the organism with the mutant DNA may have a definite survival advantage! *Because it has a duplicate set of hereditary instructions, a diploid organism can accumulate a bank account of hereditary variability, which may be cashed in at a later date if environmental conditions should change.* As you will read later in this book, the emergence of the diploid state more than a billion years ago corresponded to an explosive diversification of eukaryotic life forms.

Perspective

DNA—deoxyribonucleic acid—is the hereditary molecule of every living organism. It is the helically twisted, double-stranded blueprint inside all prokaryotes and all eukaryotes. Every DNA molecule is made of only three kinds of substances: a sugar, a phosphate group, and nitrogen-containing bases (adenine, thymine, guanine, and cytosine). Every DNA molecule is built according to the same rules: when the time comes for new bases from the cellular environment to be paired up with old bases sticking out from DNA's sugar-phosphate backbones, adenine can pair only with thymine, and guanine can pair only with cytosine.

What this means is that every living thing on earth shares the same fundamental chemical heritage with all others. Your DNA is made up of the very same kinds of substances and follows the same base-pairing rules as the DNA of those earthworms in Missouri and the flowers in Tibet you read about in Chapter One. Your DNA is replicated in much the same way as theirs; occasional mistakes in its replication are repaired in much the same way as theirs. In the evolutionary view, the reason you don't *look* like an earthworm or a flower is a result of mistakes that appeared on rare occasions during the past $3\frac{1}{2}$ billion years—$3\frac{1}{2}$ billion years of selection for beneficial mutations that led, in their unique divergent ways, to the three of you. Thus the *sequence* of base pairs along the DNA molecule has come to be different in all three of you.

Dinosaur DNA, too, was assembled from the same chemical stuff as yours. But the mutations that gave rise to the unique sequence of base pairs that specified "build dinosaurs" made those creatures unsuitable, when en-

vironmental conditions changed, for continuing their journey.

In short, DNA is the source of the unity of life, mutations in DNA structure are a fundamental source of life's diversity, and the changing environment is the testing ground for the success or failure of the mutations themselves.

If these three points are the basic concepts of this chapter, why did we bother to retrace the history of how the secrets of DNA structure and function were revealed? Because there is one more important concept to think about. Today there is an increasingly pervasive trend to single out heroes. It doesn't matter if they are heroes who save the football game, the child from the runaway car, the nation from subversive elements, the afflicted individuals who desperately need a miracle drug. Why this general tendency exists is not our concern here. What *is* of concern is its extrapolation to scientific inquiry. Science is not done by isolated heroics. It is a cumulative process, with countless individuals adding their contributions to some greater picture even though, at the time, their work may seem to be not worth mentioning at all. Today, if anyone associates any names at all with "the DNA double helix" it is usually the names "Watson and Crick." But as we have tried to point out in this chapter, many others left signposts along the road to its elucidation. Without discounting the very real contributions of Watson and Crick, we thought perhaps it might be fitting to say also, "Let's hear it for Miescher, for Griffith and Avery, Hershey, Wilkins and Franklin, and Chargaff—for all those who have, in their own way and without much fanfare, left us this legacy of profound insight."

Recommended Readings

Chargaff, E. 1970. "One Hundred Years of Nucleic Acid Research." *Experientia*, vol. 26, no. 7, p. 810. This is Chargaff's Miescher lecture—an incisive look at the politics of making history. Chargaff's writing shows him to be one who has been deeply moved by the lyrical intensity of life.

Watson, J. 1968. *The Double Helix*. New York: Atheneum. On one level, an account of how DNA structure was discovered. On another level, an intriguing revelation of the intensely competitive nature of modern research.

Chapter Eight

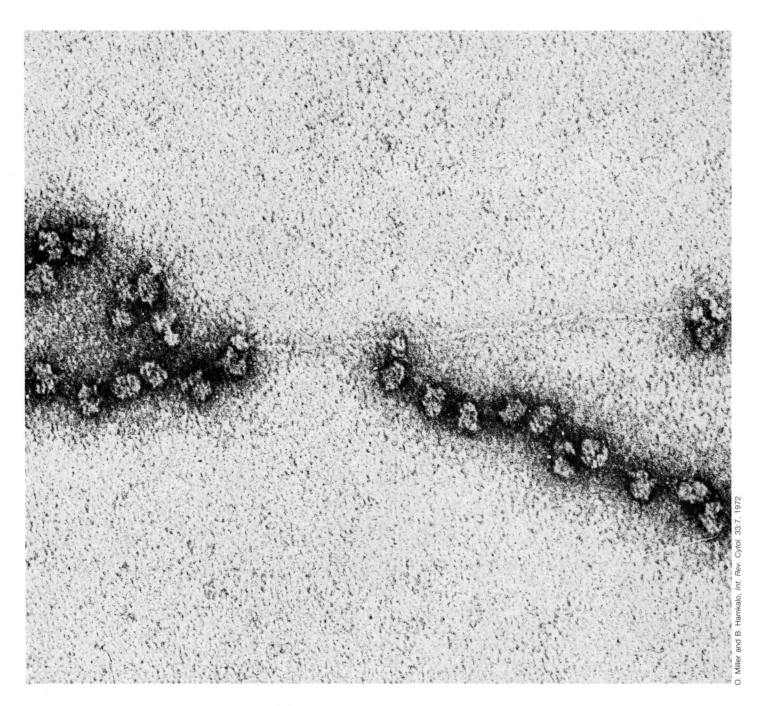

O. Miller and B. Hamkalo, *Int. Rev. Cytol.* 33:7, 1972

Figure 8.1 DNA makes RNA, which makes protein. Shown here, groups of ribosomes "reading" messenger RNA chains, which are being formed on *E. coli* DNA (the thin strand stretching from left to right). 361,200×.

How Cells Use Hereditary Blueprints

DNA, in essence, is like a book of instructions that each cell carries around inside itself. The alphabet used to create the book is simple enough: A, T, G, and C. But merely knowing what the letters are doesn't tell us how they are assembled into the language of life, with words evoking precise meaning, with meaning controlled by punctuation. More than this, the letters alone don't tell us how a cell actually reads the instructions and then puts those instructions to work in assembling all the components it needs in order to survive. And in no way do the letters alone tell us what the cell uses for an index to look up exactly the right messages it needs at each point in its life history. To see how a cell puts its genetic heritage to use, we must turn to the nature of genes and enzymes, and to the chemical link between them.

Beginnings of Research Into Gene Function

In 1908 no one knew much about genes, except that they seemed to be responsible for the traits that offspring inherit from their parents. And no one knew anything at all about the connection between genes and deoxyribonucleic acid, a substance discovered just a few decades earlier. But in that year Archibald Garrod, an English physician, reported what he thought gene function must be. After studying many cases of childhood diseases, he decided that there had to be a heritable basis for certain diseases that recur in particular patterns within some families. What did all these diseases have in common? Garrod inferred that something was blocking one of the steps in a metabolic pathway inside the body's cells. Normally, a cell converts a substance into several intermediate compounds before it ends up with the desired product. When one of the intermediate steps is blocked, there is a build-up of the compound formed just before that step, and an absence of the compounds following it:

normal pathway: $A \longrightarrow B \longrightarrow C \longrightarrow D$

blocked pathway: $A \longrightarrow B$

Eventually, the compound being accumulated reaches high enough concentrations for its presence to be detected. (For instance, in the disease alkaptonuria, concentrations of the intermediate compound alkapton turn urine samples a dark color upon exposure to air.) Garrod called such disorders "inborn errors of metabolism." He linked each of these disorders to the absence or deficiency of one essential enzyme. He suggested that each enzyme must be specified by a single gene—and if the gene is defective, then so will the enzyme be defective.

Thirty-three years later, research began that seemed to confirm Garrod's hypothesis. At that time, George Beadle and Edward Tatum, a geneticist and a biochemist working together to discover how genes function, turned their attention to a bread mold called *Neurospora*. Normally this mold can live on sugar, salts, and one vitamin; everything else it needs (other vitamins, amino acids) it synthesizes for itself. But what would happen if gene mutations were induced in *Neurospora*? Because it is a haploid organism, any effects of such mutations would show up quickly. Knowing this, Beadle and Tatum bombarded *Neurospora* spores with x-rays. They soon isolated two mutant strains. One could grow only when vitamin B_1 was added to its diet; the other could grow only when vitamin B_6 was added. Further studies showed one defective enzyme was present in one strain, and a different defective enzyme was present in the other. Beadle and Tatum concluded that (1) the ability to synthesize each kind of substance in a cell is a heritable trait linked to an individual gene, and (2) a change in one gene leads to a change in the corresponding enzyme needed to build the substance. This **one gene—one enzyme concept** was a great beginning. But further research called for its modification. For one thing, it soon became apparent that genes don't code for enzymes *directly*.

RNA: The Link Between Genes and Enzymes

By the 1940s, enzymes were known to be one class of proteins, and genes were thought to be precise regions of a DNA molecule. So the next question became this: How does a DNA molecule specify the structure of a protein molecule? The answer was startling. DNA doesn't even take part directly in protein synthesis—ribonucleic acid (RNA) does:

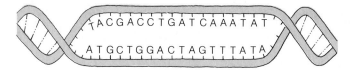

a *Strands of the DNA double helix separate in the region of a gene*

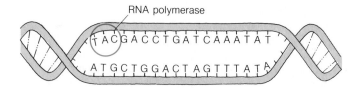

b *RNA polymerase attaches to the beginning of the gene on one strand only*

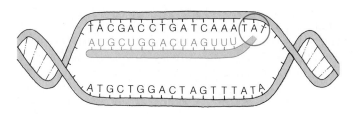

c *A strand of RNA complementary to one strand of DNA is formed on the exposed base*

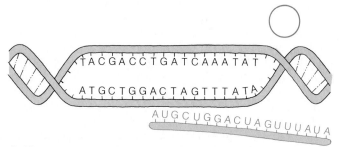

d *The newly made RNA and the RNA polymerase are released into the cellular environment*

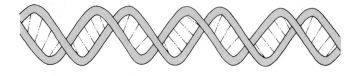

e *The double helix strands join together again*

Figure 8.2 Transcription of an RNA molecule from a DNA template, as described in the text.

A DNA molecule acts only as the template (a structural pattern) for specifying the structure of an RNA molecule.

The RNA molecule in turn is the template actually used in the construction of a cellular protein.

Base Pairing in RNA Synthesis A strand of RNA is almost exactly like a single strand of DNA. As its name implies, however, the sugar in its backbone is ribose, not deoxyribose. And instead of thymine, RNA has **uracil** (U). But uracil is like thymine, for it is capable of forming hydrogen bonds with adenine. This means RNA can form bases that match up with those found in a DNA double helix. In short, an RNA base sequence can be *complementary* to a DNA base sequence.

Whenever a certain product must be assembled within the cell, a specific gene must be called into action. That gene serves as the template for building only one kind of RNA molecule. First the two strands of the DNA double helix separate in the region corresponding to the "beginning" of the gene (Figure 8.2). Then an enzyme called **RNA polymerase** attaches to one of the separated strands. It moves along the strand, all the while picking up RNA nucleotide subunits from the surroundings. The enzyme picks up only the RNA subunits that correspond to the exposed bases. Then it links the subunits together to form an RNA strand. In this way, the DNA template and the RNA being manufactured on it form a "hybrid" double helix. But the hybrid helix is only a temporary union. The RNA strand quickly falls away and diffuses into the cytoplasm as the separated regions of the DNA double helix join together again. Each RNA molecule so formed contains what amounts to the same message—a complementary base sequence—as the DNA region on which it is assembled. That is why the process of RNA synthesis is called **transcription** ("to make a copy").

Transcription is the first step in putting a gene into action. In the next step, the transcribed copy is converted into the language of proteins. Although the message remains the same, the language is entirely different. It is written out in amino acids instead of nucleotides! (It's like writing the same sentence first in Morse code and then in standard English. The meaning doesn't change, but "words" of different structure are used.) That is why the process of protein synthesis, which proceeds as directed by a sequence of information in an RNA molecule, is called **translation** ("to change from one language into another").

The Protein Builders: Three Kinds of RNA As Figure 8.3 shows, it takes three kinds of RNA molecules to build a protein. All three kinds are created by the transcription of

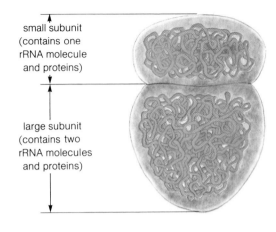

small subunit
(contains one
rRNA molecule
and proteins)

large subunit
(contains two
rRNA molecules
and proteins)

a A ribosome

b A segment of an mRNA molecule

amino
acid
1

amino
acid
2

c Two different tRNA molecules with amino acid units attached

Figure 8.3 The three kinds of RNA needed for protein synthesis. (**a**) A ribosome contains two rRNA molecules in its large subunit, and one rRNA molecule in its small subunit. Both parts also contain proteins important in ribosome structure and functioning. (**b**) Each mRNA molecule is a string of nucleic acid bases, a transcribed copy of the gene on which it was assembled. (**c**) Each tRNA molecule has a set of exposed bases at one end and an amino acid attached to the other end.

genes. But only one—**messenger RNA** (mRNA)—is then translated into protein structure. The other two (ribosomal RNA and transfer RNA) are *not* translated into proteins. Their corresponding gene simply specifies "build that particular kind of nucleic acid."

Ribosomal RNA (rRNA) acts as the physical framework for ribosomes, the workbenches on which proteins are assembled. Any protein at all can be built on any ribosomal workbench. But it is mRNA that specifies *which* protein the ribosomes must construct. The gene coding for some protein the cell needs is first transcribed to yield a unique kind of mRNA molecule, and that's the one that will be translated.

So far, we have the workbench and the RNA template. What carts in the raw materials—the amino acids—to be assembled into proteins? That's the job of **transfer RNA** (tRNA). At one end of this coiled-up molecule are a few exposed bases that are capable of matching up with a few bases on an mRNA molecule. Attached to the other end is

an amino acid. There are different kinds of tRNA molecules. Each links only to one specific amino acid out of the twenty amino acids found in living things.

What links those amino acids to tRNA molecules in the first place? It happens that the cell contains a series of enzymes, one for each kind of amino acid available to build proteins. It is absolutely vital that they hook up each amino acid in the specified order to its correct tRNA. If just one enzyme were to hook up the wrong amino acid to a tRNA molecule, it would introduce an error into the protein being assembled. Fortunately, these enzymes execute their task with astonishing fidelity. Even though tRNA molecules (and certain amino acids) are not very different from one another in structure, these enzymes are so attuned to even slight dissimilarities that they make less than one error in 100,000 operations!

We have now brought together the rRNA workbench, the mRNA template, and the tRNA delivery trucks with their load of raw materials. Protein synthesis is ready to

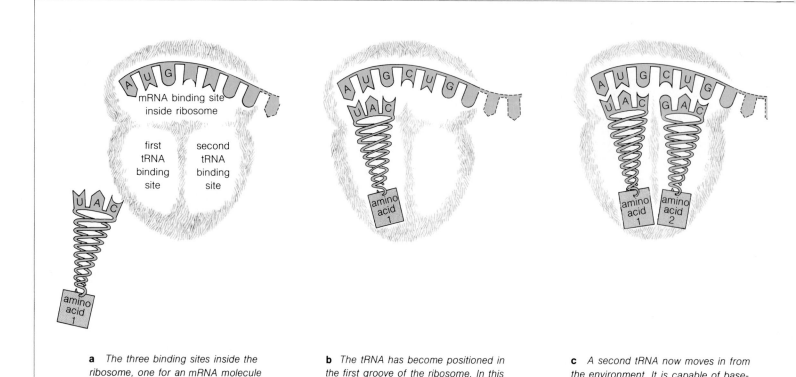

a The three binding sites inside the ribosome, one for an mRNA molecule and two for tRNA molecules. Here, the beginning of an mRNA molecule is bound in place, and a tRNA capable of base-pairing to the first segment of the message is moving in from the cellular environment.

b The tRNA has become positioned in the first groove of the ribosome. In this way, the amino acid attached to it is brought into position so protein synthesis can begin.

c A second tRNA now moves in from the environment. It is capable of base-pairing to the second segment of the mRNA's message. As it moves into the second groove, its amino acid becomes aligned with the amino acid of the first tRNA.

Figure 8.4 Protein synthesis: translation of a genetic message into protein structure.

proceed in the manner shown in Figure 8.4. First the mRNA is bound to the ribosome. Next, enzymes bring up a tRNA molecule that has exposed bases complementary to the beginning of the mRNA message. Another tRNA molecule is brought up right next to it. Once the two tRNA molecules are aligned with each other in the ribosome, an enzyme detaches the amino acid from the first one and attaches it to the amino acid of the other. Thus a dipeptide (two linked amino acids) is assembled.

At this point, the unloaded tRNA delivery truck pops off the ribosome workbench, ready to pick up another amino acid for another round of activity. In the meantime, enzymes move the mRNA and the tRNA with the attached

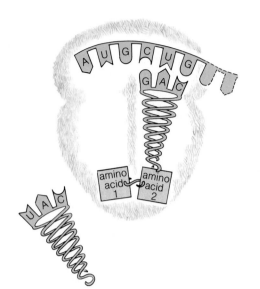

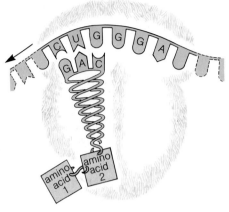

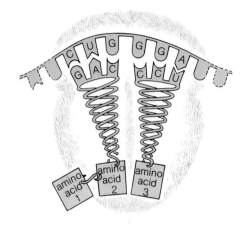

d *The bond between the first tRNA and its amino acid is now broken, and the tRNA falls away. At the same time, a bond forms between the first and second amino acid. This is a peptide bond.*

e *The groove left vacant by the first tRNA is quickly filled as both the mRNA molecule and the second tRNA molecule are moved to the left. Now a new stretch of the mRNA can be bound to the surface of the ribosome just above the second groove.*

f *Now a third tRNA molecule, which matches up with the new stretch of mRNA, is bound in place. Thus its amino acid is aligned so that it can take part in the formation of a peptide bond; in this way the protein chain grows, until the end of the mRNA message.*

dipeptide to make room for a *third* tRNA delivery truck. The dipeptide made earlier is attached to the amino acid of this third tRNA. Thus a tripeptide is formed. The process continues down the length of the mRNA template, and a polypeptide is the result. All that tRNA molecules must do is match up their bases with parts of the mRNA template, and the correct order of amino acids needed for the protein will be laid out automatically on the workbench. *Thus the tRNA reads the message, section by section, and brings amino acids to rRNA's ribosomes in the precise order specified by mRNA; hence in the precise order specified originally by the gene.*

As the protein chain grows, it begins to curl and twist in the space of the cellular environment. Now amino acids in different regions along its length bump into each other, randomly at first, but eventually taking up the positions that give the chain the greatest possible stability. By the time the end of the message is reached and protein synthesis is finished, the protein has taken on the most stable three-dimensional shape allowed by the sequence in which its amino acids have been linked together. As you read in Chapter Three, the nature and arrangement of the atomic groupings in this three-dimensional protein spell out what role the protein will play in the life of the cell. They determine where it will be located, with what other macro-molecules it will bind, what substrates it will bind together, and how it will act upon those substrates.

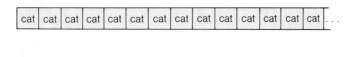

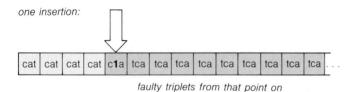

one insertion:

faulty triplets from that point on

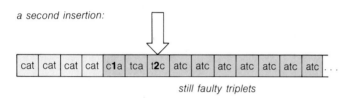

a second insertion:

still faulty triplets

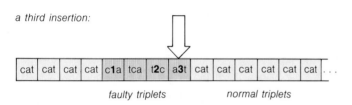

a third insertion:

faulty triplets   normal triplets

Figure 8.5 Crick's interpretation of Brenner's base insertion experiments. If translation of the genetic code depends on reading adjacent nucleic acid bases three at a time, insertion of one or two extra nucleotides will change the sequence (hence the meaning) of all subsequent base triplets. But a third addition will restore much of the normal sequence.

Cracking the Genetic Code

From time to time, a gene may change in such a way that one kind of amino acid permanently replaces another in the sequence of a protein chain. These substitutions have some effect on the protein's shape, hence function. Some substitutions are **nonconservative changes,** and they almost always spell trouble. For instance, if an acidic amino acid replaces a basic one, the resulting protein shape may be so modified that the protein simply cannot do what it is supposed to do. If the protein is vital to cell functioning, cell death is almost inevitable. But other substitutions are **conservative changes,** in that a very similar amino acid is

substituted. In such cases, the effect on protein structure and function may be extremely subtle, with the protein becoming slightly more or slightly less efficient at its intended task.

Conservative changes are extremely important, for it appears evolution proceeds largely through the gradual selection of proteins that have become permanently modified by this random process. In subtle ways, the changes adapt the protein—hence the organism bearing the modified protein—to new or changing conditions in the environment. Chapter Ten will go into the effects of these changes at the level of individuals and populations. But it is at the level of the genes themselves that we can understand how these changes come about.

Years ago, in a series of brilliant experiments, the geneticist Sidney Brenner discovered that one or two extra nucleotides inserted in the middle of the gene made the gene (and the protein it specified) completely defective. Yet, when a third nucleotide was inserted near the first two, gene function was partly restored! Why was the presence of three extra nucleotides less serious than the addition of one or two? It was Francis Crick, codeveloper of the double helix model of DNA structure, who came up with the answer. He sensed that Brenner's results probably revealed the fundamental nature of the language of protein synthesis—the **genetic code.** One would expect these results, said Crick, *if the genetic code consists of nucleotide bases that are read three at a time, with the sequence of each triplet signifying an amino acid.*

If one extra nucleotide were inserted in a gene, the "reading frame" would be modified from that point onward, and the protein to be assembled would contain the wrong sequence of amino acids (Figure 8.5). A second insertion wouldn't improve matters. A third insertion, though, would restore the reading frame. Because the region between the first and third extra bases would still be wrong, the corresponding stretch of the amino acid sequence would be defective. But the sequence coded for by the regions on either side of the insertions would be normal.

Crick's explanation was logical. RNA contains only four "letters": A, U, G, and C. If these letters were read one at a time, they could only specify four different kinds of amino acids. And yet, they somehow specify twenty different kinds. Even if each reading frame contained two letters, there could only be sixteen such words, not twenty. But three-letter words would be an entirely different matter. Sixty-four nucleotide triplets would be possible!

It has since been found that there are indeed sixty-four possible nucleotide triplets. (They are called *codons;* and the complementary triplets on tRNA molecules are called *anticodons.*) Still, sixty-four is far more than the twenty needed to assemble amino acids. What do all these triplets code for? If only twenty make "sense" (specify amino acids), are the remainder "nonsense"? Not at all. Three act like punctuation points, signifying the termination of a protein chain. The other sixty-one code for amino acids—but in

First Letter	Second Letter	Third Letter — U	C	A	G
U	U	phenylalanine	phenylalanine	leucine	leucine
	C	serine	serine	serine	serine
	A	tyrosine	tyrosine	(stop)	(stop)
	G	cysteine	cysteine	(stop)	tryptophan
C	U	leucine	leucine	leucine	leucine
	C	proline	proline	proline	proline
	A	histidine	histidine	glutamine	glutamine
	G	arginine	arginine	arginine	arginine
A	U	isoleucine	isoleucine	isoleucine	methionine (start)
	C	threonine	threonine	threonine	threonine
	A	asparagine	asparagine	lysine	lysine
	G	serine	serine	arginine	arginine
G	U	valine	valine	valine	valine
	C	alanine	alanine	alanine	alanine
	A	aspartic acid	aspartic acid	glutamic acid	glutamic acid
	G	glycine	glycine	glycine	glycine

Figure 8.6 The genetic code. Each triplet consists of three nucleotides. The first nucleotide of any triplet is given in the dark brown column. The second is given in the gray column; the third, in the light brown column. Thus we find (for instance) that tryptophan is coded for by U G G .

nearly every case, two or more different codons can specify the same thing! Usually they have the same first two letters. For instance, CCU, CCC, CCA, and CCG all specify the amino acid proline. (Using Figure 8.6, can you work out the amino acid sequence specified by the short piece of mRNA in Figure 8.2?)

The first experimental tests of Crick's interpretation of Brenner's data were done with *Escherichia coli*. Further work with other bacteria and with plants, fungi, and animals has revealed an important biological principle:

The genetic code is the universal language of protein synthesis for all forms of life.

There can be no more compelling evidence than this for the fundamental unity of life at the molecular level.

Controls for Turning Genes On and Off

If all the genes of a cell were being read all the time, the cell would waste an appalling amount of energy, for it would make far more enzymes and other proteins than it needs. Worse yet, various enzymes would be working against one another, simultaneously building up and tearing down dozens of kinds of molecules. And what would happen during the development of a multicellular organism? Here,

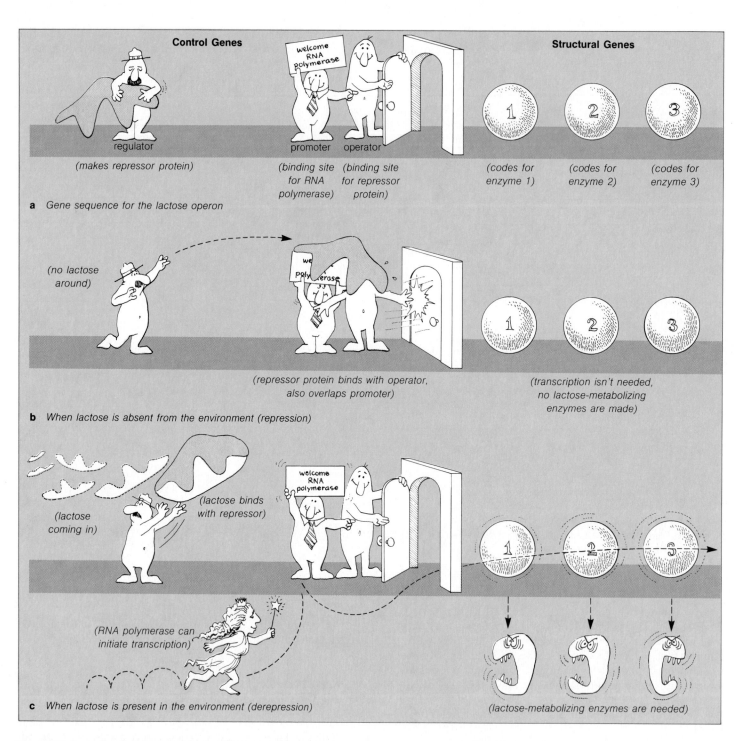

Figure 8.7 Control of gene transcription. In the lactose operon, the control processes employed are repression (when the repressor binds to the operator and blocks transcription) and derepression (when it binds to lactose instead, which allows transcription to proceed). The ''gate'' symbolizes the point of control over whether or not transcription proceeds. Hence the word ''operator'' is a little misleading because it implies some sort of busy activity—when in fact it is simply a binding site.

certain genes must be expressed in some cells even while others aren't expressed at all. For instance, in a developing red blood cell, the gene coding for hemoglobin must be read at a maximum rate and the genes coding for brain protein must be turned off. *Each cell must have a means of controlling which genes are active and which are inactive at any given moment. And it must have a means of changing the pattern of activity when environmental conditions change, or when different physical structures must be developed.* Let's turn now to the nature of these cellular controls.

Milk Sugar, a Few French Biologists, and the Operon An *E. coli* living in the gut of a baby mammal is destined to encounter an abrupt change in its diet. For a few weeks or months, the host takes in nothing but milk, which contains the sugar lactose. But once the weaning period is over, the mammalian host never takes in milk again (the only exceptions being ourselves and a few of our pets). During one part of its life, then, *E. coli* must be able to use a sugar it never will use again. For the rest of its life, *E. coli* will have no need to make the enzymes involved in breaking it down; indeed, the organism would clutter up its insides with useless enzymes if it even bothered to. Obviously a bacterium that can make lactose-metabolizing enzymes only when needed will have a considerable adaptive advantage over a bacterium that (1) can't use lactose at all, or (2) can't regulate the production of those enzymes.

Three French biologists, André Lwoff, François Jacob, and Jacques Monod, looked into this intriguing aspect of *E. coli.* Sure enough, they found the bacterial cells producing a class of three adaptive enzymes when needed. When they grew bacteria in a culture medium without lactose, they saw no sign of the enzymes. But within three minutes after switching *E. coli* to a medium in which lactose was the only sugar, they discovered the enzymes were being produced!

Lwoff, Jacob, and Monod identified three **structural genes:** genes that code for the structure of specific enzymes (in this case, the ones needed to break down lactose). They are adjacent to one another on the chromosome, and they always function as a unit. But structural genes aren't the only ones needed. There is also a category of **control genes.** They don't code for enzymes at all; instead they govern when and to what extent the structural genes may be expressed.

The term **operon** signifies *any* set of structural and control genes operating as a unit. Several different kinds of operons have now been studied in prokaryotes. Although they differ in many details, such as in the number of structural genes, they are all alike in one respect: *The coordinated activity of an operon helps conserve energy and stabilize cell functioning when environmental conditions fluctuate.*

There are three control genes in the lactose operon (Figure 8.7). One (the promoter) is the binding site for RNA polymerase, the enzyme needed to start transcription.

Another (the regulator) codes for a repressor molecule that can block transcription. (Each operon has its own regulator.) The third (the operator) acts like a control point next to the structural genes; its interaction with the repressor dictates whether transcription *can* proceed.

How does the repressor protein work? When lactose isn't around, the repressor binds with the operator. Being so bulky, the repressor overlaps the binding site for RNA polymerase. With no place for transcription to start, lactose-metabolizing enzymes are not produced. But when lactose enters the cell, it binds with and changes the shape of the repressor protein. In its changed shape, the repressor protein *can't* sit on the operator, it no longer overlaps the binding site for RNA polymerase, and transcription can start. As soon as all the lactose is broken down by the enzymes, the repressor protein is free to sit on the operator once more—thereby shutting down the production of enzymes that aren't needed any more!

(When the protein represses transcription, the control process is called "repression." That's logical enough. But when an abundance of lactose inactivates the repressor protein, the control process is called "derepression." Ask your English instructors what "derepression" means and watch them climb the walls cursing scientists who coin words.)

More is going on in *E. coli* than lactose metabolism, of course. To live and grow, organisms must also build substances such as amino acids and nucleotides. If those substances are not available, *E. coli* goes ahead and makes them itself. And it does so with a different kind of gene control process. For instance, another operon codes for the enzymes needed to build tryptophan (an amino acid). When there's more than enough tryptophan around, the tryptophan binds with and changes the shape of the repressor protein that controls this operon. But in this case, it is the *changed* shape that is able to bind with the operator to block transcription! When tryptophan is absent, the protein's shape goes back to what it was, and the repressor protein falls off the operator. Transcription then proceeds at full speed. Thus the end product helps shut down its own synthesis, because its presence *activates* the repressor protein. (This control process is called "corepression.")

Fine-Tuning the System If a cell is to make the best possible use of available materials and energy, it needs more than an "on-again, off-again" type of control. Why is this so? Raw materials for protein building (as well as for all other cellular reactions) are seldom available in exactly the right amounts at all times. The cell must do a shuffling act with materials that *are* available—sending materials down one pathway, letting them be converted to some intermediate compound, then pulling the intermediate compound out of one pathway and diverting it down another, where it is needed more urgently. The control process regulating the

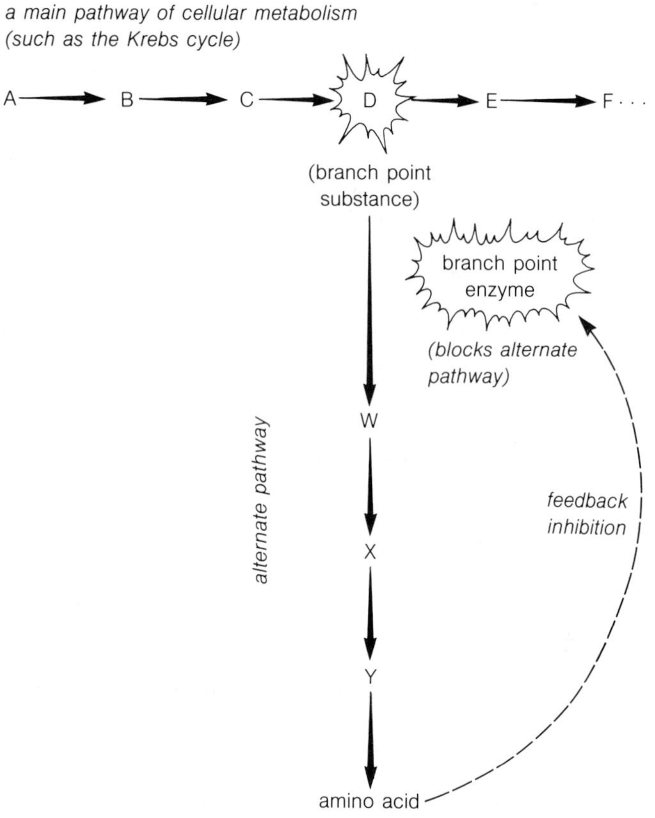

a main pathway of cellular metabolism
(such as the Krebs cycle)

Figure 8.8 Feedback inhibition of amino acid synthesis, as described in the text.

flow of compounds into more than one reaction sequence in the cell is a form of **feedback inhibition.**

As Figure 8.8 shows, the control point in feedback inhibition is at an intermediate substance found at the crossroads for two reaction sequences. The stoplight at the crossroads is a special kind of enzyme. When a cell is lacking a certain product—say, an amino acid—these enzymes latch onto molecules of the intermediate substance and divert them down the alternative pathway for use in amino acid synthesis. But as soon as there are adequate amounts of the amino acid product, the product itself gradually starts binding to the enzymes and inactivates more and more of them. Eventually all the traffic resumes in the original direction—down the main metabolic pathway.

The reason this kind of enzyme works as it does is that it has two different binding sites. One is the active site typical of all enzymes, for it binds a certain substance and speeds it along the alternative pathway. The other binds only the end product of the alternative pathway. When an end product molecule is bound in place, it distorts the shape of the enzyme so that the enzyme becomes functionally inhibited; it can't pick up the substrate when it's bent out of shape. It is an *allosteric* enzyme (which simply means it can take on "different shapes").

All life forms, from the simplest to the most complex, use feedback inhibition through allosteric enzymes to control the flow of substances down many different pathways. The important thing about this control is that it is proportional to supply and demand. *Operation of alternative pathways is not simply turned on or off; the pathways operate in proportion to the amount of product present.* This subtle control over competing cell reactions is the means by which a cell responds even more effectively to changes in the environment.

Control of Gene Expression in Eukaryotic Cells

In prokaryotic bacteria, DNA resides in the cytoplasm. It is in full contact with all cellular enzymes. As a result, all mRNA molecules in bacteria begin to be translated even before their synthesis is complete. About the only way to keep a protein from appearing in the cell is to stop the genes from being transcribed. In eukaryotic cells, the operon mechanism is not the primary mechanism for controlling gene activity; there is far more than straightforward control over transcription. *For eukaryotes, selective controls exist over specific regions of the cell. These controls depend on cell type, and on where it is in time and space.*

Part of this control is possible because of the eukaryotic cell's nuclear membrane. This membrane is a selective barrier between the nucleus (with its genetic material) and the cytoplasm, where proteins are actually constructed. As a result, the site where DNA is transcribed into RNA is kept free from the activity of the cytoplasm. Thus production of an RNA molecule doesn't have to mean the inevitable and immediate appearance of a given kind of protein; there is time for additional controls to be brought into play. For example, more than 95 percent of all the RNA produced in a typical human cell is destroyed in the nucleus within minutes of the time it is made. But it is *controlled* destruction. Some regions of one RNA molecule may be completely preserved, even while other regions of it may be completely degraded. Another RNA molecule left intact may be extensively modified, with pieces trimmed off the ends and additions tacked on before it is passed into the cytoplasm.

Once RNA is in the cytoplasm of a eukaryotic cell, still more control is exerted over the translation process. Some mRNA molecules are translated at once, and some are translated repeatedly for several days. Still others are translated a few times and then destroyed. *Thus, in addition to the control afforded by the isolation of newly forming RNA molecules in the nucleus, gene expression in eukaryotic cells is subject to cytoplasmic controls.*

For multicellular eukaryotes, the extent of this control is staggering. For instance, the egg cell from which a human life begins must divide and divide again, producing endless lines of cells that gradually develop into nerves, muscle, skin, and bone—into the total range of diverse cells and tissues found in the human body. All these cells have different capabilities from those of the single cell from which they were derived. Beyond the basic reactions of energy metabolism that all cells hold in common, the activity of a brain cell is enormously different from that of a cell in muscle, bone, blood, or kidney (Chapter Fourteen).

We have yet to discover how all the various types of cells in complex eukaryotes read the *same* genetic library in their own selective way and thereby become—and remain—unique. Somehow, these cells must respond not only to changes induced by the external environment but to the internal schedules governing their life. Somehow, they exchange chemical signals that modify the cytoplasm. In turn, the cytoplasm of a given cell may generate a new set of chemical signals to the nucleus buried within it, causing some genes to be turned off and others to be turned on. As a result, new signals may be sent out, telling neighboring cells that it's time for them to change *their* patterns of structure and behavior. In this way, eukaryotic cells move on in their journey through time and space. Through controls of a sort not yet identified, they come into existence equipped with guidelines specifying where they came from, where they are in relation to other cells, where they are headed, and what time it is as they move on their prescribed course from birth, through aging, to death.

Cancer: When Normal Controls Break Down

Even seemingly stable patterns of gene expression in a mature individual are subject to breakdown under the impact of potent external agents. Such breakdowns are extremely rare, but when they do occur the results can be devastating. For instance, at about the time of birth, brain cells always cease dividing. Each brain cell then persists for the life of the individual, carrying out its own key task in the integration of behavior in response to environmental changes. But new cells cannot be produced *after* birth to take on new assignments.

The stability of a brain cell depends on the stable state of its cytoplasm and on the constancy of cytoplasmic signals being sent into the nucleus, which regulate nuclear function. Because of these signals, all the genes governing cell growth and division are kept quiescent and the genes governing the synthesis of specific brain proteins are kept active. But if the cytoplasm becomes modified and begins sending abnormal signals to the nucleus, the pattern of gene expression may change abruptly.

Consider what happens when brain cell nuclei of a frog are transplanted into a fertilized frog's egg. In a fertilized egg, DNA replication and cell division proceed at an astonishing pace; in a mature brain cell, they normally do not proceed at all. But almost from the moment of transplantation, brain cell nuclei respond to signals from the egg cytoplasm. Within five minutes they begin replicating their DNA and preparing for division—something they never would have done for the rest of their life if they had been left in place in the brain!

Unfortunately, such gross modification of brain cell behavior is not merely a laboratory curiosity. Sometimes in human individuals, a single cell in the brain undergoes a change that modifies the behavior of its nucleus. Instead of devoting itself to the expression of the genes coding for proteins, it begins replicating its DNA and growing. And it divides. And it divides again and again and again. Suddenly the progeny of this one cell begin to crowd surrounding cells that are functioning normally. They begin exerting increasing pressure on adjacent cells and interfering with their actions. One single cell that has gone out of control has spawned a **brain tumor,** with its potentially lethal effects. Such tumors, which result from breakdown of normal cellular controls, can occur in any region of the body. If they are not removed, they inevitably lead to death of the individual.

If the only control system undergoing modification is the one that regulates cell growth and division, the situation is less serious than it might otherwise be. Tumors resulting from cells dividing faster than they should are considered **benign tumors.** If they are removed surgically, their threat to the individual ceases.

However, if changes have also occurred that modify controls over the cell surface, the situation may be far more serious. In an adult animal, most cell surfaces carry "name tags," or **surface recognition factors.** They are molecules that identify the cell as being of a certain type. It is because of such factors that cells of like type "recognize," bind to, and interact with one another. Such interactions keep cells of specific and related types bound together into tissues and organs. But sometimes changes leading to loss of control over cell growth and division also lead to loss of expression of precisely those genes coding for specific surface recognition factors. The cell loses its identity; it becomes a kind of cell that gives rise to a **malignant tumor,** or **cancer.**

Malignant cells that have lost their surface characteristics become invasive. They can slip out of the site where they arose, enter the bloodstream, or wander through tissues of the body. They may become lodged in a variety of other places in the body, there to begin multiplying and producing secondary tumors. Because malignant tumors are capable of such dispersal (called **metastasis**), they are more difficult to treat. Surgery in one site is not likely to remove all the malignant cells in the body. A tumor may be removed from one site, but another may appear elsewhere.

Other treatments must then be used—specific chemical agents, radiation, or both—that selectively destroy rapidly dividing cells wherever they are in the body. But these treatments carry hazards. They must be administered carefully in order to destroy cancer cells without destroying normally dividing cells—such as intestinal lining or blood-forming cells.

What causes such loss of normal cellular controls? There are many known factors, and often they act in concert. Viruses have been shown to cause cancer in experimental animals. And viruses are strongly implicated in many types of human cancer. Certain viruses are called **tumor viruses;** they seem to have little effect on an organism other than to cause specific kinds of cancer. But cancer formation may not be the exclusive property of tumor viruses. Under certain circumstances, in rare cells, and in rare individuals, common viruses normally associated with mild disease symptoms can somehow initiate the loss of normal cellular controls. Nonviral agents may also be involved. Environmental pollutants of many types, continuous exposure to sunlight, and continuous physical irritation can increase susceptibility to (if not cause) some form of cancer.

Many types of cancer are known. In all cases, something happens to cause a drastic change in conditions in the cell, which disrupts the signals being sent into the nucleus. The nucleus responds by changing the patterns of genes that it permits to be expressed. And the new patterns of gene expression lead to an entirely different pattern of cell behavior. Because we do not understand in detail the control processes of the normal eukaryotic cell, neither do we understand in detail the changes occurring in the cancerous transformation. But substantial progress is now being made, to the extent that many biologists are predicting some kinds of presently unmanageable forms of cancer may be brought under control within the next few decades.

Perspective

One gene, one enzyme. That simple concept was a basic step in research that led to the deciphering of the genetic code—the language of gene expression in all life forms. Another major step was the realization that the genes carried on DNA are not *directly* involved in assembling enzymes and all other proteins. First they must be transcribed to yield mRNA intermediates—messenger molecules carrying information from the site of storage to the site of action. Another step was recognition of an exception to the one gene—one enzyme concept. Although many genes do in fact carry instructions for building specific enzymes, some genes are transcribed only to yield tRNA and rRNA—the molecules by which mRNA is translated into enzymes and all other proteins. (On the ribosomal workbench, tRNA

decodes the transcribed instructions on mRNA. Each tRNA molecule bears at one end a triplet of nucleotide bases that may match up with some triplet segment on the mRNA. Each bears at the other end an amino acid; and for each kind of triplet, there will be only one kind of amino acid attached. This matching of amino acids to tRNA triplets is so constant in the living world that all creatures, from bacteria to blue whales, can use the basic language of the genetic code in precisely the same manner. They differ from one another not in how their genetic instructions are read, but in what particular genetic instructions are carried in their DNA.) Finally, in analyzing how cells selectively call their genes into action, researchers came to recognize one more exception to the one gene—one enzyme concept. This is the class of control genes, which function to determine how much enzyme should be made, and under what conditions.

So much has been learned about the translation of genes into cellular products and about control of gene expression—yet so much remains to be learned. Particularly in the case of complex eukaryotes such as ourselves, there is no small sense of urgency surrounding the need for further research in this area. We are far more complex than bacteria, and so is the process controlling the expression of our genes. Whereas bacteria rely largely on two forms of control (regulation of transcription and allosteric regulation of enzyme activity), we rely on myriad levels and kinds of controls. Occasionally our complex control system goes awry and cells go out of control, as they do in the process we call cancer. For those individuals and families confronted with it, there is little comfort from the fact that the extreme rareness of the cells undergoing such derangement emphasizes the precision with which the interlocking network of controls normally operates. Aside from the intellectual challenges of research into gene regulation, the success of that research may directly or indirectly affect us all.

Recommended Readings

Bautz, E. 1972. "Regulation of RNA Synthesis." *Progress in Nucleic Acid Research and Molecular Biology,* vol. 12, p. 129. Well-written summary of transcriptional regulation.

Drake, J. 1970. *The Molecular Basis of Mutation.* San Francisco: Holden-Day. Extensive coverage of the basis of mutations. Drake writes well on this fascinating subject.

Watson, J. 1976. *Molecular Biology of the Gene.* Third edition. Menlo Park, California: Benjamin. Chapters 18 and 20 are outstanding overviews of problems and prospects of cancer research.

Unit III

Complex Plans for Survival

Chapter Nine

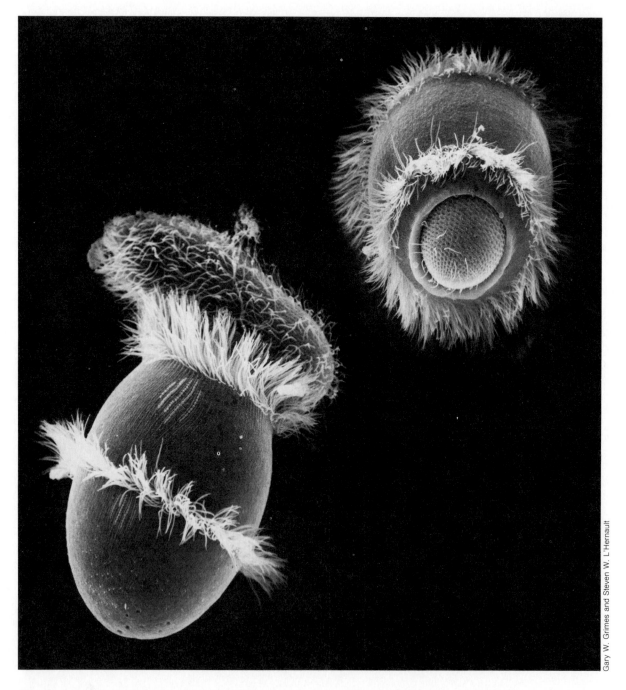

Figure 9.1 Mealtime for *Didinium*—a single-celled organism with a big mouth, and symbol of the persuasive nature of predation in bringing about evolutionary change.

Prospects and Problems of Multicellularity

Imagine a world inhabited only by the simplest prokaryotes, and you have a picture of what the earliest community of life may have been like. Our readings of the fossil record suggest those early forms were much like modern single-celled bacteria and blue-green algae, drifting side by side in the shallow sunlit zones of lakes and seas. A little photosynthesis here, a little decomposing there—for about 2 billion years, they apparently went about the business of acquiring materials and energy without interacting directly with one another. During that immense span, the body plans of these organisms remained much the same, which speaks of their relatively uneventful existence.

What pressures there were for change probably were limited to a few variable conditions in local environments—the movement of tides, strong coastal currents, the rushing flow of rivers. In such places, cells that by chance remained attached to one another after cell division could have a selective advantage. Such strings of cells could become wrapped around a pebble or rock and gain some resistance against the pull of the random swirling waters. And, in fact, sprinkled through the fossil record for this period is evidence of just such stringlike arrangements of prokaryotes. We can assume there has been persistent selection for this adaptation in certain environments. Thus today we see filamentous patterns of growth among certain kinds of blue-green algae and bacteria, which become anchored to rocks in rapidly flowing streams and hot springs. But the cells themselves remain quite simple, and the division of labor among them is rudimentary at best. It is not **multicellularity,** with the extensive division of labor and permanent interdependence among specialized cells that the word implies. Instead, these single cells merely live together in a loosely connected way. Each retains the capacity for existence and reproduction apart from the group.

About 1.2 billion years ago, however, the nature of the living community changed irrevocably. By that time the earliest eukaryotes had appeared, and among them were the world's first predatory cells (Chapter Fifteen). With their capacity for ingesting whole organisms as an energy source, they represented an evolutionary force of the first magnitude. No longer would most organisms be able to "live and let live." No longer would it be enough simply to become well adapted for tapping energy from the sun or from the waste products of other organisms. Now and for all time, survival would be bestowed largely on those life forms which could avoid predators long enough to leave descendants. There would be selection not merely for those forms which were adapted to the physical environment. There would be selection also for the swift, the elusive, and the wary—and for the ones too large to be consumed by local predators.

Increasing the Amount of Hereditary Instructions

With the emergence of predatory organisms, increased size and complexity became traits with real survival advantage. But it could not have been a simple matter for any of those first cells to have increased in size. Consider, first, the fact that all cell activities are under hereditary control. Before a cell can increase substantially in size, it obviously must have increased quantities of all the enzymes needed for building and maintaining all the extra materials. But enzymes must be synthesized from mRNA molecules, which are transcribed from their corresponding structural genes (Chapter Eight). And the structural genes coding for most enzymes occur only *once* in each haploid set of chromosomes (which is all the chromosomes a single-celled prokaryote ever has). There are no stockpiles of identical structural genes in these chromosomes. Even if there were, there is an upper limit to how fast any gene can be transcribed to produce the mRNA needed to produce more enzymes!

But what if it were possible to have extra *copies* of hereditary instructions? Among simple eukaryotes alive today, we have hints of what possibilities must have opened up for their ancient ancestors. *These organisms rely on multiple nuclei for increased size and increased complexity.* Some single-celled eukaryotes have two different kinds of nuclei. Others have many of the same kind of nucleus in the single cell body. And almost every kind of cell of a multicellular eukaryote has its own membrane-enclosed nucleus.

The single-celled eukaryotes called ciliates have two kinds of nuclei. Some species are so complex they require several hundred of one kind and up to a dozen of the other to keep all their cell parts functioning. The ciliate **micronucleus,** which contains a standard diploid complement of DNA, is concerned with reproduction. The **macronucleus** contains DNA that has been replicated several times even

a *In a small, one-celled organism, all of the cytoplasm is close to the active membrane surface, which is in contact with the outside environment on all sides.*

b *If you double that cell in size, what would have been two sides of separate smaller cells (two dashed surfaces) are no longer at the surface. The cytoplasm doubled; the surface area did not.*

c *And if you tripled the cell in size, what would have been four sides of separate smaller cells are no longer at the surface.*

d *If you kept on enlarging it you'd have a dead cell, just as you'd have a dead cell in (b) and (c). In all three cases there would not be enough surface membrane area for the efficient exchange of materials between the outside and the inside.*

Figure 9.2 Why there never will be a single *Escherichia coli* as large as an elephant: a simple portrayal of the loss in surface area as cell volume increases.

though the nucleus itself has not divided. As a result, each macronucleus often contains as many as 50 or 100 copies of each gene! Thus the macronucleus serves as an expanded reservoir of genetic instructions. This reservoir is tapped to produce the comparatively large ciliate cell, which in many cases also happens to be one of the most complex single cells the world has ever seen (Chapter Sixteen).

Another group of single-celled eukaryotes, the foraminiferans, rely on multiple copies of the same kind of nucleus. Several thousand nuclei may be spaced rather uniformly throughout the cell body. Foraminiferans have become astonishingly complex, and it is because of the abundance of hereditary material that they have been able to do so. More than this, from the point of view of a single-celled predator, they have reached discouragingly gigantic proportions. Some foraminiferans, single cells though they are, measure as wide as the palm of your hand!

Yet by far the most common strategy for increasing in size has been the use of multiple nuclei in multicellular eukaryotes, with each nucleus typically isolated within its own cell. It is the foundation for hereditary control in multicellular plants, animals, and fungi. Even with the same DNA in the nucleus of all cells, different *regions* of the DNA can be expressed in different cells (Chapter Fourteen). The extensive division of labor that exists among these cells is possible because of the selective "reading" of the same hereditary instructions.

As you read earlier, there have been few prokaryotic experiments with the division of labor, and the experimentation has been of the most uncomplicated sort. Some colonial eukaryotes (Chapter Sixteen) have evolved little beyond the prokaryotes in this respect. It is only when we come to the plant and animal kingdoms that we find the potential of cell specialization being expressed in exuberant ways.

On the Surface-to-Volume Ratio

Increasing the amount of genetic information available to an organism was by no means the only barrier that had to be hurdled along the road to increased size and complexity. To stay alive, any organism must take in nutrients, transport them through the body, and then get rid of the wastes. In the case of a small single cell, the plasma membrane is enough of an active surface to take up all the nutrients from the environment that the cell needs. Once nutrients are taken up, diffusion (through random motion of molecules) is enough to distribute them uniformly through the cell's interior. Similarly, diffusion and transport across the plasma membrane are adequate for disposing of wastes.

But when a single cell enlarges, it eventually encounters the constraints imposed by the **surface-to-volume**

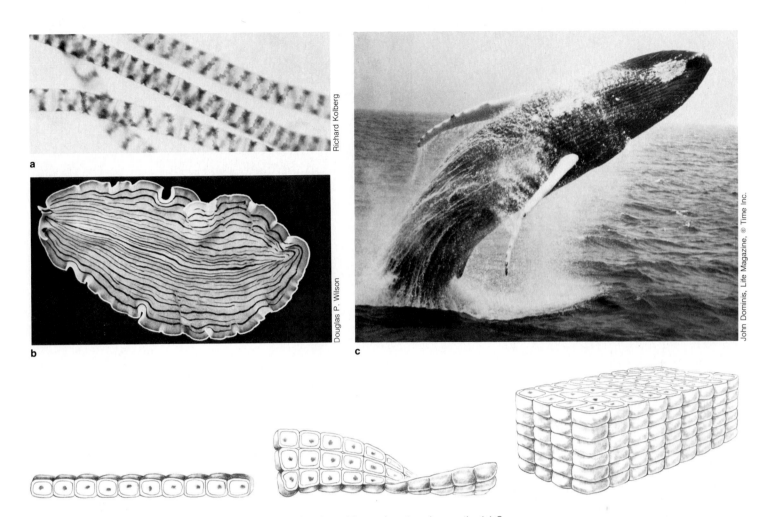

Figure 9.3 Multicellular strategies for getting around the constraints of the surface-to-volume ratio. (**a**) One-dimensional growth in *Spirogyra*. (**b**) Two-dimensional growth in a flatworm, an animal about as big as your little toe. (Actually there is some three-dimensional growth in these animals, but most of the size increase comes from sheetlike growth.) (**c**) Three-dimensional growth in a whale. Here, internal transport systems assure even the internal cells of efficient materials exchange.

Richard Kolberg

Douglas P. Wilson

John Dominis, Life Magazine, © Time Inc.

ratio: as the volume of a cell increases, the surface area does not increase at the same rate (Figure 9.2). Consider what would happen if a cell were to enlarge even four times in diameter, with no other changes. That would mean the cell volume would increase sixty-four times. To keep the expanded insides of the cell functioning, the amount of nutrients to be absorbed from the environment would also have to be increased sixty-four times. If you carry this line of reasoning one step further, you would see that sixty-four times as many waste products would have to be removed from the cell body.

The problem, however, is that a fourfold increase in diameter would only increase the cell's surface area by sixteen times. This means each unit area of the plasma membrane would now be called upon to serve four times as much cytoplasm! More than this, the cell's center would now be four times as far from the active membrane surface, so it would take four times as long for food and materials to diffuse there and for wastes to diffuse back to the surface. Clearly, enlargement alone is not what you would call adaptive; past a certain point, the cell simply would die.

You may already suspect, from Chapter Four, how the constraints of the surface-to-volume ratio were at least partly overcome during the evolution of larger cells. For one thing, complex infolding of the plasma membrane increased the membrane's surface area as the volume of cytoplasm increased. The development of pinocytosis, described in Chapter Five, was another way to help bring part of the

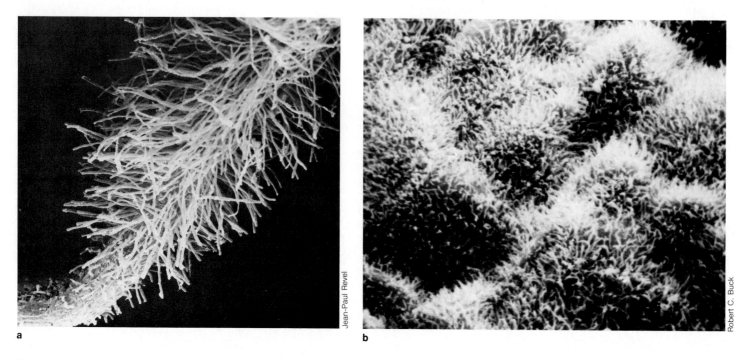

Jean-Paul Revel

Robert C. Buck

Figure 9.4 Parallel developments: (**a**) root hairs in a flowering plant (46×) and (**b**) microvilli found in parts of a mammal's digestive tract (3,000×). On the basis of appearance alone, these structures seem to hint at a common origin. But they signify only that they arose as separate responses to a similar environmental factor: both increase the membrane surface area available for absorbing materials from the environment.

environment into the cell, and to send part of the inside back out.

Such developments can work well enough for a single cell. But imagine the problems of accumulating and transporting various nutrients and wastes in a multicellular organism! If one of the cells is located too far from the surface of such an organism, it simply will not be able to obtain essential nutrients or to get rid of toxic wastes, and it will die. Under this constraint, it seems, multicellular body plans evolved according to three basic patterns: one-dimensional, two-dimensional, and three-dimensional growth (Figure 9.3).

Enlargement in one dimension leads to a long, thread-like body. This path apparently was taken by ancient colonial prokaryotes, by certain protistans, some algae, and most fungi. Its advantage is that each cell is kept at the surface, where it is capable of exchanging materials with the environment. In some evolutionary lines, such as those leading to many existing algae, long threadlike bodies became highly branched to create rather massive bodies. Still, one-dimensional growth has not shown much potential for diversification.

Enlargement in two dimensions leads to a flat, sheetlike body. This was the path taken by some colonial protistans, a

few algae, some fungi, and to some extent in one of the most intriguing groups in animal evolution: the flatworms (Chapter Eighteen). As with one-dimensional growth, each cell remains at or near the surface. Even among organisms capable of extensive three-dimensional growth, there are times when enlargement into a thin, sheetlike structure has its advantages (as it does for the leaf cells of a flowering plant, which are near the surface and therefore exposed maximally to sunlight).

Even so, it is only through three-dimensional growth that the true potential of multicellularity has been realized. With multicellularity, there is extensive cell specialization and division of labor. At the same time, though, growth in three dimensions presents the greatest problem in terms of the surface-to-volume ratio. Massive bodies are possible only if there are also internal transportation systems for exchanging materials between the environment and cells deep in the body's interior. A remarkably diverse array of multicellular organisms have such systems. In fact, you might be led into thinking some plant and animal groups are closely related, if you were to consider them only in terms of how similar their internal transport systems appear to be. But such similarities are *not* a reflection of recent common origins among these groups. Rather, they are a reflection of different groups adapting to the same kind of physical constraints on enlargement itself.

For instance, the root hairs of a flowering plant seem to resemble the microvilli lining parts of the mammalian digestive tract (Figure 9.4). But root hairs and microvilli have different origins. They are different in size, fine structure, and chemical composition. Their resemblance to each other signifies only that there are relatively few ways of adapting to certain environmental conditions. If a few cells exposed to the environment are to absorb the nutrients required by the rest of the cells in the organism, they must have as much surface area as possible. There are few ways of increasing surface area. And a series of fine, fingerlike projections into the environment happens to be a very effective way.

Similarly, a mammal's blood vessels would seem to resemble the conducting tissues of flowering plants. Both are elaborate systems of tubes connecting with all parts of the multicellular body. But when you look to the details of their structure and function, much of the similarity ends. They resemble each other only because they are both adaptations to the same physical constraint: transporting liquids through a large body. It is a constraint that, in the general sense, can be circumvented in very few ways. It is only in the details of how internal transport is accomplished that you find diversity in the multicellular body.

Even within the animal kingdom itself, some groups share certain kinds of traits, yet they are actually quite distinct. If you were simply to list the component parts of the circulatory systems of earthworms and humans, the parallels seem most uncanny: pumping hearts, major blood vessels, fine capillaries, and so on. But when you look closely at the two systems, you find substantial differences.

You might draw much the same conclusions concerning the relationship between our lungs and the tracheal systems by which insects breathe. Beyond the fact that both consist of a network of branching tubes used to conduct gases into and out of the interior of the body, they hold few other features in common. Later chapters will suggest how these two types of respiratory systems arose separately, as similar responses to similar environmental pressures. This is an important point to keep in mind when reading about parallel systems among diverse groups of organisms.

Controlling the Development of Complex Body Plans

The development of a single-celled prokaryote is somewhat straightforward. It is an extraordinarily large jump to the development of a complex multicellular body. Multicellularity never could have been realized without the establishment of a process almost never seen among less complex life forms. In this process, known as **differentiation,**

cells of identical genetic make-up become structurally and functionally different from one another according to the prescribed developmental program for the species. Differentiation requires that certain genes be expressed only in certain cells, and only at specified times.

In large part, controls over gene expression are inherited; they are an important aspect of the genetic heritage that helps define a species. For example, closely related multicellular species are not all that different in the *kinds* of materials with which they are constructed. But they do show differences in the *way* those materials are patterned as their development unfolds. As you will read in Chapter Fourteen, there are differences in the way the arrangement of parts is controlled, which leads to variations in such features as the size of the head in closely related species, or in the shape of a feather, the scalloping of a leaf. Understanding the nature of the genetic controls of development is a key to understanding how complex body plans may have evolved.

Integration of Parts Into a Functional Whole

It would not be of much use for a multicellular body to evolve unless a system of internal communication among parts evolved along with it. If one part of a complex body is to carry out its assigned tasks in helping to maintain the whole organism, it must be kept informed of the state and the requirements of other parts. Complex systems of integration exist in plants and animals. But once again, similarities between plants and animal groups are more superficial than you might initially suspect. Among complex organisms in both kingdoms, a series of chemical substances called **hormones** are produced by cells in one region of the body and are dispersed, in a controlled way, to affect the functioning of target cells in other regions. But the details of plant and animal hormone production are quite different. And in complex animals, a second communication system exists that is without parallel in plants. It is the **nervous system** of communication, integration, and control of the animal body.

In this unit of the book and the next, you will be reading about the unique aspects of multicellularity in plants and animals—the genetic bases, the plans for structural organization, the development, and the integration of specialized parts and systems of parts. Sometimes the tapestry of multicellular adaptations may seem to become so intricate that you almost lose sight of the underlying threads. But keep in mind that beneath all the myriad diverse body plans is a set of common problems: *(1) how to assure reliable hereditary control, (2) how to overcome the constraints of the surface-to-volume ratio, and (3) how to control and integrate all the separate body parts.* These common problems have shaped every multicellular organism into what it is today.

Chapter Ten

Figure 10.1 Mother's eyes, father's chin—by what means are so many heritable traits shuffled up and parceled out to a new generation of complex organisms such as ourselves?

How Multicellular Body Plans Are Inherited

With the full authority of the Spanish crown, Juan Ponce de León set off in 1513 toward parts unknown in search of *la Fontana de Juventud*—the Fountain of Youth, immortality. A futile quest, of course. But consider the ongoing saga of the immortal amoeba. It is true that ever since the first amoeba appeared, many billions of amoebas have fallen by the wayside—devoured, suddenly deprived of water or nutrients or both, generally shot down by hardships of various sorts. And yet, for the lucky ones there has been no real death. They eat, they grow, and, upon reaching a certain size, they divide their nucleus and then pinch in two. Thus they leave no dead ancestors—only two amoebas where there once was one!

Like many single-celled eukaryotes, the amoeba engages in **asexual reproduction:** the hereditary material of a single parent cell is divided by mitosis, then the cell itself divides in two (Chapter Seven). This asexual process is also the basis for the physical growth of multicellular organisms. But most existing eukaryotes rely also on **sexual reproduction.** At some stage during their life cycle, parent organisms produce haploid sex cells, or **gametes.** (Haploid cells, recall, have only one of each type of chromosome characteristic of the species, and a diploid cell has two of each.) And the fusion of two gametes (usually from two different parents, except in such cases as self-fertilizing plants) leads to the formation of a new diploid individual.

Sexual reproduction is more complex than the asexual process. For one thing, gamete formation depends on an intricate nuclear division process called meiosis, which you will be reading about in this chapter. For another thing, cycles of sexual reproduction have become molded in many ways to environmental conditions, such as seasonal changes in the availability of food, water, and living sites. Among complex plants, sexual reproduction often depends on other organisms, such as insects, to carry the sex cells from one plant or plant structure to another. Among complex animals, predictable cycles of sexual reproduction are laced with more elusive patterns that change with age, with behavioral and physiological states, with the genetic make-up of the population to which the individual belongs. Thus, throughout this book, we will be looking at sexual reproduction not only in terms of individuals but in terms of populations and community interactions. Eventually we will turn to how patterns of human sexual reproduction have come to affect the entire community of life. At this

point in our journey, however, we will first explore (1) the mechanisms by which sexually reproducing organisms sort out chromosomes into gametes, (2) how the genetic composition of two gametes becomes translated into the unique set of traits of a new individual, and (3) how change in genetic composition at the cellular level influences the character of populations, hence the species.

Mendel and the First Formulation of the Principles of Inheritance

Biology toward the end of the nineteenth century was dominated by talk of Darwin and Wallace's theory of natural selection. Many biologists remained skeptical about whether natural selection could work, however, because no one yet had a clear idea of the mechanisms of inheritance through which it would have to operate (Chapter Two). But even before the theory was made public, evidence concerning the nature of heredity was accumulating in the relative obscurity of a small monastery garden. Gregor Johann Mendel, a scholarly, mathematically oriented Austrian monk, was beginning to identify the rules governing the process of inheritance.

Mendel was active in the affairs of the surrounding agricultural community, and in fact had developed several hybrid strains of fruits and vegetables. It was through his initial interest in crop improvements that he began his experimental studies of the mechanisms of inheritance, and it was through his mathematical skills (as yet not considered very relevant to plant breeding) that he perceived patterns in the expression of traits from one generation to the next. It will be useful to retrace a few of his experiments, because the conclusions he drew from them have turned out to apply, with some modifications, to all sexually reproducing life forms.

The Concept of Dominance Mendel first simplified the question of how traits are inherited by limiting the number of traits being studied in a given experiment. For one experiment, he used pea plants that differed only in flower color. One strain always produced white flowers, the other always produced red. He crossed these "pure-breeding"

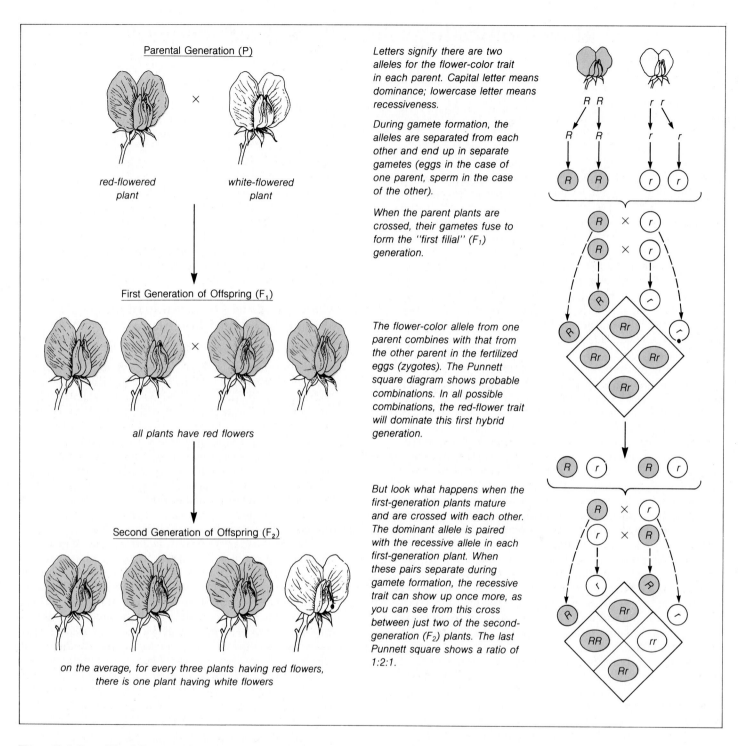

Parental Generation (P)

red-flowered
plant

white-flowered
plant

Letters signify there are two alleles for the flower-color trait in each parent. Capital letter means dominance; lowercase letter means recessiveness.

During gamete formation, the alleles are separated from each other and end up in separate gametes (eggs in the case of one parent, sperm in the case of the other).

When the parent plants are crossed, their gametes fuse to form the "first filial" (F₁) generation.

First Generation of Offspring (F₁)

all plants have red flowers

The flower-color allele from one parent combines with that from the other parent in the fertilized eggs (zygotes). The Punnett square diagram shows probable combinations. In all possible combinations, the red-flower trait will dominate this first hybrid generation.

Second Generation of Offspring (F₂)

on the average, for every three plants having red flowers,
there is one plant having white flowers

But look what happens when the first-generation plants mature and are crossed with each other. The dominant allele is paired with the recessive allele in each first-generation plant. When these pairs separate during gamete formation, the recessive trait can show up once more, as you can see from this cross between just two of the second-generation (F₂) plants. The last Punnett square shows a ratio of 1:2:1.

Figure 10.2 One of Mendel's monohybrid crosses.

(**a**) The parental (P) generation of pea plants varies only in one trait: flower color. Red flowers dominate the first generation (F₁) of hybrid offspring. But the recessive trait isn't lost; it's merely masked by the expression of the dominant one during the first generation. This and other monohybrid crosses suggested to Mendel that a sexually reproducing organism must inherit two "factors" (alleles, or alternative forms of a gene) for every trait.

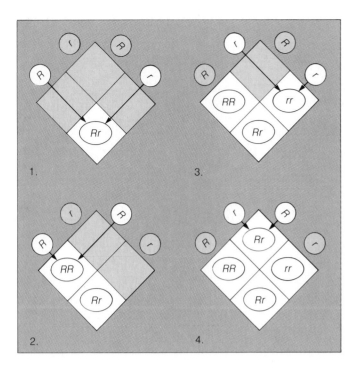

(**b**) Why does a recessive trait show up in the second (F_2) generation? We can get a good idea by filling in a *Punnett square*. Such diagrams can be used to predict the probable ratio of traits that will show up in offspring.

strains to see which color would show up in their offspring. (Whenever two parents having different forms of a trait are crossed, the offspring are called **hybrids.** If the parents differ in only one trait, the experiment is called a **monohybrid cross.**) In this case, the first hybrid generation had only red flowers. It was as if the white-flower trait had disappeared. Yet, when these hybrid plants were crossed with one another to produce a second generation, a few of the second-generation plants had white flowers! The white-color trait had not been lost. It simply had failed to be expressed in the first generation.

Mendel crossed six other pure-breeding strains of pea plants, with each pair differing only in the form of one trait (height, seed color, pod shape, and so on). In each of these monohybrid crosses, the result was the same: one of the traits "disappeared" in the first generation, only to pop up again in some of the second-generation plants. Mendel called the now-you-see-it-now-you-don't form of the trait **recessive;** in some way it was masked by the expression of the alternative form of the trait. In some way the fully expressed form was **dominant** in the hybrid plants.

Since Mendel's time, similar experiments have been performed with many different kinds of organisms, and his

description of the results is still valid. It is a straight conceptual line from Mendel's first formulation to our current view of the physical units of inheritance in sexually reproducing diploid organisms:

1. Distinct units of heredity—**genes**—are the physical basis for all traits of a new individual. Each gene occupies a stretch of chromosome that specifies some substance, or that specifies when and how a substance will be put to use. (These are the structural and control genes you read about in Chapter Eight.)

2. There is a complete set of chromosomes in each gamete. Thus, through sexual fusion of two gametes, a new individual inherits *two* genes for each trait.

3. For any trait, it might be that sexual fusion brings together two genes that code for the same form of the trait. (For instance, both genes for flower color may specify "red.") Then again, it might bring one gene together with another that codes for an alternative form of the trait. (One gene may specify "red," the other "white.") The alternative forms of a gene for a given trait are known as **alleles.**

4. If the two alleles of a given trait are identical, the individual is said to be **homozygous** for the trait. If the two alleles are not identical, the individual is said to be **heterozygous** for the trait.

5. In heterozygous individuals, the expression of one allele may mask the expression of the other (in other words, it is the dominant form of the trait). Even so, *both* alleles for a given trait retain their physical identity throughout the individual's life cycle.

Let's see what this explanation means in terms of Mendel's experiment with flower color. Assume both alleles for flower color in a diploid plant specify "red." As shown in Figure 10.2, we can represent the plant as *RR*. (A capital letter by convention means a dominant allele.) The plant is said to be homozygous dominant for the trait. Assume both alleles for flower color in another diploid plant specify "white." We can represent this plant as *rr*. (A lowercase letter means a recessive allele.) The plant is said to be homozygous recessive for the trait.

Both of these diploid plants now form gametes prior to sexual reproduction. In a diploid cell destined to become gametes, the alleles for flower color are separated from each other. Each haploid gamete formed ends up with only *one* allele for flower color. (In complex plants, as in complex animals, these gametes are called either sperm or eggs. An egg fertilized by a sperm is called a **zygote.**) Say the eggs contain the *R* allele, and the sperm the *r* allele. Thus, the zygotes turn out to be *Rr*. In this case, the alleles are not identical; the zygotes are heterozygous for the trait.

Assume these zygotes grow into mature plants. All will

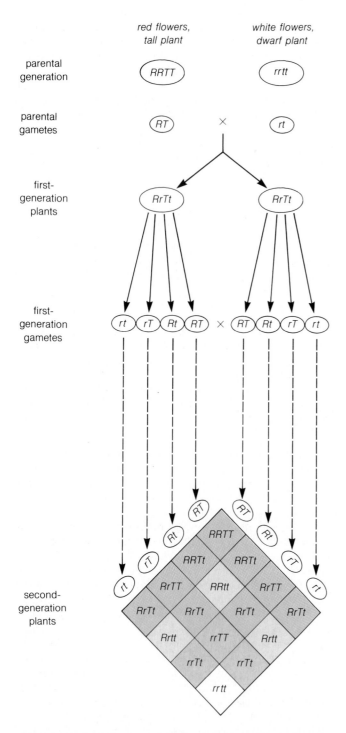

red flowers,
tall plant

white flowers,
dwarf plant

parental
generation

RRTT

rrtt

parental
gametes

RT

×

rt

first-
generation
plants

RrTt

RrTt

first-
generation
gametes

rt rT Rt RT × RT Rt rT rt

second-
generation
plants

RT   RT

Rt   Rt

rT   rT

rt   rt

RRTT

RRTt   RRTt

RrTT   RRtt   RrTT

RrTt   RrTt   RrTt   RrTt

Rrtt   rrTT   Rrtt

rrTt   rrTt

rrtt

Figure 10.3 Mendel's dihybrid cross experiment, which suggests that alleles are assorted independently into different gametes. For the alleles specifying height and flower color of the pea plants studied, these were the results: nine tall/red (shown in the dark brown squares); three dwarf/red (light brown squares); three tall/white (dark gray squares); and one dwarf/white (light gray square).

Figure 10.4 Gregor Johann Mendel. In the spring of 1865, Mendel presented the general outline of his concepts concerning the physical basis of inheritance, along with experimental data to back them up. His work was politely ignored. It was not until thirty-four years later, after his death, that the meaning of his discovery was recognized.

have red flowers, because all carry the dominant allele. But when these heterozygous plants form haploid gametes prior to sexual reproduction, half the gametes must, on the average, end up with the dominant allele, and half with the recessive. Thus, when sperm from one heterozygous plant fuse with eggs from another, three possible combinations of flower-color alleles will show up in the zygotes: *RR, Rr,* and *rr.* What's more, they will show up in a ratio of about 1:2:1 (Figure 10.2). For every *RR* plant, we can expect to find about two *Rr* plants and one *rr* plant.

(The basis for this expectation, which is valid for most cases, is that sperm and eggs from parent plants combine at random; any given sperm is as likely to fertilize one egg as another. But remember these gametes have either the *R* or the *r* allele. This means about half of all the eggs fertilized will contain the dominant allele. And about half of the sperm that fertilize them will contain the dominant allele. Thus about one out of every four zygotes will be *RR.* We could use the same reasoning to show why about two out of every four zygotes will be *Rr,* and about one out of every four will be *rr.* Even so, rarely will the ratio be *exactly* 1:2:1. To see why this is so, flip a coin a few times. We all know a coin is as likely to end up "heads" as it is "tails." But often it ends up heads, or tails, several times in a row. Only if you flip the coin an extraordinary number of times can you be assured the head-to-tail ratio will be very close to the predicted 1:1 ratio. Such random events illustrate three characteristics of **probability.** First, it's usually possible to predict the most probable ratio of outcomes of any series of randomly ordered events. Second, it's unlikely that the actual ratio will exactly match the expected ratio. Third, when only a few events are observed, the actual ratio may differ considerably from the predicted one. Mendel succeeded largely because he crossed hundreds of plants and kept track of thousands of offspring, rather than restricting his experiments to a few plants as others had done. Almost certainly his understanding of probability kept him from being confused by minor variations from his predicted results.)

But say the ratio of gene combinations does turn out to be 1:2:1 in our monohybrid cross. Even then, the ratio of second-generation pea plants with red flowers to those with white flowers is actually 3:1! The reason is that *both* homozygous dominant (*RR*) and heterozygous (*Rr*) pea plants produce red flowers. Why does the heterozygous plant do this? The function of most genes, recall, is to code for a specific protein. In this case the dominant gene codes for an enzyme (which, recall, is one kind of protein) that produces a red pigment molecule. The recessive (white-color) gene can't do this, because it has undergone a change that makes it code for a variant enzyme. But in the heterozygous plants, the one dominant gene is fully functional, so enough enzyme is produced to assure the presence of enough pigment to color the flowers red. This example illustrates why we often can't distinguish between homozygous dominant or heterozygous individuals simply on the basis of **phenotype** (physical appearance). The **genotype** (genetic constitution underlying physical appearance) of such individuals can be determined only by observing the kinds of offspring they produce.

There are a few cases in which phenotype *can* be used to identify heterozygous individuals produced in a mono-hybrid cross. For instance, when red and white snapdragons are crossed, the first-generation plants have pink flowers! When these plants are crossed, they give rise to a second generation having red, pink, and white flowers in about a 1:2:1 ratio. This is an example of **incomplete dominance,** in which a so-called dominant gene is not completely able to control phenotype. Apparently, snapdragons differ from pea plants in that one "red" gene results in the production of only half the amount of pigment needed to make the flowers red, so they end up pink.

The Concept of Independent Assortment When Mendel crossed true-breeding plants that differed in *two* traits, he began to perceive how sexual reproduction fosters diversity. Such experiments are called **dihybrid crosses.** For instance, Mendel crossed tall red-flowered plants with dwarfed white-flowered ones. Like the "red" allele, the "tall" allele was dominant, so all the first-generation offspring were red-flowered and tall. But as Figure 10.3 shows, when the first-generation plants were crossed, every combination that

could occur *did* occur. Even so, Mendel saw a pattern in the variation, for he wrote that the combinations emerged in about a 9:3:3:1 ratio. (For every 9 tall red-flowered plants, 3 were dwarf red-flowered, 3 were tall white-flowered, and 1 was dwarf white-flowered.) The pattern could mean only one thing. Each gamete from a heterozygous plant ends up with one allele for each trait. But a gamete containing the dominant allele for color will end up with either the dominant *or* the recessive allele for height! Thus Mendel perceived a principle of **independent assortment** that influences the distribution of alleles into separate gametes:

As a first approximation, the alleles coding for the different traits of an individual are parceled out independently of one another into gametes.

When independent assortment of alleles does occur, it has an important consequence. It means all combinations of parental traits that are possible may occur in the offspring. The occurrence of independent assortment provides a rich reservoir of genetic diversity. And natural selection can act to perpetuate those combinations of alleles that make their bearers better adapted to specific environments.

The Process of Meiosis

Since Mendel's time, the process by which alleles are segregated and assorted into gametes has become better understood. To gain insight into the physical basis for the patterns Mendel perceived, we must turn now to the behavior of chromosomes during gamete formation.

In a complex multicellular organism, again, the cells generally are diploid: each contains two sets of chromosomes, one set from each parent. This number must be halved before formation of gametes that are destined to fuse into a zygote. If there were no such reduction to a haploid state (a single set of chromosomes in each gamete), then the zygote would end up with twice the amount of hereditary material for its species. Even if it could survive an overload of four chromosome sets, the next generation would end up with eight, the next with sixteen, and so on until there wouldn't even be room left for anything else in the cell! The nuclear division process by which this reduction takes place is known as **meiosis:**

In the first stage of meiosis, the diploid set of chromosomes present in the nucleus of a eukaryotic cell is reduced to two haploid sets.

In the second stage of meiosis, the haploid sets are physically separated from each other for distribution into two new nuclei.

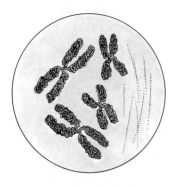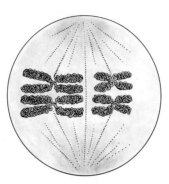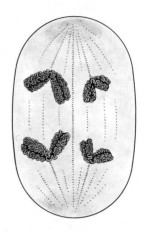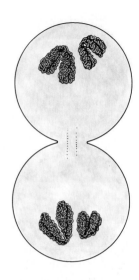

Prophase I	*Metaphase I*	*Anaphase I*	*Telophase I*
As in mitosis, chromosomes become clearly visible. We see each one has already duplicated itself, thus forming two chromatids that are attached at the centromere.	*As in mitosis, the chromosomes line up at the equator of the spindle apparatus. But <u>unlike</u> mitosis, homologous pairs of chromosomes match up with each other. Chromosomal microtubules connect one chromosome of a matched set to one pole, and the other chromosome of the set to the opposite pole.*	*As the continuous microtubules push the poles apart, chromosomal fibers attached to the chromosomes shorten—the effect of which is to <u>unpair the homologous pairs</u> and guide them to opposite poles.*	*Typically, microfilaments begin constricting the cell at the equatorial plane, and the chromosomes unwind as they do in mitosis. At this stage, meiosis I is complete.*

Meiosis I

Figure 10.5 Meiosis: halving the hereditary material in a reproductive cell of a sexually reproducing organism. To keep things simple, only two kinds of homologous chromosomes are shown here. The gray chromosomes are derived from one parent; the brown ones are their equivalents from the other parent.

Meiosis works something like mitosis, and may even have evolved from it. Prior to division, all the chromosomes present in the nucleus are replicated, just as they are before mitosis begins. A microtubule spindle and microtubule attachments are used to move the chromatids of chromosomes in similar ways. But as Figure 10.5 shows, meiosis differs from mitosis in three fundamental ways.

First, meiosis consists of *two* divisions, which usually occur in rapid succession. First the number of duplicated chromosomes is halved in a special manner, then the duplicated chromosomes themselves are pulled apart.

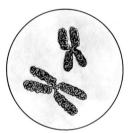
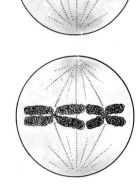
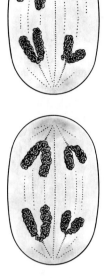
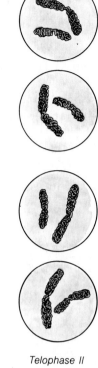

Prophase II	*Metaphase II*	*Anaphase II*	*Telophase II*
From this point on, the picture really resembles mitosis. Each cell (or each polar region, if the original cell has not divided) now has one of each kind of chromosome. Each chromosome still consists of two chromatids. The spindle apparatus quickly begins to re-form.	The centromere of each chromosome now becomes attached to microtubules, which draw it to the equator. Chromosomal microtubules attach at the centromere to connect each chromosome to both poles of the spindle.	Each centromere splits; sister chromatids are separated, and they are moved to opposite poles.	Each nucleus that forms is haploid; it has one chromosome of each type.

Meiosis II

(There is no interphase, and there is no DNA replication between these two divisions. Often the cell doesn't even divide in two at this point.)

Second, in the first division (meiosis I), all the duplicated chromosomes are lined up at the equator of the spindle in **homologous pairs:** each chromosome derived from one parent lies side by side with the equivalent chromosome from the other parent.

Third, the two sister chromatids of each duplicated chromosome don't separate during meiosis I. Instead, the paired homologous chromosomes separate from each other.

It is only during the subsequent division process (meiosis II) that sister chromatids are pulled apart, as they are in mitosis, with the result being two haploid sets of chromosomes.

With this reduction–division process, a haploid complement of hereditary instructions is faithfully parceled out for each of the forthcoming gametes. Thus meiosis works to preserve the hereditary instructions needed to build and maintain a new individual in the general image of its parents (and its species). But as you will now see, meiosis also works to scramble things up!

Independent Assortment of Nonhomologous Chromosomes Although Mendel knew nothing of meiosis, he understood that each allele of a given trait is segregated from its partner during gamete formation. This is a result of homologous chromosomes being pulled away from each other during meiosis. He also perceived that these alleles are assorted randomly into separate gametes. This is a result of the way *nonhomologous* chromosomes are arranged and moved about during meiosis:

During meiosis I, all the chromosomes from one parent become lined up on the spindle's equator with their partner from the other parent.

Either chromosome of a homologous pair can end up on one side of the equator or the other. Its alignment has nothing to do with the alignment of other homologous pairs.

There is nothing that says all the chromosomes derived from one parent must stay on the same side of the spindle's equator and their homologues must stay on the other. Assume we're talking about just three homologous pairs. Any one of the following combinations can occur as they line up on the equator:

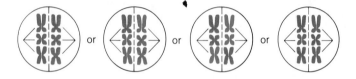

(Here, the brown chromosomes are derived from one parent, and the gray are derived from the other parent.) During the formation of gametes, these four arrangements can occur with about equal frequency. This means that when sex cells with three chromosome pairs are forming during meiosis, eight different combinations are possible. Imagine the number of possible combinations when there are not just three homologous pairs but twenty-three, which is how many you have in *your* cells!

Thus, during meiosis, the genes on one homologous pair of chromosomes are always sorted out independently of the genes on the other homologous pairs. Because chromosomes are moved about in this way, you might expect that the genes on a single chromosome always travel together; after all, they are physically linked to one another. But the tendency of genes located on the same chromosome to travel together is *not* a hard-and-fast rule. Let's now see why there are exceptions.

Recombination During Meiosis During early prophase I of meiosis, a submicroscopic, thin band of RNA and protein is synthesized between the joined sister chromatids of each chromosome. The sister chromatids come to lie on one side

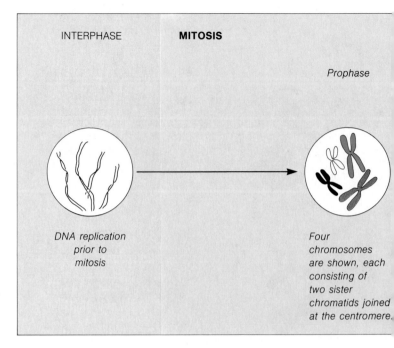

INTERPHASE   **MITOSIS**

Prophase

DNA replication prior to mitosis

Four chromosomes are shown, each consisting of two sister chromatids joined at the centromere.

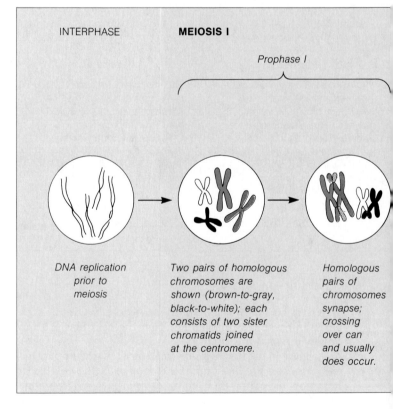

INTERPHASE   **MEIOSIS I**

Prophase I

DNA replication prior to meiosis

Two pairs of homologous chromosomes are shown (brown-to-gray, black-to-white); each consists of two sister chromatids joined at the centromere.

Homologous pairs of chromosomes synapse; crossing over can and usually does occur.

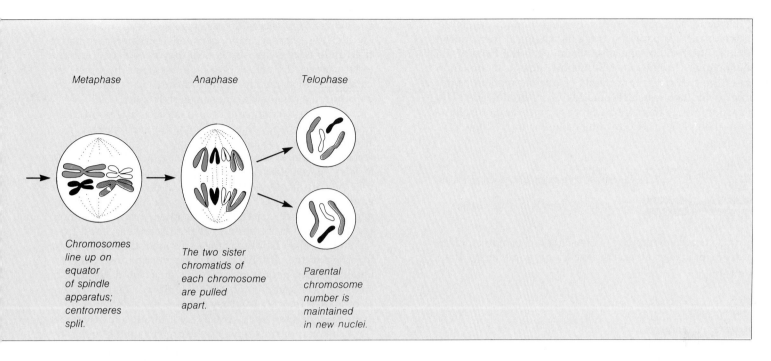

Metaphase

Chromosomes line up on equator of spindle apparatus; centromeres split.

Anaphase

The two sister chromatids of each chromosome are pulled apart.

Telophase

Parental chromosome number is maintained in new nuclei.

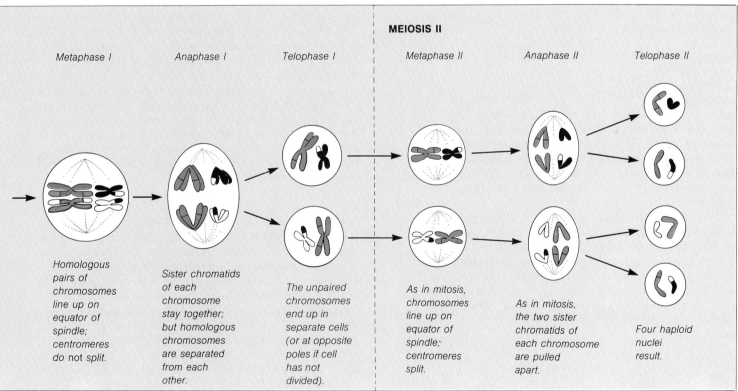

MEIOSIS II

Metaphase I

Homologous pairs of chromosomes line up on equator of spindle; centromeres do not split.

Anaphase I

Sister chromatids of each chromosome stay together; but homologous chromosomes are separated from each other.

Telophase I

The unpaired chromosomes end up in separate cells (or at opposite poles if cell has not divided).

Metaphase II

As in mitosis, chromosomes line up on equator of spindle; centromeres split.

Anaphase II

As in mitosis, the two sister chromatids of each chromosome are pulled apart.

Telophase II

Four haploid nuclei result.

Figure 10.6 Summary of chromosome movements in mitosis and in meiosis. Remember that mitosis works to maintain the parental chromosome number; it is the basis for asexual reproduction in many single-celled eukaryotes and the basis for growth of multicellular eukaryotes. Meiosis works to reduce a parental diploid number of chromosomes to the haploid state; it is the basis for the formation of gametes, which are involved in the sexual reproduction of most eukaryotic organisms.

of this band, so that the other side is exposed to the nuclear environment. Apparently there is chemical recognition between the band of one chromosome and the band of the homologous chromosome. Somehow, chemical attraction draws the two sister chromatids of one chromosome very close to the two sister chromatids of its homologue. The attraction brings them into a precise, point-by-point alignment along their length. This alignment is known as **synapsis:**

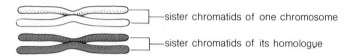

—sister chromatids of one chromosome

—sister chromatids of its homologue

In this arrangement, two "nonsister" chromatids often undergo breakage at exactly the same point along their length:

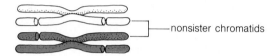

—nonsister chromatids

These four broken ends do not simply stick back together; they stick back together crosswise:

crossing over
at two sites

This event is known as **chromosomal crossing over:** the chromatids of homologous pairs of chromosomes actually exchange corresponding segments at breakage points. Its consequence is **genetic recombination**—the introduction of new combinations of genes into a chromosome (hence into gametes).

During the last stage of prophase I, the two sister chromatids of each chromosome remain close to each other. But now the chromosomes of a homologous pair seem to repel each other. Even so, these homologous chromosomes are held together at one or more regions as they are separating along most of their length. Such a region of apparent contact is called a **chiasma** (plural, chiasmata), which means a "cross." This sketch depicts two chiasmata:

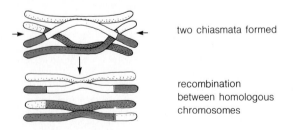

two chiasmata formed

recombination
between homologous
chromosomes

Each chiasma observed at this stage seems to indicate that an exchange has occurred between nonsister chromatids that underwent synapsis. It does not necessarily indicate exactly *where* the actual breakage occurred, however, for chiasmata tend to slip down toward the ends of the chromosomes as the chromosomes pull apart.

Crossing over apparently can occur at any point in the linear array of genes on each chromatid. *But the probability of crossing over and recombination occurring at any point between two genes located on the same chromatid is directly proportional to the distance between them.* If two genes are located very closely together, in nearly all cases they will end up in the same gamete; they will act as if they are tightly linked. If two genes are farther apart from each other, they will undergo recombination more often than tightly linked genes. Even so, in more than half the gametes formed, they will still be together on the chromatid. Such genes are loosely linked. Finally, if two genes are really far apart on a chromatid, crossing over and recombination can occur between them so often that they act as if they are assorted independently—even though they are located on a single chromosome!

The relationship between the organization of genes on chromosomes and their segregation patterns during meiosis is so regular that it can be used to determine the positions of genes relative to one another. Plotting their positions is called **linkage mapping** of chromosomes. Assume we find out that two mutant alleles located on the same chromosome end up together in the same gamete about 95 percent of the time. We know they must be so close together that crossing over can occur between them about 5 percent of the time. We could conclude they are closer together than two genes showing 10 percent crossing over—and farther apart than two genes showing 1 percent crossing over. When enough genes on a chromosome have been studied, two or three at a time, it's possible to draw a detailed map of their relative positions.

Sex-Determining Chromosomes By now, you have a general idea of how chromosomes and genes may be shuffled and reshuffled, during meiosis, into combinations that determine what the characteristics of offspring will be. Let's now see what the shuffling means in terms of a specific example: the combinations that determine whether offspring will be male or female. Most animal and some plant species have **sex chromosomes,** which differ either in number or in kind between males and females. For instance, humans have different sex chromosomes: X and Y. For us, XX specifies a female, and XY specifies a male. (All other chromosomes in the nucleus are called "autosomes.")

Normally, all the eggs from a human female contain one X chromosome. Because the X and Y chromosomes in males act as a homologous pair during meiosis, half the sperm contain an X chromosome and half contain a Y. If an X-bearing sperm fuses with an X-bearing egg, the zygote

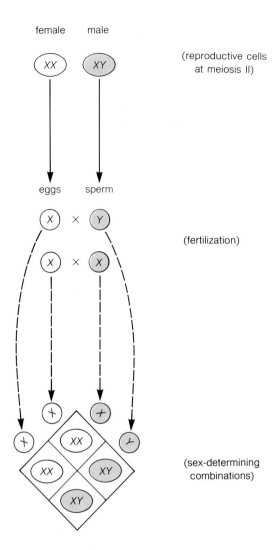

female   male

(reproductive cells at meiosis II)

eggs   sperm

(fertilization)

(sex-determining combinations)

Figure 10.7 The chromosomal basis for sex determination in humans. White indicates female; brown indicates male. (Each human has twenty-two additional sets of chromosomes, not shown here, for a total of twenty-three. But only the X and Y chromosomes determine what the sex of human offspring will be.)

develops into a female. If a Y-bearing sperm fuses with an X-bearing egg, the zygote develops into a male (Figure 10.7).

It happens that the Y chromosome in humans does not carry many genes other than those determining maleness. *But the X chromosome carries many genes besides the ones involved in sex determination.* Such genes, which are said to be **sex-linked genes,** were discovered by the American geneticist Thomas Hunt Morgan. Because males have only one X chromosome, genes carried on the X chromosome—whether dominant or recessive—are expressed in males.

For instance, one gene on the X chromosome codes for

one of the proteins needed to clot blood. A mutation in this gene means it will code for a defective protein. In a woman who is a "carrier" of (heterozygous for) the mutant gene, blood clotting time is essentially normal, because the nonmutated gene on her second X chromosome codes for enough of the normal protein. But if that woman gives birth to a son, there is a 50-percent chance he will have the X chromosome bearing the mutant gene. If he does, he will be unable to produce functional blood-clotting protein; he will suffer from the disease **hemophilia.** With a cut, bruise, or especially internal bleeding, the results may be fatal.

The mutant gene for hemophilia is rare in the human population. And many more males than females are afflicted by it. Only when a female who is a carrier of the defective gene marries a hemophilic male can daughters be born with the disease. Because the gene occurs so rarely, such births are not likely to occur, unless hemophiliacs marry close relatives.

Hemophilia was a recurrent disease among royal families of Europe during the nineteenth century. Queen Victoria of England was a carrier of the mutant gene. Unfortunately, one of her sons was hemophilic and two of her daughters were carriers of the mutant gene. Particularly for Russia, the mutant gene and the cast of characters it indirectly drew together—Czar Nicholas II, Czarina Alexandra (a hemophilic carrier who was a granddaughter of Victoria), their hemophilic son, and the power-hungry monk Rasputin who turned the disease to his political advantage—helped catalyze events that brought an end to dynastic rule in the Western world.

Abnormalities of a different sort can arise not from mutation but from the improper separation of chromosomes during meiosis. For instance, male children suffering from **Klinefelter's syndrome** have testes that develop only to about half their normal size, and the deposition of body fat follows female patterns (for example, there may be some breast enlargement). The afflicted individuals are always sterile. In 1956 it was discovered that such individuals have one Y and two X chromosomes; they have a total of forty-seven chromosomes instead of the normal diploid complement of forty-six. In another sex chromosome abnormality, called **Turner's syndrome,** the secondary sex characteristics of adolescent females fail to develop at puberty. Such individuals received only one X chromosome; hence they have only forty-five chromosomes instead of the normal diploid complement.

Genes That Govern the Translation of Genotype Into Phenotype

When research into the nature of genes began in earnest in 1900 (the year Mendel's hitherto-unappreciated work was suddenly exhumed), it was not yet possible to study anything more than the inheritance of simple features. Of necessity, geneticists had to focus on such superficial traits as eye color and bristle shape in fruit flies (*Drosophila*), coat color in mice, and kernel color in corn. With research so narrowed, it's no wonder that many believed genes in the nucleus were associated only with superficial traits. Eventually, however, genes came to be recognized as the master architects of inheritance, the source of all features essential for life and all traits that distinguish one species from another.

For instance, it was known for some time that all insect eggs contain a substance called "polar plasm." It is needed for setting aside special cells that will develop into gametes in the adult insect. Such substances in the egg cytoplasm were once held as proof that genes in the nucleus don't control all traits. Later it became clear that it takes a certain gene in the female producing the egg to produce polar plasm in the first place! If that gene undergoes mutation, a female produces eggs in which polar plasm is missing. Her offspring mature into normal-looking individuals, but they aren't able to produce gametes. Without gametes, she won't have "grandchildren" (hence the name "grandchildless mutation").

Or consider the gene that governs a species-specific trait in all true flies: their single set of wings. A mutation in this one gene affects the developmental program in such a way that the offspring end up with two sets of wings. The double-wing trait is the usual ("wild-type") feature of butterflies and beetles—but certainly not of flies!

Such mutations tell us something about which structural genes are needed to produce the central features of the species phenotype. If more have not been identified, it is because mutations in such genes are often fatal. They are **lethal mutations:** they affect some key step in structural development, and without that step, the rest cannot occur and the individual dies. Most lethal mutations in mammals cause death before or right after birth. But some are late-acting. For instance, there is a recessive form of a gene governing fat metabolism. At birth, a child who is homozygous recessive for the gene appears normal. But gradually a certain lipid builds up in the myelin sheaths of the child's nerves (Chapter Twelve). The accumulated fat blocks transmission of nerve impulses, which leads to loss of muscle coordination. Brain cells cannot function properly, and mental acuity begins to deteriorate. Within a few years, the afflicted child inevitably dies from this so-called **Tay-Sachs disease.**

Control genes (which regulate when, where, and how structural genes will be expressed) are also important in the translation of genotype into the species phenotype. Although control genes have been thoroughly studied only in prokaryotes, there is evidence for their existence in eukaryotes as well. In Chapter Fourteen, we will return to the role that control genes play in shaping the heritable traits of a species.

The Genetic Basis for the Evolution of Populations

The whole advantage of sexual reproduction is the constant reshuffling of genetic material in novel ways. Thus natural selection can constantly sift through a species for those individuals whose genotype makes them better adapted to prevailing conditions of the environment. But it doesn't take a huge leap of the imagination to perceive that "the individual" cannot be the unit of evolution; after all, an individual is stuck with the same genes for as long as it lives. Instead, it must be the *population* of individuals that is the unit of evolution. In Chapter Two, we outlined the general principles of evolution; now we are prepared to understand its genetic basis.

A **population** is a group of individuals of the same species that interact, interbreed, and thereby pool their genetic variability. The sum total of all their genes, which potentially is available to new generations of the population, is called (appropriately enough) a **gene pool.** Usually, but not always, a population is confined to one geographic region, so each new generation is derived from the same pool of potential variability. The most notable exception is the human population. Its members are found almost everywhere, yet they all retain the potential to interbreed with one another. And even putting aside the question of whether geographic barriers set apart a population, behavioral barriers (Unit V) can be another way in which its boundaries are defined.

Natural boundaries vary so much from one population to the next that it hasn't been easy to analyze how groups of sexually reproducing organisms might evolve over time. Early in this century an English mathematician, G. Hardy, and a German physician, W. Weinberg, studied the problem and independently arrived at the same place in their thinking. They both concluded that *if all individuals in a population have an equal chance of surviving and reproducing, the frequency of all alleles in the population should remain constant from generation to generation.* This concept came to be known as the **Hardy–Weinberg rule.**

Originally the Hardy–Weinberg rule was described in mathematical terms. But you can test its validity in a simple way. Return to the illustration of Mendel's monohybrid cross (Figure 10.2). Notice that as you move from the original true-breeding parents to the first hybrid generation, then to the second hybrid generation, the frequency of the two alleles (R and r) remains the same. In each generation, half the alleles are R and the other half are r. You could go on following the interbreeding of pea plants until you ran out of paper, or patience. As long as you adhered to the one stated condition—that in every generation *all* individuals have equal probability of reproducing—you would end up with the same result. The ratio of the two alleles in the gene pool would remain fixed at 1:1. You would get the same results if you dealt with more than one gene at a time. And you would get the same results even if you started out with a gene pool in which the ratio of alleles was not 1:1. For example, if you started with a gene pool of 90 percent R alleles and 10 percent r alleles, the ratio would remain 9:1 in every generation.

Such stable allele ratios in a population are known as **genetic equilibrium.** Whenever all individuals have equal probability of producing offspring, you will have genetic equilibrium. The meaning of the Hardy–Weinberg rule, then, is that a population tends to conserve its pool of genetic diversity as long as nothing disturbs survival and reproduction. Obviously, however, things do "disturb" real populations, otherwise they never would have acquired genetic diversity in the first place! Let's take a look at four ways in which they can slip out of equilibrium.

1. *In any population, gene mutations add new genetic variability.* Gene mutations arise constantly. But when a new mutant gene appears, it is by definition present only in low frequency: it is present only in the individual carrying it. If the Hardy–Weinberg rule were applicable, all mutations should remain at very low frequency levels in all future generations. But as you will read below, they don't.

2. *Population size is not always large enough to assure that the statistically probable ratios of genotypes are actually produced each generation.* If a population is relatively small and/or if it produces relatively few offspring, the numbers of individuals having one combination of traits as opposed to another combination may not correspond at all to what the predicted genotypic ratios would be on the basis of random matings alone. Consider a human family. When a couple has a child, there is a 50-percent chance it will be a girl. But it's not unusual for couples to have three children who are all girls, or all boys. In any local population, you might find more girls than boys—or more boys than girls. Only when the offspring of a large number of families are counted will you repeatedly find the sex ratio approaching the theoretical ratio. And even then it seldom will be exactly 1:1. As a consequence of small population size, certain alleles can change markedly in their frequency from one generation to the next simply by chance alone. One allele may even be lost altogether in one generation's time, again by chance. Such random fluctuations in the relative allele frequency of small breeding populations are known as **genetic drift.**

3. *There can be varying degrees of **gene flow** into and out of a population.* New members join a population, some of the old members leave for one reason or another, and again the theoretical state of "equilibrium" is disturbed.

4. Finally, and by far most importantly, *not all individuals have an equal likelihood of surviving and reproducing.* In real populations, genetic variants are not functionally equivalent, and individuals of different genotypes are not equipped in the same way to meet the challenges of the environment. Most mutations are harmful: individuals in which they are expressed have less of a chance of surviving to reproductive age. So mutant alleles are not always retained in the low frequency the Hardy–Weinberg rule would predict; they may be quickly lost. At the same time, occasional mutations enhance the probability that reproduction will occur. Such variant alleles increase in frequency in each generation because their carriers leave proportionally more offspring than others in the population.

What this all boils down to is that the Hardy–Weinberg rule is simply a baseline against which we can measure changes in *gene frequency,* or more precisely, in **allele frequency:** the number of individuals in a population that are carrying any given allele, divided by the total number of individuals in that population.

As an example of how natural selection may act to change allele frequencies in a population, let's look at a specific gene variant. For most adult humans, the red oxygen-carrying pigment in the blood is hemoglobin A (or HbA). But many individuals carry a variant known as HbS. It is the result of a mutation in the DNA that codes for the hemoglobin molecular structure: at one point, an A–T base pair is reversed to T–A. The mutation affects only one amino acid. The resulting hemoglobin is still able to carry oxygen. In fact, heterozygous (HbA/HbS) individuals show few symptoms at all. But homozygous (HbS/HbS) individuals suffer from a serious, often fatal disease called **sickle-cell anemia.**

As Figure 10.8 shows, when HbS molecules give up their cargo of oxygen to other cells in the body, they interact chemically with one another and stack up like long, rigid rods. This causes the cells to become distorted into "sickle" shapes. The deformed cells clump up in the capillaries, which blocks the normal exchange of oxygen and the removal of carbon dioxide waste products from the affected area. Severe damage to internal organs soon follows.

If the HbS allele causes such severe abnormality, why didn't the action of natural selection remove it from the population long ago? *Variant alleles are seldom advantageous or disadvantageous in themselves—their survival value can be weighed only in the context of the environment.* And in regions where the death rate from malaria is high (the environment in which the mutation has been retained), the HbS allele in the heterozygous state provides a survival advantage. Individuals carrying this allele have increased resistance to malaria. Thus there is a **heterozygous advantage,** for the heterozygous individual has a greater chance of surviving than either kind of homozygote.

In Central Africa, there was a clear relationship between the incidence of one type of malaria and the fre-

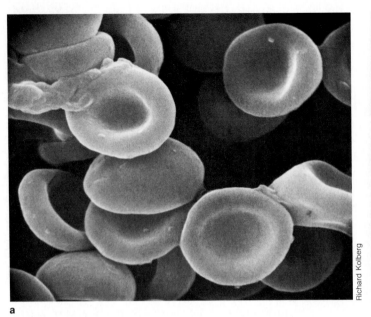
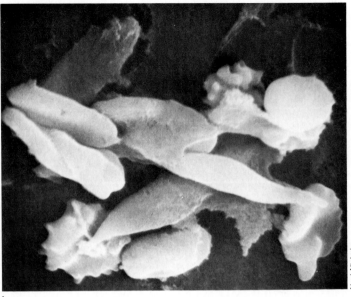

a   b

Figure 10.8 (**a**) Normal red blood cells and (**b**) cells characteristic of the disease sickle-cell anemia. Because of their abnormal, asymmetrical shape, sickle cells do not flow smoothly through fine blood vessels. Instead they pile up in clumps that block blood flow. As a result, the tissues that the blood vessels serve become starved for oxygen and nutrients even as they become saturated with waste products. These tissues soon begin to die.

quency of the HbS allele. As malaria became more prevalent, more HbA/HbA individuals died of the disease before reaching reproductive age. Thus they contributed fewer genes to the next generation—which meant the *frequency* of the HbS allele rose proportionally in the population. Soon, deaths of HbA homozygotes suffering from malaria were balanced by deaths of HbS/HbS individuals suffering from sickle-cell anemia. Therefore, in terms of this allele, the population eventually stabilized. As long as the frequency of malaria remained constant, the HbS allele would prevail at its higher frequency level through the generations.

Genetics and the Potential for Improving the Human Condition

Sickle-cell anemia is a **genetic disease**—a heritable metabolic abnormality. Today three possible ways of lowering the incidence of such diseases are being widely discussed. As you will now read, these approaches are called phenotypic cures, genetic counseling, and genotypic cures.

Preventing and Curing Genetic Diseases In a **phenotypic cure,** something intervenes to prevent the expression of mutant genes. If those genes cannot be expressed, neither

will the symptoms of the disease be expressed. Intervention has been successful for certain genetic diseases, such as **phenylketonuria** (PKU). When children are homozygous for the PKU allele, their body can't perform one essential step in the breakdown of the amino acid phenylalanine. All cells need this amino acid. But if the diet contains more than can be used, the excess must be broken down and then discarded. If the gene coding for the enzyme needed to do this is not functioning, phenylketone (an intermediate substance formed during the normal breakdown process) accumulates in the bloodstream in toxic amounts. In children afflicted with PKU, the breakdown proceeds only part way. Phenylketone builds up to high levels, and eventually its presence can be detected in the urine. In high concentrations, phenylketone can lead to damage of the central nervous system and to mental retardation. But if PKU is diagnosed early enough, the symptoms of the disease can be eliminated simply by placing the child on a special diet—one that provides only as much phenylalanine as the body needs for protein synthesis. In this way, the body isn't called upon to dispose of excess amounts. Aside from their restricted diet, children with the disease can lead normal lives. Most hospitals in the United States routinely check all newborns for PKU, so the outward signs of the disease are rapidly disappearing. The mutant gene, of course, remains in the population.

Several other diseases have yielded to phenotypic cures. **Galactosemia** leads to blindness and mental retardation as a result of a defect in the enzymes needed to break down lactose. But if children with the disease are detected early enough and placed on a synthetic (milk-free) formula, they grow up symptom-free. Similarly, children with diabetes (Chapter Thirteen) can usually lead a relatively normal life with a combination of controlled diet and regular insulin injections.

For many genetic diseases, there is no phenotypic cure. Sometimes one or both potential parents come from families with a history of one of these diseases. They may worry about giving birth to an afflicted individual. In **genetic counseling,** tests are made available to such individuals to help them determine whether they are heterozygous carriers, even though they themselves are disease-free. If both are found to be carriers, they may elect to employ birth control measures (Chapter Twenty-One) and adopt their children. If the possibility that they will have a diseased child is established and if they still wish to proceed—or if the suspected disease is one for which the presence of the heterozygous state can't be detected—another option remains. Through **amniocentesis** (sampling the fluid in the uterus), it's possible to obtain a few skin cells from the fetus at an early stage of pregnancy. The cells can be grown in sterile culture dishes and tested for the presence of any one of more than a hundred genetic diseases. If the fetus is found to have a severe disease, the parents may elect to request an abortion. Genetic counseling is bound by ethical considerations. The role of the medical community must be to provide information that the prospective parents need in order to make their *own* choice, which must be consistent with their own moral values.

Genotypic cures (modifications of the DNA in gametes or embryos) are not yet possible. But research is moving rapidly in this direction. DNA from different organisms can now be broken apart and reassembled in new combinations to form **recombinant DNA.** Recombinant DNA produced from genes of two entirely different organisms probably occurs rarely (if ever) in nature. In the laboratory, the following techniques are used.

First, special enzymes (restriction enzymes) break the DNA into pieces about the size of genes, at the same time leaving "sticky" ends on each piece of DNA. As long as two pieces of DNA have matching sticky ends, they can be attached together by another enzyme (DNA ligase). And the sticky ends do match whenever the same restriction enzyme has been used to break apart both kinds of DNA.

In this way, it is possible to isolate an allele of a gene and make copies of it. First the allele is attached to a *plasmid,* a small, self-replicating circle of DNA found in bacteria. The plasmid is then introduced into a bacterial cell. As the bacterium grows and divides, the plasmid is copied. Overnight, billions of copies of the allele can thus be produced. The gene coding for insulin in rats has been treated this way

with great success. It may become possible to provide human diabetics with a functional gene coding for insulin (diabetics, as you will read in Chapter Thirteen, cannot produce the insulin they need). There is not yet any method for delivering the functional gene to the diabetic. Besides, a far more difficult obstacle is getting the exact sequence of DNA that is required—no more and no less. The restrictive enzymes do not clip DNA precisely at gene borders. A vital regulatory signal for insulin might be lost, or the DNA used might include extra information which could have unforeseen effects.

Recombinant DNA production itself raises questions of scientific ethics. Some molecular biologists feel that the release of new combinations of DNA may have deleterious effects on human health or the environment. For instance, the main bacterium used for research into cancer-causing viruses is *Escherichia coli,* which thrives in the human gut. It would not appear to be to our advantage if *E. coli* carrying recombinant DNA were to escape from the laboratory.

Such risks cannot be accurately estimated. But neither can the benefits—and the potential benefits are great. A genotypic cure of a genetic disease would not increase the frequency of a debilitating allele in the human population, the way that phenotypic cures often do. By using recombinant DNA, plant geneticists have made great progress in modifying strains of corn and wheat so that they can obtain nitrogen from the air instead of requiring fertilizers (Chapter Five). The regulation of such human genes as those which may be implicated in cancer (Chapter Eight) could alleviate terrible human suffering. Most scientists agree that recombinant DNA research is safe as long as proper safeguards are used. For example, weakened strains of bacteria can be used, which should not be able to survive at all in the absence of special laboratory conditions. Strict physical isolation is essential, as it is now for research into highly pathogenic microorganisms. Both scientists and nonscientists have a strong interest in the ethical issues raised by recombinant DNA research. Efforts to regulate this research must carefully balance between protecting public safety and improving our ability to cure terrible diseases, and allowing as much freedom as possible in scientific inquiry.

"Good" Genes, "Bad" Genes, and the Question of Intelligence Since the beginnings of research into human genetics early in this century, some people have argued for the use of genetics to modify the human gene pool. In its most restricted sense, such modification can mean preventing the perpetuation of genes responsible for devastating diseases. In its most extreme sense, it has been touted as rationalization for genocide as the Nazis practiced it. Some genetic theories, even if they do not call for genetic intervention, may call for intervention in social or educational policies. Before any intervention at all is considered, two questions must be answered. Which genes are "the best"? What is heritable? The first question is obviously subject to the bias of the individual who gives the answer. The second question requires that we consider the relative contribution of both the genes *and* the environment, *for expression of a trait is influenced in complex ways by the environment.*

Throughout this century, some have argued that differences in intelligence, as measured by so-called IQ tests, are mainly hereditary. They point to statistics that are said to suggest American whites, on the average, are more intelligent than American blacks. They take studies of IQ scores for white identical twins, siblings, and unrelated children to suggest that intelligence is largely a heritable trait, and then conclude that the differences in intelligence *between* blacks and whites must also be heritable.

There are serious questions whether the data cited really do establish any heritable differences in intelligence. Aside from that, whether IQ tests measure anything other than the ability to perform well on the particular test being studied is questionable. Can there in fact be a "culture-fair" test that measures the same abilities in different groups? Some believe a test is "culture-fair" if parts of it are nonverbal. But cultural differences are not confined to words; they cut deeply into basic perceptions and motivations. (To give a simple example, a member of a remote New Guinean tribe may not be able to perceive the difference between cartoon sketches of Porky Pig and Donald Duck, because he may not know our graphic conventions. Does that mean he is unintelligent? Now ask yourself if you can perceive the difference between a New Guinea canoe ceremonially decorated for a marriage feast and one decorated for war—even though your life may depend on knowing the difference. If you can't, does that mean you are unintelligent? Those who argue that American blacks should be able to perform the same as whites on IQ tests because they are familiar with our cultural conventions are ignoring the fact that blacks have been kept out of the mainstream of American culture for almost the entire 200-year history of this nation.) It is outside the scope of this chapter to address these aspects of the issue. But it is not outside its scope to point out a major fallacy in the *genetic* basis for the argument.

The point is, it does not necessarily follow that if differences *within* each of two separate groups are heritable, then the differences *between* two groups are also heritable. We can show this with a simple experiment. Let's take a large sample of seed corn from each of thirty different strains of corn. Then let's mix them up until the seeds are randomly distributed—in other words, so there is genetic variation through the entire sample. We can take half the seeds and grow them in a complete, balanced medium that provides all the minerals the corn needs for optimum growth. The remaining seeds can be grown in the same medium, but we will leave one trace mineral out of it. Both batches of corn can be grown under the best possible conditions of light, moisture, and temperature.

When all the plants in such an experiment have matured, sure enough, the plants of the first group are taller on the average than plants of the second group. And sure enough, within both groups, there is substantial variation in height from plant to plant. With appropriate genetic tests, we can show that within each group, differences in height have a genetic basis. (Remember we started out with thirty different strains.) Are we then justified in concluding that the differences in average height *between* the two groups are also genetically determined? Of course not. In this case we know—because of the way we controlled the variables in the experiment—that the differences between the groups are due entirely to an *environmental* difference (the presence or absence of one mineral in the culture medium).

Race and the Channeling of Gene Flow in the Human Population In the evolutionary view, humans (like all other organisms) must be subject to the action of natural selection. But if that is so, why among all others has such a highly dispersed population remained a single species? After all, humans occupy isolated geographic pockets as well as broad stretches of almost all the diverse environments on earth! If not only geographic barriers but social barriers exist, why didn't the human species long ago split into separate species?

In fact, variations in the human species *have* existed for hundreds of thousands of years as adaptive responses to different environments. Phenotypic variations in such traits as skin color and hair color are testimony that natural selection has been at work at the genotypic level. Such variations suggest great potential in adaptive responses. Many of our traits were tested out in the forerunners of our species—in creatures that lived millions of years ago in the sun-drenched tropics (Chapter Eighteen). Consider the profuse sweat glands now characteristic of the human species. Surely they developed as a response to a need for cooling off the body of a creature that ran about upright during the heat of the day. Or consider skin color. Because cancer **can** develop in skin that is exposed for prolonged periods to intense sun, we can guess that natural selection long ago favored a gene coding for melanin, a dark pigment that screens out ultraviolet radiation. But even though too much sunlight has harmful effects, some sunlight can be used to advantage. Its ultraviolet rays are capable of reacting with lipids in the skin to produce D vitamin precursors, which the body needs for normal functioning. When some humans began moving out of the tropics, they encountered a new kind of selective pressure. In northern lands, where the amount of sunlight was relatively diminished, there would have been reduced selective pressure in terms of maintaining high melanin levels in the skin. And perhaps the need for vitamin D synthesis was enough to select *against* abundant melanin production in those environments.

At first this supposition might seem not to explain how dark-skinned populations have existed for many thousands of years in regions of low light intensity. Why, for instance, don't Eskimos have bleached-out skin in order to absorb all the sunlight they can get? It happens the Eskimo diet is based largely on fish—and fish fats are rich in vitamin D. It may be these people haven't yet experienced any selective advantage to changing from the ancestral melanic heritage.

In the adaptive sense, "black" is beautiful, and so are "yellow," "brown," "white," and "red." In the social sense, the beauty of our phenotypic variation doesn't come up very often in the conversation. People who might marvel over the variations in parakeet plumage become distressed when a human bird of a different color moves into the neighborhood. They come to attribute major differences to superficial traits. The consequences—disparagement, the self-fulfilling prophecy of economic discrimination, even war—are a commentary on our social evolution as a species.

The outcome has indeed been a partial isolation of some segments of the human population from others into what is perceived as "races." But notice the word "partial." It's true that if you were to be dropped into the middle of China, then plucked away and dropped successively into Norway, New Guinea, Arabia, and Indiana, you would certainly notice that people don't all look alike. But if you could walk step by step through every country in the world, you would probably be more aware of the gradual overlapping of phenotypic traits as one local population gives way to the next—in other words, *with the way subtle diversity is gradually laid down on the* continuum *of basic similarities as environments change.* Even then, in the midst of all this subtle gradation, you are sure to encounter individuals who "look different" who are in fact simply expressing part of the potential variability of the gene pool for the very same region! And even though there may be mountain barriers and oceans and social sanctions to bar the way, intermarriages and encounters between members of different local populations take place all the time, bridging genetic variability between isolated regions and keeping the human species what it is today: *one species,* with one gene pool of rich potential diversity.

Perspective

In this chapter, we have covered patterns of inheritance among sexually reproducing organisms. We began with Mendel's probing into the ways that units of heredity must be assorted into sex cells (gametes). We looked at a nuclear division process called meiosis, whereby the diploid number of chromosomes in the nucleus is reduced by half (to a haploid number) prior to gamete formation. For each trait,

then, a new individual formed by fusion of two gametes receives two alleles (alternative forms of a gene), one from each parent gamete. We have seen how this process assures constancy in the number of chromosomes inherited by each new generation of offspring—even as it provides opportunities for introducing novel combinations of traits. Some combinations may prove harmful. Others may be typical of the species adaptation to a given environment. Still others may somehow make their bearers better equipped to respond to prevailing conditions or to changing ones. Thus meiosis is a major source of genetic variability—and variability in the environment is the testing ground for natural selection, the process by which evolutionary change is thought to occur.

Of course, any adaptive advantage is lost if an individual doesn't reproduce—if its genetic potential does not become represented in the next generation. So it is not the individual but the *population* to which the individual belongs that is the unit for measuring constancy and change. It is not only that alleles are shuffled and reshuffled during meiosis. Gene mutations also occur and introduce new variability. The size of a population affects allele frequency, as do geographic and behavioral barriers to or channels for the movements of individuals. Not all individuals have an equal chance of surviving and reproducing, so the frequency of some alleles again may increase or decrease. And on the horizon are variations that will accrue from deliberate gene modifications. In all such ways, patterns of inheritance may vary—sometimes more, sometimes less, depending on the species and where it is in time and space.

Recommended Readings

Cavalli-Sforza, L. 1974. "The Genetics of Human Populations." *Scientific American,* vol. 231, pp. 80–89. Very good article in an issue devoted to human population genetics.

Lerner, I. and W. Libby. 1976. *Heredity, Evolution, and Society.* Second edition. San Francisco: Freeman. Excellent, intelligent introduction to genetics principles as they relate to evolution and society.

McKusick, V. 1975. *Mendelian Inheritance in Man.* Fourth edition. Baltimore: Johns Hopkins Press. Catalogs inherited human traits, and gives references under each entry.

Winchester, A. 1975. *Human Genetics.* Second edition. Columbus, Ohio: Merrill. Paperback on the important aspects of human genetics, written to be understood by the average reader.

Chapter Eleven

Figure 11.1 Birch leaves unfolding beneath the spring sun.

From Cells to Organ Systems in Plants

At the margin of a pond, in the standing water of a drainage ditch or alongside a curb, at the edge of a mountain stream or a lowland lake—in all these places you are likely to find cells of bright green algae clumped together. For all these places have the same things in common: sunlight, open air, a few dissolved minerals, and plenty of water to bathe each cell in the colony. Like simple aquatic algae, all complex multicellular plants on land also must secure these basic resources to stay alive, grow, and reproduce. But the conditions under which they must do so are far different! For one thing, abundant water is not always available to all land plants all of the time. For another, air is an entirely different medium that places different constraints on the directions of plant growth. More than this, complex land plants have not evolved in a biotic vacuum to become what they are. Not long after the first species of one-celled plants began to radiate into terrestrial zones, hungry heterotrophs began creeping out of the water after them. Land plants have been locked into competitive and cooperative relationships with heterotrophs ever since. And these relationships have had profound effect on the structure and function of both kinds of organisms (Chapters Seventeen and Twenty-Two).

Before we get into those relationships, however, it will be useful to become acquainted with plant parts and processes. In this chapter, we will focus on the tissues and organs of multicellular plants that are adapted to variable conditions of life on land. Examples will be selected from "seed plants," for they include the flowers, grasses, trees, and shrubs with which you are most familiar. This focus means we must also consider plant hormones, which have developed concurrently as a means of integrating and controlling the growth and development of multicellular parts. To do this, we will follow events by which a single seed makes the transition to a complex land plant, for the unfolding of its tissues and organ systems is a process that goes on throughout its life cycle. For instance, most of the tissues of a large tree are fully mature and are functioning in an integrated way to maintain life processes. Yet that same tree has zones in which embryological events—mitotic divisions leading to tissue formation, to the formation of new shoots, roots, leaves, and reproductive structures—are still proceeding!

For now, we will be setting aside our survey of the reproductive structures and functions that are part of this developmental process. It is largely through strategies of reproduction and dispersal of offspring that "immobile" plants were able to move onto and across the land many hundreds of millions of years ago, and that part of the evolutionary story is reserved for Chapter Seventeen.

The Seed: From Dormancy to the Beginnings of Active Growth

When you cut open a seed from any flowering plant—say, a bean seed or a kernel of corn—you find a similar array of structures in all of them. Inside is an *embryo,* an early developmental stage of the new individual (Figure 11.2). The embryo is already well developed, for it followed a limited program of growth and differentiation even while the seed was still attached to the parent plant. At one end, the embryo has already developed either one or two *cotyledons* (seed leaves). If the embryo has only one, the seed is a **monocot** (short for "monocotyledon"); if it has two, the seed is a **dicot** (short for "dicotyledon"). These seed leaves may contain food reserves for the embryo. They are often destined to become the first light-trapping structures for the young plant when it begins active growth. Where the cotyledons attach to the axis of the embryo, there is a mass of cells that will become the growing tip of the stem or shoot. At the opposite end of the embryo is another mass of cells that will become the growing tips of young roots. The rest of the seed's interior is filled with a food storage tissue called the *endosperm.* Finally, surrounding all these structures is the *seed coat,* a protective layer controlling moisture loss from the interior and offering the contents some protection against injury. Once the seed coat is completely formed, the mitotic divisions that brought the embryo to this stage of development cease; the seed enters a state of **dormancy,** a period of suspended activity preceding its germination.

How long can a plant embryo remain locked within the seed coat and still retain its ability to give rise to a mature plant? The length of time varies from species to species, for plant development is fine-tuned to seasonal fluctuations and to the prevailing conditions in different environments. Some delicate seeds will survive only for a few weeks or months, yet the seeds of some common weeds can endure for decades. In one instance, two long-quiescent seeds of a

Figure 11.2 Generalized seed structure of a monocot (**a**) and of a dicot (**b,c**), shown sliced lengthwise.

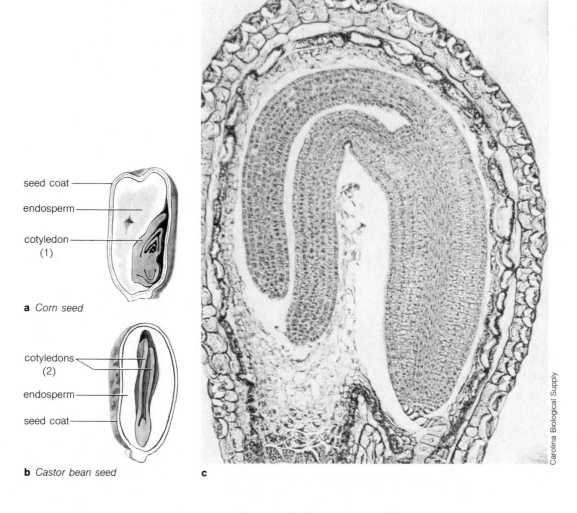

seed coat
endosperm
cotyledon (1)

a *Corn seed*

cotyledons (2)
endosperm
seed coat

b *Castor bean seed*

c

Carolina Biological Supply

pink lotus managed to sprout after 2,000 years. The seeds were found in an ancient boat buried deep below the surface of a Japanese peat bog. Lotus seeds characteristically have a thick seed coat and a large amount of endosperm, which undoubtedly had something to do with their survival—as did the cool anaerobic conditions of their burial. But perhaps the most tenacious seeds of all were found in the frozen silt of a lemming burrow in the Yukon Territory. Ten thousand years ago these seeds became frozen into the soil after glacial sheets descended from the north—and ten thousand years later they sprouted into arctic lupine flowers!

Such cases, however, are rare. Environmental cues normally merge on time, with the cycling of seasons, to shift an embryo from the dormant state to germination. Temperatures change, days lengthen, more moisture becomes available, as do certain nutrients. Consider the seeds of such

familiar plants as cherry and pine trees. For them, one of the most important cues is the cold weather of the winter months. In autumn these seeds are dormant, and it doesn't matter when they fall to the ground. They simply will not grow even if the season brings heavy rains or unusually warm days. They first must experience temperatures only slightly above the freezing point; only then will built-in controls go to work, slowly bringing about internal changes that will ensure germination with the onset of the warm spring rains. As another example, the seeds of many short-lived desert plants have built-in controls that synchronize germination with the brief rainy season characteristic of desert zones. Such seeds contain a water-soluble substance that inhibits germination. For instance, four-winged saltbush seeds contain sodium chloride. A spattering of brief showers is not enough to wash away the inhibitor from the immediate vicinity of the seed. But a heavy thunderstorm—one that drops enough water to sustain plant growth until maturity—will wash it deep into the soil, and the plant will begin rapid growth.

In all such ways, control mechanisms adapt various seed plants to unique conditions in different environments. Interestingly, our vegetable crops are plants that rarely show dormancy. Farmers have selected against dormancy, sowing the seeds and harvesting crops in season. But nearly all native seed plants depend on some internal control that governs when embryonic roots will break through and begin moving down into the soil, and when the shoot will begin its upward surge.

Compressed in the Seed: The Promise of Cell and Tissue Specialization

Once a seed germinates, its cells embark on their course of divisions and development into specialized structures of the mature plant form. The tissues composing these structures are of two general types: meristematic and permanent. In **meristematic tissues,** individual cells retain their ability to divide. These cells remain relatively small and unspecialized. They form *apical meristems,* the delicate, domelike growing tips of stems and roots. They also form *vascular cambium,* a type of lateral meristem that increases stem or root diameter because of the circular arrangement of its cells and the direction in which they divide. *Cork cambium* is another type of actively dividing lateral meristem; it is found in the outer bark of trees and shrubs.

In contrast to meristematic tissues, **permanent tissues** are composed of cells that are no longer dividing; instead they have differentiated according to the patterns for division of labor that are characteristic of plants. For instance, surface cells of leaves become waterproofed, and act as a protective barrier against insects and fungi. Surface cells of roots remain permeable to water, which permits them to absorb water and minerals from the surrounding soil. Other cells become modified in ways that help strengthen the tissue in which they are found. Still others undergo modifications that enable them to function in the conduction of water and nutrients.

For the complex land plant, protection, support, and transport are essential aspects of the acquisition and efficient use of available sunlight energy and resources.

These three requirements are met in large part with three basic systems: surface, ground, and vascular tissue systems. As you will see, some are simple tissues, composed of cells having quite similar structure and function. But some are complex, in that many specialized cell types are required in carrying out a common task.

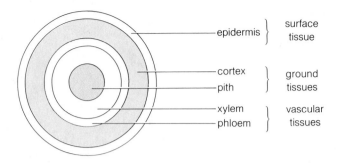

Figure 11.3 General arrangement of main tissues as they might occur in cross-section through the stem of many vascular plants.

Surface Tissue The outermost tissue on young stems and roots as well as leaves is called the **epidermis.** It is the plant's interface with the outside environment and, as you might expect, its cells are subject to considerable specialization. Usually the epidermis is only one cell layer in thickness. Above ground, the outer walls of most of these cells are typically impregnated with cutin, a waxy substance that retards water loss, and protects the plant against injury and disease-causing organisms. Various specialized cells embedded in this barrier layer act as regulatory passageways; they are essential in the absorption of carbon dioxide, which is necessary for photosynthesis to proceed. Below ground, some epidermal cells send out long extensions, which enhance absorption of the water necessary for photosynthesis.

Ground Tissue There are three main kinds of ground tissue: parenchyma, collenchyma, and sclerenchyma. As Figure 11.3 shows, ground tissues lie between the epidermis and vascular tissue and are called the *cortex.* Sometimes they also are found in the center of vascular tissue, in which case they are called the *pith.*

Characteristically, the cells of **parenchyma** remain alive at maturity. These cells have thin walls of cellulose. They are found throughout the plant body, they vary in size and shape, and some have chloroplasts but others do not. The parenchyma cells in leaves or in growing tips of a shoot are photosynthetic. With their thin walls, they are well adapted for capturing light energy and for absorbing carbon dioxide from the environment. Parenchyma cells deeper inside the plant body usually lack chloroplasts, but they may contain other kinds of plastids, including other chromoplasts and leucoplasts. Thus parenchyma serves a storage function by holding reserves of water or photosynthetically derived food. The thin walls of its cells permit

Figure 11.4 Variations in wall thickness among three basic kinds of plant cells: (**a**) parenchyma, (**b**) collenchyma, and (**c**) sclerenchyma. The upper sketches are cross-sections through tissue made up of these cells. The lower sketches are longitudinal sections (except in the case of the many-sided parenchyma cells).

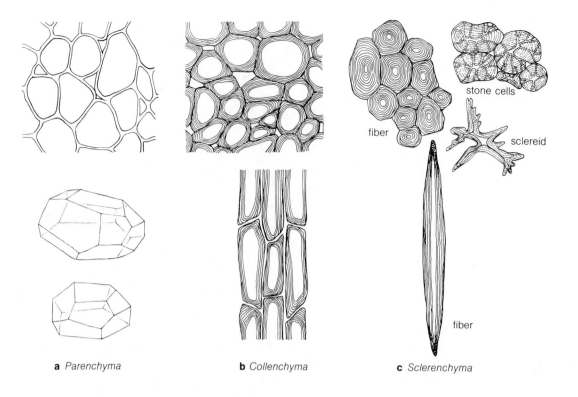

a *Parenchyma*   **b** *Collenchyma*   **c** *Sclerenchyma*

rapid transfer of these resources among cells, depending on the needs of the plant (Figure 11.4).

The cells of **collenchyma** have walls that are uneven in thickness (usually the corners are built up). Because it is somewhat more rigid than a parenchyma cell, a collenchyma cell can provide support in young stems and leaves, and in nonwoody stems. **Sclerenchyma** cells have even thicker walls. At maturity, they are dead and their contents have disappeared. All that remains is the relatively rigid cell wall, which supports and stiffens parts of the plant.

Vascular Tissue   There are two kinds of vascular tissues in plants: xylem and phloem. In **xylem,** water and dissolved minerals taken in from the surrounding soil are transported upward. In **phloem,** the products of photosynthesis are selectively distributed from the green leaves to the main stem, then up to the growing tips and down to the roots.

Both xylem and phloem are conducting systems, but they are different in composition and function. Xylem is composed mainly of three specialized cell types: fibers, tracheids, and vessel elements. *Fibers* are a type of sclerenchyma (Figure 11.4). They function primarily in support; they have no direct role in water transport. *Tracheids* and *vessel elements* are water conductors (Figure 11.5). Tracheids are long cells having thickened walls impregnated with lignin, a substance that helps strengthen them. The cell

walls are studded with porelike structures (pits). Through these pits, water flows from the empty cell cavity of one tracheid to the next in its journey up the stem. Thus the tracheid cells form a passive conducting system; the water simply flows through them. In many cone-bearing plants (such as pine trees), the xylem is composed of fibers and tracheids. In woody flowering plants (such as maple trees), the xylem also contains vessel elements, which are more advanced water conductors. Long, tubelike structures open at their ends, joining cells end to end like the pipes of a plumbing system. Although vessel elements also have pits in their walls, the main flow of water is directly through the cavities of adjacent cells, with comparatively little restriction on water flow.

In contrast to xylem, phloem is composed of living cells that actively expend energy to move photosynthetically derived nutrients through the plant. A phloem tissue system usually contains elongated cells called *sieve-tube elements,* which are lined up end to end. The walls at the ends of these cells are perforated, which allows the nutrient-laden contents from the vacuole of one cell to be passed to the next in line. Strangely enough, even though these cells are alive and are actively involved in moving nutrients through the plant, they don't have a nucleus. Instead they are associated with *companion cells:* nucleated cells that apparently function to provide control over the cytoplasm in adjacent sieve-tube elements (Figure 11.6).

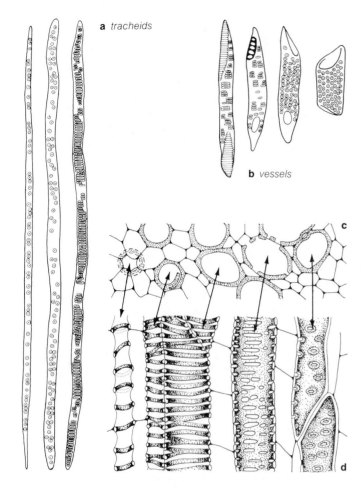

Figure 11.5 Tracheids and vessel elements shown isolated from plant tissue (**a,b**); in cross-section (**c**); and in longitudinal section (**d**).

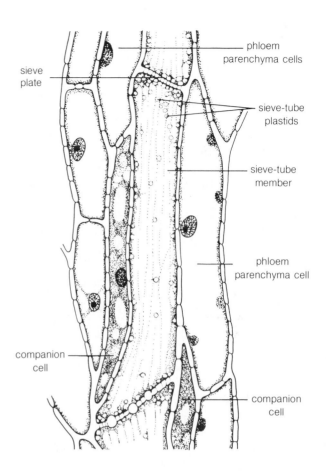

Figure 11.6 Longitudinal view of a mature sieve-tube member and a companion cell.

How phloem actually transports nutrients is something of a mystery. Phloem cells clearly make an active expenditure of energy to accomplish such transport. In a phloem tissue system, it is not uncommon to observe different substances moving in different directions—at different rates! If the metabolic activity of phloem cells is disturbed, they simply stop transporting materials. Aphids (small insects that feed on plant fluids) have been used to demonstrate that materials are transported through phloem at rates far too rapid to be attributed simply to passive diffusion. Aphids come equipped with a sharp, hollow tube (stylus) at the mouth end. This they can sink directly into phloem cells without causing any noticeable damage (Figure 11.7). Plant fluid flows into the stylus and then to the stomach, which bloats up with the incoming juices. Because the stylus is simply a tube and not a sucking device, we can conclude the contents of phloem must be under pressure. In fact, even

Figure 11.7 Well-fed aphid, exuding a honeydew droplet. The insect inserts its stylus into a sieve-tube element. The aphid does not suck out the juices; fluid pressure inside the plant does the work for it.

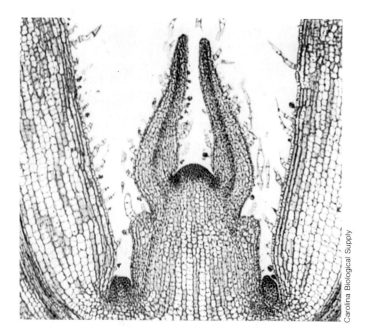

Figure 11.8 Apical meristematic tissue in a typical flowering plant, as seen sliced lengthwise.

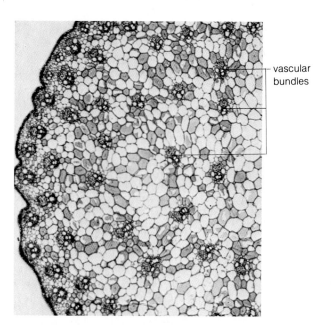

vascular bundles

Figure 11.9 Pattern of vascular bundles in a monocot stem. The bundles are scattered through the stem. Each contains xylem and phloem, usually supported by a sheath of sclerenchyma cells.

when the stylus is left in place and the rest of the insect is detached from it, fluid continues to pour out at the rate of 1 to 2 cubic millimeters per hour! Although not much may be said with certainty about the underlying transport mechanisms, complex controls obviously are at work inside vascular tissues to keep fluids moving from their place of origin to places where they will be used.

Primary Growth in Stems and Roots

Surprisingly, perhaps, differentiation of plant cells into surface, ground, and vascular tissues occurs in a region just a few millimeters (about an eighth of an inch) behind the advancing apical meristem. The term **primary growth** refers to the lengthening of stems (and roots) and the transition of the newly formed cells left behind into a complete array of tissues. The mechanism by which cells grow longer during the primary growth phase is not completely understood, although it is known to be under hormonal control. Somehow cell walls are weakened, which permits cells to enlarge as fluid pressure builds up inside. Once cells making up a stem region reach a certain size, elongation in that region comes to a halt, cell walls are reinforced, and further growth is inhibited. Regardless of whether the plant is a milkweed or a giant redwood, its increase in height or in length will

occur only immediately behind the tips of its stems and branches.

Once elongated, cells behind the growing tips undergo changes that will result in the division of labor needed to sustain the plant body. Surface cells acquire thicker walls and a waxy coat on their outward-facing side. Clusters of strandlike tissue (procambium) form behind the zone of elongation. Farther down the stem, these strands differentiate into **vascular bundles:** clusters of xylem and phloem (Figures 11.9 and 11.10). Between these bundles, parenchyma tissue forms and differentiates into various ground tissues. How many vascular bundles will actually form in the stem and what their arrangement will be depends on the type of plant examined. Cross-sections of various stems show monocot bundles are scattered through the tissues, and dicot bundles are arranged in a ringlike pattern.

Roots grow longer in much the same way that stems do. Cells behind the apical meristem lengthen, which increases root length. A special tissue forms in front of the advancing apical meristem. This *root cap* protects the tip as cell elongation drives the root deeper or farther through the soil. Immediately behind the zone of elongation, surface cells send out long extensions called *root hairs*, which give good soil contact and thus are most efficient at absorbing water and nutrients. Figure 11.11 shows the general structure of roots and root hairs.

Generally, root systems are of two kinds. In plants with

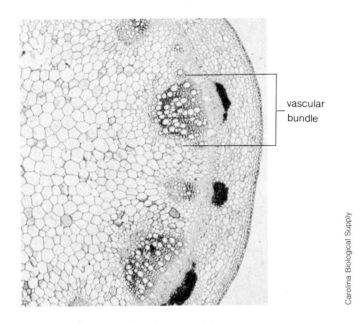

vascular bundle

Carolina Biological Supply

Figure 11.10 Pattern of vascular bundles in a dicot stem. Dicot stems are organized into three distinct regions: cortex, vascular bundles (concentric rings made of phloem, cambium, and xylem), and a central pith with its parenchyma storage cells.

a **tap root system,** there is a single main root from which smaller branch roots arise. A "carrot," for instance, is the main tap root of the carrot plant. In plants with a **fibrous root system,** many roots of about the same size originate near the soil surface. Lawn grasses have fibrous systems. If you were to pull a grass plant out of the ground, it would be easy to underestimate the size of this system. A mature ryegrass plant may have almost 14 million roots, with a total surface area exceeding 230 square meters (2,500 square feet). The system would also contain 14 *billion* root hairs—a total surface area of more than 400 square meters!

Roots grow fairly rapidly. In corn plants, they probe through tiny spaces between moist soil particles at an average rate of 60 millimeters (about $2\frac{1}{2}$ inches) each day. As they do, they encounter the thin film of water surrounding each particle. Because fluid inside roots and root hairs normally contains more salts and minerals than water on the outside, a concentration gradient exists. Thus, through osmosis, water diffuses passively into the root.

When poor drainage causes salts to build up in the soil, the concentration gradient in the root can be affected so much that plants can die of thirst even in the presence of water. That's what happens when you don't water house-plants deeply enough to dilute the salts in the soil. An occasional heavy watering does a lot more good than frequent light waterings. (At the same time, most plants can't survive in constantly saturated soil. Oxygen, recall, is

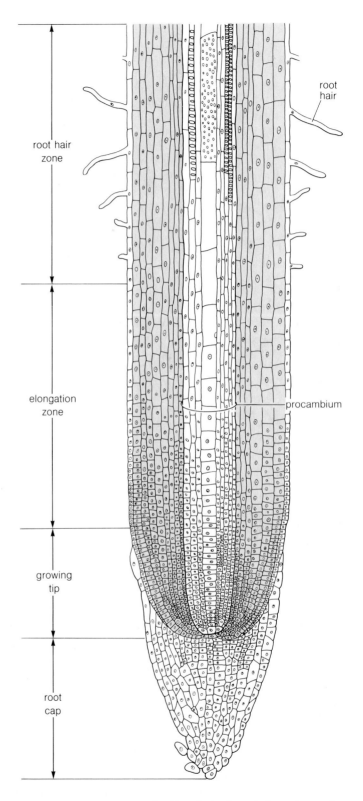

root hair zone

root hair

elongation zone

procambium

growing tip

root cap

Figure 11.11 A barley root, sliced lengthwise through the root tip and the region of root hairs. (After R. Holman and W. Robbins)

needed as an electron acceptor for cellular respiration. If the soil never has a chance to become aerated, it will contain too little oxygen to support respiration in the root cells.)

Once water reaches the inner vascular cylinder of a root, it must begin its ascent to the plant's uppermost regions. But what could possibly move water to the top of, say, a redwood tree towering 130 meters above the forest floor? It is true that the root pressure created by transport of water and minerals into the xylem can move water upward for a *few* meters. But from that point on, transport mechanisms in individual cells simply can't be invoked to account for the rest of the movement; xylem, after all, is made largely of dead cells. Perhaps we must look not to the cells of the transporting tissues but to the substance being transported.

Recall, from Chapter Three, that water shows remarkable properties because of the hydrogen bonds that form between adjacent water molecules. Water molecules inside the xylem network are hydrogen-bonded into a thin "thread" stretching in an unbroken line from roots to the uppermost stem tips. The growing stem tip cells contain salts and sugars, which means they are able to develop an osmotic concentration and draw water from the xylem threads for use in their own enlargement. This happens even when water in the xylem is under considerable tension. Later, when some of these cells mature into leaves, more water is pulled up as water is used in photosynthesis itself. Much more water is pulled up to replace that lost in **transpiration,** the inevitable evaporation of water from moist mesophyll cells when stomates are open. Even as cell growth, or photosynthesis, or transpiration pulls molecules at the top of the "threads," hydrogen bonds pull up new molecules to take their place. The idea that water is not pushed but pulled up through a plant is known as the **cohesion theory of water transport.** Experiments with small, fluid-filled tubes show the cohesive forces within vascular systems are about eight or nine times greater than is needed to carry water to the top of the tallest trees!

Secondary Growth in Roots and Stems: Increased Support for Large Plant Bodies

Obviously, a plant that depends on sunlight for its energy can survive only if it obtains enough sunlight. Some low-growing plants on the forest floor have developed ways of using what limited light they get with the utmost efficiency. Another common strategy is simply to grow taller than everything else around. This strategy has its advantages in terms of photosynthetic efficiency. But growth in height alone can proceed only so far before the weight of the plant exceeds the capacity of the stem to support it. Besides, growth in height alone doesn't offer much resistance to winds and rains and stomping animals.

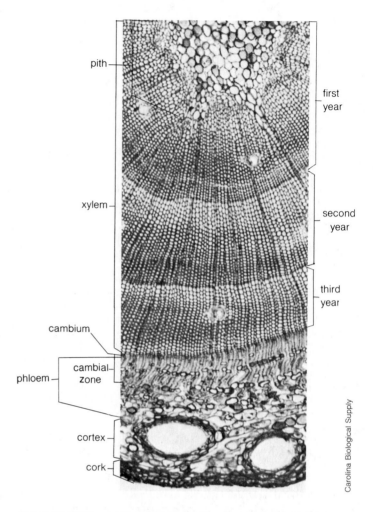

Figure 11.12 Secondary growth as seen in a cross-section of a plant stem.

Thus there have been pressures on some plants for **secondary growth:** increasing stem and root diameters. While secondary growth is occurring in stems of woody plants, essentially identical secondary growth is occurring in the roots, which even form a protective layering of bark.

Secondary growth begins with meristematic tissue left behind as the growing tip advances. In the sunflower, for instance, bits of meristematic tissue remain in each vascular bundle. Following primary tissue formation, these bits can divide mitotically to produce more xylem and phloem, which increases the diameter of the vascular bundles. It is the development of *secondary* xylem and phloem that assures the sunflower of success as it reaches for relatively dizzying heights (relative to, say, a yellow daisy).

Secondary xylem also accounts for the massive trunk of woody trees. In such plants, vascular cambium forms a

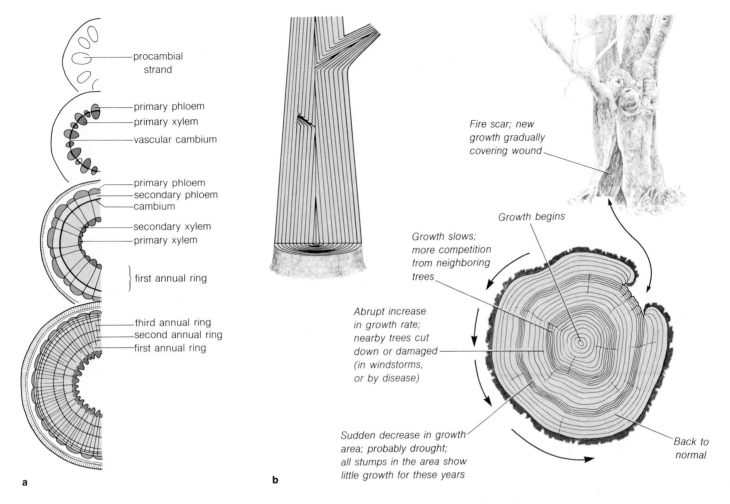

Figure 11.13 (**a**) Formation of annual rings and (**b**) life history of a tree, as suggested by its pattern of annual rings.

In the figure, labels include:

- procambial strand
- primary phloem
- primary xylem
- vascular cambium
- primary phloem
- secondary phloem
- cambium
- secondary xylem
- primary xylem
- first annual ring
- third annual ring
- second annual ring
- first annual ring

- Fire scar; new growth gradually covering wound
- Growth begins
- Growth slows; more competition from neighboring trees
- Abrupt increase in growth rate; nearby trees cut down or damaged (in windstorms, or by disease)
- Sudden decrease in growth area; probably drought; all stumps in the area show little growth for these years
- Back to normal

a   b

continuous ring, linking all the individual vascular bundles of primary tissue (Figure 11.12). Almost immediately after the primary tissues differentiate, cells of the cambium begin to divide. Those on the inside of the ring develop into secondary xylem; those on the outside develop into secondary phloem. On the outer face of the ground tissue, cork cambium gives rise to cells making up outer bark layers.

In regions having prolonged dry spells or cool winters, vascular cambium becomes inactive during parts of the year. When favorable growing conditions return, growth resumes. The cells of the xylem tissue initially produced tend to have fairly large diameters; they form spring wood. As the season progresses, the diameters become smaller and smaller; these cells form summer wood. Thus xylem comes to form alternating bands of tissue called **annual rings** (Figure 11.13). In wet tropical regions, there is a continuous growing season, and annual rings of tropical woody plants show it. They either are faint or nonexistent. The more pronounced the shifts in seasons, the more pronounced the annual rings.

As the mass of xylem increases, it tends to crush the thin-walled phloem cells from the preceding years' growth. That is why new rings of phloem cells must be produced each year, outside the growing inner core of xylem. As you can see from Figure 11.12, phloem is confined to a thin zone in the inner bark. If even a narrow band of bark is stripped all the way around the circumference of a tree, the tree will die. For the phloem cells will be stripped away and will not be able to transport photosynthetic products down to the roots, which will starve to death. (That is why many orchard owners wrap the trunks of young trees as protection against rabbits and other gnawing animals.)

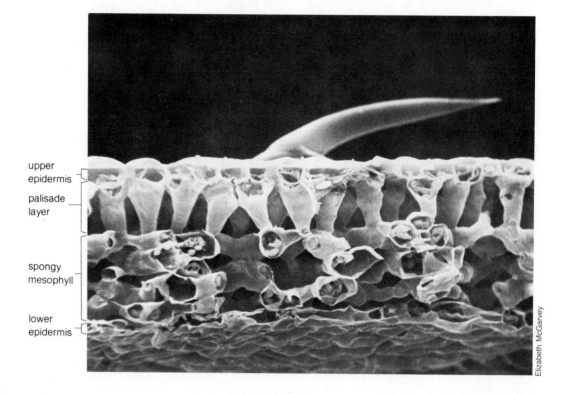

Figure 11.14 Architecture of a leaf from a privet plant.

upper epidermis

palisade layer

spongy mesophyll

lower epidermis

The Unfolding of Leaves: Gatekeepers for the Exchange of Materials and Energy

Each leaf of each flowering plant is an individual organ—an integrated, functional unit of cells and tissues working together in the performance of a common task. Each is a site for photosynthesis. With its large external surface area—and even larger internal surface areas—a leaf provides exceptional contact with the surrounding air, thereby ensuring rapid carbon dioxide intake and maximum interception of sunlight energy.

Leaves are formed from *leaf primordia,* lobelike tissue masses growing at intervals along a stem, behind the advancing apical meristem. The lobes expand into a stalked blade having highly branched vascular systems (veins), which connect with the main vascular bundles in the stem. Expansion in one direction leads to long, narrow leaves such as those of pine trees; expansion in two directions leads to broad leaf structures.

The structure of leaves gives us insight into their function. Figure 11.14 shows the tissue layers common to flowering plant leaves. Uppermost is the protective epidermis, the cutin-coated layer of surface cells. Next comes the *palisade layer,* a tissue of loosely packed parenchyma cells stuffed with chloroplasts; much of photosynthesis takes place here. Below that is the *spongy mesophyll.* Between 30 and 50 percent of an average leaf consists of air space that surrounds not only the spongy mesophyll but 80 percent of each palisade cell wall. Below is another cutin-coated epidermal layer. Even though cutin is a waterproofing agent, the leaf is not sealed off entirely. Carbon dioxide enters through many small passageways called **stomates.** These passageways usually are found on the lower surface layer of leaf cells.

Stomates open when photosynthesis in the mesophyll cells depletes the carbon dioxide present in intercellular air spaces. Thus carbon dioxide enters the leaf and is absorbed by mesophyll cells. At the same time, these cells release oxygen as a by-product of photosynthesis. These gases diffuse across the walls and membranes of mesophyll cells after first dissolving in the thin film of water adhering to each cell's exterior. Stomates close at night, and also during the day if the leaf starts to wilt in hot, dry weather. This drought closure overrides photosynthesis but, as you read in Chapter Five, it is essential for survival.

(You might think that stomate closure in hot weather would cause heat to build up to intolerable levels inside the plant. After all, water molecules converted to vapor during transpiration can leave the plant and carry away much of the

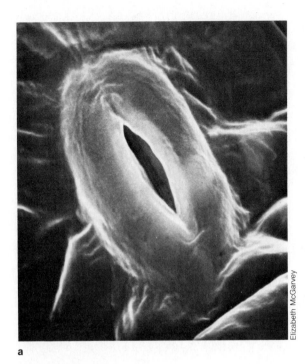

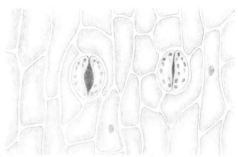

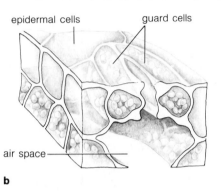

epidermal cells   guard cells

air space

Elizabeth McGarvey

a   b

Figure 11.15 Stomates. (**a**) The scanning electron micrograph reveals the shape of guard cells. (**b**) The sketch shows a stomate open (when fluid pressure is high) and closed (when fluid pressure drops).

absorbed heat. But transpiration is a consequence of leaf anatomy that functions primarily for photosynthesis, not for leaf cooling. Many desert plants such as cacti and creosote bushes keep stomates closed during the heat of the day, and may even become hotter than the surrounding air. They tolerate high temperatures without injury. During drought, even crop plants close stomates and become warmer than the surrounding air. During times when cooling is most needed, it usually fails for *lack* of water. Only oasis plants such as watermelons and palms, which have a reliable groundwater supply, seem able to afford the luxury of evaporative leaf cooling.)

As Figure 11.15 shows, stomates open and close through changes in the shape of their two guard cells. When internal fluid pressure is high, variations in cell wall thickness cause the paired guard cells to swell up like a doughnut, and gases may move through the gap between them. When their internal fluid pressure drops, guard cells become limp and collapse against each other, and the gap disappears.

What changes the fluid pressure inside guard cells? Photosynthetic depletion of carbon dioxide in the intercellular air spaces adjacent to guard cells initiates a series of events that increases the salt content (mainly potassium salts) in the guard cells. The higher salt content has the effect of lowering the concentration of water in the guard cells, which initiates an inward flow of water from surrounding cells. The increased fluid pressure in the guard

cells causes the stomates to open. At night, photosynthesis stops but cellular respiration can continue (Chapter Five). And cellular respiration increases the carbon dioxide concentration. This has the effect of lowering salt concentrations and increasing the water concentration in the guard cells. Hence there is an outward flow of water, and the decreased fluid pressure leads to closure.

As long as they are alive, leaves and other plant parts engage in cellular respiration. During the day, photosynthesis is five to ten times as rapid as respiration. But gas exchange continues at night, with leaves absorbing oxygen and releasing carbon dioxide even though stomates are closed. Because oxygen in the air is about 500 times as concentrated as carbon dioxide, sufficient amounts of oxygen can diffuse inward. The open stomates apparently have evolved only for the rapid entry of very dilute carbon dioxide.

In many plant species, leaves are more than sites of photosynthesis. Flowers, for instance, are composed of modified leaves that may function to attract insect or bird pollinators (Chapter Seventeen). Still other leaf modifications are found among insect-eating plants of the sort shown in Figure 11.16. Through natural decay or enzymatic digestion, the trapped insect body becomes a useful nitrogen supplement. Most of these plants grow in bogs and swamps, where little nitrogen is available—so who is to blame a few of them for turning the dinner table on a few heterotrophs?

Figure 11.16 Leaf modifications among insectivorous plants.

(**a**) The sundew plant has its leaves covered with small, hairlike projections, each tipped with sticky nectar. The nectar attracts insects, which promptly become trapped in the sticky goo. As the insect struggles to free itself, its movements trigger a shift in the fluid pressure of specialized cells in the leaf, which cause the dinner table to fold up into a tomb.

(**b**) The pitcher plant has leaves shaped like pitchers, and is usually partly filled with water. Halfway down the leaves, cells secrete insect-attracting nectar. As the insects climb in, they discover it is impossible to climb back out: the lips of the pitcher are lined with downward-pointing hairlike structures that prevent escape. Eventually the insect slips and drowns in the water below.

(**c**) The leaves of the venus flytrap are modified in such a way that they close up on an insect that happens to trip one of the sensitive hairs located on the leaf lobe surfaces. Spines on the leaf margin further restrain the insect.

a

b

c

Plant Hormones Governing Growth and Development

Division of meristems, cell elongation, growth of roots, stems, branches, and leaves—all these events are governed by plant hormones. A **hormone,** recall, is a chemical messenger—a substance produced in one part of an organism that delivers a chemical message affecting the behavior of cells or tissues located in some other part of the organism.

In 1926, Frits Went first identified a plant hormone; he named it **auxin.** Since that time, auxins have been found to influence cell growth and differentiation in many ways. They can inhibit as well as stimulate such events as cell elongation, budding, root initiation, fruit development, and leaf fall. As Figure 11.17 shows, they are associated with **phototropism** (growth of plants in response to light) and with **geotropism** (growth of plants in response to gravity).

Auxins are formed in growing tips of stems. They move downward, interacting with cells in some way that allows them to become concentrated in cells on the stem's

shaded side. There they stimulate enzymes into breaking down some of the carbohydrate bonds in cell walls. The walls weaken, more water and nutrients can move in, and the cell elongates. Then new cell wall material is produced, which strengthens the walls. In this way, cells on the shaded side become longer than cells on the sunlit side—which bends the growing tip toward the light.

As soon as auxin reaches certain concentrations in the tissues, however, it seems to stimulate the production of enzymes specifically designed to destroy auxin! This is a form of feedback inhibition: the accumulation of a substance in excess amounts triggers events that lower its concentration. Feedback regulation of auxin levels is vital to plant function. Without it, auxin concentrations could become high enough to damage cell walls and inhibit growth. Instead, feedback regulation permits the control of shoot growth.

In contrast, the concentration of auxins in roots *inhibits* growth! Auxin diffuses down from the growing stem tip and becomes concentrated on the underside of roots. There, the concentrations inhibit elongation of cells on the root's lower

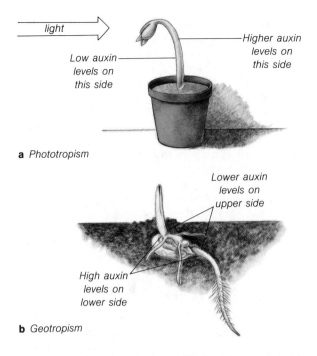

a Phototropism

b Geotropism

Figure 11.17 (**a**) Phototropism, the bending of a young stem toward light—not because of any inborn desire to do so, but because auxin tends to concentrate on shaded sides of stems, thereby causing greater cell elongation on those sides. (**b**) Geotropism, the upward bending of shoots and downward bending of roots, caused by unequal auxin and inhibitor concentrations.

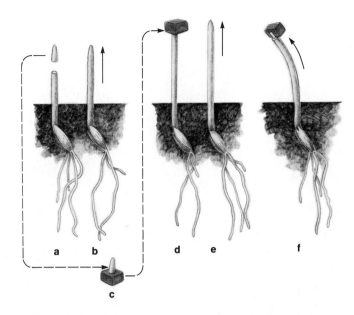

Figure 11.18 Experiment proving that auxin is a hormone that moves down from growing tips to influence elongation of the stem. In (**a**), the growing tip of an oat plant has been removed, and elongation is far less than that seen in an intact plant (**b**). If the detached growing tip is placed on a block of agar (a jellylike substance used as a culture medium), a substance moves down from the tip and into the agar (**c**). How do we know this? When agar is placed on top of the beheaded plant (**d**), the part grows as fast as or faster than its intact counterpart! If the agar block covers only half the cut tip (**f**), cell elongation will be stimulated on that half, causing the plant to bend.

surface. But cells on the upper surface continue elongating, which bends the root down. It is not known why roots show far more sensitivity to auxin than stems do, but the fact that they do helps explain the plant's integrated growth pattern in which roots grow toward water and nutrients, and shoots grow up toward the sun.

Auxins also function to control branching patterns. In many plants, growth is normally restricted to one main stem, so the plant grows more in height than in breadth. Many additional buds are often produced that are able to give rise to side branches, but they do not develop as long as the growing tip remains intact. This inhibition is called **apical dominance.** But if the apical meristem is removed, side buds develop and branching occurs. Gardeners know that if they pinch the growing tips of spindly plants, they can encourage bushier plant growth. Growth of side branches often can be inhibited experimentally if auxin is applied to a stem's cut end after the apical meristem has been pinched off. Apparently the auxin produced in the apical meristem is the natural inhibitor of branching.

The growth of side branches is a complex matter, and often hormones other than auxins are involved. But the diverse growth patterns among plant types is regulated largely through differences in the *rate* of hormone production and in differences in the *sensitivity* of various cells to those hormones. And it is this kind of growth regulation that adapts plants to different environmental situations.

Another category of plant hormones, the **gibberellins,** also promote cell elongation. They were first identified in Japan, when a chemical was isolated from the fungus *Gibberella fujikouri.* The fungus causes rice plants to elongate so rapidly they fall over and die. (The Japanese call it "foolish-seedling disease.") Many gibberellins have now been isolated, and all have amazing effects on plant growth. They have been artificially applied to plants to create radishes as large as pumpkins, and cabbages as tall as a one-story house. Many dwarf plants are deficient in gibberellin, and when they are treated with natural or synthetic gibberellin extracts, they become normal in size. Interestingly, the combined application of auxin and gibberellin intensifies plant growth in ways neither one of them can do alone. This interaction hints at the kinds of combined hormonal effects that must regulate a plant's development during its entire life cycle.

Figure 11.19 Take a pinch of auxin and one radish . . .

John Launois/Black Star

The Windfalls and Pitfalls of Artificially Stimulating Plant Growth and Development

Auxins are marvelous hormones, and they have all sorts of wonderful secondary effects. For instance, when a new and in some way desirable variety arose by chance in the midst of a food or floral crop, a long and tedious breeding program often was necessary to propagate the new strain before it could exist in great enough numbers to justify commercial production. But then someone happened to apply auxins to a branch removed from a plant—and saw that roots formed at the cut end. Just one plant of a new strain could be used to produce multiple, viable cuttings that would grow into exact replicas of the parent plant! With the application of auxin, what might take years of conventional breeding in some plants may be condensed into a few months' time.

The main effect of auxin, however, is to make cells grow longer by partially destroying cell walls in order to accommodate new growth. It happens that ways have been found to prolong auxin activity by inhibiting the enzymes which normally keep it in check. The results are disturbances in leaf growth, sometimes to the extent that leaves fall completely off the plant. This event is called **defoliation** and it can lead to plant death. It is sometimes used as a way of killing weeds. (A "weed" is any plant growing where we don't want it to.) Synthetic auxins are used to get rid of weeds that otherwise would flourish in the midst of cereal crops and seriously compete for sunlight, water, and nutrients. Auxins are used to get rid of certain kinds of weeds in lawns. (Synthetic auxin preparations—2,4D—have a potent effect on broad-leafed flowering plants, which includes most weeds, and relatively little effect on grasses and their relatives, the cereal grains.) In one year synthetic auxins were also used to clear 1,700,000 acres of foliage in Southeast Asia during military operations, thereby exposing whatever enemy camps might have been hidden therein and incidentally disrupting the ecology and soil stability of the entire region for many years.

Cytokinins also are extraordinary hormones that counteract the natural tendency of green vegetables to turn yellow as they age and lose their chloroplasts; thus they are used commercially to prolong storage life. Gibberellins, too, are used commercially. They are applied before harvesting to make grapes large and plump and celery stalks appealingly long (hence marketable). Gibberellins are used to make dwarf plants tall, and chemical inhibitors of gibberellins are used to make tall plants dwarfed. Ethylene is used to stimulate ripening in some crops, but other crops are shipped to market in tight plastic coverings to inhibit their natural production of ethylene. One synthetic hormone is used to rid bluegrass lawns of dichondra and another is used to rid dichondra lawns of bluegrass. Hormones are brushed directly onto the flowers of tomato, pepper, and cucumber plants, thereby assuring an increas-

Cytokinins are plant hormones that promote cell division. The first one discovered was called kinetin; it was not extracted from a plant at all but from fish sperm DNA. Although kinetin itself has never been isolated from plant material, many related substances with cytokinin activity have now been isolated from various plants, including one from the endosperm of corn seeds, several from the liquid endosperm (or milk) of the coconut seed, and one from pea seeds. In tissue cultures, at least, high cytokinin levels combined with low auxin levels promote bud growth—but low cytokinin levels combined with high auxin levels promote root growth.

Another plant hormone that promotes cell division (among other things) is **ethylene.** This hydrocarbon gas is produced in meristems. Without it, meristematic tissue cells would not be able to divide continually. Auxin seems to be necessary to promote ethylene synthesis and to make plant tissues more sensitive to its presence. In fact, many hormonal effects once attributed to auxins may be due to the action of ethylene. Acting in concert with other hormones, ethylene stimulates the ripening process in fruits. It also appears to inhibit the growth of old leaves, and makes them drop in response to seasonal cues. These hormones cause an **abscission layer** (a corky cell zone) to form where a leaf stalk joins the stem. The abscission layer interrupts water flow into the leaf, thereby acting as its death warrant.

ingly selective consumer of seedless tomatoes, peppers, and cucumbers. Trees are made to bear fruit faster and to postpone flowering until after the danger of frost. Fruits are made to look better by enhancing their color, to ripen faster, to slow down ripening, to keep from dropping before harvest time, to improve their storage qualities. All such applications of hormones, of hormone-releasing substances, and of hormone-inhibiting substances are designed to improve crop yields and marketability. They incidentally help stabilize prices for supermarkets by allowing fruits and vegetables to be stored past the peak growing season, a time when the abundant supplies otherwise would tend to drive prices down. They have become part of our agriculture, our eating habits, and our economy. Growth regulation of plants has come of age.

Perspective

Plant form and function are inseparable parts of plant growth and development. Even as the seed of a land plant is being formed, events inside it bring the enclosed embryo to the threshold of its birth as a new individual. In this developmental stage, the embryo waits; it develops no further until external events—the lengthening of days, the warming of the earth, and above all, the rains—converge to signal the arrival of its time to resume growth. With that signal, the masses of cells clustered at both ends of the embryo suddenly begin new growth. Even then, hormones in the young embryo are prodding away at it, causing roots to bend downward and shoots to break through the soil to the sunlit air above. This is the time when the first leaves unfold and begin capturing sunlight energy to replace the endosperm store of nutrients that was depleted during dormancy. The meristematic cells go on dividing and dividing, leaving in their wake a trail of cells that cease division and begin to differentiate into what will become permanent tissues. They give rise to vascular tissues—xylem and phloem—which will transport life-giving molecules through all parts of the plant. They give rise to new leaves, thus meeting the greater demands of the enlarging plant for nourishment. They give rise to all tissues, all structures, of the mature plant body. Underground, roots and root hairs weave past soil particles. Through diffusion and then through active transport, cells in the root tissues move water and essential minerals into the main stem, where water's cohesive properties carry them to leaves.

In all this activity, chemical messengers called hormones integrate tasks and bring parts into balance for the functioning of the whole. Moving through stems, branches, and roots, hormones trigger responses to the state of the environment, and they communicate the activities of other parts of each living cell. In this manner the plant is guided through an unbroken cycle of growth and rebirth, punc-

tuated only by the period of dormancy for the new generation—the time of the seed.

Throughout human history, the seed has been a symbol for the promise of renewed life. And rightly so. For encased in a seed is the potential of the plant upon which our own lives ultimately depend. Our ancestors learned this more than 11,000 years ago. In leaving behind the uncertainties and rigors of a life-style that had been based on the random gathering of wild plants and on the hunt, they began a move toward total dependence on the harvest. It was a move of the first magnitude. In stabilizing the food source, the foundation for explosive population growth and for its consequences of overcrowding and starvation was laid—all based on our belief in the promise of the seed.

The promise remains. Each season brings a new crop to be nurtured and manipulated in ways that will produce increasing yields to match our ever increasing demands. This chapter highlighted a few of the ways chemical substances have been used to interrupt, accelerate, and inhibit plant growth and development in attempts to meet those demands. Despite their potential for misuse, plant hormones are having increasing importance for our lives. Our current understanding of these hormones can take us in different directions, depending on how informed or indifferent we remain at the crossroads.

Recommended Readings

Fahn, A. 1974. *Plant Anatomy.* Second edition. New York: Permagon Press. An outstanding, well-illustrated book.

Jensen, W. and F. Salisbury. 1972. *Botany: An Ecological Approach.* Belmont, California: Wadsworth. Beautifully illustrated text; excellent ecological insights.

Raven, P., R. Evert, and H. Curtis. 1976. *Biology of Plants.* Second edition. New York: Worth. An exceptionally well-written, well-illustrated botany textbook.

Chapter Twelve

Figure 12.1 Ciliated epithelium in the nose of a moustached bat—an example of multicellular specialization.

James Dennis

From Cells to Organ Systems in Animals

From time to time, the multicellular body of a complex animal has been likened to a city, or a state, or some other social unit composed of separate but interdependent parts. These analogies are wonderfully optimistic about the human capacity for self-organization. In truth, there is no way a city, a state, or any other human-designed unit can approach the functional integration of parts in a complex animal. In such organisms, the lines dividing the functioning of parts at each level—from cells, to tissues, organs, and organ systems—blur to insignificance. When you come right down to it, "multicellularity" implies a unity of function for which there is no synonym in our language. This means, of course, that you are about to be swept into one of those topics about which you must first know everything before you can begin to know much about it at all. Function must be compartmentalized; structures must be pulled out of place and laid out for examination as if they have meaning apart from the whole. They don't! For a multicellular animal, it is never a question of the isolated functioning of some cell, some tissue, even a whole organ or organ system. Always the question is this: *How is the part functioning in the intermeshed, coordinated functioning of the whole?* Separation of structures and functions is necessary as a learning device for us, but it is utterly impossible for the multicellular animal body.

The Human Body: An Example of Multicellular Specialization

In approaching the topic of multicellular form and function, it's easier to begin with an organism with which you already are more or less acquainted: yourself. You undoubtedly are more familiar with your own body than you are with, say, the body of a flatworm. So it probably will be easier for you to develop the vocabulary needed for understanding the organization of multicellular animals in general by examining your own organization first. Then, when you finally get to the epic of the flatworm, you'll be able to sense not only the remarkable unity but the progressive diversity among the body plans of multicellular animals.

We will begin with the tissues in your body, then we will move on to organs and organ systems. This progression begins with chemical and structural variations in single cells. Such variations dictate which cells will bind together, in what regions they will maintain contact, and where spacing will occur between them. Thus a **tissue** is a permanent grouping of similar cells having a similar function. An **organ** is a body unit made of different tissues having a common function. From the way basic tissues are combined and arranged, organs take on their character. For instance, organs that must move constantly (such as the stomach) are typically wrapped in layer after layer of muscle tissue, and delicate organs (such as capillaries) are typically cushioned and held in place by connective tissue. As you will soon see, astonishing plans are possible with only a few tissue types.

Basic Animal Tissues

Although all animal cells are patterned according to the same basic cellular plan described in Chapter Four, in complex animals they become differentiated in these ways: as epithelial, connective, muscle, and nerve tissue.

Epithelial Tissue Sheets of cells, one or more layers thick, cover the outside of the body and line all its internal surfaces, including the lungs, gut, blood vessels, and various glands. These are the **epithelial tissues.** Epithelium is first and foremost a protective tissue. Its cells are densely packed; not much space or intercellular material is found at the cell junctions. This tissue plan provides the body and individual organs with barriers against invading microorganisms and other suspect particles. But epithelium also serves the body in the same way that the plasma membrane serves the cell: it acts as a boundary layer that regulates the flow of incoming and outgoing materials. Food enters the body by moving across the epithelium of the digestive tract. Oxygen enters and carbon dioxide leaves the body by moving across thin epithelial tissues in the lungs. Certain wastes are excreted by traveling across epithelial cells of the kidney. Special epithelial cells produce most of the body's secretions (sweat, tears, saliva, milk, mucus, and so on). As tissues they produce fluids to lubricate body surfaces; when organized into organs called glands, they have ducts to carry products to the outside (Figure 12.2).

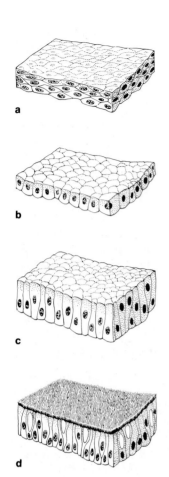

Figure 12.2 Examples of four basic types of epithelial tissue: (**a**) squamous, (**b**) cuboidal, (**c**) columnar, and (**d**) ciliated epithelium.

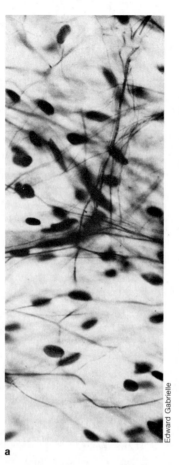

Edward Gabrielle

a

Richard Kolberg

b

Jeffrey Feldman

c

Figure 12.3 (**a**) Loose connective tissue. Cells and fibers are in a semifluid matrix. (**b**) Fibrous connective tissue, from ligaments. Parallel collagen strands give such tissues strength and flexibility. (**c**) Bone. Bone cells (one appears here) are embedded in a collagen-rich matrix they themselves produce. The matrix is reinforced with inorganic salts (mostly calcium salts), which give bone strength and rigidity. Bone contains three kinds of cells: osteoblasts form the bone, osteocytes maintain it, and osteoclasts tear it down. During growth, these cells are involved in changes in bone size and shape. They remain active in an adult animal, for bone acts as a calcium bank. If calcium levels in the blood fall, osteoclasts remove calcium from bone and pump it into the bloodstream. If calcium levels rise, osteoblasts pick it up and store it away in bone.

Connective Tissue Unlike epithelial cells, the cells of **connective tissue** are scattered through an extensive extracellular matrix. These tissues serve to bind together and support other animal tissues (Figure 12.3).

Loose connective tissue is found throughout the body. It contains a weblike scattering of strong, flexible protein fibers (collagen) and a few highly elastic protein fibers (elastin) embedded in a partially fluid matrix. Loose connective tissue acts like packing material, supporting and holding in place blood vessels, nerves, and internal organs even while according them some freedom of movement. It also binds muscle fibers together and binds skin to underlying tissues.

White fibrous tissue attaches and holds movable body parts together. The structural organization of white fibrous tissue is a key to its strength and function: it is typically made of parallel collagen fibers, densely packed with very few cells and little matrix between them. Tendons, which attach muscle to bone, are made of these strong tissues. So are ligaments, which hold bones together at body joints. And sheets of white fibrous tissue help keep muscles in place.

The skeleton of the animal body is essentially *supportive connective tissue* of two kinds: cartilage and bone. In cartilage, living connective tissue cells are suspended in a firm, rubbery matrix which they themselves have secreted. When the human skeleton is first formed, it is composed exclusively of cartilage. During growth, cartilage is slowly torn down and replaced with true bone (also known as osseous tissue). But until growth is completed during the teens, the growth points themselves remain cartilaginous.

Another form of connective tissue is made of large cells, each having a single fat-filled vesicle. The fat in this *adipose tissue* represents stored energy. It also cushions the body against shocks and blows, and insulates it by helping to retard loss of heat generated during metabolism.

The body's blood and lymph are forms of connective tissue (Figure 12.4). *Blood* circulates through the body and bathes all other tissues. It equalizes body temperature by distributing heat generated during metabolism. It takes part in body integration by transporting hormones that inhibit or stimulate organ activity, and it transports infection-fighting cells. It also transports nutrients to cells, then carries away metabolic by-products. Fluid sometimes seeps out from tiny vessels that transport blood through the body's tissues. This fluid, called *lymph,* is collected and returned to the blood by a separate system.

Blood has four components: platelets, plasma, erythrocytes (red blood cells), and leukocytes (white blood cells). Red blood cells, the hemoglobin-rich carriers of oxygen, have an average life of about four months. But replacements are continually synthesized in the soft, highly cellular interior of bones (the marrow). Platelets, also produced in marrow, function in clotting (Figure 12.4). More than 90 percent of plasma is water, which functions as a solvent. Within that fluid are sixty or so different proteins. Each carries out a specific task, such as assisting platelets in clotting when an injury occurs. Also carried in the plasma are sugars, fats, amino acids, hormones, vitamins, and various inorganic salts (some of which help maintain the acid–base balance of blood).

There are three types of white blood cells: granulocytes, monocytes, and lymphocytes. They function in the immune system (the physiological responses by which the body recognizes and eliminates foreign substances), as well as in the elimination of damaged, worn-out, and abnormal cells. The first two types are produced in bone marrow, then distributed throughout the body. Precursors of lymphocytes are also produced in marrow, but most become housed in the thymus, lymph nodes, and spleen.

Muscle Tissue In vertebrate animals, there are three kinds of **muscle tissue**—smooth, skeletal, and cardiac—which are used for movements of body parts.

The organs of the digestive system, the reproductive system, the bladder, and the arteries contain layers of *smooth muscle* (Figure 12.5). Smooth muscle movement is not consciously controlled. It is under the control of nerve impulses that travel through the autonomic nervous system, which will be described later.

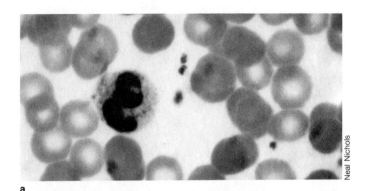

a

b

Figure 12.4 (a) Human blood. Plasma, erythrocytes (the oxygen-carrying red blood cells), leukocytes (the infection-fighting white blood cells), and platelets make up this circulating tissue.

(b) When platelets are damaged (which can happen, for instance, when a blood vessel is punctured), they release their chemical contents. One substance, called thromboplastin, reacts with two proteins that are normally dissolved in the plasma. The reactions lead to the formation of a stringy net around the puncture. When red blood cells rushing outward become tangled in the net, a *clot* begins to form, which eventually seals up the wound.

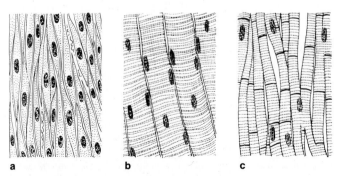

a   b   c

Figure 12.5 Muscle tissues: **(a)** smooth, **(b)** skeletal, and **(c)** cardiac.

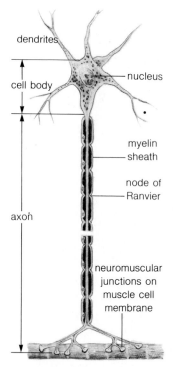

dendrites

cell body ——— nucleus

myelin
— sheath

node of
— Ranvier

axon

neuromuscular
junctions on
muscle cell
membrane

a *A motor neuron*

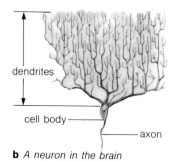

dendrites

cell body ———

——— axon

b *A neuron in the brain*

Figure 12.6 Two examples of diversity in nerve cell appearance: (**a**) a motor neuron, and (**b**) a neuron in the brain.

Epidermis

blood vessel

Dermis

hair follicle

sebaceous gland

loose connective tissue

fat cells

Evelyn Clausnitzer

Figure 12.7 Section through the intricate tissue layers of human skin.

In contrast, most *striated muscle* (or *skeletal muscle*) can be controlled voluntarily. Nerve impulses stimulating its movement originate in the conscious centers of the brain and are channeled through the nervous system to trunk and limbs, the head, the eyes, the mouth, and so on. What we call "meat" is really striated muscle. As shown in Figure 12.5, its cells are shaped like long fibers having many nuclei.

Cardiac muscle, the tissue of the heart, is made of intricately branched and fused cells, with conspicuous disks acting as boundaries between them (Figure 12.5). Cardiac muscle cells do not move autonomously. When one contracts in a tissue, they all contract in quick succession. The disks between them cause a slight delay between the contraction of adjacent cells. The contraction passes as a wave over the surface of the heart, thereby squeezing blood from the heart smoothly but forcefully.

Nerve Tissue Functionally inseparable from muscle tissue is **nerve tissue,** which is highly specialized for conducting electrochemical messages in response to stimuli. As you might expect, nerve tissue exists throughout the body, connecting and coordinating all its varied parts. It is made of billions of individually insulated nerve cells, or **neurons** (Figure 12.6). These insulated cells are held together in brain, spinal cord, and nerve fibers by a special connective tissue (the neuroglia). In general, each neuron has three parts: a central **cell body** (membrane-enclosed cytoplasm and nucleus), usually one or more **dendrites** (branched cytoplasmic extensions), and an **axon** (an extremely thin, threadlike cytoplasmic extension). Certain axons are wrapped in the plasma membrane of connective tissue cells. These multiple wrappings are called the myelin sheath. They not only protect and insulate the axon, they permit it to conduct messages much more rapidly. (In **multiple**

Table 12.1 Organ Systems in Humans

System	Some Component Organs	Main Functions
Circulatory	Heart, arteries, veins, capillaries, lymphatics	*Internal transport of materials to and from cells; protection against disease*
Respiratory	Nose, nasal cavities, pharynx, larynx, trachea, lungs	*Gas exchange between atmosphere and blood*
Digestive	Mouth, esophagus, stomach, small and large intestines, anus, liver, pancreas	*Breaking down of food molecules for their absorption into the bloodstream; elimination of indigestible and undigested residues*
Urinary	Kidneys, ureters, bladder, urethra	*Elimination of metabolic wastes of cells; regulation of fluids in cellular environment*
Skeletal	Bone and cartilage structures	*Support, protection of some organs; muscle attachment; determination of some body shapes; production of blood cells*
Muscular	Muscles, tendons, ligaments	*Movement of internal body parts; movement of whole body through the external environment*
Reproductive	Testes, penis (in male); ovaries, uterus (in females); accessory glands	*Production of new individuals; production of sex hormones*
Nervous	Brain; spinal cord; nerves; ganglia; sensory organs such as ears, eyes, and nose	*Together with endocrine system, integration of body functioning; detection of stimuli in external environment; control of behavioral responses to stimuli*
Endocrine	Pituitary, thyroid, adrenal, pineal, parathyroid glands; gonads	*Internal chemical control; together with nervous system, integration of body functioning*

sclerosis, this sheath deteriorates in places, and scar tissue forms in its absence. The scar tissue interferes with the conduction of messages that are vital to body coordination.)

Examples of dendrites and axons are shown in Figure 12.6. They are collectively known as nerve cell "processes." Their structure and organization relative to one another are the basis for integration of *all* the body's activities. For this reason, you may want to take an extra minute to study this figure in anticipation of the discussions of nervous integration (Chapter Thirteen).

Formation of Organs From Basic Tissues

When you look at your hand, you might think that "skin" is simply a sheet of tissue stretched over the bones of your palm and fingers. But as Figure 12.7 shows, human skin is a complex organ made up of all five basic tissues as well as specialized glands. The outside layer is the epidermis, the form of epithelium that covers the outside of the body. The epithelial cells making up the outermost layers are dead, but their dry toughness provides good protection. The deeper epidermal layers are the site of active cell division. Here, new cells form and move toward the surface layer as dead cells are continually sloughed off. The process goes on all the time, but we usually aren't aware of it unless we are confronted with a peeling sunburn or with dandruff.

Beneath the epidermis but tightly connected to it is the dermis, a dense layer composed mostly of connective tissue. Here also are hair follicles (epithelial shafts in which hair grows), and sweat and oil glands. Smooth muscle tissue is involved in the expansion and contraction of blood vessels to regulate blood flow and the movement of hair follicles. (When "goose pimples" form on your skin, smooth muscles have contracted and hair follicles, which normally are positioned at shallow angles to the plane of the skin, stand on end.) The subcutaneous layer, made of loose connective tissue and adipose tissue, binds the upper layers of skin to skeletal muscles below. Nerve cells wind through the dermis, reaching sweat glands, blood vessels, and hair follicles to assure responsiveness to changing conditions. Here, too, sensory nerve cells receive information about touch and temperature.

Like skin, all the diverse, specialized organs of the vertebrate body are made of some combination of two or more basic types of tissues. In the remainder of this chapter, we will look at more of these organs in terms of how they function at an even more specialized level of integration— that of human organ systems, as summarized in Table 12.1.

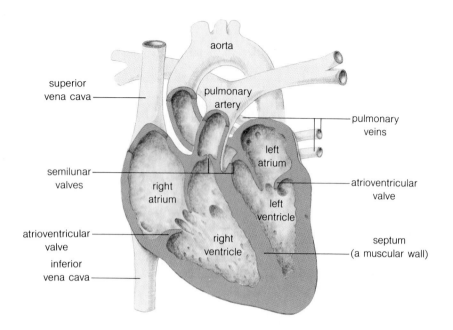

Figure 12.8 Cutaway view of the human four-chambered heart.

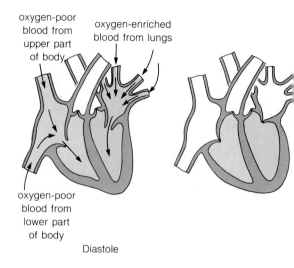

a The cardiac cycle of relaxation (diastole) and contraction (systole)

Figure 12.9 (**a**) The cardiac cycle and (**b**) the pacemaker.

Circulatory System

Almost three-fourths of the weight of an adult human is fluid—water that contains suspended or dissolved materials for nourishing all the body's cells. Most of the fluid is within cells, but about one-third is extracellular. Some of this extracellular fluid bathes the outside of the cells and acts as the transfer medium in the diffusion of materials from tissues to blood and back again. The remainder is the plasma portion of blood. Blood is contained within and fills a network of conduits called blood vessels: the arteries, arterioles, capillaries, venules, and veins. Driven by the action of a muscular organ called the heart, blood courses through this network for as long as we live. During a human life span of about seventy years, a heart will beat about $2\frac{1}{2}$ billion times to keep blood on the move! Let's take a look at this remarkable organ before we turn to the circulatory system of which it is a part.

The Heart The human **heart** is a four-chambered pump about the size of a clenched fist (Figure 12.8). A two-layered membrane sac (the pericardium) encloses the heart and protects it from injury. A muscular wall divides the heart into functionally separated halves. Within each half are two interconnecting chambers: the upper *atrium* and the lower *ventricle*. A special *atrioventricular valve* regulates the flow of blood between the upper and lower chamber in each half.

Even with the functional separation, the upper chambers expand and contract in synchrony, just as the two lower chambers expand and contract in synchrony.

The heart works in two alternating stages known as the **cardiac cycle** (Figure 12.9). In *diastole,* the heart is relaxed. This means oxygen-enriched blood from the lungs can enter the left atrium, and oxygen-depleted blood from the rest of the body's cells can enter the right atrium. All the while, the atrioventricular valves are open, so the blood flows down into the ventricles. But as these two lower chambers become distended with blood, a wave of contraction begins at their base and sweeps up toward the two atria. This contraction period is called *systole*. The upward movement of blood forces the atrioventricular valves to snap shut. At the same time, oxygen-depleted blood in the right ventricle is forced into the main artery leading to the lungs, and oxygen-enriched blood is forced into the main artery leading to the rest of the body. When the heart relaxes for the next diastole period, semilunar valves (Figure 12.8) prevent the outgoing blood from pouring back in.

The rate at which cardiac muscle fibers contract and relax is regulated by nerve impulses, but the capacity to contract rhythmically is not. The moment heart cells first appear in the developing embryo, rhythmic contractions begin. As more heart cells form, they move about until they establish contact with one another. Once they do so, they fall into the rhythmic pace established by the strongest of them: the *sinoatrial node,* or the *pacemaker* (Figure 12.9). In a fully developed heart, each beat begins with a signal from

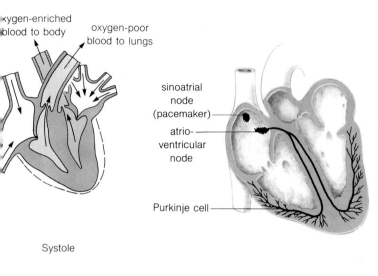

oxygen-enriched
blood to body

oxygen-poor
blood to lungs

sinoatrial
node
(pacemaker)

atrio-
ventricular
node

Purkinje cell

Systole

b *The pacemaker controlling cardiac muscle fiber contraction*

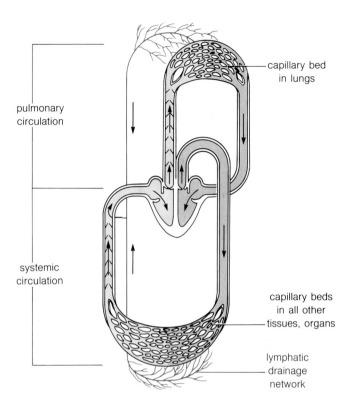

pulmonary
circulation

systemic
circulation

capillary bed
in lungs

capillary beds
in all other
tissues, organs

lymphatic
drainage
network

Figure 12.10 Generalized picture of systemic and pulmonary circulation.

the pacemaker that stimulates the atrial tissue, causing it to contract. The spreading excitation reaches cells making up another region, called the *atrioventricular node.* These cells channel the impulses into bundles of Purkinje cells, which stimulate ventricular contraction.

Systemic and Pulmonary Circulation The two halves of the heart are only pumping stations for two major conduits of internal transport: **pulmonary circulation,** which leads to and from the lungs; and **systemic circulation,** which leads to and from the rest of the body. As Figure 12.10 shows, in systemic circulation the blood is forced from the left half of the heart into arteries, arterioles, then capillaries in all body tissues. There it distributes its load of oxygen and other materials to the cells. At the same time, it picks up carbon dioxide and other wastes. Then it is sent back through veins to the right half of the heart. In pulmonary circulation, this oxygen-depleted (and carbon-dioxide loaded) blood is forced from the right half of the heart through the pulmonary artery that leads to capillaries in the lungs. Here the blood dumps its load of carbon dioxide and takes on more oxygen. It returns through pulmonary veins to the left half of the heart—where it starts the trip over again.

The heart is the main pump of this circulatory system. Blood forced out of the contracting heart flows first into arteries, then into finer branchings called arterioles. The arterioles branch into capillaries, which have such small diameters that red blood cells must flow through them single-file. Right before some branch points are precapillary *sphincters* (rings of smooth muscle) that help control blood flow in response to the changing needs of individual tissues.

The meshwork of capillaries supplying each region of tissue with nutrients is called a *capillary bed.* Because the increase in total cross-sectional area of the capillaries slows down blood flow, it's easier for materials to be exchanged between body cells and blood cells. The cells making up the single epithelial layer of capillary walls are joined tightly together, and they are selective filters. They permit small molecules such as sugars, oxygen, and carbon dioxide to pass but they prevent blood proteins from leaving. (In some unknown way, though, white blood cells can squeeze between adjacent epithelial cells and leave the bloodstream. *Why* they leave is another story, which is reserved for the next chapter.) Capillaries come together again in the veins. Unlike arteries, veins are lined with valves that are really folds of tissues pointed toward the heart. The flow of blood presses the valves open against the vein walls. If for some reason the blood starts moving backward, the flow shuts the valves, preventing its reentry into the capillary bed.

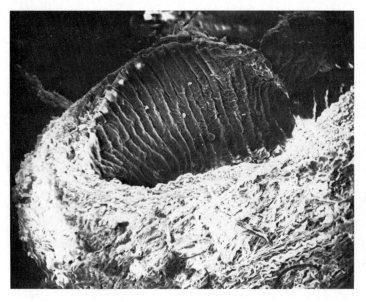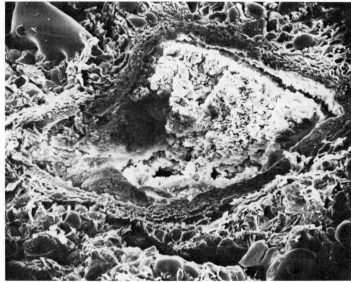

Robert LaPorta

Figure 12.11 Normal artery, and an artery clogged with lipids.

It is through the circulatory system that each cell of the vertebrate body in general (and the human body in particular) receives materials and energy needed to maintain itself, to grow, and to remove wastes. And it is through controls over the circulatory system that the volume of blood flow can be varied according to cellular needs. The more active the cells in any given region, the more blood must reach them every minute. As you will read in the next chapter, circulatory adjustments to changing metabolic needs are met through both endocrinological and nervous control.

Blood Pressure When your physician takes your **pulse,** he is measuring the difference between systolic and diastolic pressure. During systole, the volume of blood being forced out of the arteries equals only about a third of the volume leaving the heart. The volume of blood remaining distends the arteries (which have thick, elastin-rich walls), and arterial pressure increases. This arterial pressure is commonly called "blood pressure." When ventricular contraction ends, the arteries recoil passively to their original diameter, driving blood through the arterioles. The greatest pressure exerted on the arteries occurs when blood is forced out of the heart; it is called *systolic pressure.* The least pressure, which occurs just before blood is forced out of the heart, is *diastolic pressure.* The difference between the two is the pulse pressure; it is usually measured in terms of how much it can raise a column of mercury (Hg). For an adult male, the average pulse pressure is 50 ($125 - 75 = 50$ millimeters Hg). For an adult female, it is 40.

Arteriosclerosis During times of increased physical activity, systolic pressure rises because of the increased volume of blood leaving the heart. Thus pulse pressure rises. As you might well imagine, factors that decrease the ability of the arteries to distend also increase systolic pressure. That is what happens in **arteriosclerosis,** or hardening of the arteries.

This disease is associated with the natural process of aging, but other factors promote its occurrence. A diet rich in saturated fats or excess carbohydrates, lack of physical exercise, and chronic tension seem to be by-products of modern society. And the combination of these three factors seems to promote hardening of the arteries. The ravages of this disease are intensified by cigarette smoking and by air pollution. Circulatory failure caused by arteriosclerosis is the major cause of death in the United States.

In Chapter Three, you read that lipids such as fats and cholesterol are insoluble in water. Such lipids are transported in the bloodstream, where they normally are bound to protein carriers that keep them suspended in the plasma. In arteriosclerosis, lipids are deposited in the arterial walls. As the depositions build up, the arteries become clogged—which cuts down blood flow to the regions the arteries supply (Figure 12.11). The passageways may become so blocked and roughened that platelets are caught and break open. When platelets are broken, they initiate the formation of a blood clot, or *thrombus.* If the clot occurs in an artery that supplies blood to the heart itself, cardiac muscle cells begin to die within thirty seconds after being deprived of their supply of oxygen.

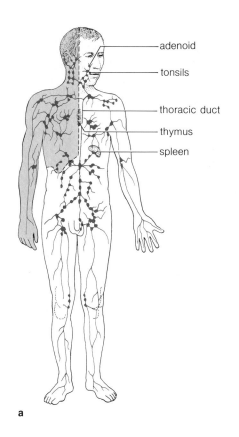

- adenoid
- tonsils
- thoracic duct
- thymus
- spleen

a

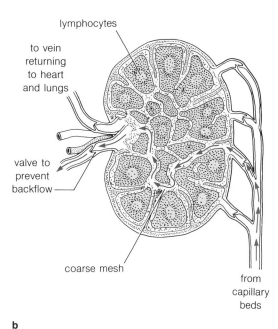

lymphocytes

to vein
returning
to heart
and lungs

valve to
prevent
backflow

coarse mesh

from
capillary
beds

b

Figure 12.12 (**a**) Lymphatic system, as described in the text. The shaded region is the area drained by the right lymphatic duct; the thoracic duct drains the rest of the body's tissues. (**b**) Close-up of a lymph node. (Refer back to this figure as you read Case Study 5 in the next chapter. It gives an idea of the lymphatic system's vital role in the body's defense against invading cells and particles.)

The heart will stop beating in this region, and that is why death from **coronary thrombosis** can be immediate. When similar clots form in the arteries that supply the brain, then brain cells die at once, which is one source of a **stroke.** If the stroke occurs in brain regions governing the coordination of body movements, paralysis may follow. If it occurs in regions controlling vital functions such as breathing, death will follow.

The Lymphatics Although red blood cells can carry carbon dioxide back to the lungs, many substances such as some proteins become intermixed with water in the body tissues. Such material is returned to the bloodstream through supplementary "drainpipes." In the capillary beds, tiny lymph vessels take up fluids diffusing from the capillaries as well as fluids and proteins from tissue. These vessels converge into larger and larger vessels until they become two major conduits: the *thoracic duct* and the *right lymphatic duct.* Both ducts have valves to prevent backflow, and both depend on muscular contractions for pumping fluid into two major veins near the heart. Together with special organs called the *thymus, spleen,* and *lymph nodes,* this network makes up the **lymphatic system.** This system is shown in Figure 12.12.

The lymphatic system is concerned with more than returning seeped-out fluid to the blood. It also takes up fats absorbed from the intestinal tract and transports them to the blood. Most important of all, it is essential to the body's defense against invading cells and substances. Special blood cells manufactured in bone marrow are housed in the thymus, spleen, and lymph nodes. The nodes are small, bean-shaped bodies that act as filters and as defense stations. Fluids entering a lymph node follow a twisted path through its meshlike interior, percolating past countless specialized blood cells (the lymphocytes). If the fluid has been drained from a tissue suffering from infection, these cells trap bacteria from the lymph and begin making special proteins called antibodies. These proteins are released to kill the bacteria and abolish the infection. (If you've ever felt painful, swollen bumps beneath your jaw where it meets your ears, you know the location of at least two lymph nodes in your body. At that time they were encountering a throat infection, and they had become enlarged with the foreign material and with lymphocytes during the battle.) More will be said about this process in Chapter Thirteen.

Respiratory System

In cellular respiration, oxygen is brought into the cell and carbon dioxide leaves it through the metabolic pathways described in Chapter Six. In multicellular animals, most cells are too far removed from the outside environment for simple exchange to occur across the plasma membrane. You have just finished reading about how oxygen is circulated to all body cells once it has entered the bloodstream. Another system—the **respiratory system**—gets oxygen to the blood in the first place, and gets rid of the carbon dioxide that blood transports away from the cells. Key organs in this system are the pharynx, trachea, bronchi, lungs, chest muscles, and the diaphragm (Figure 12.13).

Air normally enters the body through the nose and nasal cavity. Hairs at the entrance of this passageway and cilia on its thick epithelial lining filter out dust and other foreign particles. Numerous blood vessels embedded in the lining help warm the incoming air, and mucous cells moisten it before it flows into the lungs. From the nasal cavity, air moves through a region in the back of the mouth (the pharynx) then down to tubes that have walls reinforced with cartilage rings (the larynx and the trachea). These tubes are lined with mucous epithelial membrane that is also covered with cilia. As inhaled particles become stuck against the surface membrane, the upward-beating cilia sweep debris-laden mucus back to the mouth or nasal cavity, where it can be expelled from the body.

The trachea branches into two conduits, the **bronchi** (singular, bronchus), in the region of the lungs. **Lungs** are a pair of elastic, spongelike sacs that house the structures in which gas exchange between blood and air occurs. Like trachea, bronchi are ringed with cartilage and are lined with mucous, ciliated epithelium that works to keep the passages clear. Within the lungs, each bronchus branches into smaller tubes and then bronchioles (epithelial conduits in which cartilage is not present and which therefore are more elastic). Mucous membrane is absent from the smaller bronchioles, for it could interfere with air passage.

Bronchioles terminate in air sacs from which thin-walled pouches—the **alveoli**—protrude like bunches of grapes. Each one of the 300 million alveoli in human lungs is a functional exchange point. At any moment, about one-fourth of your body's entire blood supply is concentrated in fine capillary nets that surround the alveoli, giving up carbon dioxide to them and taking oxygen from them.

Tidal Ventilation The respiratory channel just described is highly branched, with oxygen moving inward and carbon dioxide being expelled through the same channel. The exchange of these two gases is possible because of rhythmic changes in the pressure gradient between the lungs and the atmosphere during **tidal ventilation:** the alternating expan-

sion and compression of the lungs. The lungs are expanded when the pressure in the thoracic cavity falls below the pressure within the lungs, so air flows down the gradient *(inhalation).* Once the lungs are filled, the chest cavity relaxes, which increases pressure in the cavity and compresses the lungs. Thus the pressure gradient is reversed, and gases are forced outward *(exhalation).*

This alternating air flow occurs by the muscular expansion and contraction of the diaphragm (Figure 12.13) and the highly elastic muscles attached to the ribs. The diaphragm and the rib muscles are under control of neurons housed in the brain. These neurons monitor how much carbon dioxide is present in the blood coursing through the arteries. When the carbon dioxide level rises, those neurons act to increase the rate of inhalation and exhalation. The diaphragm moves downward and flattens, while the rib cage moves outward. This action expands the thoracic cavity and lowers the pressure in the lungs. Thus air rushes in to fill the pressure deficit. When this muscle system relaxes, it returns elastically to its normal position. The lungs are compressed, and air flows outward.

Gas Exchange Not all of the air leaves the alveoli during exhalation. If it did, the structures would collapse. But as the lungs expand once more, atmospheric air (which is low in carbon dioxide but rich in oxygen) comes swirling in, mixing with the air left in the lungs. The oxygen concentration rises and the carbon dioxide concentration falls relative to the levels present in the blood inside the lungs' capillary nets. Immediately these two gases begin to flow from the region of high concentration to the region of low concentration (oxygen into the blood and carbon dioxide out of it).

As the oxygen moves into the plasma, it is taken up at once by red blood cells. There it combines with hemoglobin to form oxyhemoglobin. Through this reaction, free oxygen levels are kept low; hence more oxygen keeps diffusing inward. If hemoglobin were not present, the blood would be able to carry only about 2 percent of the oxygen that it does carry—and the heart would have to beat 50 times faster to supply all the body tissues with oxygen!

When oxygen-rich blood reaches a capillary bed within the body, it enters a region of high carbon dioxide concentration and low oxygen concentration. Once again, there is diffusion down the gradient, with oxygen moving out of the blood and carbon dioxide moving into it. With a new cargo of carbon dioxide, the red blood cells are ready to make another round trip.

Bronchitis and Emphysema In urban environments, in certain occupations, even in the microenvironment surrounding a cigarette smoker, airborne particles and gases exist in abnormal amounts, and they put an extra workload on the ciliated mucous membrane of the respiratory pas-

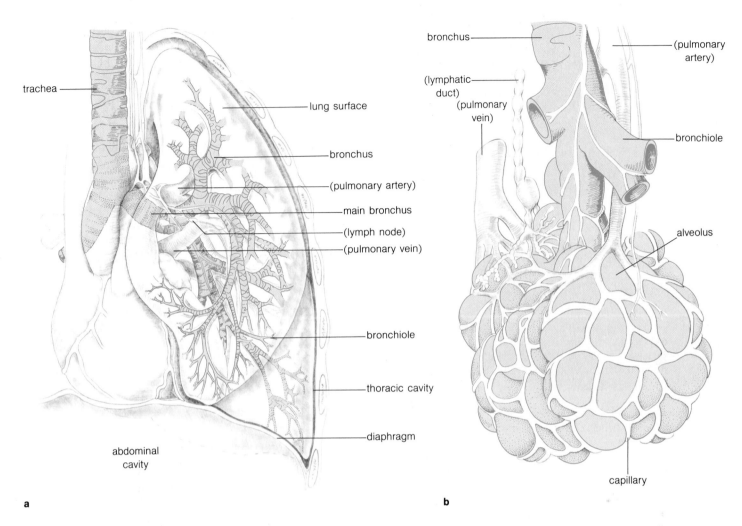

trachea

lung surface

bronchus

(pulmonary artery)

main bronchus

(lymph node)

(pulmonary vein)

bronchiole

thoracic cavity

diaphragm

abdominal cavity

bronchus

(lymphatic duct)

(pulmonary vein)

(pulmonary artery)

bronchiole

alveolus

capillary

a

b

Figure 12.13 (**a**) One of the paired lobes of the human lung, shown in position next to the heart. (Lungs are located in the thoracic cavity, which is separated from the abdominal cavity by the diaphragm—a layer of muscular tissue that acts as a partition.) (**b**) Close-up of alveoli, terminal points of the bronchioles and sites for gas exchange between blood and the atmosphere. (After Avery, Wang, and Taeusch, Jr., "The Lung of the Newborn Infant," *Scientific American,* April 1973. Copyright © 1973 by Scientific American, Inc. All rights reserved)

sageway. The membrane is highly sensitive to cigarette smoke, probably because it is so concentrated. Smoking and other forms of air pollution interfere with ciliary movement in this passageway. As a result, mucus—and the particles it traps, which include bacteria—begins accumulating in the trachea and bronchi. Coughing sets in as the body tries to clear away the mucus. If irritation continues, the coughing response persists. It aggravates the condition because it expands the bronchial walls, thereby providing pockets for the accumulation of various moist substances in which bacteria may grow. The bronchial walls become inflamed

with infection; tissue is destroyed by bacterial activity; cilia diminish in numbers; and mucus-producing cells increase as the body works to fight against the accumulating debris—all of which leads to the formation of fibrous scar tissue. Such are the characteristics of **bronchitis.**

A person suffering from an acute attack of bronchitis who is otherwise in good health responds to antibiotics. But what happens if the irritation persists—if, for example, a chain-smoker continues to smoke? As fibrous scar tissue begins to obstruct the respiratory passageway, the bronchi become progressively clogged with more and more mucus.

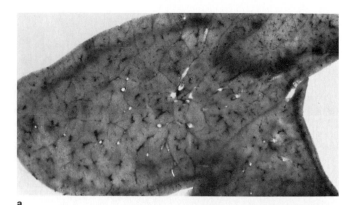

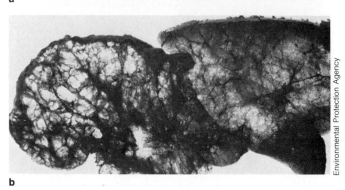

Environmental Protection Agency

Figure 12.14 (**a**) Healthy human lung and (**b**) lung from a person who suffered from emphysema.

Carbon-dioxide-rich gases then become trapped in the alveoli, which stretch and tend to disintegrate under the pressure. The lungs fill with these gases, which can't be expelled efficiently. The normally pink, elastic lung tissue becomes dry and perforated (Figure 12.14). The outcome is **emphysema**—the distension of lungs and loss of gas exchange efficiency to the extent that running, walking, even exhaling become painful experiences.

Why don't all cigarette smokers get emphysema? There is evidence that early environmental conditions—poor diet, chronic colds, other respiratory ailments—can create in some persons a predisposition to this disease later in life. Moreover, many who suffer from emphysema have a hereditary deficiency in antitrypsin, a substance that inhibits tissue-destroying enzymes produced by bacteria. These people may therefore be at a disadvantage in fighting off respiratory infections when they do strike. When such people have a smoking habit, their prospects unquestionably are grim. Part of the problem is that the potential threat seems exaggerated—what's so terrifying about coughing up mucus now and then? But emphysema is insidious, for it can develop slowly over twenty or thirty years, and by the time the disease is detected, the damage to lung tissue is

irreparable. The threat is not exaggerated: about 1,300,000 people in the United States alone now suffer from this disease.

Digestive System

If the ultimate purpose of food intake is to sustain the single cell, the multicellular animal needs a pathway by which food can be broken down into molecules small enough to move across the epithelial boundaries separating the internal body from the outside world. (Appearances to the contrary, food chunks inside your body aren't really inside your body. They are merely part of the external environment that your body "surrounds," which concentrates food for processing. Only when they have crossed those epithelial boundaries are they truly inside you.) The organ system that takes food chunks—primarily proteins, carbohydrates, and fats—and dismantles them chemically into usable parts (which are absorbed) and nonusable parts (which are discarded) is called the **digestive system.**

Smooth muscle is constantly at work throughout the length of the digestive canal wall. Its churning motions cause food to be mixed with digestive substances, and its wavelike motions cause food to be transported along at a pace slow enough for absorption to occur. The canal itself is some 5 meters (15 feet) long from mouth to anus in an adult human. In between are pharynx, esophagus, stomach, intestines, and rectum (Figure 12.15). Attached to the digestive tract by tubelike ducts are a series of digestive glands, including the salivary glands, the liver, and the pancreas. The enzymes and other substances these glands secrete aid in digestion.

It is in the mouth that food begins to be prepared for its journey through the digestive tract. Here the food is chewed to break it into manageable chunks and to mix it with saliva. The saliva moistens and softens the food, lubricates it so that it can pass smoothly into the stomach, and stimulates the taste buds that trigger secretions in the stomach in anticipation of the oncoming meal. Saliva also contains *amylase,* a hydrolytic enzyme that breaks down starch. Hydrolysis of starch is the only digestive step that takes place in the mouth. (Which is fortunate, because a constant supply of gastric enzymes in your mouth would tear down your mouth along with the food.)

The chewed and softened lump of food (bolus) is then sent through the pharynx and the esophagus, a tube of circular muscles. These muscles push the bolus downward by squeezing behind it in successive waves of contraction, called **peristalsis** (Figure 12.15). From the esophagus, the food enters the **stomach**—a muscular, elastic sac designed as a temporary holding station for food. Here food is partially broken down before moving on through the digestive tract. A membrane of epithelial and connective

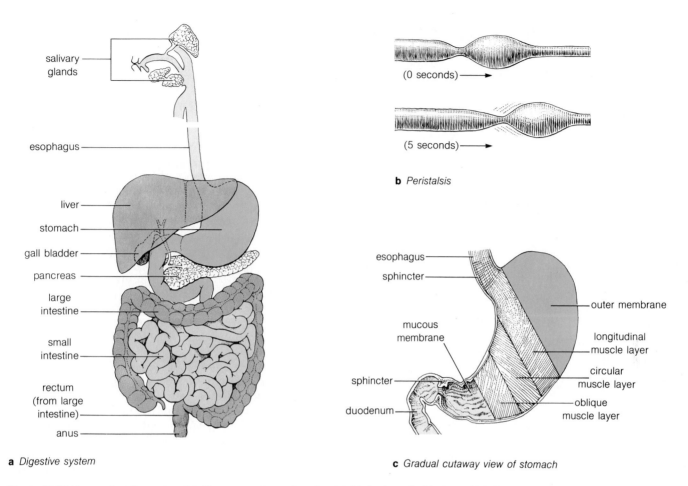

a *Digestive system*

b *Peristalsis*

c *Gradual cutaway view of stomach*

Figure 12.15 Human digestive system (**a**). The progressive action of peristalsis is shown in (**b**); the multiple layers of muscle in the stomach are shown in (**c**).

tissue surrounds the outside; within the membrane is a coat of criss-crossed, fibrous muscle tissue, then a layer of smooth muscle tissue. Lining the inside is a mucous membrane thickened with myriad digestive glands that secrete not only mucus but *gastric juice,* a watery substance containing hydrochloric acid and various enzymes. The hydrochloric acid dissolves various salts and destroys most of the bacteria coming in with the food. Its presence is also needed to activate the enzyme *pepsin,* which is one of the enzymes that split the peptide bonds holding amino acids together into proteins. Pepsin isn't released and activated until food is present in the stomach. (The entire digestive tract lining is mostly protein, hence a potential target for pepsin. In fact, if the protective mucous coating on the epithelium breaks down, pepsin eats away at the lining, with the subsequent perforation being called an *ulcer.*)

Once the partially digested food leaves the stomach, it moves into the **small intestine,** an epithelial–muscular tube. The small intestine is divided into three zones: the duodenum, the jejunum, and the ileum. It is in the first two zones that digestion is completed. The intermediate molecules that left the stomach are reduced here by various enzymes, which are present in the *intestinal juice.* (The components of the intestinal juice are secreted from cells lining the tract, as well as from the liver and pancreas.) It is here—not in the stomach—that organic molecules are made small enough to be transported into the bloodstream. This transport occurs through dense outfoldings of the mucous membrane, called villi. Villi are covered with millions of microvilli—fingerlike projections that are most efficient at increasing the surface-to-volume ratio of cells concerned with absorption (Figure 9.4). Beneath its one-cell-thick epithelium, each villus houses a capillary network and a lymph vessel lying in a bed of connective tissue. As the small intestine moves, the microvilli sway slightly, thereby coming into contact with more small molecules that can be

Table 12.2 Some Enzymes of Digestion

Enzyme	Source	Substrate	Main Breakdown Products
Amylase	Salivary glands	Polysaccharides	Disaccharides (glucose, maltose)
Amylase	Pancreas	Polysaccharides	Disaccharides (glucose, maltose)
Pepsin	Stomach	Proteins	Polypeptides
Trypsin, Chymotrypsin	Pancreas	Proteins	Polypeptides
Lipase	Pancreas	Fats	Fatty acids and glycerol
Ribonuclease	Pancreas	RNA	Nucleotides
Deoxyribo-nuclease	Pancreas	DNA	Nucleotides
Dipeptidases	Small intestine	Dipeptides	Amino acids
Carbohydrases	Small intestine	Disaccharides (glucose, maltose, sucrose)	Monosaccharides

absorbed. Once they have crossed the surface boundary layer, amino acids, glucose, and water molecules enter the capillaries, and glycerol and fatty acids enter the lymph vessels.

The **liver,** the largest gland in the body, has many functions. As you will read in the next chapter, it is central in glucose metabolism, and in converting various molecules (such as amino acids) into glucose as part of the body's program for maintaining blood sugar levels. Its main digestive function is the continual secretion of *bile,* a solution containing bile salts, bile pigments, cholesterol, and lecithin. Bile is needed in the digestion of fats in the small intestine. Fats, recall, are insoluble in water. Once fat leaves the stomach, its hydrophobic "tails" clump together in large globules. Some of the substances in bile break down the globules to an emulsion (small, suspended droplets) in the liquid contents of the intestine. In this way, the surface area of the fat components is increased, and enzymes can break them down more effectively.

Bile enters the duodenum through a bile duct, which is ringed by a sphincter where it joins the duodenum. When this sphincter is closed, bile is temporarily stored in the **gall bladder** (a small sac branching off the bile duct). You probably have heard about **gallstones.** What happens is that something changes the concentrations of bile salts, lecithin, and cholesterol in the bile. Bile salts and lecithin normally keep the cholesterol from forming fat globules. But if their concentrations are lowered, or if the cholesterol concentration rises, the cholesterol molecules aggregate into "stones." Large stones may become stuck in the bile duct, which effectively shuts down bile secretion, hence fat digestion. The disruption of the digestive system is extremely painful, and surgery is typically called for.

Another gland associated with digestion is the **pancreas.** It secretes enzymes that digest proteins, lipids, and carbohydrates in the duodenum (Table 12.2). The main body of the pancreas produces the digestive enzymes. But small patches of pancreatic cells (the *islets of Langerhans*) secrete the hormones insulin and glucagon into the bloodstream. They are a vital link in body functioning. In fact, the secretions of the pancreas and the liver are so linked in feedback relationships for the entire body that a special section has been set aside in the next chapter to describe them, after you have read about the control systems governing all metabolic activity.

Within two to five hours after eating, the undigested and undigestible residues of the meal enter the **large intestine,** or *colon.* Unlike the small intestine, the colon doesn't secrete enzymes, only mucus. The reason is that the colon is a holding station not for digestion but for the absorption of water and minerals—and by the time the residues leave the small intestine, they are in a highly liquid state. As the water is being absorbed, bacteria living in the colon convert the remaining intestinal contents to a semi-solid state *(feces).* The mucus secreted in the colon facilitates movement of feces to the rectum and then out through the **anus,** the terminal point of the digestive tract.

A symbiotic relationship exists between part of your digestive tract and resident bacteria. From the first few hours of your life until after your death, millions of bacteria (mostly *Escherichia coli*) live in your large intestine. They are acquired during the birth process or immediately thereafter. Once they enter the relatively benign environment of your colon, they grow and multiply in proportion to the amount of food traveling past. They have come to form a necessary link in digestion, for they help convert the intestinal contents to a bulk form that stimulates the intestinal lining and triggers peristalsis. Their diligent attention to the intestinal slush assures the elimination of wastes from your body in an expeditious manner. These and other bacteria in your gut also synthesize certain essential vitamins (such as vitamin K), which are absorbed by the epithelial walls of your colon. Whatever toxic by-products the bacteria create are normally eliminated through the intestinal tract or are absorbed into the bloodstream, then carted off to the liver for detoxification by special enzymes. The detoxified molecules are then carted off to the kidneys, which excrete them through the urine.

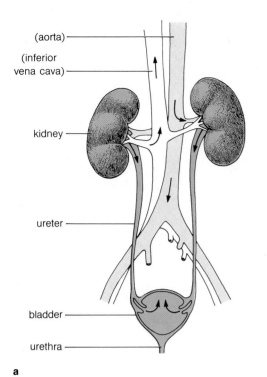

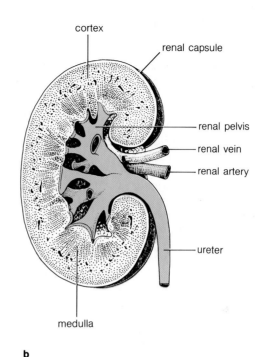

Figure 12.16 Human urinary system (**a**) and longitudinal section of a kidney (**b**).

a

b

Urinary System

Digestion of food molecules and oxygenation of blood provide resources for cell growth and repair, for building new substances, for energy. But in all these reactions, certain toxic by-products are formed. If allowed to accumulate, they would lead to widespread cellular death.

The lungs eliminate carbon dioxide, the intestinal tract eliminates solid residues, and the skin eliminates some water, salts, and excess heat generated during metabolism. All these organs represent exchange points with the environment. In other words, there is no single "excretory system" for the body; different organs rid the body of different types of wastes. What is often labeled the "excretory system" is really the **urinary system:** two kidneys, two ureters, one bladder, and one urethra (Figure 12.16). Its concern is not only elimination of certain substances but the recycling of others (including wastes).

The two **kidneys** are cushioned on a bed of adipose tissue near the diaphragm. Each looks like a kidney bean in shape and in color; each is 12 centimeters (over $4\frac{1}{2}$ inches) long. A million slender tubules, the **nephrons,** are found in each kidney (Figure 12.17). These structures filter waste materials from the blood. At one end of a nephron is a **Bowman's capsule:** a cuplike structure housing a cluster of

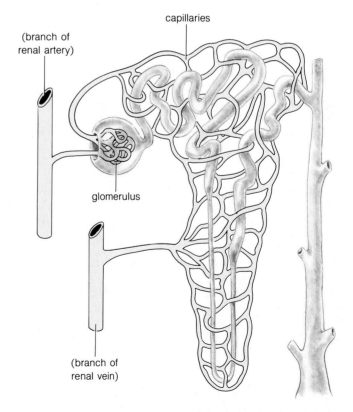

Figure 12.17 A nephron tubule and its associated capillary network.

capillaries (a **glomerulus**). These capillaries lead into the system from a branch of the renal artery. Every twenty-four hours, 180 liters (190 quarts) of fluid passes through them!

Filtration, the first step of urine formation, begins at the glomerulus of each nephron. The tiny capillaries have thin walls that allow part of the fluid to escape. The force driving the fluid into Bowman's capsule is the pressure exerted on blood by the heart's pumping action. This fluid is the **kidney filtrate:** it contains salt, sugar, and other small molecules. But essentially the kidney filtrate has no blood protein molecules, which are too large to move across the capillary walls.

During **reabsorption,** nearly all substances removed during filtration are returned to the bloodstream. Reabsorption begins as the filtrate flows from each Bowman's capsule through the tubular regions of the nephron. In this tubule system, the blood and the filtrate that parted company earlier are brought side by side. The two streams are separated only by two cellular layers: that of the tubule wall, and that of the capillary wall. As you will read in the next chapter, much of the fluid leaves the tubule and reenters the capillaries through osmosis at these regions. In fact, more than 99 percent of the water and nearly as much of the dissoived solids that entered the system are returned to the bloodstream.

The remaining 1 percent of the filtrate (now called urine) eventually flows into a cavity that leads to the **ureters:** thick-walled ducts that empty into a urinary reservoir (the **bladder**). The ureter walls are made of muscular layers that propel urine into the bladder by peristaltic movements. Ureters enter the bladder at an angle, so that when the bladder becomes distended with urine, the pressure closes them off, which prevents backflow. Leading out of the bladder is the **urethra,** the duct that empties to the exterior.

But filtration, reabsorption, and elimination are only the basic processes of this system. For the kidney is not merely a passive filter for the blood. *The kidney is highly selective in the accumulation of the blood's waste products that will leave the body in urine, and it is highly selective about what it sends back to the bloodstream.* Its action depends on the state of the organism as a whole. The kidney helps regulate the flow of various salts, conserving them when they are in short supply, and removing them when they are present in excess amounts. In times of water shortage, the nephron system operates at full capacity to reabsorb water; only a tiny volume of highly concentrated urine leaves the body. In times when there is excess water, a large volume of highly dilute urine leaves. In this regulatory behavior, the kidney cannot be viewed as an organ separated from the whole; neither can the urinary system be viewed as separated from the whole. In this chapter we simply are describing organs and systems. In the next chapter we will look at the marvelous control processes that bring about the integration of these interdependent systems in the state we call being alive.

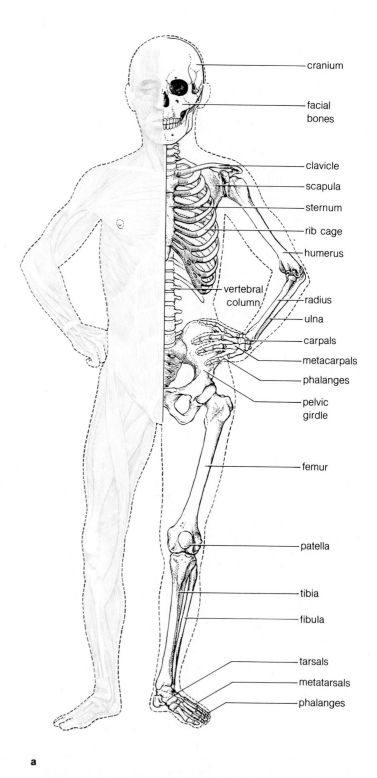

a

Figure 12.18 (**a**) Arrangement of muscles and bones in the human skeletal–muscular system. (**b**) Antagonistic muscle movement. The sketch shows how two muscles of a pair contract and relax in opposition to each other.

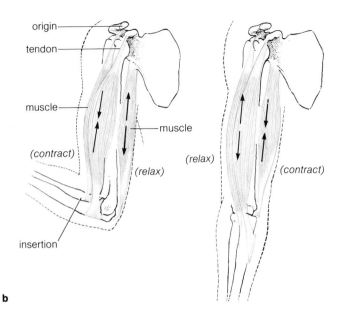

origin

tendon

muscle

muscle

(contract)

(relax)

(relax)

(contract)

insertion

b

Skeletal–Muscular System

The human **skeletal–muscular system** is the framework that supports and protects all the other organ systems, and it is the means of moving them as a unit through the environment. Bone and cartilage make up the internal framework, the *endoskeleton* (Figure 12.18). Bone and cartilage are at once strong and light in weight. Generally they take on a cylindrical shape in bones—a shape that affords more strength without taking up too much room. The endoskeleton serves three functions. The *axial* portion (vertebral column, skull, and rib cage) supports and protects vital organs. The *appendicular* portion (arms, legs, and associated appendages) supports the body; and, in conjunction with special skeletal muscle tissue, it acts as a system of levers for moving the body. Webs of muscles stretch over the framework, attached tightly to the skeletal bones by tendons. Ligaments form flexible connections between adjacent bones, bridging them like door hinges and using them to bring about movement in response to signals from the nervous system.

The human body requires movements in multiple directions. Muscle cells happen to contract and relax in one axis only—which would be fine only if you wanted muscles to lengthen in a straight line and pop back again. Thus muscles in the limbs of the body are arranged in antagonistic pairs that pass from one bone to the next on the opposite sides of a joint. When one member of an antagonistic pair is

contracted, the limb is bent. As it relaxes and its partner is contracted, the limb straightens out again (Figure 12.18). In the next chapter, we will look at the mechanisms underlying movement, after we have had the opportunity to consider certain aspects of neural activity that are related to coordinated muscle movement.

Sexual Reproductive System

The strategy of sexual fusion between cells derived from two individuals has given rise to an astonishing array of reproductive structures in the plant and animal kingdoms. All function to promote successful fertilization. All are linked to the environment in which they developed. For some organisms living in seas and lakes, sperm may be released into the surrounding medium to swim about until they bump into receptive eggs. Such a strategy obviously poses some problems for land-dwelling organisms. How these problems are circumvented by the stationary plants is fascinating, as you will read in the next unit of the book. Certain land-dwelling animals such as ourselves have developed a very different reproductive strategy. Within the female body is a portable microenvironment conducive to the successful fertilization of eggs, and to protecting and nurturing the embryo. Figures 12.19 and 12.20 show the reproductive systems of the human male and female.

In both males and females, the primary reproductive organs are **gonads;** the remainder are **accessory reproductive organs.** The gonads not only produce gametes, they secrete important sex hormones. Accessory reproductive organs include gland-lined channels and structures through which gametes are transported. In both males and females, these organs don't become functional until the onset of puberty (usually between the ages ten and fourteen).

Male gonads are known as **testes** (singular, testis). Within testes are highly coiled *seminiferous tubules,* in which sperm are formed. At any given time, cells lining some segment of these tubules are undergoing division to become mature sperm. In fact, gamete production in adult males is a continuous process by which several hundred million sperm are manufactured every day. In its final form, a mature sperm cell consists of a DNA-packed head, a mitochondrial powerhouse, and a tail. That tail is so powerful it can propel the sperm 90–360 millimeters ($3\frac{1}{2}$–14 inches) in 90 seconds!

Also in the testes are *interstitial cells,* which produce and secrete sex hormones. The most important is the hormone *testosterone.* It is involved in sperm production, in maintaining normal sexuality, and in the growth and maintenance of secondary sex traits in males. These traits include the form of external sex organs, deepening of the voice, skin texture, overall body size, the distribution of hair, fat, and skeletal muscle masses—even certain behavior patterns, which can

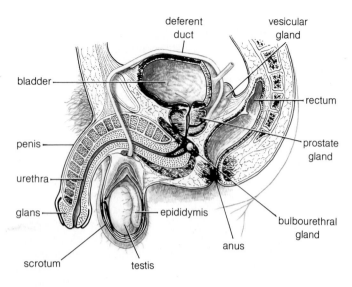

Figure 12.19 Primary and accessory reproductive organs of the human male.

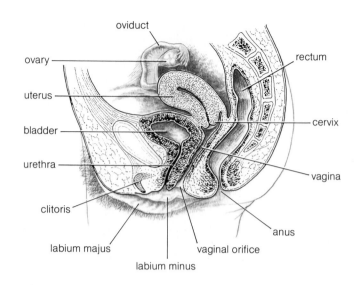

Figure 12.20 Primary and secondary reproductive organs of the human female.

be correlated with increasing and decreasing levels of testosterone secretion. (Testosterone is also present in females; for them, it stimulates overall body growth and helps maintain normal sexuality.)

Mature sperm leave the seminiferous tubules and pass through a long, coiled tube (the *epididymis*) and a thick-walled duct (the *vas deferens*), which either store most of the sperm or, during sexual activity, move them along by peristaltic action. As they are transported, the sperm become mixed with secretions from glands such as the *seminal vesicles* and the *prostate*. These secretions make up most of the *semen*, the sperm-carrying fluid that is ejaculated from the male body during sexual intercourse, or *coitus*.

The organ from which sperm are ejaculated is the *penis*. It consists of three cylinders of spongy vascular tissue arranged around the urethra. During normal activity, the blood vessels leading into these cylinders are constricted, and the penis is limp. During early sexual excitation, blood flows into the cylinders faster than it flows out. Blood collects in the spongy tissue and makes the organ rigid, at which time blood inflow equals blood outflow. Rhythmic muscular contractions force the contents of the seminal vesicle and the prostate into the urethra (which, recall, also serves to excrete urine from the body), and then expel the semen from the penis. During ejaculation, a sphincter closes off the bladder so that sperm cannot enter it; and neither can urine be excreted at that time. The muscular contractions, ejaculation, and the associated sensations of release, warmth, and relaxation constitute the event called *orgasm*.

Female orgasm is characterized by similar physical events (uterine and vaginal contractions, skeletal–muscular activity) and similar sensations. A male or female may or may not reach this state of excitation during sexual intercourse. But if the male ejaculates, the female can become pregnant whether she experiences orgasm or not.

In the female, the primary reproductive organs are a pair of **ovaries,** where eggs mature prior to fertilization. Ovaries produce the important sex hormones estrogen and progesterone, the effects of which will be described in Chapter Fourteen. Accessory reproductive organs include the oviducts (or fallopian tubes), uterus, vagina, external genitalia, and mammary glands. The *oviducts* are passageways that channel mature eggs from the ovary into the uterus. The *uterus* houses the developing individual during pregnancy. It is also the source of menstrual discharge (the monthly sloughing off of blood-enriched uterine lining when pregnancy does not occur), and it is the organ of muscular contraction during labor (when the fetus is expelled from the body at birth). The *vagina* is the organ that receives sperm from the male, and it forms part of the birth canal. It also acts as a channel to the exterior for uterine secretions and for menstrual flow. The external genitalia, or *vulva*, include organs for sexual stimulation and organs lined with fat tissues for protection of these external parts. *Mammary glands*, or breasts, function in lactation (the secretion of milk for offspring).

The actual timing of reproduction, and the intricate controls governing the female reproductive tract during the span from fertilization through development to birth, are a subject of Chapter Fourteen. But in this context, we can say that all the structural and behavioral features just outlined represent the products of natural selection; they promote reproduction and help perpetuate adaptive genotypes.

Systems of Control

Each human body comes equipped with systems for circulation, respiration, digestion, metabolic waste disposal, movement, and reproduction. Each of these systems has evolved as a functional unit for accomplishing one highly specialized task. But one does not operate independently of the others. Two organ systems exist that integrate the activities of all the body's separate systems. The effects of one are rapid and of short duration; the effects of the other are more prolonged.

The **nervous system** is designed to receive signals from the external (and to some extent) internal environment, and then to stimulate the body into response. Special cells and organs act as *receptors* for stimuli, which may be chemical, olfactory, tactile, auditory, or visual in nature (Chapter Nineteen). Nerve cells transform the messages into electrochemical signals called *nerve impulses* that travel rapidly to the body's control centers—the *spinal cord* and *brain*, which are essentially networks and clusters of nerve cells. At this point, the incoming messages are interpreted, and appropriate responses are set in motion by sending new signals outward on other nerve tracks to appropriate body parts.

The **endocrine system** helps govern body functioning through the production and secretion of chemical messengers called *hormones*. Hormones travel through the bloodstream to effect the functioning of body parts in a coordinated way. Activities of the endocrine system are linked closely with that of the nervous system. Like the nervous system, these activities are linked to external conditions as well as to the internal state of affairs. Together these two systems bring us to a new level of biological organization—that of the multicellular organism, and how its parts function as a whole. That is the subject of the next chapter.

Perspective

Of all the topics covered so far in this book, why have we singled out human anatomy and physiology—the form and function of parts of the human body—as a vocabulary-building subject? There are two reasons, one philosophical and the other somewhat practical.

First, familiarity with the structure and function of your own body is a gateway to discovering how much you belong to the rest of the living world. The division of labor in your own body is only one strategy among countless multicellular strategies. Yet it is like all others in its solutions to the same general problems associated with increased size and complexity. In later chapters, when you read about animal evolution, you will come to see your place in an evolutionary story of magnificent dimensions.

Second, it may be that you are not going to study further in the biological sciences, but no matter where you go from here, you must take your body with you. Learning about it can give you a better sense of how it functions—and what might be going on when its functioning becomes impaired. Having read about the kidney's central importance in filtration of the blood and in water regulation, for example, you might sooner recognize that a persistent pain in the region of your kidneys is a warning signal that ought to be heeded at once. Having read about the delicate, ciliated structures designed to keep your respiratory passageways clear, you might sooner admit that a persistent coughing-up of mucus tells you something is interfering severely with their functioning. And learning about your body can also demystify a doctor's diagnosis of what might be going wrong—which might help make you an active participant rather than a passive receptacle for probes and pills in reinstating your physiological health.

Recommended Readings

Anthony, C. and N. Kolthoff. 1975. *Textbook of Anatomy and Physiology.* Ninth edition. St. Louis: Mosby. Straightforward introduction to human body parts and processes.

Beck, W. 1971. *Human Design.* New York: Harcourt Brace Jovanovich. A comprehensive but accessible coverage of human physiology.

Crouch, J. and J. McClintic. 1976. *Human Anatomy and Physiology.* Second edition. New York: Wiley. Well-written text, with excellent illustrations.

Schmidt-Nielsen, K. 1970. *Animal Physiology.* Third edition. Englewood Cliffs, New Jersey: Prentice-Hall. Brief, well-written comparative approach to animal physiology. An excellent book.

Chapter Thirteen

Figure 13.1 Olympic medalist Nadia Comaneci in her hour of total control and integration of the human body.

Animal Systems of Integration

That all your cells, tissues, organs, and organ systems show such unity in the way they function is simply extraordinary. At every instant of your life, some parts of your body are being called into action in a coordinated manner, even as other parts are being actively suppressed. At every instant, every body part is constantly monitored and evaluated—not for its own sake, but for how it fits into and will influence the working pattern of the whole. This sweeping integration of bodily activities is a never ending process that begins at or before the moment of conception and continues until the moment of death. Whether you are asleep, awake, relaxed, or alert to danger, integration demands participation of all systems in your body. It depends utterly on the functioning of two interrelated networks for control: the human nervous system and endocrine system. These systems interact with all other body systems to maintain a state of **homeostasis**—a constancy of internal conditions even as the environment changes. In terms of sheer complexity, they are without parallel in the living world.

In this chapter, we will look first at the electrical and chemical nature of the **neuron** (Figure 12.6), the nerve cell of all complex, multicellular animals. We will see how messages are received, how they are propagated within a neuron, and how they are passed from one neuron to another. We will then see how messages influence **effectors** (the body's muscles and glands) to produce movement and chemical changes that counterbalance changes in external and internal conditions. We will move on to the manner in which messages are coordinated by the nervous system and endocrine system. Finally, we will look at some case studies of neural and hormonal interactions—not as an exercise in memorization, but as a way of conveying the awesome complexity on which even the simplest events in your life are based.

The Neuron "At Rest"

Like all cells, neurons have a plasma membrane, a lipid bilayer structure in which proteins and other molecules are embedded. This membrane separates the components of the cytoplasm from the extracellular fluid surrounding the neuron. You know from Chapter Four that the membrane is differentially permeable. It passively permits some molecules and ions to move along their concentration gradient across the membrane, and it passively prevents others from doing so. Sometimes it actively transports molecules and ions across the membrane, so that they move *against* their concentration gradient. Thus you already know something about the nature of the neuron, which is the basic building block of the nervous system:

The plasma membrane of a neuron is differentially permeable, and this property is essential in the propagation of messages from one part of the neuron to the other.

The nerve cell membrane happens to be a region of high electrical resistance: its lipid components by themselves show little or no tendency to permit the passage of charged particles across the membrane. But within the neuron are high concentrations of positively charged potassium ions (K^+) as well as many fairly large organic molecules (proteins, for instance) that contain negatively charged groups. And in the extracellular fluid surrounding the neuron are high concentrations of positively charged sodium ions (Na^+) as well as other kinds of ions. Thus we have a situation in which the nerve cell membrane separates two highly charged regions.

A neuron "at rest" (in other words, when it is not being stimulated) has far more K^+ ions concentrated inside than out, and far more Na^+ ions concentrated outside than in. The resting neuron is constantly engaged in maintaining these concentration gradients across its plasma membrane. Potassium ions tend to diffuse along their concentration gradient (out of the neuron), but enzymes in the membrane actively pump them back in. Similarly, sodium ions tend to diffuse down their concentration gradient (into the neuron). But the membrane is not nearly as permeable to Na^+ as it is to K^+. Furthermore, the sodium ions that do diffuse in are actively pumped back out.

Even with these sodium and potassium "pumps," the fact remains that the membrane is *more* permeable to outgoing K^+ than it is to incoming Na^+, so there is a net outward, passive movement of positively charged ions. And remember that large, negatively charged organic molecules inside a cell cannot cross the membrane at all. As a result, *the inside of a resting neuron has a negative charge with respect to the outside.* Thus there is an electric gradient across the membrane, which is known as the **resting membrane potential.**

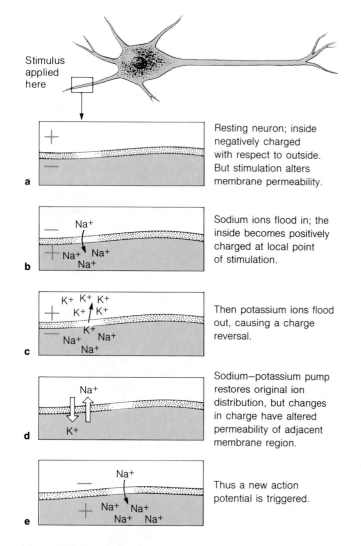

Stimulus applied here

a Resting neuron; inside negatively charged with respect to outside. But stimulation alters membrane permeability.

b Na+ Na+ Na+ Na+ — Sodium ions flood in; the inside becomes positively charged at local point of stimulation.

c K+ K+ K+ K+ K+ K+ Na+ Na+ — Then potassium ions flood out, causing a charge reversal.

d Na+ K+ — Sodium–potassium pump restores original ion distribution, but changes in charge have altered permeability of adjacent membrane region.

e Na+ Na+ Na+ Na+ Na+ — Thus a new action potential is triggered.

Figure 13.2 Propagation of a nerve impulse.

Propagation of Nerve Impulses

When a neuron is stimulated, its resting membrane potential changes. Not only is there a change in the ionic components on either side of the membrane, there is an associated movement of electrical energy (a current). First sodium ions flood in at the point of stimulation, which makes the interior positively charged with respect to the outside. Then potassium ions move out, so that the interior rapidly reverts to the negative state. The rapid, cyclic alteration in the electric gradient across a nerve cell membrane is called an **action potential.** This is the "message," or nerve impulse, that travels along a neuron.

But it is not a *single* action potential that travels along. When an action potential occurs at a point of stimulation, the local current triggers a *new* action potential at an adjacent point on the membrane (Figure 13.2). The new action potential is no different from its predecessor. This means a nerve impulse does not dwindle away as it travels along, and neither does it build up in intensity. An action potential arriving at one end of a neuron is exactly the same as the first one that was triggered. A nerve impulse transmitted in each neuron is sometimes called an "all-or-nothing" event.

What triggers each new action potential in line? At the initial point of stimulation, the electric current flow somehow causes a local change in membrane permeability. The membrane momentarily becomes *600 times* more permeable to sodium ions than it was before! The sodium ions carry in more net positive charge than the potassium ions are carrying out, which reverses the charge. A fraction of a second later, the membrane becomes less permeable to sodium ions and more permeable to potassium ions, so that the interior of the neuron reverts to the negative state. The sodium–potassium pump then works to restore the original distribution of sodium and potassium ions. In the meantime, the electrochemical activity has altered the permeability of the adjacent membrane region (Figure 13.2). Sodium ions flood in, potassium ions move out. The current flow affects the permeability of the *next* membrane region, and so on down to the end of the neuron.

What determines the direction of an action potential? It depends on the location of the stimulus. For most neurons, action potentials are triggered at one end and move to the other. For muscle cells, which will be described later, action potentials are triggered midway between the two ends and are propagated in both directions. The direction of propagation is always away from the point of stimulation, for it takes a certain period of time before the membrane has been restored to the resting state following its stimulation. (Most nerve cells may produce up to 100 action potentials every second.)

Receptors

Any change in an organism's external environment or within its body is known as a **stimulus.** All stimuli are one form of energy change or another. Some are changes in light or heat energy, sound wave energy, or chemical energy. But no matter what the form of energy may be, it must be translated into the electrochemical energy of action potentials, for there is no other way that neurons can communicate information about the internal and external world to different regions of the body. The task of translating the energy of stimuli into electrochemical energy falls to the body's **receptors.**

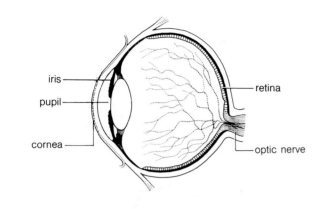

a *A human eye*

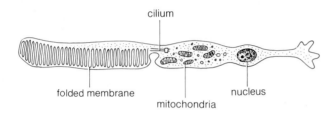

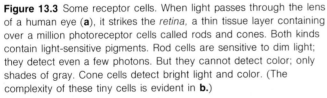

b *A photoreceptor cell (rod cell)*

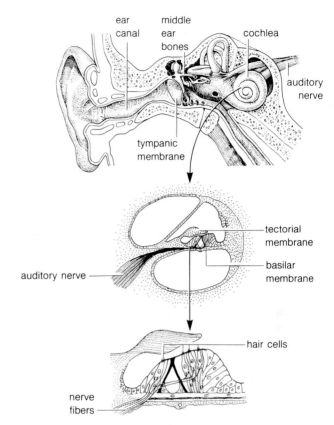

c *Membranes, compartments, and pressure receptor cells in the human ear*

Figure 13.3 Some receptor cells. When light passes through the lens of a human eye (**a**), it strikes the *retina,* a thin tissue layer containing over a million photoreceptor cells called rods and cones. Both kinds contain light-sensitive pigments. Rod cells are sensitive to dim light; they detect even a few photons. But they cannot detect color; only shades of gray. Cone cells detect bright light and color. (The complexity of these tiny cells is evident in **b.**)

Absorption of photons changes the arrangement of pigment molecules, which changes the cell's membrane permeability at the point of stimulation. Thus action potentials are triggered. Each rod and cone cell synapses on a sensory neuron, which sends an electrochemical message on its way to the *optic nerve,* a bundle of nerve cells that extends into the brain. Many rod cells synapse on sensory neurons, so the summation of their activity can trigger an action potential even in dim light. Only a few cone cells synapse on each sensory cell, so the threshold value for triggering an action potential is higher (the light must be more intense).

When sound energy passes into the ear canal (**c**), it disturbs air molecules. The molecules become compressed in some regions and more spaced in others, creating pressure waves. The waves bounce back and forth and strike the tympanic membrane. Three ear bones transmit the vibrations across the middle-ear cavity, then to pressure receptors positioned in the fluid-filled cavities of the inner ear (cochlea). The inner ear is a coiled passage divded down its length by a basilar membrane, on which receptor cells (hair cells) are arranged. When this membrane moves up and down in response to pressure waves, the hairs rub against the tectorial membrane. The friction becomes transformed into an action potential. The message moves through sensory neurons to auditory centers in the brain.

Receptors at or near the body's surface are specialized for receiving stimuli from the outside—changes in sunlight or temperature, sights, noises, smells, the presence of sweet- or sour-tasting things, touch, and pain. Receptors embedded within the body receive signals concerning balance, movement, pressure of blood and other body fluids, temperature, and the chemical balance of substances in specific regions. Figure 13.3 gives two examples of the degree of cellular specialization involved in the reception of stimuli.

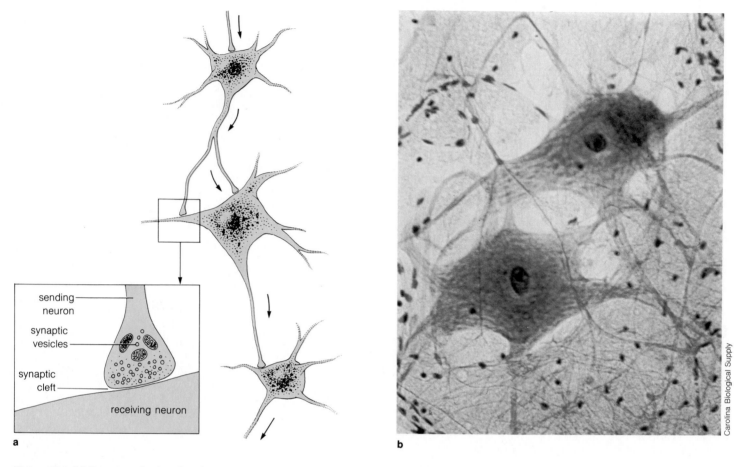

a

b

Figure 13.4 (**a**) Synapses, the junctions between neurons (**b**). The boxed insert is a close-up of a synaptic cleft. The beginnings of information control occur as signals are passed from one neuron to the next at synapses.

Labels in figure: sending neuron; synaptic vesicles; synaptic cleft; receiving neuron

Neuron-to-Neuron Junctions

Receptors are special kinds of nerve cells or glands. Adjacent to their endings are **sensory neurons,** which carry the information received to the spinal cord and brain, then to the body's muscles and glands. Some carry outgoing messages to skeletal muscle, hence they are called **motor neurons.** A third and by far the largest group contains the **intermediate neurons,** which are concentrated in the brain and spinal cord.

Each junction between all these neurons is a special region called a **synapse.** Here, the activity of one neuron may influence the activity of another. As Figure 13.4 suggests, most synapses are junctions where the axon of one neuron terminates near the cell body or dendrites of another neuron.

It is not the action potential that bridges the small cleft between a sending neuron and a receiving neuron. Rather, when an action potential reaches a synapse, it causes a chemical substance to be released from small, membrane-enclosed vesicles within the end of the neuron. This substance, called a **neurotransmitter,** diffuses across the synaptic cleft to the receiving neuron.

A receiving neuron may respond to neurotransmitter by generating an action potential of its own. But other factors may intervene to prevent it from doing so. A neuron sending an impulse usually has many branches and may synapse with more than one other neuron. Similarly, each receiving neuron usually has many hundreds or thousands of synapses from different neurons acting upon it. The patterns of synapses that these branchings create are important control networks. Some sending neurons release *inhibitory* chemicals at their synapses. Their message to the receiving neuron is this: "Don't listen to what others are telling you; more important things are going on elsewhere that require you to be quiet now." Other sending neurons release *excitatory* chemicals, but by themselves they may not cause a receiving neuron to generate an action potential. Instead their message is this: "Stand by to receive a message from other neurons, and if you do, fire at once."

Each neuron must receive a certain minimal intensity of chemical stimulation, a **threshold value,** before it will fire off. Some neurons are activated easily, but some will not respond unless they are hit with a barrage of incoming messages. Each neuron sums the incoming inhibitory and excitatory signals, and it generates an action potential only when the net stimulation passes its threshold value.

Thus the state of the receiving neuron, and the nature of all incoming messages, dictate whether or not it will generate a nerve impulse in response to a given signal.

Synapses are known to be vulnerable to many drugs and to toxic substances. The synthesis, storage, or activity of neurotransmitter substances can be modified by these substances. Sometimes the neurotransmitter can be prevented from interacting with the membrane of the receiving neuron or muscle cell. The bacterium that causes **tetanus** produces a toxin that blocks inhibitory messages that normally would reach motor neurons. Thus excitatory messages flow incessantly to the motor neurons, which can cause uncontrollable muscle spasms in the jaw muscles, then spasms and seizures throughout the body.

Neuron-to-Muscle Junctions

Muscle Contraction Like nerve cell membranes, muscle cell membranes also can transmit an action potential. Muscle cells may be stimulated in three ways. As you read earlier, cardiac muscle activity is initiated during embryonic development and continues spontaneously during the life of the organism. Hormones influence the activity of all three types of muscle (smooth, skeletal, and cardiac), as do neurons that terminate upon them. Here, we will focus on the manner in which skeletal muscles contract under signals from motor neurons.

Figure 12.5 showed an illustration of skeletal muscle tissue. Each cylindrical muscle cell in this tissue is called a **muscle fiber.** What we call a "muscle" is really a number of muscle fibers bound together with connective tissue, through which blood vessels and neurons thread. Some muscles contain only a few hundred muscle fibers, others contain hundreds of thousands.

Figure 13.5 is a closer view of skeletal muscle. Each fiber can be seen to be composed of smaller filaments

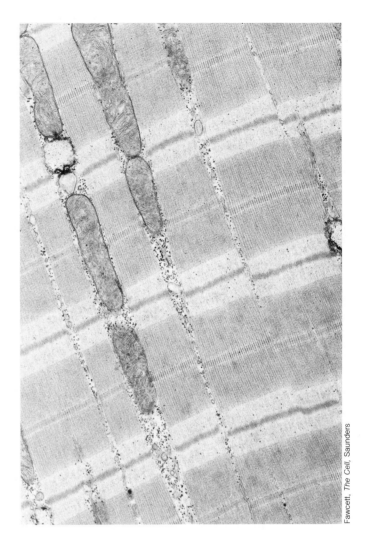

Figure 13.5 Electron micrograph of skeletal muscle.

(*myofibrils*), which contain smaller filaments still. These *myofilaments* are of two kinds: one is made of myosin (a thick protein), and the other is made of actin (a thin protein). Myosin makes up the dark bands visible in Figure 13.5, and actin makes up the lighter bands. Each light-dark-light pattern is a **sarcomere,** the fundamental unit of contraction.

The process of contraction may begin where a motor neuron forms a junction with skeletal muscle (Figure 13.6). When a nerve impulse reaches the small synaptic vesicles in the end of the motor neuron, they are stimulated into releasing a chemical substance (in this case, acetylcholine). The interaction of this substance with receptor molecules on the muscle cell membrane increases that membrane's permeability to sodium and potassium ions. The resulting

Figure 13.6 Transmission electron micrograph (**a**) and sketch (**b**) of a neuromuscular junction. The axon of this motor neuron terminates in a slight depression on the muscle cell surface. A narrow space (the synaptic cleft) separates the nerve ending from the muscle cell. Notice how the part of the nerve cell facing this space is not wrapped in a protective myelin sheath. Notice also how channels from the main synaptic cleft penetrate the region of the muscle cell membrane. An as-yet unidentified substance fills all these channels. And it is through this substance that chemical messages may travel from the nerve to bring about muscle contraction.

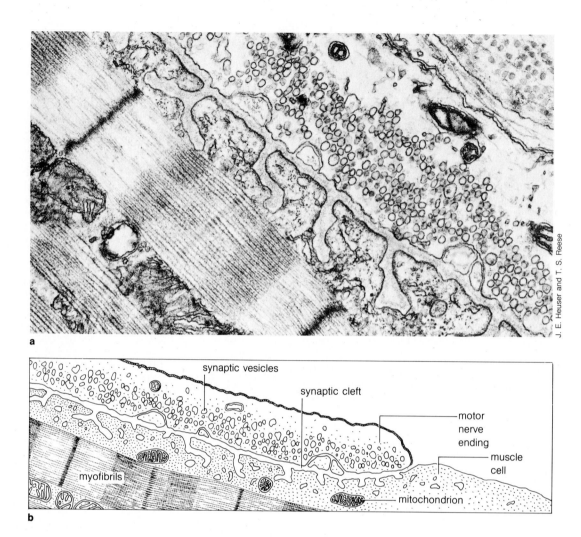

a

synaptic vesicles

synaptic cleft

motor nerve ending

muscle cell

myofibrils

mitochondrion

b

changes in ion levels form the basis of propagation of a muscle action potential. (Acetylcholine does not accumulate in the cleft, because enzymes present on the muscle cell membrane rapidly break it down. Thus a single impulse can't trigger multiple contractions.)

At each sarcomere, the propagating muscle action potential leads to the release of calcium ions from a special membrane (the sarcoplasmic reticulum) that surrounds the myofibrils. The calcium ions bind to the actin in such a way that the actin is free to interact with adjacent myosin filaments. These interactions involve the activation of enzymes in the myosin that split ATP molecules, which releases energy that is used to produce the following movements. "Cross-bridges" projecting from the myosin filaments attach to actin, bend, release their hold, and attach again (Figure 13.7). In this way the actin is propelled past the myosin, much like a boat being propelled down a narrow stream by oarsmen alternately digging their oars into the bank and pushing on them, all in the same direction. Each cross-bridge movement uses up one ATP molecule. When you think of the tremendous number of cross-bridges involved, you can seen why muscle activity consumes tremendous amounts of ATP. More ATP is used to run a "calcium pump," which transports calcium ions back to the sarcoplasmic reticulum. With the removal of calcium ions from the myofibrils, the actin no longer can interact with myosin and the sarcomere returns to the relaxed state.

The Reflex Response So far, we have described action potentials in individual neurons and muscle cells. Let's now see how such action potentials can be coordinated to produce an integrated response to a simple environmental stimulus. A relatively simple response pattern, which trav-

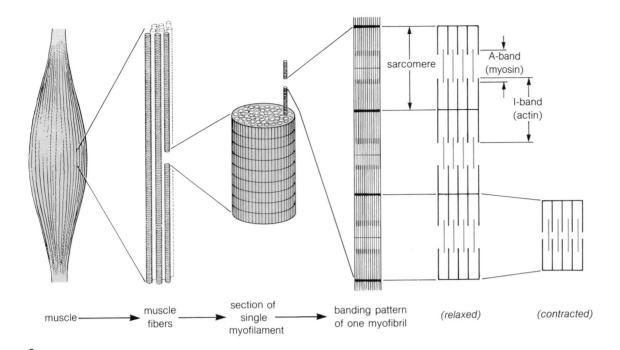

muscle ⟶ muscle fibers ⟶ section of single myofilament ⟶ banding pattern of one myofibril

(relaxed) *(contracted)*

A-band (myosin)

I-band (actin)

sarcomere

a

els directly from receptor cells to a sensory neuron, then to only a few intermediate neurons, then to a motor neuron, is called a **reflex response.**

Consider what happens when you step barefoot on a thorn. Even before you have time to think about what has happened, your foot is raised by the **flexion reflex.** Receptors in your foot stimulate a sensory neuron. Some branches of the sensory neuron synapse on an intermediate neuron that transmits information to the brain. Other branches synapse on a motor neuron, which synapses on certain flexor muscles that bend the ankle (causing the foot to be lifted). If it is an especially big thorn, and the pain is very intense, a number of motor neurons may be stimulated so that the knee and thigh also bend. At the same time, intermediate neurons synapse on extensor muscles in the other leg, which cause the leg to straighten and take more of the body's weight off the injured foot.

Simple reflex action serves the body well enough in some situations. But often, changes in the inner and outer environment demand far more complex responses. Myriad parts of the body must be called into action in a coordinated way. Thus even the neurons involved in simple reflexes may have their activities modified by signals coming from neurons in the brain. If you had stepped on that thorn while running to catch the last boat out of Pompeii when Mount Vesuvius blew its top off in AD 79, impulses sending information to the brain about the exploding mass and the advancing wall of lava behind you, the falling ashes and cinders all around you, and the last boat pulling away from

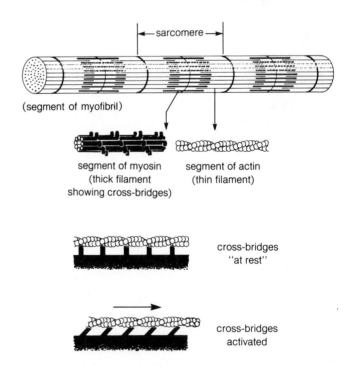

sarcomere

(segment of myofibril)

segment of myosin (thick filament showing cross-bridges)

segment of actin (thin filament)

cross-bridges "at rest"

cross-bridges activated

b *Movement of myosin cross-bridges. Several actin filaments are arranged around each myosin filament. For clarity, only one is shown here.*

Figure 13.7 Proposed mechanism of skeletal muscle contraction, as described in the text.

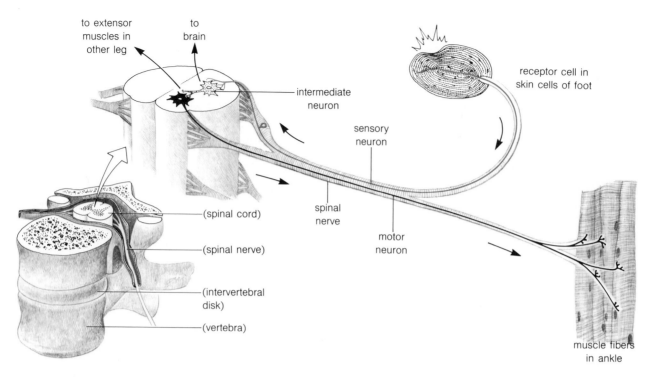

Figure 13.8 The flexion reflex, one of the simpler examples of nervous coordination, as described in the text.

the dock in front of you would have resulted in the generation of other streams of impulses that would have modified the behavior of your leg motor neurons. They would not have responded to signals from the sensory neurons in your foot, your progress would not have been interrupted as it otherwise would be by a reflex lifting of your foot, and you probably would not even have been conscious of the pain (Figure 13.8).

The Nervous System: Increasing Refinements on Basic Reflex Activity

Billions of neurons make up your nervous system. But they don't weave haphazardly through your body like so many limp noodles. Instead they are organized in what we perceive as two interconnected zones: the **central nervous system** (the spinal cord and brain) and the **peripheral nervous system** (all the sensory and motor neurons leading to and from the spinal cord and brain). In the peripheral nervous system, the nerve processes of large numbers of individual neurons are combined into bundles, which we call "nerves." Depending on what kind of neurons are bundled up in them, there are "sensory" nerves, "motor" nerves, even "mixed" nerves.

The Spinal Cord Pair after pair of nerves lead into and out of different levels of the **spinal cord**—a conduit of nerve tissue encased in the protective bones of the vertebral column. If you were to cut out one of these levels and look at it in cross-section, you would find it looks something like a gray butterfly mounted on a white background (Figure 13.8). The **white matter** is mostly made of the myelin sheaths that surround the axons of neurons running up and down the length of the spinal cord. The **gray matter** is mostly the cell bodies of unmyelinated intermediate neurons that lie between sensory and motor neurons. It is in this zone of intermediate neurons that integration of nerve impulses begins.

In simple reflex activity, sensory neurons terminating in the butterfly zone can synapse directly on dendrites of motor neurons, so appropriate responses *can* be activated directly. (The word "reflex" is derived from the Latin word meaning "to turn back.") But such one-to-one connections are not all that common. More often, many intermediate neurons intervene, coordinating all the interrelated events (such as antagonistic muscle movements) needed to bring about reflex action in the body's trunk and limbs. Nervous coordination of these body parts is the simplest level of control; no other part of the nervous system need be involved. *Even so, information about every single reflex action is sent, through upward-directed neurons, to the brain.* The brain

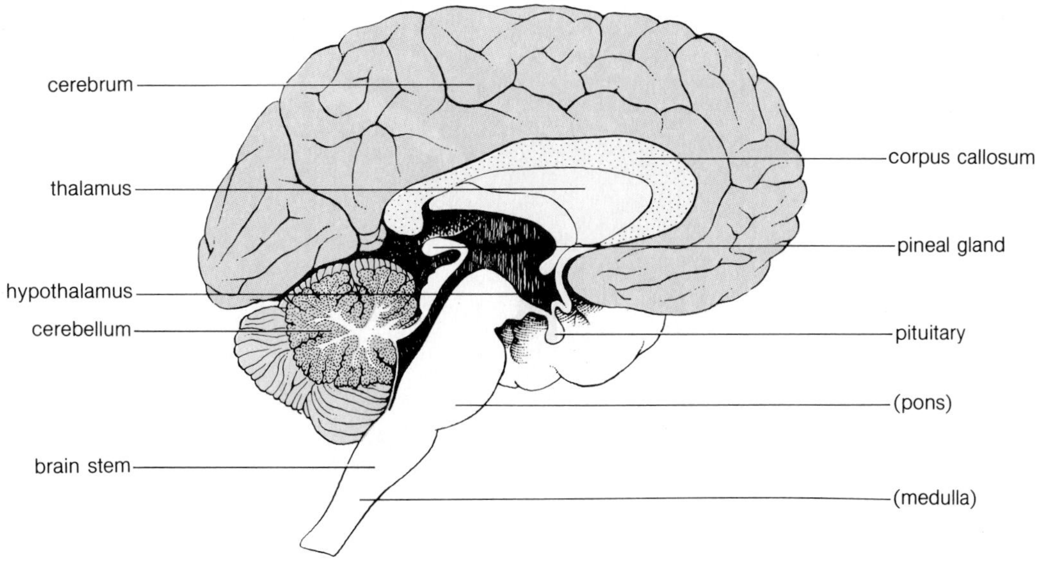

Figure 13.9 Parts of the human brain, shown cut through its midsection.

cerebrum

thalamus

hypothalamus

cerebellum

brain stem

corpus callosum

pineal gland

pituitary

(pons)

(medulla)

interprets the information, and stores it away for future reference. And if more modifications to the response are needed at once, then the brain sends further instructions to the appropriate levels by way of downward-directed neurons.

Parts of the Brain The human **brain** is about 1,360 grams (about 3 pounds) of concentrated nerve tissue. In both the structural and functional sense, the brain acts with the spinal cord as a single, coordinated unit. But as impulses ascend the central nervous system—from spinal cord through brain stem, cerebellum, hypothalamus, thalamus, and finally reaching the cerebrum—there is increasing integration of all messages that are being received and acted upon.

Figures 13.9 and 13.10 show the functional zones of the brain. The function of the spinal cord, again, is to integrate trunk and limb movements at the reflex level, and to relay information about those activities to the brain. The **brain stem** is basically an extension of the spinal cord, for it integrates the movements of the sense organs and the head at the reflex level. One region of the brain stem, the *medulla,* governs the automatic reflexes of breathing, heartbeat, blood pressure, sucking, and swallowing. These are the most basic of all vital functions. The **cerebellum** is an outgrowth of the brain stem, a crossroad for all neurons

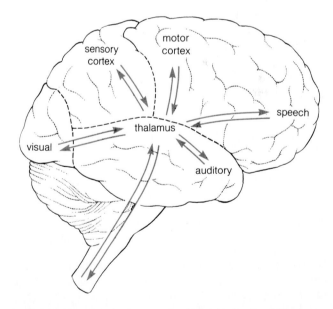

Figure 13.10 Relationship between the human thalamus and the cerebrum. The thalamus acts as a crossroads in the brain, routing impulses arriving from sensory neurons of the head and body down preestablished nerve channels leading into appropriate regions of the cerebrum. In Figure 13.11 the cerebral regions involved in registering tactile stimuli and generating motor outputs are shown in more detail.

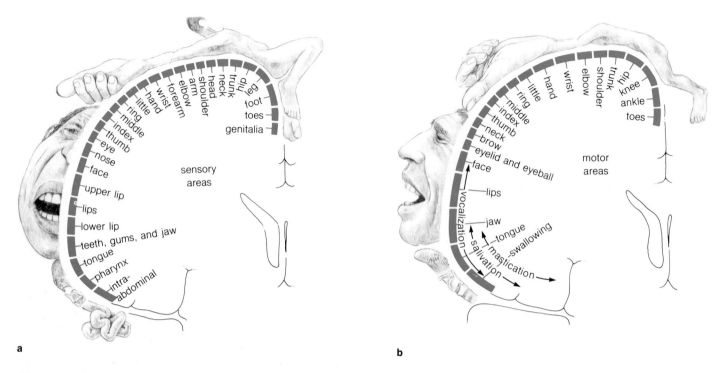

Figure 13.11 (**a**) Organization of sensory areas in the cerebral cortex. The sensory region is a strip a little over an inch wide running from the top of the head to just above the ear. This sketch is a cross-section through the right side of the cerebrum of someone facing you. It shows the proportion of sensory cortex devoted to different body parts. (**b**) Organization of motor areas in the cerebral cortex. The motor region is just in front of the sensory region. Notice in both cases that the amount of cortical tissue devoted to various body regions does not correspond to relative size. (After Penfield and Rasmussen, *The Cerebral Cortex of Man*)

concerned with the stimulation and responsiveness of muscles and skin. It smooths out gross movements, and it stabilizes movements to ensure maintenance of the body's balance in space. (We get an idea of its importance by observing individuals whose cerebellum has been damaged. All their movements—walking, speaking, even breathing—are characterized by uncontrollable jerking and tremors.) Above the cerebellum is the **hypothalamus,** a region that governs metabolism of the entire body. Embedded within it are receptors and effectors concerned with monitoring such acts as heat regulation, hunger, thirst, fear, and sex.

All major pathways of incoming neurons wind their way to the **thalamus,** the gateway to the cerebrum (Figure 13.10). It is here that impulses concerning sights, sounds, tastes, smells, position, movement, and pressure converge before they are sent on. These messages are not interpreted here, for the thalamus is little more than a switching station, with roads running out in different directions to take the messages to exactly the right destination in the cerebrum.

The **cerebrum** is the brain region responsible for the highest level of integration. It takes messages and rapidly responds by shooting impulses back and forth from one of

its regions to the next in order to interpret and evaluate their meaning. The cerebrum is the most advanced layering of intermediate neurons. At least a *billion* of them radiate outward from the thalamus and form the convoluted masses we usually think of when we hear the word "brain." These masses are divided into two distinct halves: the left and right cerebral hemispheres. But the hemispheres are not functionally separate. They are connected through various bridges, the most prominent of which is the *corpus callosum* (Figure 13.9). The cerebral hemispheres themselves are made of an inner mass of white matter and an outer layer of gray matter called the *cortex* (after the Latin word meaning "tree bark"). The cortex is highly convoluted. Evolution has confined the human brain in the space of a rigid, protective skull, which has grown larger over time but at a slower pace than our increasingly complex needs for integration. The convolutions provide the maximum surface area possible within the limits of the skull's volume.

As Figure 13.10 shows, certain regions of the cerebral hemispheres can be assigned specific roles. The *visual cortex* receives and processes impulses coming from the eyes. The *auditory cortex* receives and processes impulses coming from

Table 13.1 Functional Divisions of the Human Nervous System

Central nervous system (brain and spinal cord)	*Control center for all body activity. Receives, then analyzes, coordinates, and stores information from the internal and external environments. Maintains overall homeostatic state by originating new commands as conditions change.*
Peripheral nervous system (sensory and motor neurons)	*Branches and clusters of sensory, motor, and mixed nerves leading to and away from the central nervous system. Connects all receptors and effectors to the central nervous system.*
Somatic pathways	*Two-way communication with body parts (such as arms and legs) that can be consciously controlled.*
Autonomic pathways	*Two-way communication with internal organs and accessory glands. Usually not under conscious control. Autonomic pathways are of two kinds:*
	Sympathetic nerves which turn on certain organ activities and glandular activities, and turn others off.
	Parasympathetic nerves, which work antagonistically with sympathetic nerves in turning the body's organs and glands on and off. See Figure 13.12.

the ears. Another region is involved in the uniquely human process of relating exterior objects and events to words—to controlling our use of language and speech. Near the center of each hemisphere, two adjacent bands of cerebral cortex receive sensory information from the body and send out commands for the integrated movement of body parts (Figure 13.11). The number of neurons devoted to these tasks is determined not by the size of the body part but by its functional importance. For humans, more neurons are devoted to the hands and the face than to any other body part.

Mapping out this much of the cerebral cortex has been quite an accomplishment. But we have yet to penetrate the mysteries of the greater part of it—the immense, undefined mass of neurons known as the *associative cortex*. It is in this mass that one of our most prized possessions, our intelligence, resides. It is here that associations are made between past stimuli and present stimuli, between the consequences of past responses and what future consequences may be. It is here that we may someday come to understand the basis of our capacity for learning and memory, for abstract thought, emotions, and the conscious modification of behavior.

The Autonomic Pathways

So far, we have seen how reflex responses can be increasingly modified through activities in the peripheral and central nervous systems. Taken together, the motor and sensory pathways and the controls involved in reflex responses to external stimuli are called the **somatic nervous system.** Usually, this system is under our conscious control; at the very least we can become aware that a reflex has

occurred. But there is another important aspect of integration that is also based on a portion of the peripheral and central nervous systems. It has to do with the mobilization of internal organs such as the heart, stomach, and glands. Because it generally is not under conscious control, we call it the **autonomic nervous system.**

The autonomic nervous system is subdivided into two antagonistic parts: the *sympathetic* and the *parasympathetic* systems. It is through the interplay of the sympathetic and parasympathetic pathways that the body's organs and organ systems are adjusted as a unit to prevailing conditions.

At no time does the human body exist in suspended animation, waiting for a stimulus from the outside world to spark it into some kind of action. Whether you are standing, sitting, sleeping, eating, running, digesting food, or eliminating wastes, something must be bringing all your organs into balance with one another. But this balance is subject to rapid change. Say you have just finished eating and are doing nothing more than sitting out on the porch watching the sun go down. Impulses flow outward along parasympathetic nerves, stimulating into action all the organs and body parts related to digesting your meal. At the same time, your sympathetic pathways are quiescent. Blood vessels serving your digestive tract are wide open, but those supplying muscles and brain are constricted, so you are relaxed. Suddenly three fire engines with sirens screaming and horns blatting zoom down the street, followed by two patrol cars with flashing red lights, barking dogs, and agitated children. The excitement triggers impulses that race through your body to shut down the parasympathetic pathways and turn on the sympathetic pathways. Digestion stops; your eyes open wide, your heart starts pounding as you leap up to see what's going on. Only when things quiet down do the pathways shift priorities again, and digestion resumes where it left off.

Figure 13.12 Antagonistic pathways of the autonomic nervous system.

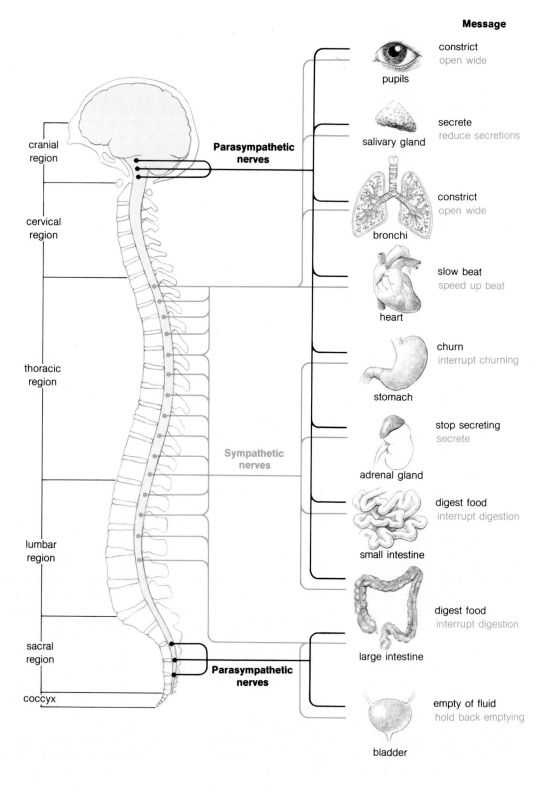

Message

constrict
open wide
pupils

secrete
reduce secretions
salivary gland

constrict
open wide
bronchi

slow beat
speed up beat
heart

churn
interrupt churning
stomach

stop secreting
secrete
adrenal gland

digest food
interrupt digestion
small intestine

digest food
interrupt digestion
large intestine

empty of fluid
hold back emptying
bladder

cranial region

cervical region

thoracic region

lumbar region

sacral region

coccyx

Parasympathetic nerves

Sympathetic nerves

Parasympathetic nerves

Generally, the parasympathetic pathways dominate muscles and organs when environmental situations permit normal metabolic functioning. But in times of stress, danger, excitement, or even heightened awareness, the sympathetic pathways dominate internal events, alerting the entire body so it is ready to respond to the need for change.

Of course, it's important to keep in mind that autonomic control isn't always an "either/or" kind of thing. Just as your life is usually filled with complicating circumstances, so also does autonomic control over organ functioning usually have gradations of antagonistic dominance. Just because your parasympathetic system signals that it's time to empty your bladder doesn't mean you are going to empty it on the front porch. If one were to isolate the principle underlying autonomic pathways, it would have to be, "There's a time and place for everything."

The Endocrine System of Chemical Control

Closely linked to the nervous system in the integration of body activities is the **endocrine system,** which consists of small, ductless glands located throughout the body (Figure 13.13). Each gland produces one or more unique chemical messengers, or **hormones.** Under appropriate conditions, these hormones are secreted directly into the bloodstream, which carries them to one or more target organs. Here they act to stimulate or inhibit key metabolic processes in the organ's cells. In most cases, the induced changes reverse the conditions that triggered the release of hormones in the first place. Hence the endocrine system is a key link in the body's negative feedback loops, helping to maintain the constancy of internal conditions even as environmental conditions change.

The main endocrine glands, the hormones they produce, their target organs, and the main changes the hormones elicit are listed in Table 13.2. As you can see, the hypothalamus also acts as an endocrine gland, for it is not only involved in nervous integration, it also secretes its own hormones. As you will read later in this chapter and in others to follow, the hormones secreted by the pituitary are central to events as diverse as migratory behavior among birds to regulation of kidney functioning to menstrual cycles in women. Another important gland is the thyroid, which produces a hormone that accelerates the rate of cellular respiration, and a hormone that inhibits the release of calcium from bone tissue. The hormones secreted by the pancreas and the adrenal cortex are extremely important in glucose metabolism.

Interestingly enough, the first endocrine glands arose as specialized regions of the nervous system. Instead of se-

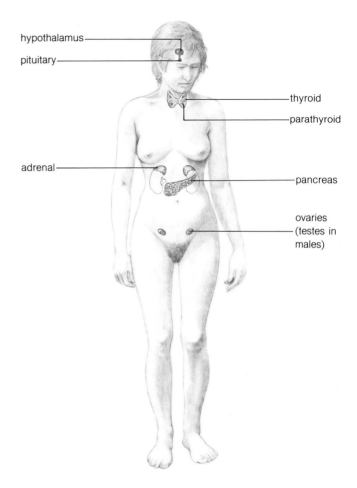

Figure 13.13 Human endocrine glands.

creting their chemical messengers at synapses, where they would affect the functioning of adjacent neurons, these specialized structures came to secrete their messages into the bloodstream, which could carry them off to influence distant parts. Even today, many of the endocrine glands retain a close functional link to the nervous system and are either directly or indirectly controlled by it. Later, in Chapter Nineteen, you will see how the endocrine system is influenced profoundly by cyclic changes in the external world, and what role it plays in adapting animal behavior to those changes.

The remainder of this chapter is devoted to a few examples of how the nervous system and the endocrine system work. Again, these examples are not presented as something you should be expected to memorize, but as a way of conveying the complex integration upon which ordinary as well as extraordinary events in your life are based.

Table 13.2 Partial List of Endocrine System Glands and Hormones

Gland	Hormone(s)	Target(s)	Main Effects
Hypothalamus	Releasing factors	Pituitary	*Regulates release of pituitary hormones*
	Oxytocin	Mammary gland	*Release of milk*
	Antidiuretic hormone (ADH) synthesis*	Kidney	*Promotes water reabsorption*
Pituitary	Adrenocorticotrophic hormone (ACTH)	Adrenal cortex	*Release of adrenal cortex hormones*
	Growth hormone	Bone, muscle	*Promotes growth*
	Gonadotropins	Ovaries, testes	*Promote gamete, hormone formation*
	Prolactin	Mammary gland	*Formation of milk*
	Thyroid-stimulating hormone (TSH)	Thyroid gland	*Release of thyroid hormone*
Thyroid	Thyroxin	Most cells	*Increases rate of metabolism*
	Thyrocalcitonin	Bone	*Lowers blood calcium*
Parathyroids	Parathyroid hormone	Bone	*Elevates blood calcium*
Adrenal cortex	Glucocorticoids	Many cell types	*Raises blood sugar level*
	Mineralocorticoids (e.g., aldosterone)	Kidney	*Promotes retention of Na+ and loss of K+ in urine*
Adrenal medulla	Adrenalin	Liver, adipose tissue	*Raises blood sugar and blood fatty acid levels*
		Heart	*Speeds heartbeat*
		Smooth muscles of blood vessels, gut	*Some contract, others relax*
Pancreas	Glucagon	Liver	*Raises blood sugar*
	Insulin	Liver, muscles, etc.	*Lowers blood sugar*
Ovaries	Estrogen	Uterus, rest of body	*Development of female traits; prepares uterus for pregnancy*
	Progesterone	Uterus	*Prepares uterus for implantation; maintains it during pregnancy*
Testes	Testosterone	Body generally	*Development of male traits*

*Secreted by pituitary.

Case Study 1: Integration of Movement

While you are bounding about on a tennis court, it's probably safe to say you aren't giving much thought to the evolutionary significance of your motor skills. Yet these are the skills that were developed not on a field of play but in contests of another sort. It was not a tennis ball but a predator stalking its human prey on some ancient savanna that once tested the human brain—tested it for its ability to analyze the predator's speed, direction, and changing pace, to assess the readiness of all parts of the body's systems of response, and to integrate all the motions needed to swing not a racquet but a hand-held ax. Even before that, in all the ancestral lines leading to the first humans, systems of integrating motor responses were developing. Just how efficient those systems have become is evident today, as you move across the court.

As the ball leaves your opponent's racquet, sunlight reflected from the ball and the racquet strikes receptors in your retinas. These modified nerve endings start tiny nerve impulses on their way to your brain. The neurons themselves converge into large nerve fibers, the **optic nerves** (Figure 13.14), then separate into two pathways. Neurons coming from the left half of each eye send messages unerringly to the left half of the brain; those coming from the right half of each eye send messages to the right half of the brain. These messages reach the thalamus, then are sent on to the visual cortex. There, specific neurons are stimulated into registering information about the vertical and horizontal angle between your eyes and the tennis ball at the time it was sent flying toward you. Impulses travel back and forth between the neurons receiving messages from the right eye and those receiving messages from the left. In comparing those impulses, your brain computes the distance between your eyes and the ball.

Meanwhile, impulses from other parts of the thalamus travel deeper into the cortex. There, each message is stored and compared in a split second to determine the direction and speed of your opponent's racquet before, during, and after it hit the ball. Impulses shoot out through your cerebral cortex to regions where sensory information has been stored on each game of tennis you have ever played in your life. These impulses are compared to information stored in memory banks to predict the ball's angle, speed, and spin.

At the same time, messages from muscles and tendons in arms, legs, trunk, feet, and hands are channeled to the spinal cord and brain stem. Messages come quickly from some muscles and tendons, slowly from others, not at all from still others—thus informing the brain stem of which muscles are taut and which are relaxed. These inputs combine with signals from the middle ear, a sensory region supplying information on the angle of your head with respect to the earth's gravitational pull.

Now the brain stem sends off new signals, this time to the sensory cortex. Using information coming from the visual cortex and memory banks, the sensory region builds up a picture of your body's position in three-dimensional space. New impulses fire off to the motor cortex, which generates signals that travel to the brain stem, down the spinal cord, and out to muscles. Some motor neurons carry impulses that cause certain muscles to contract strongly; others carry impulses widely spaced in time that call for moderate contraction. Still others are silent; the muscles they serve are left relaxed.

What your brain ignores is just as crucial as what it monitors. It sends out a steady stream of inhibitory signals to certain neurons. Regions of the cortex receiving signals from leaves rustling nearby and from traffic moving in the distance are actively suppressed by neurons that are responding to the stimulus of the ball. If those regions were to become active, *they* would send out inhibitory signals to the regions you are now employing—and your concentration would be lost.

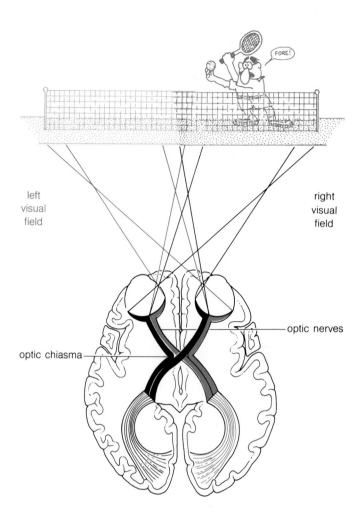

Figure 13.14 Optic nerve pathways.

As the ball approaches, your brain calculates the expected angle and speed of the bounce. Quickly you glance at your opponent, for you intend to return the ball right over the net. But he's moving forward to the center. Again memory banks are searched. Your opponent's ability to reverse direction and make a backhand return is checked against your ability to place the ball in the far corner of the court. The same instant, another region of your cortex is reassessing the ball's flight, for it is curving unexpectedly close to the outside line. Although you must be prepared to carry out an entire program of muscular contraction for your return shot, now you must be prepared to interrupt it if the ball lands outside the line! You will have only a split second to make your decision.

It strikes just inside. Impulses arriving in the visual cortex permit comparison of the actual rebound with the

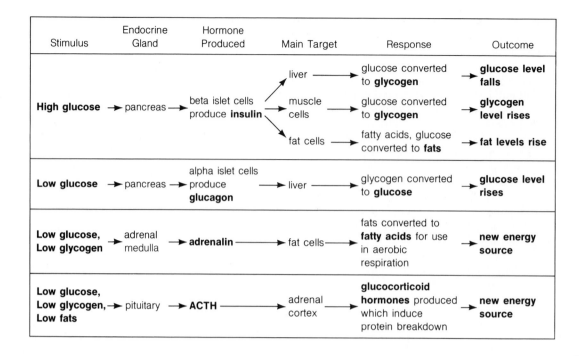

Figure 13.15 Summary of a few homeostatic feedback processes at work in energy metabolism.

Stimulus	Endocrine Gland	Hormone Produced	Main Target	Response	Outcome
High glucose →	pancreas →	beta islet cells produce **insulin**	liver →	glucose converted to **glycogen**	**glucose level falls**
			muscle cells →	glucose converted to **glycogen**	**glycogen level rises**
			fat cells →	fatty acids, glucose converted to **fats**	**fat levels rise**
Low glucose →	pancreas →	alpha islet cells produce **glucagon**	liver →	glycogen converted to **glucose**	**glucose level rises**
Low glucose, Low glycogen	adrenal medulla →	**adrenalin**	fat cells →	fats converted to **fatty acids** for use in aerobic respiration	**new energy source**
Low glucose, Low glycogen, Low fats →	pituitary →	**ACTH**	adrenal cortex →	**glucocorticoid hormones** produced which induce protein breakdown	**new energy source**

expected. Again your motor program is revised. Your body is brought closer to the ground, your right arm is straightened more, and your wrist is rotated slightly as your right foot makes one last push against the court. Again memory banks are checked to determine the speed and force with which your arm muscles must contract to hit the ball. Signals smoothed out in the cerebellum race down your spinal cord and out to your arm, your racquet swings, and the ball rockets over the net, just beyond your opponent's reach. Had it been a predator closing in for the kill, you would have lived another day. As it is, all you receive is a sense of accomplishment and your opponent's comment, "Good stroke!"

Case Study 2: Integration of Metabolic Activity

The dawn of the first day of the semester break finds you camping in the mountains, about to take off on an all-day hike. You don't want to waste time getting started, so you gulp down some cereal, cheese, and a chocolate bar as you leave camp. While you move along, your breakfast moves along, too. Inside your digestive tract, enzymes secreted by the pancreas and intestinal cells are slowly breaking it down. Starches from the cereal are degraded into glucose molecules, which move across the intestinal wall and into the bloodstream. Right now, glucose is entering the blood

faster than you can use it up, so the blood glucose concentration begins to rise. If left unchecked, it would damage every neuron in your brain. Even though neurons demand a constant supply of glucose (their *only* energy source), too much or too little can set events in motion that can lead to cell damage, loss of consciousness, even death.

But your body has a homeostatic program for storing away glucose when it's flooding in, and then distributing what it has stored when supplies are scarce. Long before glucose approaches dangerously high levels, two kinds of cells making up the islets of Langerhans in the pancreas are called into action. These cells are part of the endocrine system (Figure 13.13). Alpha islet cells make and secrete **glucagon** (a hormone). Beta islet cells make, store, and secrete **insulin** (another hormone). When too much glucose is present, molecules of glucose bind to both types of cells and change their shape. Thus changed, the alpha islet cells are inactivated and shut down—and the beta islet cells are activated and start secreting insulin.

Insulin pours into the blood, which carries it to your body's cells. (About the only ones insulin won't directly affect are your brain cells, which must remain functional largely outside the ebb and flow of hormones). As insulin reaches your liver cells, it prods enzymes of glycolysis and respiration into breaking down the glucose from your breakfast. It also stimulates enzymes needed to assemble glucose into **glycogen** (a storage starch). With the arrival of insulin in your muscle cells, glucose breakdown and glycogen storage begins here, also. On reaching fat cells throughout your body, insulin prods them into taking up more fatty acids (the legacy of oils and fats from your breakfast). These fatty acids are assembled into fats, then stored away in glistening vacuoles. At the same time, fat cells are stimulated into taking up and degrading glucose so its carbon atoms can be used in making even more fatty acids (to be converted into still more fats).

With all this insulin-stimulated activity in your body's cells, your blood glucose level stops rising. In fact, it now starts to drop. So pancreatic cells are made to act once again. At first, the imbalance affects only the beta islet cells. Without glucose to bind to them, they stop secreting insulin. Without insulin, your liver, muscles, and fatty tissues can no longer take up glucose, so the glucose level stabilizes.

But even though no more glucose is entering your body, your brain has not lessened its demands for it—and neither have your muscles, which are getting a strenuous workout! Little by little, glucose dwindles away. Finally, the alpha islet cells no longer can hold onto their bound glucose, either. The inhibitory signals *they* had been receiving disappear and they start secreting glucagon. This hormone has an entirely different effect than insulin. When it reaches your liver, it stimulates enzymes needed to break glycogen back down into glucose. In this way, glucose is returned to the blood. The blood glucose level stabilizes, even if it's at a lower level than before.

But the best-laid balance of internal conditions can go astray as new situations pop up. You happen to see a deer at the far edge of a meadow and you race toward it. Halfway across the meadow you pull up in pain: you forgot to reckon with the change in altitude, and your leg muscles cramped up. Receptors had already detected the lower oxygen concentration in your blood at this elevation, and they had already called for the production of more oxygen-carrying red blood cells. But it would take days for extra cells to accumulate. Thus your blood could not supply muscle cells with enough oxygen for cellular respiration. Your muscle cells had to switch to the anaerobic pathway of glycolysis, in which lactic acid is an end product (Chapter Six). When that happened, the sudden lactic acid build-up caused the cramping—and glucose began to be used up eighteen times faster than before!

Again controls in your body go to work to return it to a homeostatic state. Detectors in the brain note the build-up of carbon dioxide. They respond by stimulating your chest and diaphragm muscles to speed up your breathing rate. Slowly your cells are purged of carbon dioxide and are replenished with oxygen. Gradually the lactic acid leaks out of your muscle cells and into the blood, which carries it to the liver. There, it is recombined to produce glucose, which is returned to the blood.

By now, the falling levels of glucose and fatty acids in your blood have triggered activity in the appetite center of your hypothalamus. Nerve impulses initiate a churning in your stomach—hunger pangs! On checking the sun's position, you see it's well past noon. And guess what: you forgot your lunch. When you start the long walk back, your liver's store of glycogen is nearly gone and the blood sugar level drops once more. In combination with your anxiety about reaching camp before dark, the decline calls on a new agent of homeostatic control. Your adrenal medulla begins secreting the hormone **adrenalin**, which flows through the bloodstream to your body's fat cells. There, adrenalin activates fat-metabolizing enzymes, so fatty acids begin pouring into the blood. With this fresh nutrient, skeletal muscles and organs such as the heart switch metabolic pathways and begin using fatty acids. For every one fatty acid molecule they use, three glucose molecules are saved for the brain. What you call "getting your second wind" is really getting this new energy source.

It's enough to get you back to camp by sundown. In fact, your body had enough calories stored away as fat to sustain you for many more days, so the situation was far from desperate. Even then your energy supplies would not have run out. Another hypothalamus center would have been stirred into releasing a hormone, which would have prodded your pituitary gland into secreting **adrenocorticotropic hormone** (ACTH). In turn, ACTH would have prodded adrenal cortex cells into secreting **glucocorticoid hormones,** which would have acted on your muscles and organs to bring about the slow disassembly of your body's proteins. Amino acids released by their action would have been used in the liver as one more source of carbon atoms for producing glucose—and once more the brain would have been kept active. As extreme as this last pathway may be, it's a small price to pay for keeping the brain functional enough to figure out how to bring the body back to a homeostatic state.

Case Study 3: Salt and Water Regulation

In a few hours, he would be starting his summer job at the steel mill—hard work, he was thinking. Would he hold up? He was sure going to try: he could use the money when he returned to college in the fall. He absentmindedly poured too much salt on his breakfast of ham and eggs, and he winced with each mouthful. But there was no time to fix another meal, so he washed everything down with several glasses of water to rid his mouth of the bitter taste of salt.

During his hour-long drive to the mill, the water and salt (along with the first traces of digested food) were passing from his intestine to his bloodstream. Pressure receptors in the blood vessels sensed the increased water volume, and salt detectors in the brain sensed the increased salt content. And regulatory events were set in motion that would permit the kidneys to rid his body of the excess.

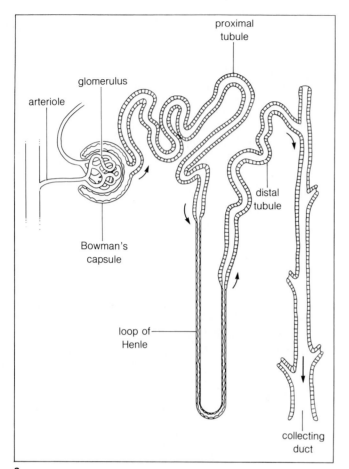

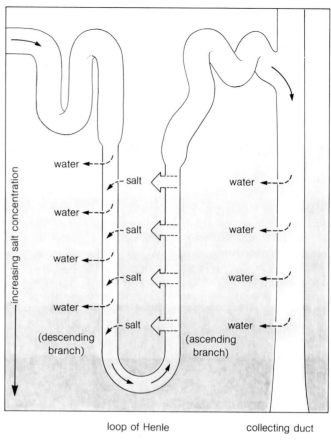

a

b

Figure 13.16 The *countercurrent system* in the "hairpin" loop of Henle. (The word "countercurrent" simply refers to a flow in opposing directions within the same system: here, down one branch of the loop and up the other.) In this system, salt and water reabsorption is regulated. Two transport mechanisms are involved: movement of water and salts (indicated by dashed arrows), and active transport of salts (large arrows). Here we will focus on sodium reabsorption.

As the kidney filtrate moves down the descending branch, water moves outward, down its osmotic gradient. On the upward part of the journey it remains inside, for the cells making up the ascending branch are impermeable to water. But depending on the activity of aldosterone (Figure 13.17), these cells can be prodded into actively transporting sodium to the *outside*. They do this against a concentration gradient: there are more sodium ions in the fluid surrounding the loop than there are inside. But the high external concentration also means simple diffusion can carry sodium back *inside* the descending branch. In short, as the filtrate flows downward, it becomes progressively more concentrated because sodium keeps getting pumped back in. And as it flows upward, it becomes progressively more diluted as sodium is pumped out.

With all this sodium circulation, the extracellular fluid surrounding the loop of Henle is perpetually salty. The farther down we measure around the loop, the saltier it gets. This has important consequences for urine formation, for water can be reabsorbed only if sodium has been reabsorbed first. The more sodium leaving the tubule, the greater the volume of water that will passively follow. In themselves, the cells of a distal tubule and a collecting duct show low permeability to water—but ADH secretions greatly increase their permeability to water. Depending on ADH activity (Figure 13.17), it is here that the volume of water leaving the body—hence final urine concentration—is largely regulated.

As you might well imagine, the concentration gradient for sodium, on which this process is based, depends largely on the length of the loop of Henle, which varies among species. The longer the loop, the steeper the gradient. Can you guess why such variations might be adaptations to different environments?

The filtration process was beginning in the glomeruli, the capillaries clustered in the entrance to each kidney nephron (Figures 12.17 and 13.16). As the fluid moved down the proximal tubule, sodium ions (Na^+) were actively transported across the tubule cell membranes and into the surrounding interstitial fluid. Chloride ions (Cl^-), being oppositely charged, were "pulled" in the same direction as the sodium ions in most parts of the tubule system. Linked to this outward movement of solutes was an outward movement of water, by way of osmosis. As a result, fluid pressure in the interstitial region began to rise, and both the solutes and water flowed into a region of lower fluid pressure—into the capillaries intimately associated with the tubule system. Thus, by this **bulk flow** process, water was reabsorbed into the adjacent capillaries. *But water was reabsorbed only because sodium had been reabsorbed first.* This simple relationship would be important in the regulation of his body fluids.

By the time he ate breakfast, the excess salts and water from his evening meal had long since been disposed of. During the evening, his kidneys had been working in ways that minimized salt and water excretion while maintaining a flow of urine, which could rid his body of toxic substances. (For instance, tubule cells were actively transporting penicillin, an antibiotic he had been taking to fight an infection, to the urine. Blood entering the kidneys with a cargo of penicillin left without it.) Also, his adrenal cortex had been secreting **aldosterone,** a hormone concerned with maintaining sodium levels in the blood. Its main effect was to cause cells of the distal tubules to actively pump sodium ions against their concentration gradient and into the interstitial fluid; from there, sodium was reabsorbed into the capillaries (Figure 13.17). Chloride ions followed the sodium. Through the night, his pituitary had been secreting **antidiuretic hormone** (ADH), a hormone that promotes water reabsorption. Under the influence of ADH, cells of the distal and collecting tubules had become permeable to water, which meant water reabsorption kept up with sodium reabsorption. (That is the reason why the urine produced during periods between meals is low in salt and in volume.)

As soon as water from his breakfast began to flow from his intestines to his blood, things began to change. Before the blood volume had increased 1 percent, ADH secretions slowed, and cells lining the distal and collecting tubules of his kidneys became impermeable to water. Thus water began to be excreted at a faster rate. His urine became more dilute, and the flow rate into his bladder rose rapidly. Changes in aldosterone secretions followed. As salt entered the blood from his intestine, the adrenal cortex responded indirectly to the rising sodium levels. Aldosterone secretions slowed, the distal tubule cells stopped expending as much energy to transport sodium out of the tubule, and the urine sodium content rose. By the time he reached the mill, his body was prepared to rid itself of much of the water and salt he had consumed less than an hour before.

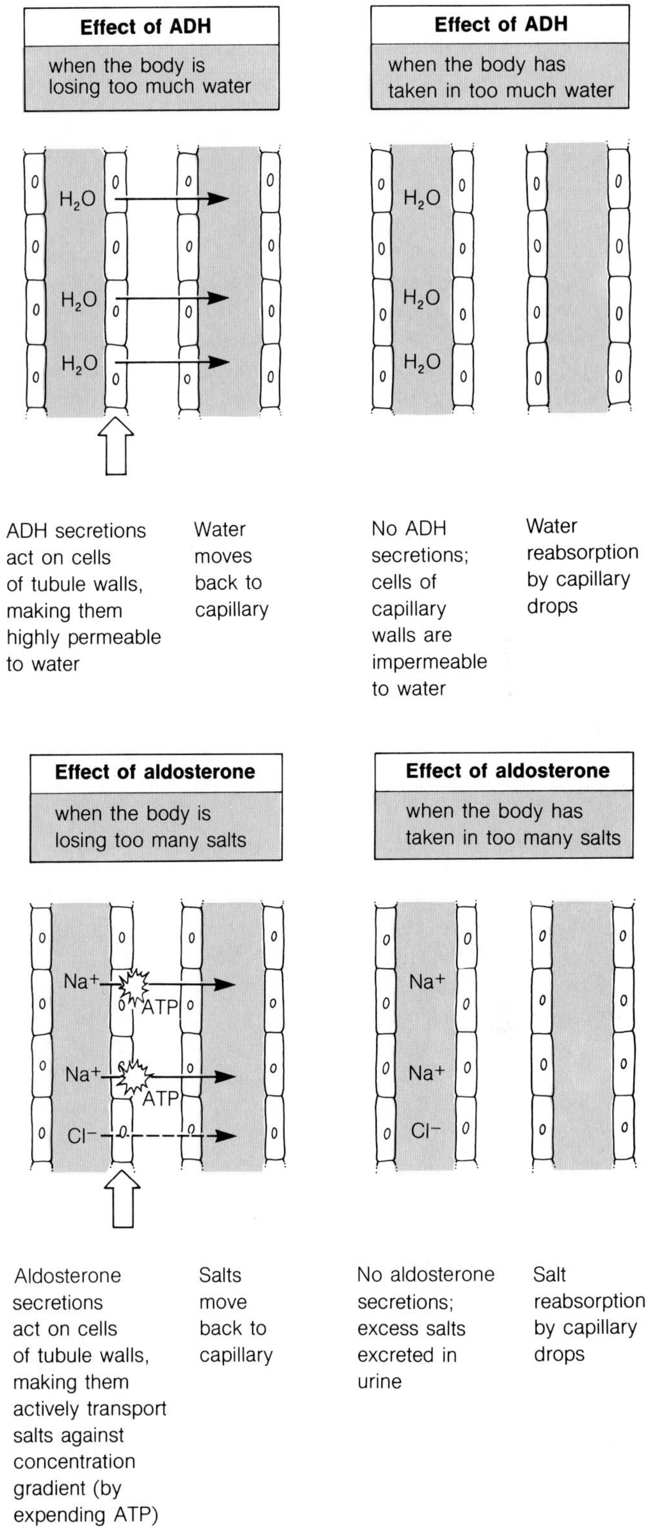

Figure 13.17 Summary of the effects of ADH (on water regulation) and aldosterone (on salt regulation) in the kidney.

As he was escorted to his work station, the intensity of the heat reminded him of a sauna and the summer desert rolled into one. Almost at once, sweat began flowing from every pore in his body. At first it burned his eyes, but after a few hours the stinging disappeared. It was a sign that his body had switched gears and was now working swiftly to conserve salt. For the profuse sweating was altering his blood volume and composition. As the water concentration decreased slightly, sensors throughout his body detected the change. Signals went out to the hypothalamus, which triggered two responses. First, his pituitary began pouring out a fresh supply of ADH. The distal and collecting tubules of the kidneys responded by absorbing water from urine. Second, his thirst center was activated. At every opportunity, he now took a drink of water, which restored the volume of body fluids. *But it did not replace the salt he was losing in sweat.* The sodium concentration in his blood began to drop, which activated receptors in his brain and kidneys. Chemical signals from the kidneys and nerve impulses from the brain reached the adrenals and prodded them into releasing aldosterone. Under the influence of aldosterone, the cells of the distal tubules and the collecting ducts began actively transporting sodium from urine to blood, and potassium from blood to urine. The aldosterone also prodded cells lining the ducts of his sweat glands into actively reabsorbing sodium from the sweat. It was the decreased saltiness of his sweat that accounted for the fact that his eyes no longer burned.

By midafternoon he felt weak, dizzy, and nauseous. He was also vaguely hungry, but he couldn't think of what to eat. His kidneys were fighting a losing battle. By drinking so much water without replacing the salts he was sweating out, he was on the verge of **water intoxication:** his internal environment was too dilute for cells to function properly; nerve and muscle cells were losing potassium and were beginning to malfunction. Fortunately, his foreman stopped him and handed him a glass of salt water. The liquid smelled and tasted salty—but it seemed appetizing. Eight hours earlier, one sip of that same solution would have made him gag. But by now, his **salt-hunger center,** a region in his hypothalamus that normally is quiescent, had been activated. Not only humans but other mammals have a salt-hunger center. When sodium levels in the blood fall dangerously low, salt solutions that normally would be revolting taste delicious. And the nose, which normally is insensitive to the smell of salt, becomes sensitive even to tiny amounts of chlorine gas escaping from salt. The selective advantage of this response is clear: with it, an animal desperately in need of salt may discover a source it might otherwise overlook. (Hence deer are drawn to a road in winter to lick salt from the pavement.) The kidneys are remarkable organs that will maintain salt concentrations in the blood within normal limits, even when excess salt is consumed. But one thing they cannot do is create salt from something else.

Case Study 4: The Fight-or-Flight Response

As you glide through the warm waters of the tropical reef, you marvel at the way your new scuba gear allows you to move through this totally different world. You dive to explore the coral formations below, and there, beneath the shadow of an overhanging rock, an extravagantly colored anemone catches your eye. You find a handhold and pull yourself toward it. Suddenly, out of the corner of your eye, you see glistening needlelike teeth—*MORAY!* Your arms and legs explode into action and your body shoots upward. Out of danger, you try to relax, but your head is still pounding and you can feel your pulse throbbing in your neck and temples. By violating a prime rule of diving—by allowing part of your body to come into contact with a place you had not yet examined carefully—you had come within a fraction of a second of losing your hand.

In one instant, your hand had been saved by one of the most ancient, rapid, and thoroughly integrated reactions of which the human body is capable—the **fight-or-flight response.** Every part of your body had been mobilized in one way or another to give you supernormal speed and strength. Your retinas registered only a glimpse of flashing teeth. As soon as those few impulses traveled from your retinas to your brain, they were processed and interpreted. Their meaning was clear: *DANGER!*

Once impulses triggered the appropriate center in your cerebral cortex, a new set of nerve impulses shot rapidly down through the spinal cord. The impulses had two direct effects on the autonomic nervous system: parasympathetic pathways were turned off, and sympathetic pathways were turned on. The secondary effects were directed toward preparing your body for fighting or escaping. Impulses shot out to every vital organ. The heart beat stronger and faster to accelerate blood flow. Artery wall muscles contracted slightly to speed blood through the capillary beds. The accelerated blood flow was not randomly directed. Sympathetic nerve impulses closed off blood vessels in the skin and in the wall of the gut, thereby diverting blood from those organs—which would play no part in the reflex response—to those that would. Simultaneously, the gut was made to go limp so it would drain off less energy during the emergency. Sympathetic nerve impulses also caused the blood vessels supplying skeletal muscles and brain to open wide. The increased blood flow prepared the brain for operating as fast as possible to sort out all the response options, select the appropriate one, and set it in motion; it prepared muscles to respond with maximum speed and strength to whatever urgent commands descended from the brain.

To assist the muscles, sympathetic nerve impulses went to work on other organs. Smooth muscle cells in the spleen were made to contract, so that any excess red blood cells being stored there for just such an emergency were

Figure 13.18 Fight or flee?

squeezed into the bloodstream. Smooth muscle cells of the bronchi were made to relax, so the respiratory passage could open wide to increase the intake of oxygen. The adrenal medulla was stimulated into pouring adrenalin into the blood. The adrenalin flowed to the liver and to fat tissues, stimulating the release of glucose and fatty acids. Glucose and fatty acids in the blood rose to supernormal levels to provide the urgently needed energy.

With all systems alerted, your cerebral cortex ran through its memory banks, searching for the appropriate responses your brain had stored away during your scuba diving lessons. "Don't panic!" "If it's a shark, freeze!" *"If it's a moray eel, get out of there!"*

Having found the appropriate general response, your cerebral cortex fired off impulses to the motor control center of the cortex, where the correct detailed program was selected and activated. From there, impulses flowed down the spinal cord and out the motor neurons in just the right

sequence and frequency needed to activate the muscles into producing a powerful swimming stroke. *And all this happened in a fraction of a second—even before you were consciously aware of the danger.*

Your body is not only prepared for constant integration of activities to maintain a homeostatic state. It is also prepared to instantly modify internal conditions. The fight-or-flight response is not unique to humans. All mammals display it in times of stress, whenever a sudden confrontation demands an immediate and unequivocal response.

Case Study 5: The Immune Response

It's a warm spring day and you are walking barefoot across the lawn to class. Abruptly you stop: you stepped on a staple that had been lying prong-upward in the grass. The next morning the punctured area is red, tender, and slightly swollen, even though you had pulled out the staple at once. The following day it is inflamed, but then the swelling begins to subside. Within a few more days your foot is back to normal, and you have completely forgotten the incident.

All that time, however, your body had been struggling against an unseen foe. You had inadvertently stepped on the place where a diseased bird had fallen the night before and had lain until a scavenging animal carried it off. Your foot picked up some of the bacteria that had caused the bird's death. As long as the bacteria remained outside your body, they were dormant. But the staple prongs had carried several thousand bacterial cells into the moist, warm flesh beneath your skin. Here they found conditions suitable for vigorous growth. And grow they did—they soon doubled in number and were on their way to doubling again. Left alone, they would have threatened your life.

There was nothing vicious in their attack; the bacteria simply were exploiting available resources, as all life forms must do. But they also were producing by-products that interfered with your own cell functioning. If the bacterial population became large enough, those by-products would begin interfering with the functioning of certain organs.

Such attacks are far more common than you might suspect. They occur hundreds of times in all vertebrate animals. But it happens the vertebrate body has ways of defending itself against them—not with fangs and claws but with molecules and cells. When the staple penetrated your skin, it tore open many of your cells. As the contents spilled out, lysosomes broke apart. Their digestive enzymes were unleashed as the body's first line of defense. They began digesting proteins, lipids, and nucleic acids from the broken cells. But they could not digest the invaders because of the tough bacterial cell wall.

Even so, as the lysosomal enzymes went to work, a second line of defense was called up. Around the wound, some blood had escaped from broken capillaries, and it now pooled and clotted. As in all blood, it contained **complement** (a collection of inactive proteins). When the enzymes partially degraded the complement molecules, the fragments began diffusing through the tissue. First these fragments went to work on capillary sphincters, causing the ones through which blood enters to relax, and the ones through which blood leaves to contract. Thus each capillary filled with blood. Next, the fragments loosened the junctions between epithelial cells in the capillary walls. The capillaries became somewhat leaky, and blood proteins normally confined within them began oozing out. But most importantly, the fragments acted as cues for white blood cells.

Usually, white blood cells remain in the bloodstream. But once complement stimulates them, their behavior changes. They attach themselves to capillary walls and crawl along, amoebalike, until they reach a weakened epithelial junction. Then they slip out into the tissue. There they creep about between cells and fibers, following the concentration gradient of complement fragments. Like bloodhounds on the trail, they move in the direction of higher concentration.

Once released in this way, your white blood cells went into a frenzy of phagocytosis. Specifically avoiding your own cells, they began engulfing any particle not having a surface marker that meant, in effect, "Leave me alone: I'm *self!*" Dirt, rust, bits of broken host cells, bacteria—all were engulfed indiscriminately. Particles were enclosed in vacuoles, then digested by lysosomal enzymes that spilled in. Every so often, in their frenzy, the white blood cells released lysosomes before a vacuole had formed completely. When these enzymes spilled outside, they activated more complement, caused more capillaries to become swollen and leaky, and attracted more white blood cells to the battleground.

This entire process is called **inflammation.** It is a general response to any tissue injury; it follows whether foreign substances are present or not. Distension of capillaries (as inlets are opened and outlets are closed) causes redness in the tissue. Blood plasma oozing out from capillaries and filling spaces between cells causes swelling. And the constant stimulation of pressure receptors once a tissue has become swollen causes pain.

If bacteria had not entered the wound or if they had been of a type that couldn't multiply rapidly in the tissue, the inflammatory response alone would have cleaned things up. Once the debris had been swept away, enzymes no longer would have been released and the inflammation would have abated. But in this case the bacteria were outmultiplying the inflammatory response! It was time for the body's main line of defense against bacteria and viruses: the **immune response.**

An immune response differs from an inflammatory response and from all other defense mechanisms because its agents respond to and "remember" each type of invading cell or particle. The white blood cells called **lymphocytes** carry out the immune response.

Each lymphocyte is able to bind only to one kind of **antigen** (a foreign cell or substance). Your body contains thousands of different lymphocytes, specific for thousands of different antigens. When a lymphocyte encounters its target, it becomes activated. It begins dividing rapidly and producing more cells like itself. Some of these new lymphocytes go on to produce **antibodies:** molecules that combine specifically with the antigen. Others go dormant; they are a reservoir of lymphocytes that can be quickly tapped if the same antigen should reappear; hence their name, "memory lymphocytes."

(If this had been the first time your body encountered this bacterial species, few lymphocytes would have been around to respond to it. Your body would have had to make a *primary* immune response—it would have to wait for days while its lymphocytes divided enough times to produce enough antibody to control the invasion. But during your childhood, your body did fight off the same type of bacterium, and it still carries vestiges of the struggle: memory lymphocytes. When the bacteria showed up again, they encountered a lymphocyte trap ready to spring; they activated a *secondary* immune response.)

As the inflammation progressed, lymphocytes were among the white blood cells creeping out of the bloodstream. Most were specific for other types of antigens and did not take part in the battle. But each specific memory lymphocyte entering the area found its target, bound a few antigens to its surface, and became activated. It moved into lymph vessels with its foreign cargo, tumbling along until it reached a lymph node and was filtered out of the fluid (Chapter Twelve). For the next few days, memory cells with bound antigens steadily accumulated in the node. Then they began dividing several times a day, so their numbers began rising rapidly. And they began manufacturing antibody molecules.

Antibody molecules began trickling out of the lymph node. By the morning of the third day, the trickling had become a torrent. On reaching the inflamed site, antibodies oozed out through leaky capillaries. Each antibody was able to bind two bacteria, so the invaders were being clumped together. And now complement was called on for the second time. A certain kind of complement protein was activated, becoming an enzyme capable of digesting a hole in each bacterial cell to which antibody was attached. Through such holes, bacterium after bacterium lost its contents and died.

For the first two days the bacteria appeared to be winning, for they were reproducing faster than the phagocytotic cells, antibody, and complement were destroying them. But by the third day, antibody production reached its peak and the tide of battle turned. For two weeks or more, antibody production will continue until every last bacterium is destroyed. When it finally shuts down, a new set of memory lymphocytes will go into the resting state, prepared for some future struggle.

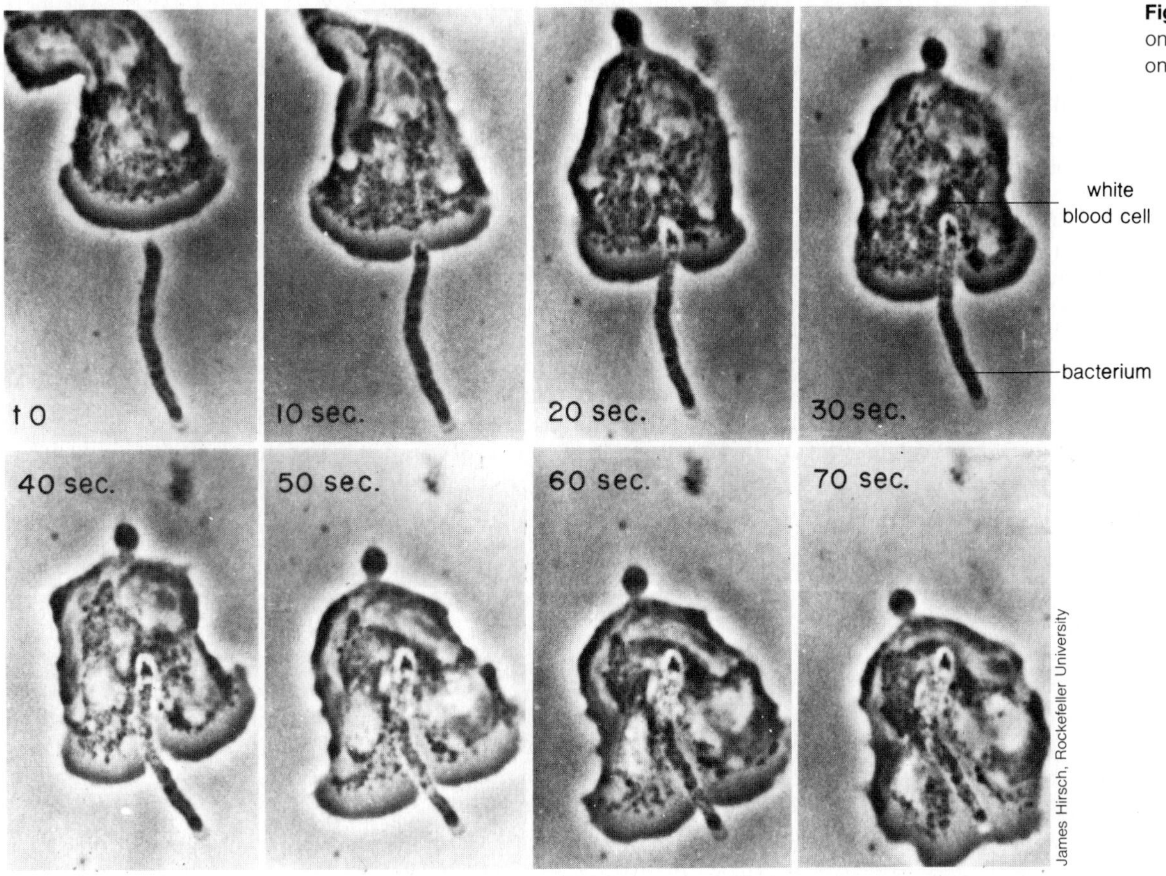

Figure 13.19 White blood cell on the prowl and the demise of one rodlike bacterium.

white
blood cell

bacterium

t 0   10 sec.   20 sec.   30 sec.

40 sec.   50 sec.   60 sec.   70 sec.

James Hirsch, Rockefeller University

Case Study 6: The Immune System Gone Awry

It started out as a simple case of measles—discomfort for a few days, but nothing to be too concerned about. The rash you saw on the child was a clear indication that her immune system was on the move, destroying the virus and the virus-infected cells. Each localized red spot was not an effect of the virus itself but of her body's inflammatory response. The fever, too, was a sign that the immune response was working. And that was encouraging, because for those rare individuals unable to develop an immune response against measles, there is no rash, no fever. The virus multiplies in the body, undetected, until it reaches and infects brain cells. The results are fatal.

But even with these encouraging signs, the work of the immune response was destined to go out of control. The child was a carrier of one mutant gene that would shift the outcome of the infection—a mutation of one of the many genes coding for antibodies. In any other immune response to other antigens (to bacterial invaders, for instance), the outcome would be favorable. And if this virus did not happen to settle in the child's pancreas, the outcome would also be favorable. But if the virus did settle there, the mutant gene somehow would prevent the body from distinguishing virus-infected pancreatic cells from uninfected ones! Thus the white blood cells tracking down the virus would also be unleashed against the normal cells of the pancreas—particularly, against the insulin-secreting beta islet cells. They would not stop until virtually all of those pancreatic cells were wiped out.

The virus did reach the pancreas before it was stopped, as many viral diseases do in all of us. Within a few days, the measles infection was gone. But a few days later the child began eating, drinking, and urinating excessively, even through the night. Having lost her ability to produce insulin, she could not regulate the glucose concentration in her blood; after each meal, glucose accumulated to dangerously high levels. In an effort to rid the body of excess sugar, her kidneys began pouring it out in the urine.

Because of the volume of urine being lost, the child's skin dried out and she craved fluids. She lost weight; dizziness and vomiting set in. Within two weeks of the end of her encounter with measles, she was in the hospital at the brink of death. The diagnosis: **diabetes mellitus,** a severe insulin deficiency brought about, in this case, by the interaction of a viral attack and one mutant gene.

Insulin injections saved her. But for the rest of her life, each day's survival will depend on carefully balancing the timing and the amount of food intake with insulin injections, which will have to do the work of the destroyed beta islet cells. If the insulin is too little, or if the injection is delayed, the blood glucose level will rise high enough to endanger her brain. If the insulin injection is too large or ahead of schedule, her blood glucose level will fall dangerously low. Thus for this child, for the rest of her life, a daily conscious effort must be substituted for the delicate homeostatic regulation that exists in the rest of us, without our conscious intervention.

Case Study 7: Drugs and the Nervous System

It was going to be one of those monumental midterms, and he was up all night studying. About two in the morning, when his inner eyelids were beginning to feel as if they were made of sandpaper, he popped a dexie—dexedrine, a **psychoactive drug,** a substance deliberately introduced into the body to affect the functioning of the nervous system.

Inside his intestine, the drug was released from the tiny pellets in the capsule, and it entered his bloodstream. Traveling through his body, it caused the release of extra neurotransmitter at thousands upon thousands of nerve endings in his brain and body. His cerebrum, which had been urging sleep for a body in need of rest, was artificially stimulated into a state of arousal. His heart began to beat faster; fatty tissues were stimulated into releasing fatty acid into the blood, and fat metabolism throughout the body was accelerated. With his new alertness, things began falling into place; for the first time he felt a growing mastery of the subject he was studying.

For the next twelve hours, right through the midterm, the tiny pellets had gone on dissolving in his gut. But the drug carried him through, he thought—although he was now beginning to feel a letdown. His skin was taking on a strange numbness. And his mouth was parched and dry. By the time he returned to the dormitory, the nerve endings in his throat were bombarding his brain with impulses calling for a drink. All day long the dexedrine had suppressed his appetite center in the hypothalamus, so for nearly twelve hours all he had taken into his body were a few sips of water. But even though the appetite center was finally able to send out signals, the heavy sensation in his gut made him feel revulsion at the thought of solid food.

Instead of eating he filled a waterglass with wine, downed it, then downed another. As the alcohol began seeping into his blood, it was carried quickly to the brain, where once again nerve endings began to be controlled artificially. But this time, instead of being stimulated as they had been by the dexedrine, they were inhibited. The first to be inhibited were special neurons in the cerebral cortex that function to suppress the activity of other cortical neurons. Thus freed of inhibitory signals, certain cerebral activities were accelerated. This accelerated activity in the mind creates the false impression that alcohol is a stimulant for the central nervous system, rather than the depressant that it is. (Hence we have the heightened jabbering and activity among the tipsy as a prelude to the slow, labored speech and uncoordinated movements among the drunk.)

In itself, occasional drinking of alcohol causes little permanent damage. It is only with a pattern of constant drinking that the body loses its ability to return to a state of homeostasis. For the body can become adapted to alcohol; and the liver can become more and more adept at burning it up. With time, it might take more than ten times as much alcohol as before to achieve the same psychological effect. By then, the body would be in a state of **addiction**—a state in which it can no longer function normally in the absence of the drug. Alcohol addiction is marked by uncontrollable trembling when the individual is sober; the trembling stops only when more alcohol is consumed.

Because alcohol can be used as a fuel, the staggering amounts consumed during a period of addiction begin to satisfy the body's need for energy. Hence the appetite center is suppressed, and the urge to eat diminishes. But even though it can satisfy the body's need for energy, alcohol supplies none of the myriad other nutrients needed to build and maintain physiological health. Faced with the absence of essential amino acids, vitamins, and fatty acids, and continuously assaulted by the drug, the body begins to degenerate. Neurons are damaged. The digestive tract can develop open wounds (ulcers) that cannot heal because of the lack of body-building nutrients. The liver, an organ that is central to maintaining body fluids in a healthy state, can begin to deteriorate. Healthy liver cells are replaced by degenerate fatty ones. At the very least, the changes can shorten life expectancy by a decade or more. And unless the cycle of addiction is broken, this **cirrhosis** of the liver can claim an individual's life at an early age.

Perspective

Somehow, during the course of time, neighboring cells in the simplest multicellular organisms managed to communicate information about themselves and the environment to one another. Somehow, neurons evolved—cells with extensions capable of propagating a message from one end to the other. This message, the nerve impulse, was not born of some unique chemical trait of neurons. Rather it was an extraordinary refinement of an attribute shared by all living cells—the capacity to regulate sodium and potassium ion concentrations so they are at different levels on the outside of a neuron than they are on the inside.

Over evolutionary time, neurons became assembled first into nerve networks and then into whole systems of integration, through which all competing signals impinging on different receptor cells could be sorted out, evaluated, and acted upon to maintain homeostasis for the whole organism. The animal way of life came to depend more and more on nervous integration to detect changing conditions and to respond to them. And the systems themselves became more complex. In one line of descent, this tendency led to an accumulation of neurons in the head—the region near the sense organs responsive to light, sound, taste, and smell. This concentration of neurons, which we now call the brain, became subdivided into special regions that could integrate nerve impulses at ever more sophisticated levels. With the appearance of the mammals, the cerebrum developed. To a greater extent than before, experiences could now be stored and used as the basis for modifying future behavior. The capacity for memory and learning began.

With the appearance of the ancestors to monkeys and apes, the cortex of the cerebrum expanded. The capacity to use the forefeet as grasping hands developed. Finally, with the appearance of human animals, there was an increasing capacity to coordinate what the hands did with what the eyes saw—to create tools as extensions of the body. With further development of the cerebral cortex, there also came the capacity for language—the use of words to symbolize objects and experiences. And now the brain could begin to store not only the experiences of the individual but the experiences of the group as a whole. Thus, through the evolution of cellular communication—the evolution of nervous systems of integration—the sense of history was born, and the sense of destiny. Through the nervous system of integration we can ask how we have come to be what we are—and where we are going from here. These are the subjects of later chapters in the book.

Recommended Readings

Nathan, P. 1969. *The Nervous System*. New York: J. B. Lippincott. This is just about the best book around for an introduction to the nervous system. Highly readable, lively examples. Covers current research into learning, remembering, and memory storage. Recommended for your personal library whether or not you go on in life sciences.

Penfield, W. and T. Rasmussen. 1952. *The Cerebral Cortex of Man*. New York: Macmillan. Fascinating account of initial attempts to map regions of the human brain.

Vander, J., J. Sherman, and D. Luciano. 1975. *Human Physiology: The Mechanisms of Body Function*. Second edition. New York: McGraw-Hill. More advanced reading, but an excellent treatment of homeostatic regulation and control of body functions.

Chapter Fourteen

Figure 14.1 Few dramas in life are more compelling than the development of single cells into complex multicellular animals.

Reproduction and Development
of the Animal Body

The Frog's Hour

With a full-throated croak that only a female of its kind could find seductive, a male frog proclaims the onset of warm spring rains, of ponds, of sex in the night. By August the summer sun will have parched the earth, and his pond dominion will be nearly gone. But tonight is the hour of the frog! Through the dark, a female moves toward the irresistibly vocal male. They meet, they dally; he clamps his forelegs about her swollen abdomen and gives it a squeeze. Out streams a ribbon of thousands of large eggs. The egg cytoplasm harbors the kinds of structures and substances found in eukaryotic cells—mitochondria, enzymes, ribosomes—as well as many substances and structures unique to itself. At the same time the eggs are released, the male's body expels a milky cloud of furiously swimming sperm. Unlike the egg, each tiny sperm is little more than a store of genetic information. (Its other parts, such as a flagellum and mitochondrion, simply get the sperm where it's going, and thereafter are quickly destroyed.) Each egg accepts and joins with only a single sperm. Not long afterward, the sperm nucleus and egg nucleus fuse. With this event a zygote is formed, and **fertilization**—the fusion of the genetic endowment of two gametes—is completed.

Within minutes after fertilization, a drama begins to unfold that has been reenacted each spring, with only minor variations, for countless millions of years. The single-celled frog zygote begins to divide rapidly into two, then four, then eight cells, and many more. These newly formed cells do not separate; they remain attached as a multicellular embryo (Figure 14.2). There is no growth in size during this cleavage. Instead, the cells become smaller and smaller with each successive division. At last there is a hollow ball of tiny cells. And now division begins to slacken. There is a rearrangement of parts that will give rise to the specific shape of the embryo. A dimple forms at a point on the ball's surface, and some cells begin sinking into the interior. This inward migration results in internal layers of cells as well as the external cell layer. Hence the development of internal and external features may now proceed separately.

Soon a groove forms along one side of the developing embryo. The embryo begins to lose its spherical shape and grows longer in the direction of the groove. The groove deepens, then the edges fold over and seal to form a hollow tube. From this tube a nervous system—brain, spinal cord, and nerves—will ultimately form. It is the first visible sign of organ formation. Now an eye begins to form on each side of the developing head. Within the trunk a heart is forming, and soon it will be beating rhythmically. Fins take shape, a mouth forms and opens to the surface. These developments, appearing as they do at different points in time, speak of a process going on in *all* the cells that were so recently developed from a single cell: they are all becoming different from one another in appearance and in function!

Soon the embryo becomes transformed into a tadpole (a kind of larva) that can swim on its own. For the first time since fertilization it feeds itself—and it grows. For many months it grows until, in response to environmental cues, the larva embarks on a new course of development. Its body now changes into a form that is in the image of its parents. Legs grow from its body. Its tail becomes ever shorter, then disappears. Its mouth, once suitable for feeding on algae, develops jaws with armored teeth suitable for devouring insects and worms. Structures called gills, which served so well in trapping oxygen from the pond environment, decrease in size; lungs have developed to replace them. Eventually a full-fledged frog leaves the water for life on land. If it is lucky it will avoid hungry predators, bacterial attacks, and other assorted threats through the many months ahead. And come spring it may find a pond, swollen with the warm waters of the new season's rains, and the hour of the frog will be upon us again.

Watching the promise of the zygote unfold into the reality of the adult, it is difficult not to view it as one of life's greatest mysteries. *How does a simple-looking zygote become transformed into all the specialized cells and structures making up the adult body of a complex animal?* Somehow, during **embryonic development,** cells first become different from one another in position, then in their developmental potential, and finally in their appearance, composition, and function. Cell division is of course needed for this process. But what underlies **differentiation**—the actual changing of cells, according to predetermined patterns, so they become chemically, visibly, and functionally distinct from one another and from their common predecessor cells? What controls are at work, coordinating the ways in which these separate events must proceed to arrive at a normal integrated adult? This chapter outlines what we know and what remains to be discovered about the process of development.

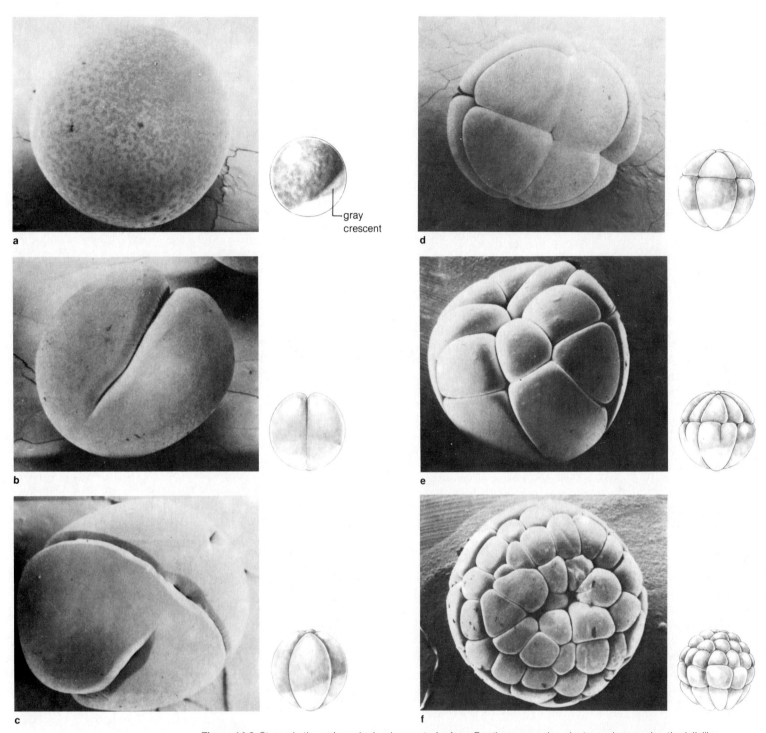

gray
crescent

Figure 14.2 Stages in the embryonic development of a frog. For these scanning electron micrographs, the jellylike layer surrounding the egg has been removed. (**a**) Within about an hour after fertilization, a gray crescent region appears opposite the site where the sperm penetrated the egg. (**b**) The first cleavage divides the cell in two. (**b–e**) Notice how the distribution of substances in the egg cytoplasm dictates their ultimate destination as cleavage proceeds. (**c**) The second cleavage divides the egg into four cells; the third cuts horizontally across all four to form eight cells (**d**). (**f**) The cells in the upper half now divide more rapidly than those in the lower half. (**g**) Successive divisions lead to a ball of cells (a blastula) in which a cavity (blastocoel) has formed. (**h**) During the second day after

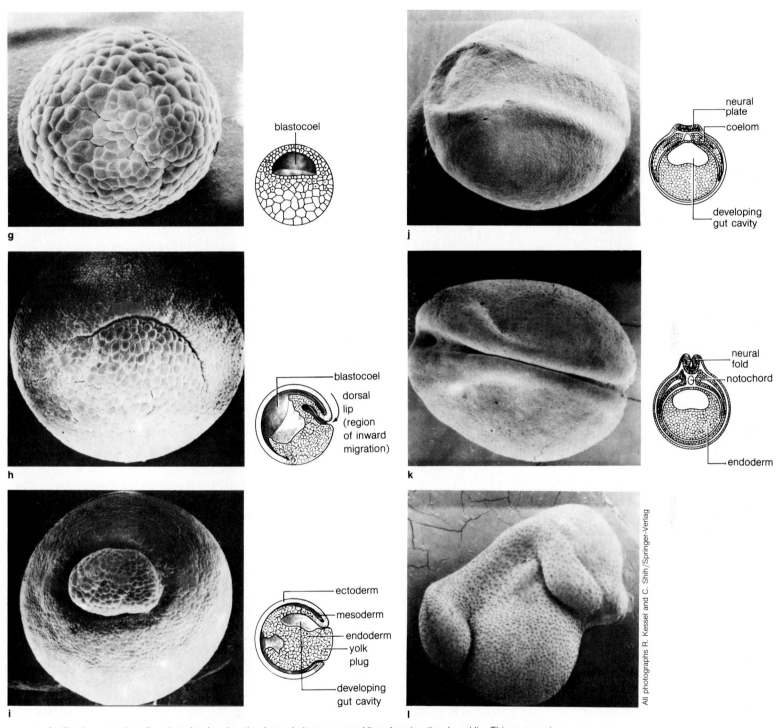

labels in figures:

blastocoel

blastocoel
dorsal lip (region of inward migration)

ectoderm
mesoderm
endoderm
yolk plug
developing gut cavity

neural plate
coelom
developing gut cavity

neural fold
notochord
endoderm

g

h

i

j

k

l

All photographs R. Kessel and C. Shih/Springer-Verlag

fertilization, certain cell regions begin migrating inward along a curved line, forming the dorsal lip. This process is called gastrulation. (**i**) Eventually the curve of inward migration extends full circle, forming the yolk plug. Three cell layers—ectoderm, mesoderm, and endoderm—are forming; each will give rise to certain body structures. (**j**) For instance, the ectoderm gives rise to the nervous system; the mesoderm, to such organs as the spinal cord; and the endoderm, to organs associated with digestion and respiration. (**k, l**) Differentiation proceeds, moving the embryo on its way to becoming a member of the new generation (Figure 14.1).

Cytoplasmic Influences on Early Embryonic Development

The sexual fusion of the genetic material carried in sperm and egg determines what combination of traits will show up in an adult individual. But even before fusion, cytoplasmic events in the egg are established that will come to regulate cell divisions and differentiation in the embryo. In fact, the diploid nucleus apparently plays only a passive role in the early developmental program.

Consider what happens when a diploid nucleus can be removed surgically from a cell, and a nucleus from another, highly differentiated cell can be transplanted in its place. Is development henceforth mucked up? Not at all. The transplanted nucleus stops whatever it was doing and starts carrying out the same metabolic tasks as the nucleus it replaced! Say the nucleus from a gut cell of a mature embryo replaces the nucleus stripped from the egg. It won't produce a large population of gut cells; it will now produce many substances that gut cells normally don't produce. In fact, in many cases a normal embryo develops! Thus, because the nucleus has been embedded in a different kind of cytoplasm, the patterns in which its genes are expressed have changed.

Such transplantation studies tell us three basic things about development:

All the genes necessary to produce a complete individual are present in the nucleus of each differentiated cell.

The fact that differentiation occurs must be the result of different cells expressing different parts of the collection of genes in their nucleus.

Which genes in the nucleus will be expressed apparently is under the control of the cytoplasm.

What is it about the cytoplasm that directs this differential development? Consider what happens when localized cytoplasmic regions are destroyed (by surgical removal, by searing with a tiny heated needle, by irradiation with a beam of ultraviolet light). In some animals, the individual that develops lacks one or more specific structures or cell types. For instance, if a certain region of a squid zygote is irradiated, the left eye of the embryo won't form. In frogs, a part of the zygote surface known as the **gray crescent** (Figure 14.2) affects the patterning of the whole embryo. When a gray crescent from one zygote is transplanted next to the gray crescent of another zygote, the resulting embryo develops into Siamese twin frogs! The zygote from which the gray crescent was removed never divides into more than a shapeless mass of cells. These experiments tell us one more important thing about development:

Localized differences in the egg cytoplasm can dictate which developmental pathways different embryonic cells are destined to follow.

Such localized differences exist because of the way certain substances are distributed inside the cytoplasm. Some of these substances are packaged into the egg while it is being formed in the female's reproductive organs. For example, mRNA molecules coding for specific proteins are included in the egg during its formation. The molecules lie dormant until fertilization. Then suddenly they become activated, one after another. With the help of ribosomes and tRNA molecules, which were also prepackaged in the egg, they initiate development without any direct intervention of the zygote nucleus. During the next stage of events, the preprogram of the egg leads directly to the development of regionally specialized cells.

Cleavage: Segregating Control Factors Into Different Cells Shortly after fertilization, a zygote enters the stage of **cleavage,** in which successive cell divisions convert the zygote into a multicellular embryo. As Figure 14.2 showed, the large, single-celled zygote is divided into smaller and smaller cells, each having its own nucleus. Among many species, cleavage results in a **blastula,** a hollow ball of cells (Figures 14.2 and 14.3).

This process carries forward the promise of the unequal distribution of cytoplasmic substances. *As cleavage proceeds, cytoplasmic differences appear among the dividing cells because some of the substances end up in certain cells but not in others.* These differences will come to exert controls over which genes will be expressed in different regions of the developing embryo.

Gastrulation: Layered Hints of Things To Come During cleavage and blastula formation, cells divide rapidly. But suddenly the pace slackens. Now, in one region of the blastula, cells grow longer and in some cases put out long cytoplasmic threads. Either one by one or while attached to each other, these cells begin a mass migration to the interior, where they create one or more layers of cells. The process of layer formation is called **gastrulation,** and the resulting embryonic form is called a **gastrula.** Figure 14.3 shows how two layers can be formed; Figure 14.2 shows a more complex three-layer formation, typical of amphibians.

Gastrulation is a developmental stage in almost all animal groups, and it is a first step in the differentiation of internal and external parts. The cells remaining on the outside are now known as the **ectoderm** ("outer layer"); they will give rise to the nervous system and the animal's skin. The cells forming the innermost layer is the **endoderm**

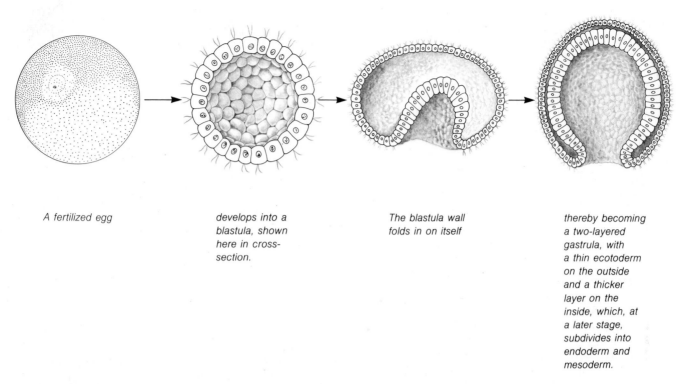

A fertilized egg

develops into a blastula, shown here in cross-section.

The blastula wall folds in on itself

thereby becoming a two-layered gastrula, with a thin ecotoderm on the outside and a thicker layer on the inside, which, at a later stage, subdivides into endoderm and mesoderm.

Figure 14.3 Blastulation and then gastrulation, as they occur in a small marine animal *(Amphioxus).* In most animals, gastrulation is a more complicated process leading to a three-layered embryo, such as the one shown in Figure 14.2. (After J. Young, *The Life of Vertebrates*)

("inner layer"). Basically, the endoderm is a hollow tube in the center of the embryo; it will give rise to the gut and to such structures as the lungs and certain glands. Usually, between these two layers will be the cells of the **mesoderm** ("middle layer"). The bones, muscles, and other internal organs of complex animals are derived from the cells of the mesoderm. But it is through cooperative interactions among the cells of different layers that most body parts of the adult animal will develop.

Factors Governing Later Stages of Embryonic Development

Embryonic Induction: Fitting Parts Together In an adult, many body parts work only if they fit precisely with adjacent parts. For instance, the retina is the light-sensitive tissue of the eye. It must be positioned in a precise way with respect to the lens of the eye, which focuses light onto the retina. The remarkable thing about this relationship is that the retina forms in a structure called the optic cup, which

develops as an outgrowth of the brain—but the lens develops as an ingrowth of the skin (Figure 14.4)!

What coordinates the development of such separate but interdependent parts? The answer came from experiments by the German embryologist Hans Spemann at the beginning of this century. Spemann used a salamander embryo in which the eye lenses were still unformed but in which the optic cups had started to grow out from the brain. He surgically removed one of the optic cups, then placed it just under the skin in the belly region. A lens never did develop where it was supposed to on the side of the head from which the optic cup was removed. But belly skin cells that had been touched by the transplanted optic cup differentiated to form a lens—which fit perfectly into the transplanted part! Spemann concluded that a lens does not develop independently of a retina. Rather, it is caused (or induced) to form wherever the optic cup makes contact with skin. *In such* **embryonic inductions,** *one body part differentiates because of signals it receives from an adjacent body part.* These signals are needed for coordinated development. They are thought to be chemical in nature, but the precise nature of the chemicals has yet to be determined.

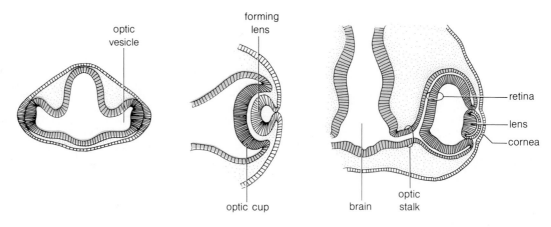

Figure 14.4 Cross-sectional view of the way the retina of a frog's eye develops as an outgrowth of the brain, and the lens develops as an ingrowth of the skin. (After John W. Saunders, Jr.)

To what extent do embryonic inductions dictate the course of development? Half a century ago Hilde Mangold, one of Spemann's students, provided an answer. For her experiments she used salamander embryos of a dark-pigmented species and of a light-pigmented species. From the light-pigmented embryo, she cut out a region known as the **dorsal lip,** which is where gastrulation begins (Figure 14.2). She inserted this piece of tissue into the body wall of the dark-pigmented embryo, across from its own dorsal lip. Gastrulation proceeded at both sites and led to Siamese twin salamander embryos, each with its own set of organs. More than this, almost all the organs of the twin embryo were dark-pigmented. In other words, the extra set of organs had developed from the host, *not* from the transplanted tissue! The transplanted dorsal lip had induced some of the host's cells to develop in ways they normally never do. Thus, although the gray crescent–dorsal lip region assures some prepatterning, induction had to be the key process underlying most of the subsequent organization of individual salamander embryos.

In other animals, such as the squid mentioned earlier, far more prepatterns are laid out as localized substances in the egg cytoplasm. For them, embryonic induction is not as important in determining the ultimate positions of body parts. *In all animals, however, the coordinated development of body parts in specific regions depends on some combination of (1) the influence of cytoplasmic substances within the unfertilized egg, and (2) inductive interactions between cells at later stages of development.*

Embryonic Regulation: Compensating for Slip-Ups If an embryo is to develop into a normal, integrated adult, its cells must function on schedule and in their proper place. But sometimes cells become damaged, or they divide too slowly or too fast. Fortunately, embryonic cells are able to modify their behavior in ways that compensate for missing or extra parts. This capacity is known as **embryonic regulation.**

The control mechanisms underlying embryonic regulation have not yet been discovered. But there is impressive evidence of their existence. For instance, half a century ago, the American biologist Ross Harrison identified the region on a salamander embryo where a foreleg was destined to form. He cut out the entire block of cells from this region and transplanted them to the head. Would a leg form on the head or where it was supposed to form? Or would some parts form in both places? Surprisingly, a complete, normal leg developed at *both* sites. Cells in the leg region weren't committed merely to forming parts of a leg but to forming an entire leg. And they did so despite Harrison's intervention. But apparently *more* cells had become committed than are normally needed in leg formation. Just as apparently, each cell's behavior was somehow modified by its neighbors in such a way that missing parts could be replaced.

To cross-check his findings, Harrison grafted the leg-forming block of cells of one embryo right next to the leg-forming region of another. In some cases, he even grafted two, three, and four of these cell blocks onto the same place. In each case, only *one* leg of normal size and shape developed. Just as the cells modified their behavior to compensate for missing parts, so did they compensate for extra parts.

We see the most dramatic evidence of the capacity for embryonic regulation when a zygote is split, either naturally or experimentally, into two equal parts. The result is not two half-embryos but **identical twins:** two complete, normal individuals. (In contrast, non-identical twins occur when two different eggs are fertilized by two different sperm.)

Differentiation: Emergence of Diverse Cell Types

An adult animal contains dozens or hundreds of differentiated cell types. Few generalizations apply to all the pathways by which they develop. Some cells, such as those of the vertebrate nervous system, acquire their final form and function before birth; if they are later destroyed, no other cells can replace them or assume their function. Other cells, such as those in skin, blood, and gut, undergo constant renewal because a population of dividing cells (stem cells) gives rise to replacements. (For instance, every two days, stem cells replace all cells lining the harsh environment of a mouse intestine.) For other cells, differentiation begins early but isn't completed until some stage of adulthood. (Eggs are one of the first cell types to be set aside in a human female embryo. The first completes differentiation a decade or so after birth—and the last may not do so until four decades later!)

Cell types undergoing differentiation do have one thing in common: they can be identified by the proteins they produce. For instance, in the final stages of differentiation, 95 percent of the protein that red blood cells are making is hemoglobin. Although all the cells *contain* a gene coding for hemoglobin, the gene is expressed only in red blood cells. *Differentiation is based on selective expression of genes in certain cells at certain times.*

Control of gene expression was discussed in Chapter Eight and won't be repeated here. It's enough to say that the cytoplasm is the primary control agent. Certain cells in the dividing zygote end up with localized cytoplasmic substances, which modify their gene behavior. When certain genes become active, their products modify the cytoplasm. These modifications presumably cause further changes in the patterns of *which* genes are expressed. Inducer substances from one cell somehow stimulate another cell, and the pattern changes again. Such interactions between the nucleus and cytoplasm are so complex that no one has yet detailed all the steps in the differentiation of even a single cell type!

Morphogenesis: Beginnings of Functional Body Units

It's one thing to ask how a bone cell becomes different from a muscle cell. It's another thing to ask how one bone becomes different from all other bones. All the bones of your hand and arm are practically identical in cellular composition, yet they differ in size, shape and arrangement—and all the parts fit together. How do coordinated differences arise in clusters of cells, all of which are committed to making bone?

It is through the process of **morphogenesis** that groups of similar cells become spatially coordinated and produce structures of predefined shapes. Morphogenesis requires localized cell movements, localized controls over rates of cell division and differentiation, and localized contractions of some cells and extensions of others. We are only beginning to understand how the controls work.

Perhaps the most fascinating example of morphogenesis is one you might not associate with a vigorously growing animal. It is called **controlled cell death.** Essentially what it means is that, so the rest may live, some cells must die. Preprogrammed cell death is part of many morphogenetic processes. The folding of parts, the hollowing-out of tubes, the shaping of bones, the separation of one part from another, the opening of eyes, mouth, nostrils, and ears—all involve plans of cell death. Perhaps you've noticed that kittens and puppies are born with their eyes sealed shut. From the time their eyes form until just after birth, the eyelids are an unbroken layer of skin. But then, certain cells stretching in a thin line across each eyelid respond to some internal clock and die on cue. As the dead cells degenerate, a slit forms in the skin, the upper and lower parts of the lid separate, and light from the outside world pours in.

Often, cell death is a response to an inducer signal from a nearby developing structure. But there may be a long interval between receipt of a death warrant and cellular execution. In one experiment, John Saunders isolated two blocks of apparently similar cells from an embryo. One block came from a region in which cell death normally occurs, the other came from a region in which it does not. Both were placed in culture dishes containing a nutritionally rich medium. For days, both flourished. Then, at precisely the moment it was scheduled to die in the embryo, the predoomed block of cells suddenly died! The other block continued to thrive.

Let's look at one more kind of controlled cell death to see if we can derive a general principle about morphogenesis. Originally, the hands and feet of developing vertebrates are shaped like paddles. In many species such as our own, skin cells between the lobes of the "paddle" die on cue, leaving separate digits. In other species such as ducks, cell death normally does not occur; that's why a duck has webbed feet. In some mice and some humans, a certain gene mutation blocks cell death, and hands and feet remain webbed. Experiments in which skin was grafted from normal mice to mutant mice (and vice versa) showed the mutation doesn't affect the ability of the inner parts of hands and feet to *generate* the death signal. Rather, it changes the capacity of skin cells to *respond* to the signal. All other events of controlled cell death in all other parts of the developing body proceed on cue. Apparently the gene controls the response to one signal needed for morphogenesis. On the basis of such studies, we can make this generalization: *In complex animals, control genes regulate many of the responses of specific cells to specific developmental signals.*

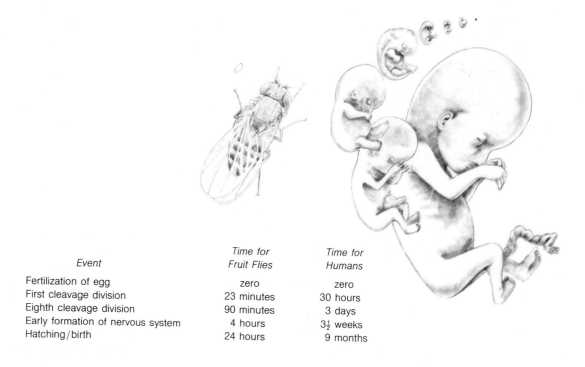

Event	Time for Fruit Flies	Time for Humans
Fertilization of egg	zero	zero
First cleavage division	23 minutes	30 hours
Eighth cleavage division	90 minutes	3 days
Early formation of nervous system	4 hours	$3\frac{1}{2}$ weeks
Hatching/birth	24 hours	9 months

Figure 14.5 Time table of the development of a fruit fly and a human embryo. By the time a fruit fly larva has emerged and is crawling about in search of food, the human zygote has not even divided for the first time into two cells. By the time the infant is born, a fruit fly conceived at the same instant would have been dead many months— but in theory it could have left behind many *billions* of great, great grandchildren!

Human Reproduction and Development

So far, we've isolated some principles governing animal development. Let's see how these principles apply to human development, from the time of fertilization to birth.

The Nature of Placental Development Among many animal groups, it's common for sexual reproduction and development to follow the cycling of seasons. In this way, the young are born when environmental conditions (such as food availability) converge in the most favorable way for the species. To a large extent, differences among animal groups relate to differences in how the developing young are nourished. In most cases, the embryo develops outside the female parent's body. Often eggs are produced in great numbers, and each is provided with only a meager store of

nutrients. Such eggs must develop rapidly into **larvae:** immature forms that look quite different from the adult but are able to live and feed on their own. Such rapid development means the eggs must be extensively preprogrammed. Larvae eventually undergo **metamorphosis** (change in form) to become adults. Tadpoles undergo metamorphosis to become frogs; caterpillars undergo metamorphosis to become butterflies. A program that progresses through a larval form, followed by metamorphosis into the adult form, is known as **indirect development.**

In other cases, large eggs are produced in small numbers. Such eggs contain abundant **yolk** (food reserves for nourishing a more slowly developing embryo). A shell protects the eggs of many land animals from predators. And by providing the egg with its own "pond," so to speak, a shell protects the embryo from harsh environmental conditions. Usually the parents protect, hide, or bury the eggs, which improves chances for survival. When the embryo

Table 14.1 Major Hormones Dominating the Female Reproductive Cycle

Hormone	Where Produced	Controlled By	Main Effects
Follicle-stimulating hormone (FSH)	Pituitary	Hypothalamus (via estrogen and progesterone levels)	*Stimulates growth and development of ovarian follicles; stimulates estrogen secretion and maturation of eggs (in males, FSH stimulates sperm production)*
Luteinizing hormone (LH)	Pituitary	Hypothalamus (via estrogen and progesterone levels)	*Stimulates progesterone secretion; stimulates mammary glands (in males, LH stimulates testosterone secretion)*
Estrogen	Ovaries	FSH, LH	*Causes development of secondary sex characteristics in teenage women; stimulates development of uterine wall early in each menstrual cycle*
Progesterone	Ovaries	LH	*Stimulates growth of glandular tissue in breasts; causes further development of uterine wall late in the menstrual cycle*

hatches, it hatches essentially as a young adult; it undergoes **direct development.** Birds are the most familiar example of this strategy.

We, as mammals, are among the few groups of animals that rely on **placental development.** Our developing young are retained within the mother's body, where they attach to the lining of the reproductive tract and are nourished by absorbing food from the maternal bloodstream. In this way they are protected for the duration and their development can proceed at a less rapid pace, without extensive preprogramming. Figure 14.5 shows just how leisurely our development proceeds.

The Time of Fertilization in the Female Reproductive Cycles In mammals generally, the onset of the female's sexual receptivity to the male (a period called **estrus**) is controlled by the hypothalamus and the pituitary, which secretes certain sex hormones. On the surface, human sexual activity seems to have broken away from the cyclic mammalian plan. Tempered only by social standards (as well as by inclination, opportunity, and endurance), a sexually mature human male is capable of producing and dispersing viable sperm on a more or less continuous basis. A human female, too, has the potential to be physically and emotionally receptive to the male's overtures at any time. But there has been no escaping the cyclic processes that correspond to estrus in other mammals. During each cycle, the female's reproductive system and the gametes maturing within it are physically prepared for fertilization. But fertilization can occur only within a span of about three days of that cycle. Once fertilization does occur, it cannot occur

again for another nine months (assuming embryonic development runs its full nine-month course).

The structural and functional changes during these events are under the cyclic influence of pituitary and ovarian hormones (Table 14.1). Together, these hormones orchestrate the recurring events of the ovarian cycle and the menstrual cycle. The **ovarian cycle** consists of the ripening of an immature egg in a fluid-filled cavity inside the ovary (Figure 14.6), the transformation of the ruptured cavity into an endocrine gland when the ripened egg is expelled from it, and the concurrent production of ovarian hormones. The **menstrual cycle** consists of profound cyclic changes in the uterine lining, which are synchronized with the events of the ovarian cycle.

Ovaries do not constantly produce new eggs. A human female is born with about 400,000 immature eggs, which is all she ever will have. Of those, only about 400 will mature; the rest start to degenerate from birth onward. Usually between ages twelve and fourteen, the female begins to produce mature eggs, which usually are released, one at a time, on a monthly basis. (Occasionally hereditary defects occur among children born to women approaching **menopause,** the end of the period of reproductive potential. These defects are thought to arise from changes in the remaining immature eggs, which may be forty to fifty years old by then. Menopause usually doesn't begin until the forties or fifties.)

Each maturing egg is contained in a spherical chamber called a **follicle** (Figure 14.6). A ripening follicle grows in size, and the cells around the egg secrete a fluid about it. Soon there is so much fluid that a mature follicle balloons outward from the ovary's surface. At **ovulation,** the follicle

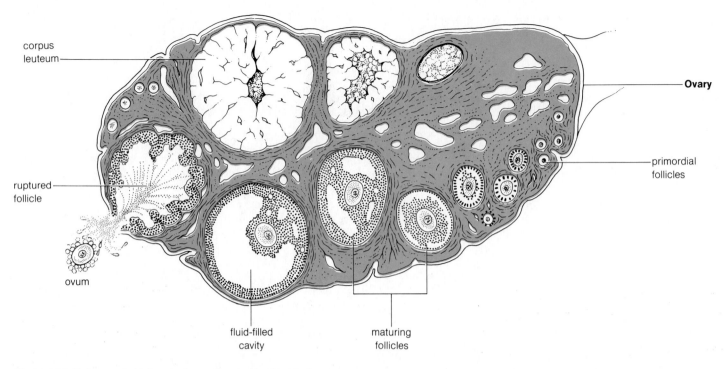

corpus leuteum

ruptured follicle

ovum

Ovary

primordial follicles

fluid-filled cavity

maturing follicles

Figure 14.6 Major events in the ovarian cycle, as described in the text.

ruptures and the egg escapes from the ovary. It is then swept by ciliary action into the oviduct, the road to the uterus. It takes about two weeks for a follicle to develop fully to the stage of ovulation. (The exact time varies from one female to the next, even from one month to the next in the same female.) Following the egg's expulsion, the ruptured follicle develops into an endocrine gland: the **corpus luteum.** This gland becomes a major hormonal control agent during subsequent reproductive events.

All this activity is influenced by subtle feedback relationships among an array of hormones (Figure 14.7). At the start of the ovarian cycle, the hypothalamus stimulates the pituitary into secreting FSH, which in turn stimulates the growth of several young follicles. (Only one of these follicles normally reaches maturity; the rest collapse and degenerate during the course of the cycle.) Cells of the follicles secrete estrogen, which triggers the release of a second hormone, LH, from the pituitary. It is the midcycle peak of FSH and LH levels that triggers ovulation. After the egg escapes, the high levels of LH in the blood stimulate cells in the ruptured follicle to differentiate into the corpus luteum and increase their rate of progesterone production. For the next ten or twelve days, the corpus luteum secretes estrogen and progesterone. Together these hormones affect

the uterus along with the pituitary gland's secretion of LH and FSH.

The changing estrogen and progesterone levels throughout the cycle cause profound changes in the uterus. Estrogen causes the uterine muscles and epithelium (the **endometrium**) to thicken; progesterone goes to work on the thickened tissues to stimulate development of glands that secrete various substances. The uterus becomes rich in blood vessels. These events prepare the uterus for conception. Under the influence of estrogen, the cervix secretes a clear, thin mucus, which is an ideal medium through which the motile sperm can travel rapidly on their journey to the oviduct.

But as the progesterone level continues to rise, it begins to inhibit the pituitary's secretions of FSH and LH. Eventually the flow of FSH and LH is shut off altogether by progesterone. This completes the feedback loop; without the stimulation of FSH and LH, the corpus luteum stops secreting estrogen and progesterone. Without estrogen, the blood vessels in the endometrium become constricted, which means oxygen and nutrients no longer nourish the thickened lining. The endometrium begins to disintegrate and is gradually cast off, which marks the first day of the menstrual cycle. And now the blood vessels open wide. The

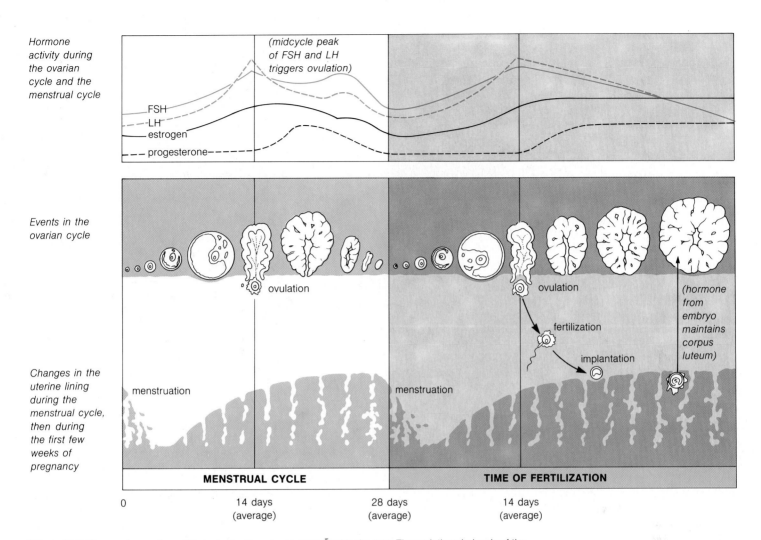

Figure 14.7 The ovarian cycle and the menstrual cycle, as described in the text. The variations in levels of the major hormones involved in these cycles are indicated. The shaded areas to the right show the ovarian and uterine events that occur from the time of menstruation to fertilization, implantation, and the formation of the corpus luteum during the first few weeks of pregnancy.

increased flow causes the weakened capillaries in the endometrium to hemorrhage. For about three to six days, blood mixed with the epithelial lining is cast off. On the average, 50 to 150 milliliters of blood is lost. During this menstrual period, the level of all the sex hormones is low. But then the pituitary begins to secrete FSH once more, because it is no longer inhibited by progesterone. Now a new follicle is stimulated to develop and secrete estrogen. The estrogen shuts off the menstrual flow and stimulates the uterus to begin rebuilding for a new cycle. It takes about ten days from the end of the menstrual flow for hormonal activity to again prepare the uterus for ovulation.

From Fertilization to Birth If sexual intercourse coincides with ovulation—if it occurs at any time between about two days before ovulation and about fifteen hours afterward—pregnancy can result. Within thirty minutes after ejaculation from the penis, sperm are in the oviduct, where fertilization usually takes place (Figure 14.8). Although many are produced, few sperm enter the oviduct. Of the several hundred million that are expelled into the vagina, only a few thousand complete the journey, and only one succeeds in penetrating the ovum. Once the sperm nucleus and the egg nucleus have undergone fusion, fertilization is completed.

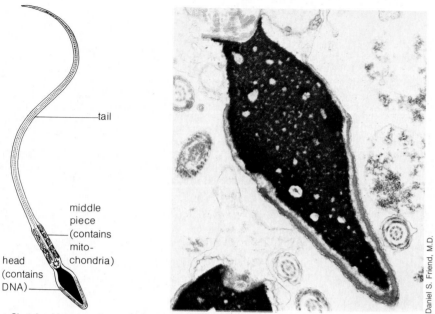

head
(contains
DNA)

middle
piece
(contains
mito-
chondria)

tail

a Sketch of human sperm, and micrograph of head. 21,000×

Daniel S. Friend, M.D.

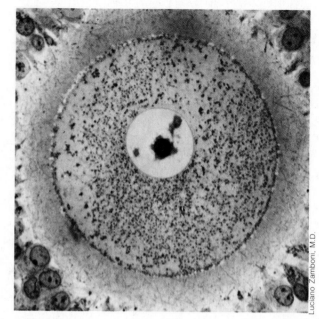

Luciano Zamboni, M.D.

c Human egg, ripening in follicle. 800×

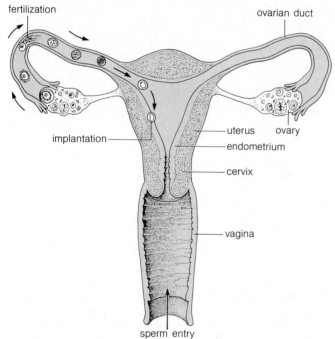

fertilization

ovarian duct

implantation

uterus

endometrium

cervix

ovary

vagina

sperm entry

b Fertilization and implantation.

Figure 14.8 Moments in the continuum of human embryonic development. The diagrams corresponding to the photographs point out some of the major events going on at each stage. This illustration continues over the next few pages. (Unless otherwise indicated, these illustrations have been provided by The Carnegie Institution of Washington and Ronan O'Rahilly, M.D., Director of the Carnegie Embryological Collection)

For three or four days the fertilized egg travels down the oviduct, picking up nutrients from maternal secretions and undergoing division. By the time the egg reaches the uterus, it is a solid cluster of cells. But now the surface cells separate from those inside to form a **blastocyst,** a hollow sphere of sticky surface cells with inner cells massed at one end (Figure 14.8f). The surface cells, collectively called the *trophoblast,* are important in implantation; they will quickly establish vital links between the embryo and the mother. The *inner cell mass* will give rise to three primary layers of cells—endoderm, ectoderm, and mesoderm—from which all organs will be derived.

Five or six days after conception, the blastocyst contacts and adheres to the uterine wall. Enzymes secreted by trophoblastic cells destroy some of the uterine lining, allowing the blastocyst to burrow inside. This enzyme action also ruptures maternal blood vessels being encountered. The entry site heals over, leaving the blastocyst embedded inside in a pool of the mother's blood. About this time, two membranes begin to form. The inner **amnion** arises from the inner cell mass; this membrane will hold a fluid in which the embryo will develop freely (Figure 14.8g). Later, the outer surface of the amnion and the inner surface of the trophoblast acquire a lining of connective tissue. Blood vessels develop in the connective tissue associated with the trophoblast, at which point this structure forms the **chorion,** the second major membrane.

The chorion secretes a hormone, much like LH, that maintains the corpus luteum (hence the endometrium).

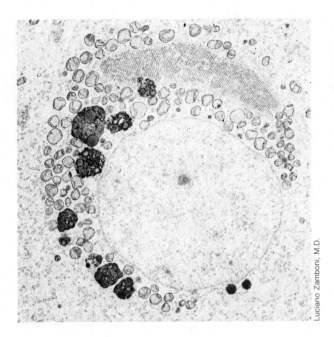

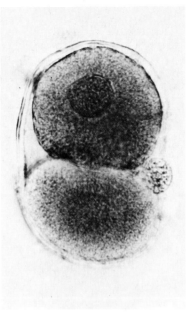

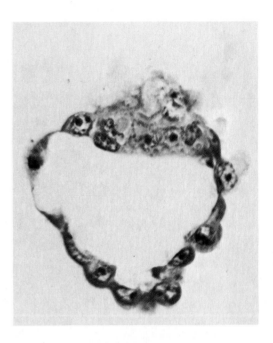

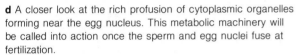

d A closer look at the rich profusion of cytoplasmic organelles forming near the egg nucleus. This metabolic machinery will be called into action once the sperm and egg nuclei fuse at fertilization.

e Between the second and third day following fertilization, cleavage begins, progressing from the two-cell stage to about the sixteen-cell stage.

f Between the fourth and sixth days, a blastocyst forms, with an outer cell layer (the trophoblast) and an inner cell mass (the epiblast).

Extensions from the chorion fuse with the endometrial layer of the uterus to form the **placenta,** an organ specialized for mediating the interchanges between mother and fetus. (At childbirth, it will weigh about a pound and will be cast off as the "afterbirth.") In the placenta, maternal blood vessels and embryonic blood vessels lie side by side—separate circulatory systems, but close enough for the transfer of materials through diffusion. Here, the embryo absorbs nutrients needed for growth; here, its metabolic wastes are cast off, to be disposed of through the mother's lungs and kidneys. As the embryo grows, it remains firmly attached to the placenta by the umbilical cord, a flexible strand of tissue through which blood vessels extend.

By the time the placenta is forming, the inner cell mass has begun to differentiate. Its development proceeds fairly rapidly along the same general course described earlier for all complex animals, except that the human embryo is not spherical. Prior to gastrulation, the embryo itself consists simply of two flat layers of cells, the *embryonic disk* (Figure 14.8h). During the second week, the onset of gastrulation is marked by the appearance of a groove on the upper surface of the embryo. This groove, the *primitive streak,* is the site where cells from the surface move inward to form the mesoderm. When gastrulation is completed, the remaining surface tissue is the ectoderm, which gives rise to the nervous system and skin. Inside, the endoderm forms the respiratory and digestive systems; and the mesoderm develops into internal organs such as the heart, muscles, bone, and blood. By the end of the first month, the embryo has grown 500 times its original size and has begun taking on recognizably human characteristics. This rapid spurt of growth gives way to two months of relatively slow development for the main organs. This three-month period is the **first trimester.**

By the beginning of the **second trimester,** all major organs have been formed (Figure 14.8n). The embryo now resembles an adult in miniature—it is about 7.5 centimeters (3 inches) long. Even through the **third trimester** the individual grows considerably, but few new parts form. That is why the term "embryo" usually is reserved for the first trimester of human life, when body parts are being formed. In the remaining time inside the uterus, the developing individual is called a **fetus.**

Not until the middle of the third trimester will the fetus be sufficiently developed to survive on its own if born prematurely or if removed surgically from the uterus. By the seventh month, fetal development appears to be relatively complete, but fewer than 10 percent of infants born at this stage survive, even with the best medical care. In most cases they are not yet able to breathe normally, swallow, or maintain a normal body temperature. By the ninth month, survival chances increase to about 95 percent.

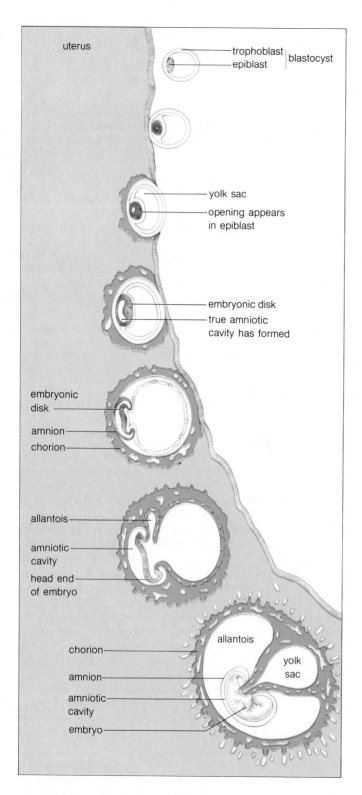

g A current view of developmental events, from the time of implantation (about eight days) to about the end of the first month.

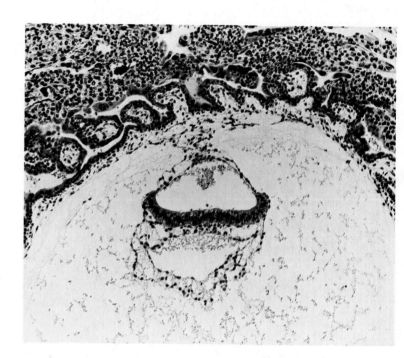

h Cross-section of the embryonic disk at about sixteen days. The amniotic cavity has formed.

The Mother As Protector, Provider, and Potential Threat

With the shift to internal fertilization and development, the offspring of mammals have secured a survival advantage. But with this greater protection for offspring has come a greater responsibility for the prospective mother. No longer does her material contribution end with producing the egg. From conception to birth, the developing individual is at the mercy of her diet, her health habits, and her life-style.

Over evolutionary time, many controls and safeguards have been built into the human reproductive system. The placenta, for example, is a highly selective filter. By screening substances present in the maternal bloodstream, it prevents many noxious substances from gaining access to the embryo and, later, to the fetus. More than this, if the mother's diet is in some way deficient, the placenta often can (at her own body's expense) preferentially take in scarce nutrients for the fetus.

But millions of malformed, malnourished infants born each year are testimony to the fact that human development is far from an infallible process. Even in the United States, where women generally have enough food and adequate medical care, between 250,000 and 500,000 infants are born each year with physical and mental handicaps severe enough to warrant special care. Only about half of these handicaps can be corrected by intensive medical care immediately following delivery. Although it is not known

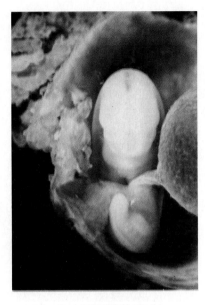

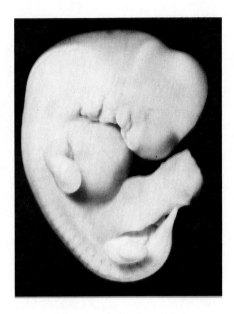

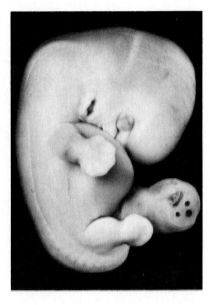

i Eighteen days. Neural folds appear.

j Twenty-eight days. Neural folds have fused; optic vesicles have already formed.

k Thirty-nine days. Limb buds and foot plates have appeared.

l Forty-two days. Head begins to enlarge, trunk begins to straighten. Finger rays appear.

how many birth defects could have been prevented in the first place, the number is certainly significant.

Particularly during the first six weeks after fertilization, the embryo is vulnerable to certain damaging influences from the maternal bloodstream, because that is the critical period of organ formation. If, for instance, the mother contracts German measles during this period, there is a 50 percent chance the embryo she is carrying will be severely malformed. The probability of damage diminishes after this stage; the same disease, contracted during the fourth month or thereafter, has no discernible effect on the embryo.

Throughout pregnancy, the developing individual is well protected from all but the most severe bacterial diseases. But certain viral diseases may have their effects if they are contracted during the sensitive first developmental period. During that time, the mother must maintain her body in the best of health and avoid becoming exposed to individuals with virus infections.

During the first trimester, the embryo is also sensitive to drugs—to new, manufactured agents against which natural selection has not had the opportunity to build defenses. The most shocking example of drug effects came during the first two years after the tranquilizer *thalidomide* was introduced on the market. Women using this drug gave birth to infants with missing or horribly deformed arms and legs. As soon as these deformities were traced to thalidomide, the drug was immediately withdrawn from the market. But there is evidence suggesting that various other tranquilizers as well as sedatives and barbiturates still in use may cause similar, albeit less severe, damage.

Even though certain drugs may cause embryonic malformations only if taken during the first trimester, the embryo does not become impervious to drugs in the maternal bloodstream at any stage. Many drugs pass freely across the placenta and have the same kind of effect on the fetus as on the woman who takes them. For example, infants born to heroin or alcohol addicts are themselves addicted. Without expert detection and prompt medical care, they are likely to die shortly after birth. It is true that certain kinds of medicines and drugs may be necessary in certain cases, but the decision to use them at any time during pregnancy must be made by a skilled physician.

As birth approaches, the growing fetus places more and more demands on the mother for essential nutrients. Thus it is during the last phase of pregnancy that her diet profoundly shapes the course of development. Poor nutrition affects all organs of the fetus, but it is most damaging to the brain. It is in the weeks just before and just after delivery that the human brain undergoes its greatest growth. And normal brain growth is assured *only* if there are adequate amounts and kinds of amino acids for building brain proteins.

Obviously nutrients other than amino acids are needed for normal development. But a balanced diet that provides

m Two months old. Only the major organs are sketched here. Even so, it is possible to get an idea of the complexity of the fetus at this stage of development.

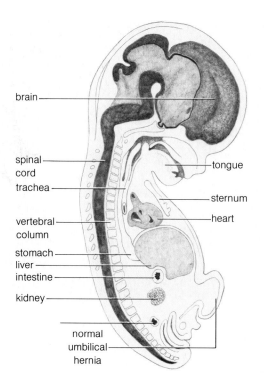

brain

spinal cord

trachea

vertebral column

stomach
liver
intestine

kidney

normal umbilical hernia

tongue

sternum

heart

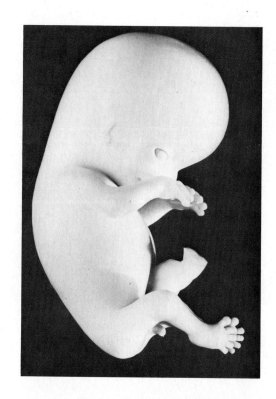

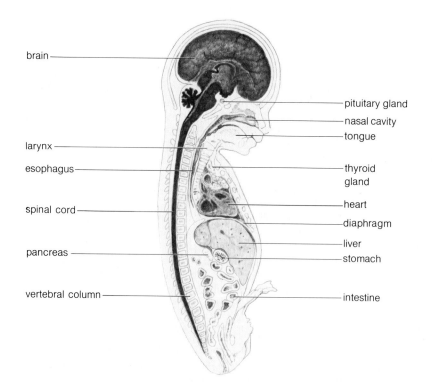

brain

larynx

esophagus

spinal cord

pancreas

vertebral column

pituitary gland

nasal cavity

tongue

thyroid gland

heart

diaphragm

liver

stomach

intestine

n About four months old—the adult in miniature. In this median section, organs such as the lungs and kidneys cannot be seen. (After Hans Elias, 1971.)

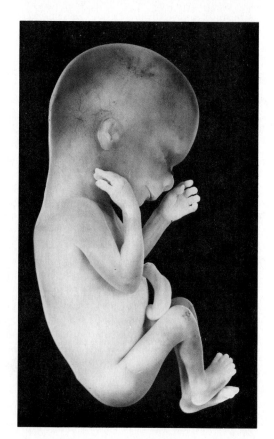

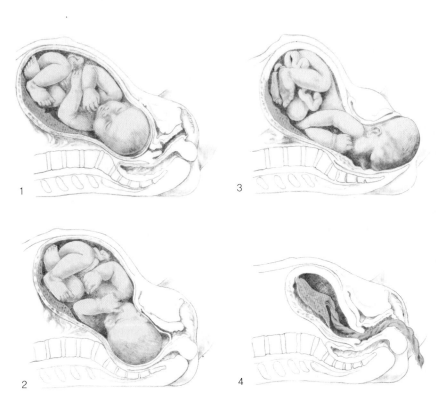

1

2

3

4

o Childbirth.

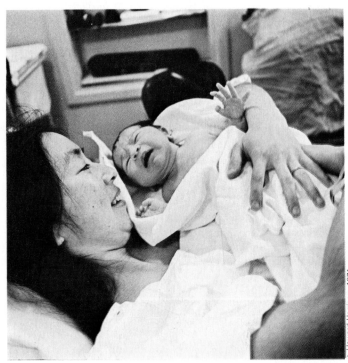

© Howard Harrison, 1971

p The newborn, five minutes old.

enough calories and amino acids normally provides all other nutrients in sufficient amounts. One of the most common fallacies of Western culture is that if you pop a vitamin pill into your mouth every day, the rest will take care of itself. Particularly with respect to pregnancy, nothing could be further from the truth. Because the placenta preferentially absorbs vitamins and minerals from the mother's blood, the developing fetus is more resistant to vitamin and mineral deficiencies than she is. And in cases where the diet is marginal, the money spent on vitamin pills and other food supplements usually does the fetus more good if spent on wholesome, protein-rich food.

There is one other important aspect of food intake during pregnancy that has recently come to light. A few years ago, it was fashionable for a pregnant woman to keep her total weight gain to 10 or 15 pounds. But it is now clear that a woman who limits her weight gain to this level does so at the expense of her child. Particularly during the last trimester, restricted food intake leads to restricted fetal development and to the birth of an underweight infant. Significantly underweight infants face more postdelivery complications than infants of normal weight. They also face a much higher incidence of mental retardation and other handicaps later in life than do infants of normal or somewhat above-level birth weight. In most cases, it is now recommended that a pregnant woman manage her diet to assure a total weight gain of between 20 and 25 pounds.

Factors other than diet during the last trimester also affect birth weight, hence the state of maturity of the newborn. Two factors are the mother's height and her long-term nutritional status. Generally, the taller she is and the better her nutrition has been (from before her own birth to the time of conception), the larger and healthier her own infant is likely to be. In developing countries, the effects of marginal or inadequate nutrition persist for more than a generation. Recent studies in Asia show that children of malnourished women are less capable of using food efficiently. And when the female children mature, they in turn tend to be small and to have smaller, less well-developed infants even if their own diet is improved. Studies in Japan suggest that only over the long term will improved diet diminish the effects of malnutrition.

One other factor known to have an adverse effect on fetal growth and development is cigarette smoking. A woman who smokes every day throughout pregnancy has smaller infants even when her weight and nutritional status and all other variables are identical with those of pregnant women who do not smoke. Often it has been stated that such infants are normal in all other ways. But a recent study in the British Isles suggests otherwise. In England, Scotland,

and Wales, records were kept for seven years on all infants born during one particular week. Those infants born to women who smoked were indeed smaller. More than this, they had a 30 percent greater incidence of death shortly after delivery and a 50 percent greater incidence of heart abnormalities. Most startling of all, at age seven such children had an average "reading age" nearly half a year behind that of children born to nonsmokers.

In this last study, the critical period was shown to be the last half of pregnancy. Children born to women who had stopped smoking by the middle of the second trimester were indistinguishable from those born to women who had never smoked. Although the mechanism by which cigarette smoking exerts its effect on the fetus is not known, the fact that it does affect the fetus is further evidence for the inability of the placenta—marvelous structure though it is—to prevent all the assaults on the fetus that the human mind can dream up.

Perspective

Few processes in the living world are more inspiring than the gradual transformation of a single zygote into an intricately detailed, coordinated adult of the species. And few processes evoke more profound questions. *When does development begin?* As you read in this chapter, many key aspects of development have already emerged before the union of sperm and egg. But this answer leads to a much more basic question: *When does life begin?* Each human female can produce, during her lifetime, as many as four hundred eggs—and all those eggs are alive. Each human male can produce, during one ejaculation, a quarter of a billion sperm—and all those sperm are alive. Even before they merge by chance to establish the genetic constitution of a new individual, sperm and egg are as much alive as any other life form. In no sense, then, is it tenable to suggest that "life begins" when they fuse. *Life began only once, billions of years ago; and each sperm and egg, each zygote and mature individual, is only a fleeting stage in the continuation of that beginning.* This fact cannot diminish the meaning of conception, for it is no small thing to entrust a new individual with the gift of life, wrapped in the unique threads of our species and handed down through an immense sweep of evolutionary time. Child, man, woman—who among us can witness the birth of a living creature and not know the profound force of the life process? For an instant time stands still; and a sense of past and future descends on us. For an instant our pulse beats in synchrony with the newborn and unseen predecessors.

How can we reconcile this compelling moment of individual birth with our growing awareness of too many births in the biosphere? At the time this book is being written, an average of 2.2 infants are being born each second—132 each minute, 7,920 each hour. By the time you go to bed tonight, there will be 190,080 more humans on this earth than there were last night at that hour. Within a week, the number will reach 1,330,560—about as many people as there are now in the entire state of Massachusetts. *Within one week.* This astounding birth rate has outstripped our resources, and each year millions face the horrors of starvation. Living as we do in one of the most productive lands on earth, few of us can know what it means to give birth to a child, to give it the gift of life, and have no food to keep it alive. Few of us can know what it means to a mother thirty years old, with eight children and with the knowledge that they face poverty and starvation. From a photograph in a magazine her dying children look out at us, and, uncomprehending, we turn the page.

Living as they do with these realities, people over much of the world have practiced abortion, often by the most primitive and dangerous methods imaginable. Long before birth control became a global issue, individuals resorted to infanticide. These are some of the practices followed in some countries to assure that some individuals, at least, have enough food to live.

Such practices mean the deliberate termination of life. Just as surely, indifference to rampant population growth means the deliberate termination of life. Many believe it is wrong to deny life to an unborn embryo; and just as many believe it is no solution to withhold compassion from those—child and adult—who have no alternative but to resign themselves to starving to death. For our species, reproduction and development and maturity flow on, through years, through centuries. But there is no escaping the principles underlying this flow of life, and in some way the birth rate for our kind must be reconciled with available resources. How we decide to control population size is one of the most volatile issues of this decade. Yet the decision must be made if we are to assure our kind of long-term stability in the biosphere. We will return to this issue in Chapter Twenty-One, in the context of principles governing the growth and stability of all populations.

Recommended Readings

Ebert, J. and I. Sussex. 1970. *Interacting Systems in Development.* Second edition. New York: Holt, Rinehart & Winston. Readable account of control patterns in development.

Rugh, R. and L. Shettles. 1971. *From Conception to Birth: The Drama of Life's Beginnings.* New York: Harper & Row. Magnificent illustrations of human embryonic development.

Saunders, J. 1970. *Patterns and Principles of Animal Development.* New York: Macmillan. An excellent, lucid book.

Unit IV

Adaptations in Form and Function: Organisms Evolving

Chapter Fifteen

Figure 15.1 The present is often key to the past, if we know where to look. The view northward across Oregon's Crater Lake, a collapsed volcanic cone aligned with Cascade volcanoes reaching into Washington—and all paralleling the Pacific coast.

Diversity:
Evolutionary Force, Evolutionary Product

By the close of the nineteenth century, "the fixity of species" was crumbling as a scientific concept. By the middle of the twentieth, "the fixity of continents" met a similar fate. Not only was life seen to be an ever changing drama, the stage itself was now seen to be changing in its most fundamental prop: the solid earth beneath our feet. It was not just that the earth's crust has been buckling upward and eroding downward through time. It was not just that the earth has gone through at least four immense glacial epochs and thaws, in which fully a third of the land has been alternately drained and drowned. It was not even that the earth has been steadily spinning slower and slower about its long axis, thereby making something as "basic" as the length of day variable through time. *It was that the whole crust has been divided into vast plates, which have been moving about uneasily on top of a plastic mantle and carrying the continents with them!*

There is no question that this conceptual upheaval is making a shambles of many previously entrenched scientific theories. And yet its implications are exhilarating. Here we are, living at a time when we can anticipate a cohesive theory in which the flow of earth history and life's history are inseparable. In this theory, the chemistry of life, its molecular and cellular architecture, and its myriad multicellular forms and behaviors—all these things have unfolded as parts of the same drama, which has been linked since the very beginning to the stage of a restless chemical earth.

The story you are about to read is sketched in broad outline, simply because there isn't space enough to do much more than this. But these broad strokes should not belie the impressive evidence from diverse fields that has been converging in support of it. This evidence comes from investigators with such far-ranging interests as the nature and evolution of the stars, the solar system, and the earth. It comes from centuries of increasingly precise research into the history of the earth's changing landscapes, oceans, climates, its fossils, its flow of life. We encourage you at the outset to explore and evaluate this evidence for yourself by continuing with the survey books recommended at the chapter's end. At the same time, we encourage you to view this story for what it is—a *tentative* undertaking in our attempts to explain all the observations currently available. The excitement of discovery is there, and so must be the legitimate tests of the validity of these newest attempts to find meaning in what is unfolding before us.

Origin of Life: A Transition From Chemical to Biological Evolution

From the remnants of titanic stellar explosions that ripped through our galaxy billions of years in the past, our solar system was born. And through the accretion and gravitational compression of dust and debris swirling turbulently about the primordial sun—a process that continued until about 5 billion years ago—our planet took form. At first it was a cold, homogeneous mass. But through contraction and radioactive heating, it soon had a core that was growing increasingly dense—and hot. By 3.7 billion years ago, it was hurtling through space as a thin-crusted inferno.

How could life have been destined for this forbidding place, with its violent volcanic outpourings and fountains of gases, with its surface quaking from the tumult below and the bombardment of meteorites from above? Yet, in the solidification of the lighter crust over the hot, denser layers below, and in the orbit that the primitive earth settled into around the sun, conditions that could favor the origin of life were assured.

Long before life originated, gases that had been trapped beneath the slowly forming crust or that were released as by-products of reactions in the molten interior were being forced to the outside. Although much of this early atmosphere was lost to space, after millions of years the earth's surface finally retained what is thought to have been a dense, gaseous atmosphere of hydrogen, nitrogen, methane, ammonia, hydrogen sulfide—and water.

Water is so essential for all the reactions and properties of all living things on this planet that it's difficult to imagine life originating here without it. But without enough gravitational mass to retain its atmosphere, the earth would have been as devoid of water as the moon is today. In addition, had the earth come to orbit the sun much closer than it did, its surface would have remained so hot that water vapor never would have been able to condense in liquid form. Had the orbit been much more distant, the earth's surface would have become so cold that any water formed would have become locked up as ice. *Thus the size of the earth, its composition, and its distance from the sun must be considered in identifying conditions that would be favorable for the origin of life.*

At first, however, the promise of liquid water could not be fulfilled. As fast as water vapor condensed in the cool,

Table 15.1 Main Elements Needed To Build Molecules of Life, and Their Likely Location in the Early Environment

Element	The Atmosphere	The Oceans
Carbon	Methane (CH_4)	
Hydrogen	Hydrogen gas (H_2)	
Oxygen	Water vapor (H_2O)	Water (H_2O)
Nitrogen	Ammonia (NH_4)	Ammonium hydroxide (NH_3OH)
Sulfur	Hydrogen sulfide (H_2S)	
Phosphorus		Dissolved salts
Potassium		
Iodine		
Calcium		
Iron		
Magnesium		
Sodium		
Chlorine		

upper reaches of the atmosphere and began falling earthward, it evaporated in the intense heat blanketing the rumbling crust. Millions upon millions of years passed and the surface cooled. And then it began to rain. For centuries it rained, with interminable torrents pouring over the parched rocks and stripping them of minerals and salts, then streaming through crevices and canyons and thundering downward to the lowest points on the crust to form the hot primeval seas.

Somewhere between 3.7 billion and 3.2 billion years ago, living systems appeared in these mineral-rich waters. Precisely how and where we may never know, for there are few fossil records of the event that are left for us to interpret. Most of the rocks from that period have been melted, solidified into new rocks, and remelted many times over as a result of ongoing movements in the earth's mantle and crust. Some are buried far beneath more recently formed rocks, where they have been so subjected to heat and compression that whatever clue they might have held of this critical period is now altered beyond recognition.

But here is what we do know: with three basic conditions met, life could have originated as a product of chemical evolution. What are these conditions? *First,* the environment would have to contain those elements which make up living things. *Second,* the chemical environment would have to promote reactions between these basic elements so that they could form organic molecules—the building blocks of life. *Third,* there would have to be some physical means for keeping certain molecular building blocks right next to one another. Otherwise, it would be impossible for a self-contained system to arise that would have the chemical and structural complexity characteristic of life, and it would be unlikely that the system ever would be reproducible.

Let's consider the first condition. As Table 15.1 indicates, the ancient atmosphere and oceans probably contained all the elements needed for the synthesis of complex organic compounds. So the question becomes not whether the required elements were available, but whether there was some way for them to combine into the building blocks of life. If the chemical environment of the early earth had been the same as it is today, there would be no way for that to happen. The main deterrent would be the abundance of free oxygen in the existing atmosphere. Organic compounds, recall, are spontaneously broken down into simpler molecules in the presence of free oxygen. It's true that the process appears to occur slowly in terms of human life spans. But over the millions of years it would have taken for the first cells to evolve, even a slow rate of decomposition would have been notable: organic molecules simply could not have accumulated faster than they were being broken down if free oxygen had been present.

Analysis of the chemical composition of the earth's early rocks as well as the atmospheres of other planets does indeed suggest the early atmosphere had little or no free oxygen, but that it contained abundant hydrogen as well as nitrogen, methane, ammonia, and water. Hydrogen, a reducing substance, would have encouraged reduction reactions in which complex organic molecules could be built up. That is why the early atmosphere is thought to have been a "reducing atmosphere" instead of the "oxidizing atmosphere" of today (Chapter Three). Any organic molecules built up would be less likely to undergo degradation in a reducing atmosphere. In fact, polymerization of small molecules into macromolecules would tend to occur spontaneously. All that would be required would be a source of environmental energy to activate molecules and make them collide with enough force to assure chemical interaction.

What energy sources might have been available? Ultraviolet radiation from the sun undoubtedly was bombarding the earth's surface more strongly at that time than it does today. For in the absence of free oxygen, there would have been none of the ozone (O_3) layer that now envelops the earth and screens out much of the intense ultraviolet wavelengths. Aside from that, levels of light and heat energy reaching the earth would have been higher during that period, when solar reactions were somewhat more intense than they are today. Energy might also have been provided by volcanic activity and by the radiation being released from the abundant radioactive elements making up the rocks in the earth's crust—elements that have largely decayed now to nonradioactive forms. And surely electric storms in the hot, dense atmosphere would have sent lightning crackling incessantly to the earth.

Thus the early earth could well have had the elements, the chemical environment, and a variety of energy sources. With that combination, organic molecules could arise spontaneously. We know this from the pioneering experiments of Stanley Miller and of many other chemists since. In 1953, while a graduate student at the University of Chicago, Miller set up a reaction chamber (Figure 15.2) containing a reducing mixture of hydrogen, methane, ammonia, and water. For a week he kept the mixture recirculating. All the while, he bombarded it with a continuous spark discharge to simulate lightning as an energy source. By the week's end, he found organic molecules had formed—including many kinds of amino acids!

Such experiments have been repeated many times, with variations in elements, in gas mixtures, and in the types of energy sources used. The results invariably show that all the building blocks required for life—including lipids, carbohydrates, proteins, and nucleotides—can form under abiotic conditions. More than this, if inorganic phosphate is present in the starting mixture, even ATP—the energy molecule used by all living systems—will form!

But now we come to the most puzzling question of all. By what mechanism could independently formed organic molecules have combined into a permanent, reproducible living system? There is no universally accepted hypothesis here, although two lines of experiments have given us some interesting things to think about. According to the first hypothesis, abiotic synthesis of increasingly complex molecules led gradually to a DNA molecule. With its double-stranded structure, DNA is intrinsically reproducible. And its formation would be a key step on the road to self-reproducing organisms. But even though DNA is central to life, it must be kept in mind that *a DNA molecule is incapable of reproducing itself.* To the best of our knowledge, DNA can be replicated only as part of a metabolic system that can provide (1) nucleotide subunits in great enough concentrations, and (2) an enzyme for holding the reactants in place while they combine. It's difficult to envision how a fully self-replicating molecule could have arisen without the support of a relatively complex and functionally isolated metabolic system. Even supposing there had been enough nucleotides concentrated in the primordial seas to permit the first DNA molecules to self-reproduce spontaneously, what would it have meant? DNA reproducing in isolation would have been like a book with no words on the page, and no one to read it.

According to the second hypothesis, a closed metabolic system must have appeared first. Metabolism, recall, means chemical control—*and chemical control is possible only with chemical isolation from the random ebb and flow of materials in the environment.* If we assert that chemical evolution alone led to a metabolic system, we must show that molecular boundaries can arise spontaneously, thereby isolating and concentrating the essential molecules from the rest of the environment. Laboratory experiments again suggest how such partitioning might have occurred. When Aleksandr Oparin mixed together proteins, polysaccharides, and nucleic acids, he discovered that these molecules can assemble into small spheres. These **coacervate droplets** are intriguing, for they seem to be differentially permeable. They can

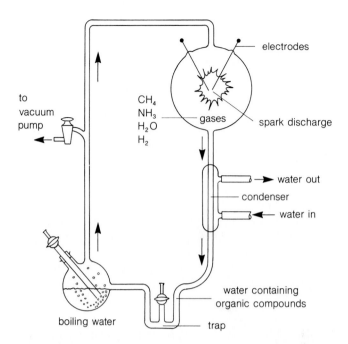

Figure 15.2 Stanley Miller's apparatus used in studying the synthesis of organic compounds under conditions believed to have been present on the early earth.

selectively accumulate certain substances in greater concentrations than are found outside. There is a tendency also for droplets to pinch off the main spheres, creating smaller subsystems from them.

In a similar kind of experiment, Sidney Fox and his coworkers began by heating amino acids under dry conditions, which produced a number of long polypeptide chains. Perhaps, Fox suggested, such polymers were formed in hot volcanic regions and then were washed by rains into the seas. Whatever the case, when these chains were added to hot water solutions and then allowed to cool, they assembled into small, stable **microspheres** (Figure 15.3). Like Oparin's droplets, these microspheres tended to accumulate specific substances in their interior. Depending on the substances present in the solution, the spheres could shrink or grow in size. In fact, given the presence of enough polypeptide chains, they could grow to the point of producing budlike growths that could break off into new microspheres. More than this, when lipid molecules were also present in the solution, they tended to combine with the protein microspheres. *And the outcome was the formation of a lipid–protein film surrounding each droplet that was remarkably like a simple cell membrane.*

It's important to keep in mind that neither coacervate droplets nor microspheres are alive. They are no more than

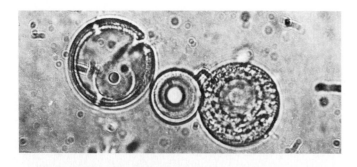

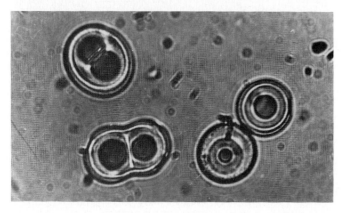

Figure 15.3 Micrographs of microspheres. (Sidney Fox, Institute for Molecular and Cellular Evolution, University of Miami)

the outcome of relatively simple chemical interactions between water and complex organic molecules. They do demonstrate that some boundary layers, at least, can come to exist spontaneously. But even though a selective boundary layer is one of the prerequisites for life, in itself it does not demonstrate the property of "being alive." Similarly, even though microspheres can grow by simple chemical acquisition of molecules and then fragment after a certain volume is achieved, they still cannot be called alive—for there is a long way to go from passive, random fragmentation to a self-reproducing cell.

At present, we can only speculate that the formation of such structures *could* have marked the beginning of a long series of chemical developments, with individual systems accumulating various environmental resources and carrying out such reactions, probably at comparatively low rates. For if such spheres existed that could grow in size, more molecules could be forced together in their interior. And because of chemical interactions of the sort outlined in Chapters Three and Seven, the molecules inside them would tend to interact spontaneously and become spatially *organized* relative to one another. If the spheres grew large enough to break in two, or if the pounding against some

ancient shore broke them in two, the fragments could continue to gather materials selectively and thus grow in their turn—only to be broken apart into fragments that again would grow.

Each new fragment would enter a slightly different part of the environment and would accumulate a slightly different array of molecules. Perhaps some sphere even acquired a molecule such as a polypeptide that acted to speed reactions involved in the formation of new membranelike films. The molecule essentially would be a catalyst; it would permit the sphere to grow more rapidly, hence to accumulate materials more rapidly, than neighboring spheres. Such a polypeptide catalyst may have represented the beginnings of enzyme action. Gradually, microspheres containing enzymes that speeded their own growth would succeed at the expense of less efficiently organized ones. In the nondirected chemical competition for resources in the seas, there would automatically be selection for microspheres having enzymes.

Where does this line of speculation take us? During the many millions of years that lifeless microspheres were accumulating organic molecules, suppose they were also gathering in various nucleic acids and nucleotides from their surroundings. Suppose some of the nucleic acids were no less than short molecules of deoxyribonucleic acid— DNA. If the nucleotides became concentrated near the DNA molecules, there would be a greater likelihood that the DNA could be replicated, given the presence of an enzyme capable of unzipping the double helix. In itself, this step would not represent a selective advantage. But among all the millions of tiny spheres that contained nucleic acids, perhaps there was one that came to acquire a nucleic acid having a base sequence that was, by chance, no less than a template for a membrane-forming enzyme. It would be a self-enclosed system for promoting the replication of a nucleic acid: the nucleic acid would promote formation of the enzyme, and the enzyme would promote formation of new membrane. Thus the sphere could grow faster and accumulate more resources, which would speed the growth—replication cycle. The spheres into which it fragmented could come to differ slightly in the interdependent cycle of membrane, nucleic acid, and enzyme formation. And at each stage, variants capable of functioning somewhat more efficiently would grow faster and produce more microspheres like themselves. Given the replicative nature of DNA, the perpetuation of new modifications would have been assured. The evolution of a living cell, with its gene-action system and its battery of gene-coded enzymes, would have begun.

It is true that the entire sequence of events just described is indeed speculation. But it is also true that between 5 billion and 3.2 billion years ago, there was time enough and resources enough for chemical evolution of this sort to occur. And by 3.2 billion years ago, we know now that clearly recognizable forms of life had appeared.

A Time Scale for Patterns of Diversification

In itself, the notion that living systems not much different from prokaryotic bacteria could evolve into forms as complex as ourselves might elicit disbelief. Part of the problem is that we are used to thinking in terms of minutes, hours, days, and years, certainly never much beyond centuries. Besides, until recently there was no way to judge the true age of the earth, hence the span of time in which evolution might have taken place. We could only observe the relative sequences of rock formations and the fossils they contained without knowing how to assign dates to the boundaries between them. Thus the **geologic time scale** (Figure 15.4) initially was no more than a progression of four broad eras: the *Proterozoic* ("very first life"), the *Paleozoic* ("ancient life"), the *Mesozoic* ("between-ancient-and-modern life"), and the *Cenozoic* ("modern life"). Within the past three decades, however, we have been able to assign fairly firm time boundaries to these geologic intervals by using radioactive dating methods on an enormous number of rock samples taken from all over the earth. All these methods are based on comparing the known, invariant decay rates of such radioactive isotopes as uranium, thorium, potassium, and strontium to the measured amounts of these isotopes and their decay products in different kinds of rocks (Figure 15.5). Some of the most ancient fossilized cells taken from carefully dated rocks are shown in Figure 15.6. These fossils have been found in certain regions of Africa, North America, and Australia, where large land masses have remained fairly stable through episode after episode of geologic unrest.

As we move from these earliest signs of life, the fossil organisms preserved within younger and younger rocks become progressively more abundant, more diverse, and more like modern forms. In many cases, it's possible to trace gradual modifications of form among apparently related organisms. And often there appears to be gradual divergence of related forms into distinct species. Modern evolutionary theory explains this progression in the following way. Within each population of organisms, there is always a certain amount of genetic variability. This variability constantly increases as new mutations occur. In each generation, natural selection favors those variants best adapted to a given environment, and they will be the ones most likely to reproduce. As environments change, so do selective pressures change—and so does the population change in character. If one population somehow expands into two different settings, different traits will be selected for in each setting. With time, the two populations will accumulate more and more structural and/or behavioral differences until they reach the point of *speciation:* members of the two populations can no longer interbreed even if they are brought back together.

Era	Period	Epoch	Age (Millions of Years)
Cenozoic	Quaternary	Recent	30
		Pleistocene	
	Tertiary	Pliocene	
		Miocene	
		Oligocene	
		Eocene	
		Paleocene	
Mesozoic			65
	Cretaceous	Late	100
		Early	130
	Jurassic		185
	Triassic		
Paleozoic			230
	Permian		265
	Carboniferous		355
	Devonian		413
	Silurian		425
	Ordovician		475
	Cambrian		
Proterozoic (Precambrian)			570
			1,000
	oldest definite fossils known		3,500
	oldest dated rocks		4,000
	origin of the earth		5,000

Figure 15.4 Geologic time scale, with dates based on radioactive isotopes from rocks of each era. Figure 15.5 describes the naturally occurring process that makes radioactive dating possible.

Figure 15.5 Radioactive dating. For many years, scientists tried to measure the age of rocks by assuming erosion, uplifting of mountains, and so on occurred at a constant rate. All such attempts were frustrated because these processes simply don't occur at a constant, invariable rate. In this century alone, volcanoes have suddenly popped out of the sea, only to crumble and disappear in a matter of months!

More recently, radioactivity has proved to be an accurate timekeeper. A *radioactive element* has an unstable combination of protons and neutrons in its nucleus. It breaks down spontaneously, and it releases radiation (such as x-rays or electrons) when it does. Some combinations are inherently more unstable than others. Thus some radioactive elements break down rapidly, others slowly. *But each radioactive element has its own characteristic decay rate.* And that rate simply can't be modified. Radioactive elements have been subjected to intense heat and pressure, suspended in high vacuum, cooled to nearly absolute zero, rearranged chemically into myriad chemical forms, charged with electricity, enclosed in a vast array of environments, and studied as gases, liquids, and solids. And no one has ever found any evidence that such changes have any effect whatsoever on the rate of radioactive decay. *All radioactive elements go on ticking off the seconds in total disregard for the environment.*

Each radioactive element has a *half-life:* the time required for half the atoms in a sample of the element to decay. How can half-life be used to date ancient rocks? Consider a radioactive isotope of potassium. When it breaks down, it's converted to a lighter element: argon. Argon is stable, but it is a gas at normal earth-surface temperatures. The half-life for the decay of radioactive potassium is 1.28 billion years. In that time, half of a pure, radioactive potassium sample becomes converted to argon. (Since a given rock first solidified, all the argon present in it has been derived from potassium. Any argon present before that time would have escaped to the atmosphere as gas. But once the rock became solid, argon formed by radioactive decay of potassium would have no way to escape.) If the measured ratio of potassium-to-argon is 1:1, then the rock solidified 1.28 billion years ago.

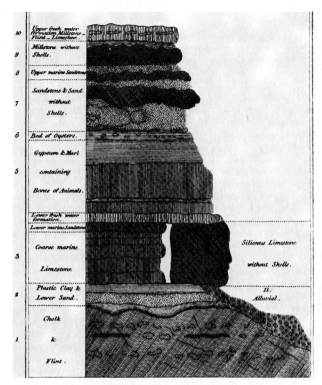

G. Cuvier, in Dott and Batten, *Evolution of the Earth*

Main Radioactive Elements Used in Rock Dating			
Radioactive Isotope	Half-life	Stable Product	Age Range (Years)
Rubidium 87	50 billion years	Strontium 87	>100 million
Thorium 232	14.1 billion years	Lead 208	>200 million
Uranium 238	4.5 billion years	Lead 206	>100 million
Uranium 235	0.71 billion years	Lead 207	>100 million
Potassium 40	1.28 billion years	Argon 40	>100,000
Carbon 14	5,730 years	Nitrogen 1	0–60,000

Data from *Handbook of Chemistry and Physics*, 1975.

But not all transitions in the fossil record are gradual, for certain major groups of organisms often appear relatively abruptly—and they are already highly developed and diverse when they do. What could account for such dramatic entrances of whole groups onto the evolutionary stage? First, not all organisms become fossilized when they die. The fossil record represents only a tiny fraction of all the organisms alive during any span of the remote past. Second, the probability that representatives of any group will be preserved depends partly on how many individuals there are in the group. Thus small numbers of organisms might be successfully adapted to some restricted environmental pocket—even though the chance of finding fossil evidence of their existence would be extremely low. Third, suppose the range of the environment to which an isolated group was adapted suddenly expanded as a result of a comparatively abrupt climatic or geographic change. Then its populations could be expected to expand proportion-

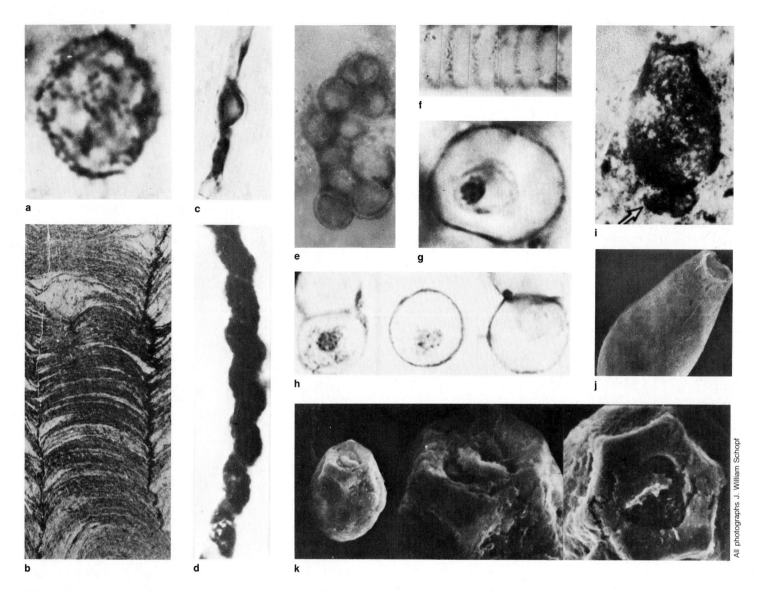

Figure 15.6 A sampling of the oldest known fossils. (**a**) From South Africa, bacteriumlike form about 3.2 billion years old. (**b**) Limestone stromatolite from Rhodesia, about 3.1 billion years old. The stacked organic (dark) and calcium-containing (light) layering is identical with the one laid down by modern communities of photosynthetic microorganisms. (**c**) Cellular form from South Africa, about 2.25 billion years old, and similar to modern blue-green algae. (**d**) From central Australia, a spiral form 900 million years old. (**e**) Colonial cells found in stromatolites in the USSR, about 650 million years old. (**f**) Filamentary form, about 650 million years old. (**g**) Eukaryotic cell from Central Australia, about 900 million years old; the granules and spots may be remnants of organelles. (**h**) Fossil eukaryotes containing cytoplasmic remnants, about 900 million years old. (**i**) Eukaryotic cell 750 million years old. (**j, k**) Notice the well-developed mouth region in these fossil eukaryotes, also 750 million years old.

ately. Not only would the change mean an increased probability that fossil evidence would become more abundant, it could well set the stage for rapid diversification. For as the group of organisms radiated into an expanded environment to which it was already adapted, local populations would have new opportunities to exploit existing genetic variability, and they would embark on a rapid course toward speciation. As you will now read, sprinkled through the fossil record of gradual modifications in form over vast spans of time are just such episodes of abrupt environmental change—with a corresponding acceleration in the emergence of new species.

The Age of Prokaryotes

Until about 3.7 billion years ago, the earth's crust may have been too unstable or too thin—and perhaps the heat flow from the interior was too great—for permanently stable land masses to form. (It would be at least another 1 billion years before there would be stable crustal plates, and even then they would be fringed with mobile, volcanically active zones.) Nevertheless, there probably were myriad volcanic islands rising above the primeval seas. And it may have been in the shallow, nearshore waters of these islands that life appeared. Reeflike formations, perhaps formed by calcium-secreting algae, date from about 3.5 billion years ago. Some of the oldest rocks, formed from lime deposits and black mud (in which oxygen is absent), contain organic compounds such as amino acids and parts of the chlorophyll molecule—and as far as we know, only life forms can organize a chlorophyll molecule. Some organisms apparently were developing the means for photosynthesis (Chapters Three and Five). Even so, fossils of any sort from the Precambrian are scattered sparsely in limited regions, and they speak only of the most limited diversification. For about 2 billion years, it seems, organisms resembling modern bacteria and blue-green algae had the world much to themselves. If there was little or no oxygen in the atmosphere, anaerobic pathways must have been the main style of the day.

Between 2 billion and 1 billion years ago, both heterotrophic and autotrophic prokaryotes grew steadily more abundant, if not more diverse. Apparently there were no pressures prodding life toward greater structural and functional variation. It was only with the convergence of new environmental conditions that new evolutionary roads opened up.

The Rise of Eukaryotes

Between 1.4 billion and 570 million years ago, the first eukaryotes arose. If we are to assign some meaning to the array of fossilized shells, spines, and armored plates they left behind, it would have to be that predatory forms were among them. That being the case, simple coexistence could no longer be possible; change would be the order of the day. Fossils of prey organisms from that time onward are increasingly larger. Fossils of predators are correspondingly larger, and some even sport devices for tearing off pieces of larger prey. In some cases, apparently, the only survivors among certain prey groups were the ones too large to be swallowed. With their selective advantage, they would be the ones more likely to leave progeny. Thus, subsequent generations became larger in size. At the same time, other groups may not have included enough variants capable of

withstanding predation, and they became extinct. One of the groups that seem to have been hit hardest by predation were the blue-green algae of the seas. An important group, whose fossilized, matlike remains form stromatolitic rocks (Figure 15.6b), had been expanding steadily in numbers for many millions of years. But with the rise of predatory eukaryotes they began to decline. Perhaps it's no coincidence that living representatives of the group are confined to marine environments having low levels of diversity and low levels of predation.

By 570 million years ago, astonishingly diverse eukaryotes dominated the scene. Why such rapid diversification *then,* and not before? Consider the possibilities. During the period between 1.5 billion and 570 million years ago, oxygen had begun to accumulate to significant levels as a result of the combined activity of billions of photosynthetic microorganisms. (These increased levels have been determined by measuring the extent of oxidation of surface rocks which were solidifying at that time.) The primitive eukaryotes were mobile, and they were heterotrophic; they chased after their prey. The use of cellular respiration (Chapter Six) would clearly be advantageous in providing an energy source for their movements, and it seems inevitable that the gradually increasing atmospheric levels of oxygen would foster organisms capable of cellular respiration. Whatever the case, by the beginning of the Cambrian, oxygen levels were high enough to allow considerable cellular energy to be devoted to building mineralized shells, spines, and the other hard protective parts we see for the first time in the fossil record. In addition, the rising oxygen levels meant that an effective ozone (O_3) barrier against incoming ultraviolet radiation was forming. Thus shallow offshore waters became more hospitable environments for eukaryotes—which are highly sensitive to these potentially deadly wavelengths.

Correlated with these rising oxygen levels was a geographic change of the first magnitude. By the dawn of the Cambrian, the marine environment expanded dramatically as the climate warmed up in what appears to have been the sudden aftermath of a world-wide glaciation. As the ice melted, shallow seas inundated much of the land. Into these vast seas the photosynthetic algae and eukaryotes radiated.

Before this transitional period in the history of life—a comparatively brief span of about 800 million years—the eukaryotes did not leave behind much evidence of their early adaptations. But with the convergence of conditions favoring the development of hard parts around fragile bodies, the eukaryotes would henceforth commit evidence of their journey to the sediments of time. And so today we pick up the remains of that past age and watch them fall silently through our fingers; and we begin to suspect at last the meaning of the relatively abrupt appearance of ever more elaborate shells and spines that once harbored vulnerable as well as voracious life in the crashing surf of the Cambrian shores.

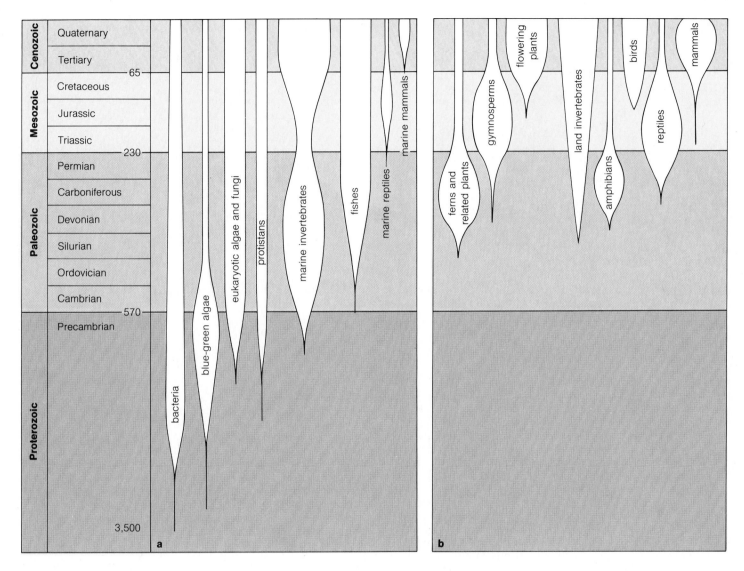

Figure 15.7 Diversification of life in the oceans (**a**) and on land (**b**). The varying width of each evolutionary line represents periods of relative dominance and decline for each group of organisms.

Up Go the Mountains, Out Goes the Sea

If there is one thing that characterized the Paleozoic, it was the number of massive inundations of continents and retreats by shallow seas. By the late Cambrian, there apparently were three major land masses. Two would eventually become North America and Europe. The third, **Gondwanaland,** was a supercontinent destined to become fragmented into Africa, South America, Australia, Antarctica, even parts of Asia and the eastern North American coast. But the three land masses were covered in large part by shallow seas. Because uplifted land had been eroded almost to sea level, not far below the water's surface were mud, sand, and more mud. Perhaps that's why this was the golden age of trilobites—of mud-crawling, mud-burrowing scavengers that eventually were 600 genera strong. They, along with algae and with myriad other shelled organisms, thrived in this marine setting.

By the Ordovician, even low-lying land masses seem to have become submerged. Abruptly, there was a new burst of adaptive radiation into vacant marine environments. This

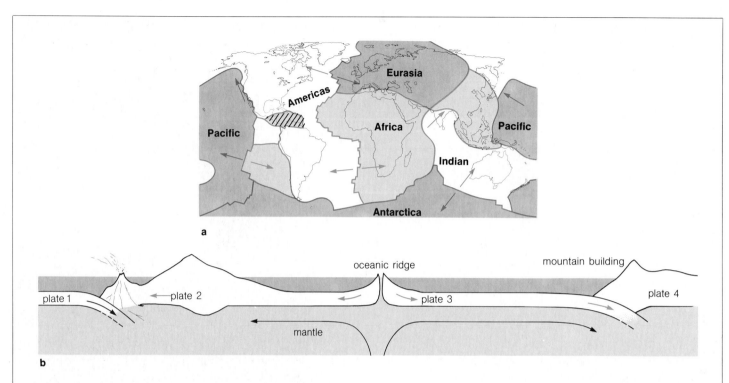

Figure 15.8 Plate tectonic theory. The earth's surface is seen to be broken up into rigid plates. Today these plates are drifting toward or away from each other in the direction of the arrows in (**a**). Boundaries between plates are marked by recurring earthquakes and volcanic activity. (Recent eastward movement of the Caribbean plate, the cross-hatched area in **a,** caused the devastating 1976 earthquake in Guatemala, which, like much of Central America, is part of this plate. Explosive volcanic activity and extensive surface faulting suggest this region of the crust is slowly being torn apart.)
(**b**) Sea-floor spreading and continental drift. Plate tectonics is based on observations that the sea floor is slowly spreading away from sites called oceanic ridges, and on measured displacements of the continents relative to these ridges. Thermal convection in the mantle is proposed as the mechanism underlying these movements. More heat is seen as being generated deep beneath oceanic ridges than elsewhere. The hotter material slowly wells up, then spreads out laterally beneath the crust (much like hot air rising from a stove, then spreading out beneath the ceiling). Oceanic ridges are places where the material has actually ruptured the crust (as it continues to do in the volcanic Hawaiian Islands). As the cooler material moves away from the ridges, it acts like a conveyor belt, carrying older oceanic crust along with it. Thus plates grow and spread away at oceanic ridges. As the plates push against a continental margin, they are thrust beneath it, which causes the crumpling and upheaval we call mountain building.

was what we might call the age of invertebrate experiments. All but one invertebrate phylum had arisen by the close of the Cambrian, but now new varieties were decked out in the most astonishing adaptations—including well-developed eyes for locating predators or prey. Many of the innovations of this period would lead nowhere, but others would survive even into the present (Chapter Eighteen).

For several million years, life flourished in the seas. Then, in the mid-Ordovician, the crustal plates bearing the three land masses began moving toward one another. As they did, they began edging over the crustal plate that formed the basin of the ancestral Atlantic Ocean. At first, the land masses were raised slightly and the shallow seas were drained off. The sudden shrinking of the immense marine havens must have severely tested existing populations, and the fossil record does indeed reflect a corresponding explosion of new forms apparently vying intensely for food and space near the continental margins. By the late Ordovician, the seas inundated the land once more, and a new variety of life forms appeared in vast numbers.

The jawless fishes—organisms at the crossroads leading eventually to reptiles, birds, and mammals—seem to have been among them, but they were not yet populous enough to leave behind much more than traces of their appearance. No matter. Toward the end of the Ordovician, storms of mountain building were brewing as the continental plates continued on their collision course. Volcanic outpourings in the mobile zone along the eastern edge of the North American land mass created the immense ancestral Appalachian Mountains as the European plate closed in.

During late Silurian times, some life forms by chance must have developed traits that would eventually permit them to escape from the seas. In the fossil record are transitional forms that appear to have been evolving in ways that would lead to the invasion of land. Regardless of when the first invasions occurred, they could have occurred only when certain challenges were met. Before some organisms could leave the seas, they would have to be equipped with the potential for tapping into new water supplies or carrying water around with them. They would require tough or waxy surface layers to keep from drying out. No more could aerobic organisms absorb oxygen from the surrounding water—they would live on land only with systems for taking in oxygen from the surrounding air. Without the buoyancy of water, different frameworks would be needed to support the body. For animals, new modes of moving about would be required. And how could the balance of salts in body fluids such as blood be maintained once salt-water environments were left behind? There would be selection for new structural and behavioral adaptations. And what about reproduction? Gametes could not be simply dispersed for fertilization in the waters of the sea; there would be selection for new modes of reproduction and offspring dispersal. Adaptations that permitted radiation onto the land are shown in Chapters Seventeen through Nineteen.

As the Silurian gave way to the Devonian some 400 million years ago, the continents collided. Even as the proto-Atlantic Ocean disappeared, there was a dramatic increase in dry land area. Plants and then animals, equipped by chance with variations that permitted them to survive at the land's edge, began their tentative adaptive forays into this vacant environment. The first land plants were single stalks no taller than your little finger, and they were anchored in wet mud. Crabs and related forms, with their protective shells, were scuttling about in the alternating wet–dry tidal zones. Many groups of fish perished. But certain fish, left behind in drying tide pools or in evaporating fresh-water ponds, were able to use their swim bladders (air sacs) as lunglike organs, and their fins as simple limbs (Figure 15.9). With such adaptations, they had the means of crawling from pond to pond. Over prolonged periods of alternating wet and dry conditions, those individuals best able to breathe air and crawl about in the mud survived even as their water-dwelling relatives perished. Thus there apparently began an evolutionary trend that gave rise to the

American Museum of Natural History

Figure 15.9 Reconstructions of the lobe-finned fish (**a**) and their descendants, the amphibians (**b**) of Devonian time.

amphibians: organisms adapted to life both in water and on land. Yet, despite the flourishing diversity, the plants and animals of this period had one thing in common: a total dependence on a constant supply of water. The transitional plant forms still needed a wet environment to complete their sexual cycles. Devonian amphibians, like their descendants, had to return to water to lay eggs, for their eggs had no protective shell to keep from drying out.

The Carboniferous was the heyday of the amphibians. And why not? From beginning to end of this period, land masses were submerged and drained no less than *fifty times!* Besides these major inundations, there were ever-changing sea levels in between. And imagine the conditions along the flat, low coastlines. Immense swamp forests became established; over time the seas moved in and buried them in sediments and debris; they became reestablished as the seas moved out, and then they were submerged again. (The organic mess left behind has since compacted to become the world's great coal deposits.) But some groups survived the inundations, for adaptations had appeared among them that allowed movement onto higher (and drier) land. The gymnosperms (Chapter Seventeen) were one such group.

Figure 15.10 (**a**) A Jurassic reptile capable of running swiftly on two legs. Such reptiles are thought to be the ancestors of dinosaurs and birds. During the age of dinosaurs, some forms took to the air; (**b**) shown here, pterosaurs having wingspans of 12 meters (40 feet) or more. Other forms became adapted to the water and became impressively large, swimming predators—the long-necked plesiosaurs (**c**) reached 15 meters (50 feet) from head to tail. These forms were doomed to extinction when the vast inland seas dried up. On land, dinosaur diversity reached its peak, with herbivores and carnivores ranging from the size of gophers to behemoths three stories high and half a block long (**d**).

These seed-bearing plants were not restricted to the water's edge; they could complete their reproductive cycles without free-standing water. The reptiles were another: these animals could break away from an aquatic existence because of shelled eggs and internal fertilization. These adaptations meant development could proceed within the eggshell, a moist and (compared to the outside) dependable setting.

Swamp forests and ancestral frogs—as magnificently adapted as many species were to these alternating environmental conditions, many of their adaptations would prove disadvantageous when the environment changed. And change it did, for as the Carboniferous gave way to the Permian, collisions of crustal plates brought all land masses together into one vast supercontinent, called **Pangaea.** The changes in land area and in elevation brought pronounced

the water, others came to develop means to take to the air. The ones remaining on land diversified in the most spectacular ways and were unchallenged for the next 125 million years—the golden age of dinosaurs (Figure 15.10).

Their ultimate challengers, though you might never have believed it at the time, would be the little ratlike mammals scurrying through the shrubbery of the Jurassic. These mammals were food for many dinosaurs, so they had to be clinging most precariously to their place in the community. But cling they did, only to explode into the biotic vacuum left by their predators at the end of the Cretaceous. For at that time all dinosaurs disappeared suddenly from the earth. This total, abrupt extinction has yet to be explained. There were no profound shifts in climate. Although there is evidence of extensive mountain building, there apparently were no sudden, massive upheavals in the crust. Other kinds of organisms living in the same places managed to survive. Why not the dinosaurs?

Perhaps the answer lies with changes in the community of life itself. For example, in the midst of gymnosperm dominance, the flowering plants began to take hold and diversify (Chapter Seventeen). The plant-eating dinosaurs were adapted to feeding on gymnosperms. Perhaps, with brains no larger than a golfball, those giants never figured out that flowering plants might be food, too. At least, the mummified remains of hadrosaurs (large and widely distributed plant eaters of the late Cretaceous) show the stomach contents to be entirely gynmosperm needles and twigs—even though gymnosperms were relatively rare by that time. With plant-eating dinosaurs starving themselves out of existence in the midst of plenty, the populations of dinosaurs that preyed on the plant eaters would have starved, too, and the whole community structure might have toppled. As you will read in Chapter Twenty-One, local disruptions can have that irreversible effect. Whatever the reasons, with the land essentially cleared of major predators, the mammals eventually diversified along with the flowering plants.

Cenozoic Upheavals and the Further Diversification of Life

It was the best of times, the worst of times; it was the dawn of the Cenozoic and the continents were on the move.

Beginning in the Mesozoic, the supercontinent began to break up. Widespread volcanic activity marked the birth of the Atlantic basin as North America, Europe, and Africa began moving their separate ways. But by Cenozoic times, major reorganization was going on among all the crustal plates. Unbelievable amounts of lava began pouring out through immense faults and fissures that penetrated to the very basement of the earth's crust. Brittle fragmentation occurred along the coasts; severe volcanism and uplifting

differentiation in world temperature and climate. To the north, arid lowlands and humid uplands emerged. To the south, glaciers built up and ice sheets spread over the land that had become positioned there. As shallow seas were drained from the massive continent, life forms radiated throughout the land. And everywhere in the shrinking seas: massive extinctions.

After the great Permian marine extinctions, the survivors in the seas had 165 million years of relatively stable environments in which to diversify. On land, the character of the living community changed profoundly. By Jurassic times, the climate was warm and humid; mountains had emerged, along with plains and vast lagoons. In these new settings, gymnosperms and reptiles had the competitive edge. Some groups of reptiles became readapted to life in

Figure 15.11 Reconstruction of North American mammals of the Eocene (**a**) and Pliocene (**b**).

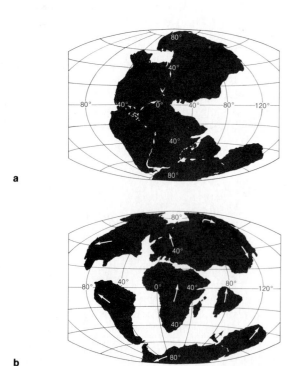

Figure 15.12 Restoration of the supercontinent Pangaea (**a**) and its breakup, showing the impending collisions of Afro-Arabia and India with Eurasia in Cenozoic times (**b**).

gave rise to mountains along the margins of massive rifts and along zones where some plates were thrust under others. The immense volcanoes dominated by Kilimanjaro in Africa, the Sierra Nevada and the Cascade ranges paralleling North America's Pacific coast, the Alps, the Andes, the Himalayas—never before had there been such world-wide mountain-building as crustal plates collided with and jockeyed for position relative to one another.

Correlated with this latest redistribution of the land was the onset of the world's most recent glaciation. Sea levels changed, as did the earth's overall temperature, which brought about widespread extinctions among some groups even as it opened the door to expansion for others. The separation of continents encouraged diversification of life in divergent ways. In much of the world, the concurrent shifts in climate led to the emergence of extensive, semi-arid, cooler grasslands into which plant-eating mammals and their predators radiated. Now the ungulates—hoofed mammals of the sort shown in Chapter One—roamed in vast herds. The ancestors of such modern forms as deer, giraffes, horses, pigs, rhinos, camels, and elephants appeared, as did carnivores such as the saber-toothed tiger. And as the climate continued to change, the vast tropical forests dating from the preceding epoch began to be fragmented into a patchwork of new environments. Many of their inhabitants were forced into new life styles in mountain highlands, in deserts, in the plains. One such evicted form was destined to give rise to the human species—with all our problems, with all our promise. The events surrounding that emergence will be described in Chapter Eighteen.

Perspective

All living things are composed of molecules made of the same chemical elements—primarily carbon, hydrogen, nitrogen, and oxygen. We can reasonably assume that all these elements were present, in one form or another, in the crust and atmosphere of the primordial earth. It has been demonstrated that these elements can combine spontaneously into carbohydrates, lipids, proteins, nucleic acids, even ATP—into the stuff of life—given the right chemical environment and some source of activation energy. It has been demonstrated that, given a plausible set of environmental conditions, such macromolecules can form differentially permeable structures much like simple cell membranes. It is possible that further experiments into the molecular and biochemical nature of life will throw light on how such structures might have evolved chemically into the first organized, self-reproducing systems—into the first living cells. Whatever the details of such chemical evolution might have been, we suspect from fossil signposts along the way that the progression flowed in this manner to the threshold of biological evolution and the subsequent explosive diversification of life.

And so today you are sharing the earth with some 10 million different kinds of life forms. They, and you, share allegiance to the same principles of energy flow and chemical interactions. They, and you, share the same molecular and cellular heritage. All these things speak of the underlying unity of life. But more than this, they speak eloquently of its subsequent diversity. For if the environments of the first living things had never changed, if there had never been different horizons waiting for the vanguard of inadvertent explorers, perhaps the world today would hold little more than testimony to life's unity. Perhaps there would be little more than single cells of the sort preserved in ancient rocks—cells matted against rocks or suspended in the waters of the sea, soaking up nutrients.

But the record of earth history tells us environments *have* changed, and that organisms within a given population either have been equipped to respond to those changes or have perished. The record also suggests that just as the diversity of life has been a product of evolution, so has it been an evolutionary force of the first magnitude. "Diversity" not only means adaptations to some combination of temperature, chemical balance, available water, light, dark, and living space. "Diversity" means adaptations to different kinds of predators, different prey, different competitors or cooperative groups after the same resource, different behaviors and patterns of fur and feather for attracting mates and assuring reproductive success.

It is possible, then, to view all existing species as the product of interactions with the environment and with one another. These interactions are the focus of paleo-ecology; they continue to be the focus of modern-day ecology. And this brings us to a concept of profound importance. By the very fact of their continuing existence, all 10 million kinds of organisms now on earth can claim adaptive success. However, as you will read in this unit and the one to follow, *success is assured only as long as there is responsiveness to the environment, and only as long as there is dynamic stability between the requirements and the demands of organisms making their home together.* Both the environment and the community of life change, shifting as imperceptibly as shifting winds over the centuries or abruptly obliterating all trace of forms that have gone before. And therein lies the story of evolution, the story of random chemical competition leading to the first organized, self-reproducing forms of life, of dinosaurs and continents on the move, of simple strategies unchanged since the dawn of life, and of the complex human strategy—as yet unresolved—that can hold a world together or rip it apart. But must we predict gloomily that such unresolved activity on our part will end this magnificent story for all time? We doubt it. For if the record of earth history tells us anything at all, it is that life in one form or another has survived disruptions of the most cataclysmic sort. That life can evolve tenaciously through tests of flood and fire suggests it has every chance of evolving around and past our transgressions, too. *Viva Vida!*

Recommended Readings

Dott, R. and R. Batten. 1976. *Evolution of the Earth.* Second edition. New York: McGraw-Hill. Findings from myriad lines of research are distilled into a stunning picture of earth and life history. The authors are masters at initiating the reader into the excitement of discovery. This second edition continues to weather the conceptual storms of a rapidly changing field; you can almost see the authors bracing for the unexpected with wit and good humor. We enthusiastically recommend this book for your personal library.

Barghoorn, E. and J. Schopf. 1966. "Microorganisms Three Billion Years Old From the Precambrian of South Africa." *Science,* vol. 152, pp. 758–763.

Cox, A. 1973. *Plate Tectonics and Geomagnetic Reversals.* San Francisco: Freeman. Excellent overview of the evolution of plate tectonic theory.

Kay, M. and E. Colbert. 1965. *Stratigraphy and Life History.* New York: Wiley.

Marquand, J. 1971. *Life: Its Nature, Origins, and Distribution.* New York: Norton.

McAlester, A. 1977. *The History of Life.* Second edition. Englewood Cliffs, New Jersey: Prentice-Hall.

Schopf, J. W. 1975. "The Age of Microscopic Life." *Endeavor* (May), vol. XXXIV, 122:51–58.

Chapter Sixteen

Figure 16.1 Scanning electron micrograph of the magnificent glass house of a diatom. 5,800 ×.

Monerans and Protistans

Lake bottoms, decaying plants and animals, hot springs, snow, swamps, sewers, soil, rivers, oceans, your gut—all these environments are the kinds of places that single-celled monerans and protistans call home. Just as their environments are diverse, so are monerans and protistans diverse, although most are so small you might never even suspect their presence. A handful of rich, moist soil may contain myriad different kinds of these microorganisms, for it is really a composite of microenvironments, each having slightly different conditions that encourage unique adaptive responses. Thus diversity means more than increasing in size and complexity; diversity also means unique single-celled adaptations to worlds that can fit on the head of a pin. The principles of evolution apply to these microworlds, too, of course. But where there have been no pressures to increase in size—*where resources are plentiful and there is no advantage to be gained by expending the energy needed to maintain an increase in size*—then there has been no trend to become much larger or much more complex than the first living things. In such places, the most successful organisms are small and capable of extremely rapid reproduction. Their reproductive exploits are mind-boggling: given enough resources, they may double in numbers twice an hour and increase many millionfold in just one day!

In this chapter we will consider the monerans, the tiny single-celled prokaryotes that have done well enough by remaining small and organizationally simple. We will also consider the puzzling protistans—simple yet successful eukaryotes that in many ways seem to mirror the ancient ancestors of the multicellular plants, animals, and fungi.

At the Boundaries of Life: Bacteria, Mycoplasmas, and Viruses

At the boundary with the nonliving world are 15,000 known species of bacteria and their stripped-down cousins, the mycoplasmas, along with some truly stripped-down entities called viruses. All **bacteria** are monerans built according to the same general body plan: a rigid cell wall surrounding a plasma membrane, which encloses the cytoplasm. Bacteria are prokaryotes: they lack a true nucleus and the varied organelles typical of eukaryotes (Figure 4.3). Sometimes the bacterial cell wall is encased in a slimy capsule, which affords protection against such threats as bacterial viruses.

Even though all bacteria have the same general body plan, they differ somewhat in shape. Some look like rods, others look like balls or spiral noodles (Figure 16.2). These three shapes speak of functional adaptations to the environment. Bacteria must have a moist environment in order to grow and reproduce. Ball-shaped bacteria, or **cocci,** are less apt to shrivel up if their living quarters dry out. Rodlike forms, or **bacilli,** have more exposed surface area per unit volume than cocci, which means they are less resistant to drying out—but which also means they have more plasma membrane for soaking up nutrients when conditions are favorable. Many spiral forms, or **spirilla,** can move more rapidly than either cocci or rodlike forms, much like a corkscrew twisting through a cork.

Although each bacterial cell is a functionally independent unit, bacteria are sometimes clustered together as an outcome of the way they reproduce. Many cocci reproduce by dividing in only one direction. Others divide in two directions and adhere to form a sheet, and still others divide in three directions and form a cube. At one time, bacteria were classified according to the way they remained attached to one another after cell division. But variations in the environment often have a profound effect on what the resulting colonies look like. For instance, if the environment is deficient in just one mineral that some bacterial species must have, the chains that form are far shorter than what you would otherwise observe.

If you can imagine what a bacterium would look like without its rigid cell wall, then you have an idea of the appearance of free-flowing **mycoplasmas,** disease agents that are the smallest organisms known. Mycoplasmas have no cell wall, although they typically have a tough plasma membrane which helps keep their contents from dribbling out. They can get along without a cell wall because they live in the osmotically protected sanctuary of an animal's body. When grown in a rich solution in the laboratory, the unwalled mycoplasmas fail to take on any characteristic shape: they form random threads, balls, and blobs (Figure 16.3). When grown on a solid medium such as agar, their options are fewer and they end up looking like tiny fried eggs.

Despite the structural simplicity of bacteria and mycoplasmas, these organisms are capable of metabolism and of transforming energy into usable forms. They also have a

a

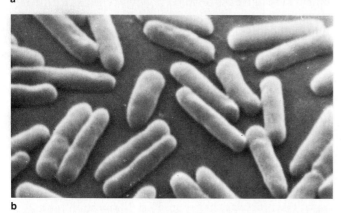

b

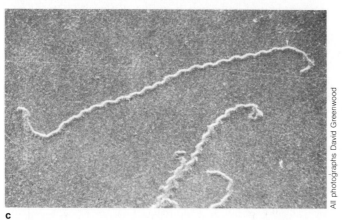

c

All photographs David Greenwood

Figure 16.2 Bacterial body plans. (**a**) The ball-shaped forms are represented here by *Streptococcus*, the causative agent of strep throat which, if left untreated, may lead to rheumatic heart disease and irreversible heart damage. (**b**) Rod-shaped forms are represented by *Pseudomonas aeruginosa*, an opportunistic pathogen that can infect animals whose natural resistance has been lowered. (**c**) Spiral shapes are represented by *Spirochaeta stenostrepa*. A related spiral bacterium causes syphilis, one of the most common infectious diseases in the United States. Although many bacteria are agents of human disease, the benefits derived from their activity far outweigh the harmful effects some bacteria have on our lives. Because bacteria recycle resources, life without them would be impossible for more complex organisms.

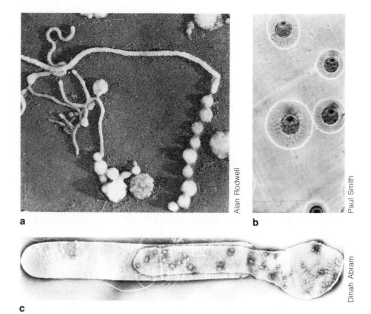

a

Alan Rodwell

b

Paul Smith

c

Dinah Abram

Figure 16.3 (**a,b**) Mycoplasmas, and (**c**) a bacterium undergoing lysis. Three things point to an evolutionary link between mycoplasmas and bacteria. First, under certain conditions, cells resembling mycoplasmas can be made from bacteria. Most bacteria undergo lysis when their cell wall is degraded by an enzyme, which allows water to rush into the cell, expand it, and make it burst. But lysis can be prevented if the wall is removed in a concentrated sugar solution. (The sugar lowers the concentration of free water on the outside and prevents it from diffusing rapidly into the cell.) The organisms so stabilized end up looking rather like mycoplasmas (**c**). Second, although under certain conditions these naked bacteria may revert to their former clothed selves, some never do: they may go on growing like mycoplasmas. Third, the DNA composition of mycoplasmalike organisms closely resembles that of certain bacteria. If past environments somehow counteracted wall-degrading agents, stripped-down bacteria could have given rise to mycoplasmas.

complete genetic system for maintaining and reproducing themselves. Hence they unquestionably are alive. Sometimes **viruses** (Chapter Seven) have been said to be the simplest of all living things. But viruses are a paradoxical group that many biologists would not consider to be alive. They do have a set of nucleic acids sheathed in a protective protein coat, and their nucleic acids do contain the blueprint for making more virus. But they are not capable of metabolism; neither are they capable of reproducing themselves on their own. It is the infected host cell, *not* the virus itself, that assembles new virus particles. Because viruses can be perpetuated only inside the cells of another living organism, they probably developed at some point long after cellular life originated. It is likely that they are the noncellular remnants of some ancestral bacterial cells that became parasitic. Invariably, when one organism becomes parasitic, it becomes simplified. If the host will do some task for it, there is no longer any selective pressure demanding that the parasite retain the ability to perform the task for itself. Viruses thus may be the ultimate expression of parasitic simplification. They retain only two features the host will not provide: a mechanism for recognizing and infecting suitable host cells, and the hereditary blueprints for making new viruses.

Even if viruses are not alive, their influence on the living world is staggering. On the one hand, they are the causative agents of many human diseases such as smallpox, influenza, polio, and possibly cancer; they also cause extensive damage to livestock and crops. On the other hand, they play an important role in keeping certain bacterial populations within bounds in some environments.

Bacterial Life-Styles The entire bacterial way of life—growth, reproduction, even going "dormant"—centers on the energy extraction process. When nutrients are plentiful, bacteria reproduce rapidly. They do so through **binary fission,** a simple division process that can be completed every twenty to thirty minutes in some species. With bountiful resources, there seems to be little advantage to the longer process of sexual reproduction. However, under certain conditions some bacterial cells engage in **conjugation:** the transfer of short segments of genetic material between two cells (Figure 16.4). Under other conditions, some bacteria are capable of **genetic transformation:** picking up intact DNA molecules from dead relatives and incorporating parts of the DNA into their own genetic library! And in some cases, **genetic transduction** takes place: viruses released from one bacterial cell carry genes from that cell to the next one they infect. Conjugation, transformation, and transduction are far simpler processes than the sexual reproduction process seen among eukaryotes. Yet they, too, result in genetic recombination—the production of individuals having new combinations of heritable traits. Undoubtedly such primitive methods of exchanging genetic information helped bring about diversification of the bacteria.

When the environment becomes hostile, many bacteria respond by forming an **endospore,** a spore structure within the bacterial cell body (Figure 16.5). The spore coat offers amazing resistance to moisture loss, boiling, radiation, disinfectants, even acids. Once the endospore forms, the rest of the bacterial cell disintegrates. Thus for many bacteria, survival means following this dictum: When resources are abundant and environmental conditions benign, grow explosively—but when resources dwindle or conditions become harsh, wait it out.

The part about waiting it out causes humans all sorts of problems. Exposure to high temperatures for a few minutes is often enough to rid various foods and equipment of most

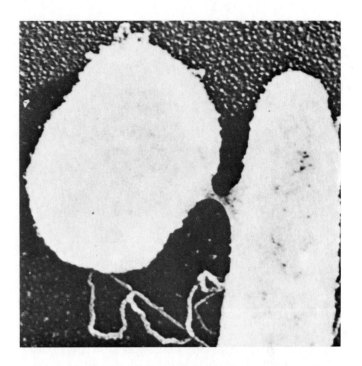

Figure 16.4 Conjugation between female strain (left) and male strain (right) of *Escherichia coli,* showing the conjugation bridge. (T. F. Anderson, E. L. Wollman, and F. Jacob)

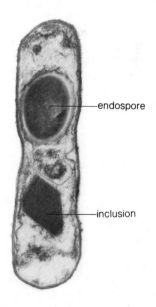

endospore

inclusion

Figure 16.5 Endospore formation in *Bacillus thuringiensis.* The crystalline inclusion below is of a kind that may be "harvested" and used as a natural insecticide for certain insect species. (David Vitale and George B. Chapman, Georgetown University)

actively growing bacteria. But if the environment is slightly alkaline, endospores can live through several hours of boiling! That is why hospitals use autoclaves (steam pressure devices that raise the temperature above that of boiling water) for killing all microorganisms. In home canning, the risks are ever present. *Clostridium botulinum,* for instance, is a bacterium that thrives in the anaerobic environment of canned foods. You know that the metabolic activity of any organism leads to certain by-products. In the case of *C. botulinum,* one by-product is extremely toxic to humans. It gives rise to the often-fatal food poisoning, **botulism.** Unless the canning solution is quite acidic, simply boiling the jars of food is not enough to kill these endospore-forming bacteria. If the foods are neutral or slightly alkaline (such as peas, beans, and corn), the bacteria can survive endless boiling. Either acidic substances must be added or the jars must be processed in a pressure cooker, which is clearly the simplest course. (Every now and then you read about commercially canned foods being recalled because of improper sterilization, but that is a relatively rare event. What is less rare is the accidental denting of cans or jar lids. Because even a small break in a seal may be enough to let undesirable bacteria inside, dented cans of food that are on sale may prove to be no bargain.)

Bacterial Agents of Venereal Disease Some bacteria, such as the photosynthesizers, are not adapted for growth in the animal body. But many heterotrophic bacteria can live *only* within animals. Some produce toxic by-products that cause disease in humans. Cholera, diphtheria, bubonic plague, leprosy, tuberculosis—long ago, these and other diseases were traced to such bacteria. Most have been brought under control, at least in industrialized countries. But one type of bacterial disease has reached epidemic proportions even in countries with the highest medical standards. And it has done so in spite of the fact that it can be diagnosed, treated, and quickly cured if it is reported promptly. This is **venereal disease** (VD), of which syphilis and gonorrhea are the most rampant forms. The bacterial disease agents are transmitted to uninfected persons through sexual intercourse, kissing, or intimate body contact with the sexual organs of infected persons. There must be intimate contact for the disease to spread, for the bacteria don't form endospores; they can't live long when away from moist, warm parts of the human body. They die on exposure to light and air.

Syphilis is caused by a spiral bacterium, *Treponema pallidum.* The first warning of infection occurs between two to six weeks after exposure. It is a single **chancre,** a painless bump under the skin that soon becomes an open ulcer at the site where bacteria entered the body (typically the genitals, anus, or mouth). Because the chancre is not sore or itchy and is often internal (especially in females), it may not even be noticed. Even without treatment it soon disappears. But this only means the bacteria have gone deeper into the body.

Two to six months later, the infected person may experience a round of headaches, enlarged lymph glands, a sore throat and fever, aching joints, and a skin rash. These symptoms come and go over a four-year period. The problem is that too many other common diseases have the same symptoms, and without proper diagnosis the danger may still be ignored. Thus the bacteria have time to become more deeply entrenched. For five years or more, *T. pallidum* may slowly attack every organ in the body. The final stage of syphilis may last for as long as twenty years. By then, the damage is irreparable—nothing can save the victim from a combination of heart damage, blindness, and insanity. In all but the final stage, an infected woman who becomes pregnant can transmit syphilis to the fetus during the second and third trimester. The bacteria move across the placenta and into the unborn child's bloodstream. And the newborn will be either stillborn or syphilitic.

Gonorrhea is caused by the bacterium *Neisseria gonorrhea,* which thrives in the mucous membranes of the genital tract (and the eyes, if carried there). With males, there is a greater chance of detecting the disease in its early stages. Within a week, yellow pus is discharged from the urethra. Urination becomes more frequent and painful, because the infection leads to inflammation of the urinary tract. If left untreated, gonorrhea can cause severe bladder infections and sterility. With females, there may or may not be a burning sensation in the genital region; there may or may not be a slight vaginal discharge—and even if there is, it is often mistaken for a simple, common genitourinary infection. As a result, the disease often goes untreated. But the bacteria are there. They spread into the oviducts, eventually leading to violent cramps, fever, vomiting, and often sterility. An infected pregnant female can transmit gonorrhea to her unborn child as it moves through the birth canal during delivery.

The grave consequences of prolonged infection from these diseases *can* be avoided with prompt diagnosis and treatment. But as a preventive measure, those who engage in sexual activity with more than one partner should recognize how important it is for the male to wear a condom (Chapter Twenty-One) to prevent the spread of infection. They should undergo medical examinations, complete with blood tests for the diseases, at least once a month. As a long-term measure, school and community programs of sex education should be encouraged. Part of the problem is that the initial stages of the diseases are so uneventful that the true dangers are masked. Another part is an unfortunate tendency to dismiss the problem as being the deserved curse on a few isolated sex freaks. More than this, it is a common misconception that a person cured once of venereal disease is immune for life, which is simply not true. An individual may contract venereal disease on exposure to an infected person no matter how many times he or she has been infected and cured. The result is that every twelve seconds, somebody—young and old, male and female, of any social class—becomes infected. Gonorrhea, in short, is out of control. With 3 million people known to suffer the disease (public health workers estimate the number may be seven times higher), it has become the leading communicable disease in the United States; syphilis ranks third. Indifference to these appalling statistics can only encourage widespread ignorance and casual attitudes that perpetuate the vicious cycle of venereal disease.

Fifteen Thousand Species of Bacteria Can't Be All Bad It is tempting, after reading about some of the horrors of bacterial infections, to render all bacteria guilty by association. But the vast majority are not guilty of terrible deeds, and in fact they are an essential underpinning not only of human life but of the biosphere in general. From earlier discussions, you know they are a vital link in the recycling of carbon, nitrogen, and other materials in natural communities (Figures 3.15 and 5.4). But bacteria do more than decompose organic debris in the forest and the sea; they do the same in the digestive tract of most animals, ourselves included. The digestive tract is home to a wide variety and tremendous numbers of bacterial cells, of which *Escherichia coli* is the most prevalent. They secrete enzymes for breaking down food materials, and they build up vitamins, which are absorbed and used by their host. Just how important they are becomes apparent during intensive antibiotic therapy for serious infections. The broad-spectrum antibiotics often wipe out the beneficial bacteria as well as the bad, which leads to digestive and intestinal disorders!

And can you imagine what the world would be like without bacteria to help dispose of human wastes, as well as the pathogenic organisms lurking therein? Modern wastewater treatment centers rely heavily on bacteria for restoring water to some semblance of its former self before it is dumped into the waterways (Chapter Twenty-Three). Waste water from domestic and industrial sources includes solid and liquid substances from sewers, sugar mills, textile and paper mills, food processors, dairies, slaughterhouses, oil refineries, chemical manufacturers, and mining operations. It takes aerobic and anaerobic bacteria to break down the organic residues of this stuff. (That is also why we now have "biodegradable" laundry detergents. At one time, soaps were replaced with synthetic detergents, which brighten clothing. As it turned out, the detergents contained chemical groupings that microorganisms couldn't break down. Streams, rivers, lakes, even parts of the sea began to foam up. Only by redesigning detergents have we started to remove suds from the waterways.)

Finally, consider the way crops are sprayed each year with insecticides and herbicides. For instance, some chlorinated hydrocarbons sprayed on crops may linger in the fields for more than a decade. Thus they accumulate to levels high enough to harm not only insects and weeds but humans (Chapter Twenty-Two). Certain factors such as

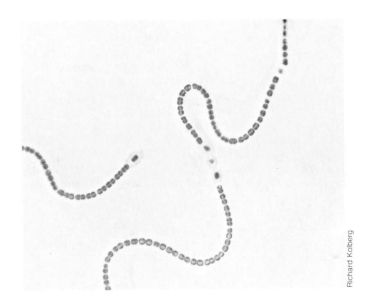

Figure 16.6 Blue-green algae, showing a filamentary growth pattern and special nitrogen-fixing cells called heterocysts.

Richard Kolberg

leaching of the soil help keep these compounds to tolerable limits—but in some cases, so do bacteria out for a meal of carbon. In their energy extraction processes, they can convert some toxic compounds to harmless ones. As we are pressured into using new kinds of insecticides when old ones become ineffective, we must determine in advance whether they are biodegradable by soil bacteria.

Success in Simplicity: The Blue-Green Algae

Bacteria are one major kind of prokaryote; **blue-green algae** are another. Aside from their photosynthetic activities, the blue-green algae (like some of their bacterial relatives) are nitrogen fixers of no small importance. If deprived of their activities, the biosphere as we know it would probably collapse for lack of enough usable nitrogen compounds for protein synthesis.

Although blue-green algae may exist as single cells under some conditions, most of the 1,500 known species look like filaments because cells remain attached to one another after division (Figure 16.6). In many blue-green algal filaments, we see the first inklings of a division of labor: a process not seen among bacteria, but a feature of more complex organisms. Frequently the filaments consist of two cell types: one specialized for photosynthesis, the other for nitrogen fixation.

Like photosynthetic bacteria, blue-green algae don't have chloroplasts (light-trapping organelles found in some eukaryotic cells). Instead their plasma membrane is folded inward and studded with light-trapping pigments, some of which give each species its typical blue, red-green, or almost black color. Unlike those bacteria, blue-green algae have the *same* kind of chlorophyll pigment found in all true algae and plants. Yet they are so much like their bacterial predecessors in cellular organization, it would probably be less confusing to think of them as "blue-green bacteria." Whatever the name, they appear to be living representatives of a long-since vanished evolutionary bridge between bacteria and the plant kingdom.

Blue-green algae are found throughout fresh-water, salt-water, and land environments. Many species thrive in places where other organisms would be hard-pressed even to stay alive—in near-boiling hot springs, in extremely salty lakes, on desert rocks, in permanent snowfields. Two factors underlie their adaptability to such a broad range of settings. First, they have very simple requirements for energy and raw materials. Almost any environment has sunlight, carbon dioxide, some water, and simple mineral salts, and that is all the blue-green algae need in order to grow. Second, they reproduce by the simplest means possible. When conditions are favorable, they divide rapidly by binary fission. When conditions are unfavorable, some species form thick, sporelike walls or a gelatinlike coat to keep from drying out. Like bacteria, the blue-green algae grow in good times and wait out the bad. Thus most of the energy-rich nutrients they take in can be used at once for rapid growth and reproduction—which means they can compete quite well with other organisms. It is a time-tested adaptation. Both bacteria and blue-green algae were among the earliest forms of life on earth, and it's probable they will be among the last.

The Boundary-Straddling Protistans

For the earliest prokaryotic monerans, survival was bestowed on those forms which were adapted for extracting energy-rich molecules from certain environments, for reproducing rapidly, and for being able to go dormant when conditions deteriorated. But by 1 billion years ago, some had evolved into eukaryotic forms adapted for feeding on other organisms. What were those first eukaryotic organisms like? Their membrane-enclosed nucleus had to mean the potential for more complex developments (Chapter Four). One such development was the frequent, more elaborate genetic recombination possible with sexual reproduction, which must have accelerated the process of diversification. The fossil record suggests ever more diverse forms evolved rapidly in several different directions, in what must have been intense competition for energy and

a

resources. Thus today we have the animals, which move about actively in search of food energy. We have the plants, which generally stay put, trapping nutrients and sunlight in their immediate surroundings. We have the fungi—nonmotile, nonphotosynthetic, and specializing in absorption of organic nutrients. But more revealing for our question of what the first eukaryotes were like, we also have a diverse collection of single-celled eukaryotes. Some are actively prowling predators without photosynthetic pigments, and in this they resemble animals. In fact, they are known as "protozoans" (meaning "first animals"). Others are somewhat immobile and have photosynthetic pigments; they resemble multicellular plants. Still others resemble fungi in their life-style. But also in this hodgepodge are many motile, photosynthetic forms such as *Euglena*, which resemble plants, animals, *and* fungi! How do we go about classifying "little green funguslike animals"?

But taxonomic schemes, recall, are a little like fences dividing up the expanse of nature. With observation and with educated guesses, we can make our fences more or less follow the terrain. But they are still artificial boundaries, subdividing a continuum that extends from the distant past. We are left with trying to discern what these eukaryotic microorganisms might represent in the total evolutionary scheme. And that is what the five-kingdom scheme, which we use in this book, has attempted to do. By viewing all the single-celled eukaryotes as a separate kingdom—the Protista—they are recognized not as insignificant cousins left behind in the great flow of life history but as successful adaptive lines in their own right.

b

R. F. Head/National Audubon Society/PR

Figure 16.7 Dinoflagellates. (**a**) The scanning electron micrograph shows the plates making up the cell wall. Assorted accessory pigments give these organisms a yellow-green, to brown, to red color. (**b**) Dinoflagellates such as *Gonyaulax* are a source of episodic red tides. They produce a toxin that affects the central nervous system of fish. If their population levels are high, hundreds of millions of fish may be killed and washed up along coasts, as shown here. Sometimes dinoflagellate blooms occur in lakes and reservoirs, which makes the water foul-tasting and often toxic. (Many dinoflagellates exhibit *bioluminescence,* the production of light caused by the metabolic reaction of ATP with specific compounds. This light is usually a pale blue. It often is seen at night.

Protistan Experiments in Photosynthesis

Early in their life history, certain eukaryotes gained the capacity for photosynthesis. Perhaps, as some biologists have speculated, it was when a predatory animal cell happened to engulf a photosynthetic, bacteriumlike alga. Instead of succumbing to digestion, the organism so engulfed had the means to stay alive—indeed, to thrive in the presence of the carbon dioxide being given off by its engulfer. Why not take up permanent residence, dividing and reproducing even as the diner-turned-host divided and reproduced itself? The animal cell would benefit from the energy-trapping abilities of the smaller cell; the smaller cell would gain protection as well as raw materials for photosynthesis. (Such mutually rewarding dependence is one form of **symbiosis.** As you will read in Chapter Twenty-Two, it occurs today among many organisms.) Over hundreds of millions of years, evolutionary forces worked on this relationship, the outcome being the retention of adaptive structures and the weeding-out of nonadaptive ones. And gradually the photosynthetic organism lost its ability to live outside its host, even as its host came to depend on the energy-trapping guest.

Although this scenario might seem difficult to prove, detailed comparisons of chloroplast DNA and metabolism with those of bacteria and blue-green algae are compatible with such a theory of origin. Chloroplasts have their own complement of DNA and an apparent ability to reproduce parts of themselves with the instructions it contains—in partial independence of the division of nuclear material of the cell in which they reside!

If this was the route by which some eukaryotes acquired the capacity for photosynthesis, it must have been traveled more than once. Probably several different lines of photosynthetic protistans arose long ago, and have persisted into the present. Today, they resemble one another only in the fact that they are photosynthetic. Their chloroplasts are different and resemble different types of blue-green algae. They have different body forms and life-styles. Their accessory pigments are different, so they differ in color. All these lines are called "algae," but they are often distinguished from one another on the basis of color: "golden algae," "red algae," "green algae," and so on. There is a problem in trying to confine these lines to the protistan kingdom, for several have multicellular as well as single-celled members! The green algae are a case in point; they almost certainly gave rise to the multicellular green plants. Thus the green algae will be described in the next chapter as part of the story of plant evolution, even though they include some members that belong with the protistans as much as anywhere else. In this chapter we will look only at some of the algal lines that are largely or completely protistan in their make-up.

Dinoflagellates The **dinoflagellates** (Figure 16.7) are what you might call primitive photosynthetic eukaryotes. True, they have a membrane-enclosed nucleus. But the composition of their chromosomes is much like that of prokaryotes.

Their mode of DNA replication seems vaguely prokaryotic, for the two DNA molecules formed prior to cell division attach to different parts of the nuclear membrane (just as bacterial DNA molecules attach to different parts of the plasma membrane), then they are separated as the membrane pinches in two.

The dinoflagellates are photosynthetic *and* motile. Most are found in the seas as members of **phytoplankton:** communities of microscopic photosynthesizers that are suspended in the water, where they form the basis for much of the region's productivity. In both fresh water and sea water, populations of dinoflagellates can grow explosively under certain conditions. During these so-called "blooms," there may be several million microorganisms in each liter of water. The sheer numbers usually play havoc with the environment. At night, when photosynthesis is impossible, the cellular respiration of such large populations depletes the dissolved oxygen in the water. The depletion causes the death of many organisms such as fish, which are sensitive to oxygen levels. Some dinoflagellates of the oceans produce toxic by-products. Nearly every year, destructive **red tides** occur somewhere in the world. They are the result of blooms of such dinoflagellates as *Gymnodinium* and *Gonyaulax*. As Figure 16.7 indicates, many dinoflagellates exhibit bioluminescence. Other marine organisms show bioluminescence, often because dinoflagellates are living inside them or in close association with them.

Golden Algae and Diatoms Somewhere around 570 million years ago, oxygen had accumulated to high enough levels in the atmosphere and enough salts had accumulated in the seas to permit the formation of hard body parts (Chapter Fifteen). There apparently was selection for those forms having the genetic blueprints and the capacity to expend energy for building shells, spines, and armored plates that provided some defense against predation. Two lines of descent from those adaptive forms are the armor-plated **golden algae** and **diatoms.** The golden algae have armor made of cellulose, pectin, and sometimes silicon. Their close relatives, the diatoms, have what is literally a glass shell, made mostly of silicon dioxide. The diatom shell is made of two pieces, much like a glass petri dish, and is often elaborately sculptured (Figure 16.1).

Today, the golden algae exist mostly in fresh water. At one time they numbered in the billions; the Cretaceous fossils of one group make up tremendous chalk deposits in various parts of the world. In terms of sheer numbers, however, the diatoms are the truly long-standing success. Immense diatomaceous deposits date from 70 to 10 million years ago—and these organisms are *still* abundant in almost all aquatic settings. Areas rich in diatoms are rightfully called "the pastures of the sea," for the diatoms, despite their glassy shell, remain the prey on which many marine communities are built.

The Puzzle of Little Green Funguslike Animals

Several hundred species of **photosynthetic flagellates** exist, and they clearly are related to both plants and animals. *Euglena* belongs to this group. Each *Euglena* contains about ten or so chloroplasts. But its cell body is not surrounded by a cell wall, as plant cells are. Instead, it is enclosed in a semi-rigid layer intimately connected with the plasma membrane. *Euglena* is not completely autotrophic, even with its array of chloroplasts. It needs certain nutrients such as vitamins—which it absorbs from its environment in a funguslike manner. This photosynthetic flagellate and its relatives are most successful in places where organic molecules are concentrated—in polluted barnyard pools, for instance, which are heavily contaminated with urine and fecal material. They are also abundant in waste-water treatment centers (Chapter Twenty-Three).

Except for their chloroplasts, *Euglena* and its relatives are virtually identical with a number of heterotrophic flagellates (Figure 16.8). In fact, if some species are raised in the dark, they rapidly change into forms having no chloroplasts at all. If these bleached-out forms are later grown in sunlight, they never regain chloroplasts. From then on, they survive only if they can absorb a complete diet of food molecules. Other heterotrophic flagellates are even more puzzling. Consider a variety known as *Ochromonas* (Figure 16.8). They ingest whole particles through their gullet, much like an animal would do. But they also have chloroplasts and are capable of some photosynthesis. They can't grow in the dark, no matter how rich their surroundings are in nutrients. Even so, they are incapable of producing enough food by photosynthesis to support themselves. They must have both light for photosynthesis *and* food molecules to survive. They belong to a rare group of organisms known as **photoheterotrophs.** The resemblances between photosynthetic and heterotrophic flagellates, together with the existence of photoheterotrophs, could signify divergence of many modern flagellates from a common ancestor. It is a strong argument for establishing a protistan category instead of attempting to differentiate in every case between plantlike and animallike forms at this level.

Protistan Experiments in Predation

There are four main types of animallike protistans: the heterotrophic flagellates discussed above, the amoebas and their relatives, the sporozoans, and the ciliates. All are unwalled single cells that live by pinocytosis and phagocytosis (Chapter Five). Some species are nearly as small as bacteria; others can be seen with the naked eye.

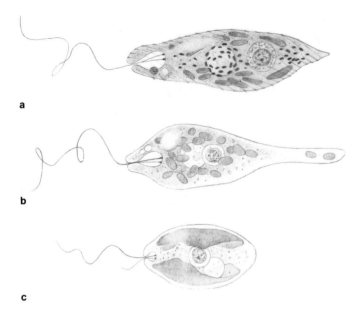

a

b

c

Figure 16.8 Organisms with features reminiscent of plants, animals, and fungi. (**a**) *Euglena*, a photosynthetic flagellate. (**b**) *Astasia*, a heterotrophic flagellate. (**c**) *Ochromonas*, a photoheterotrophic flagellate.

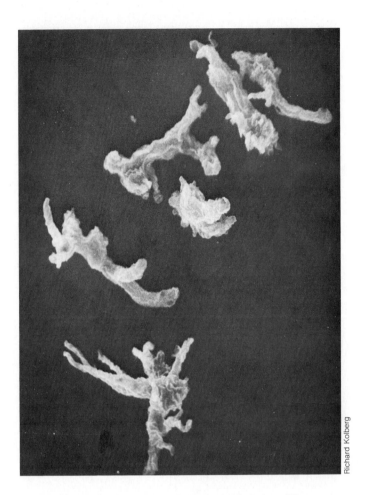

Richard Kolberg

Figure 16.9 Not a dance of the amoebas but a clustering of these single cells in all their pseudopodial potential.

Amoebas and Their Relatives As primitive as the **amoebas** seem to be as they ooze formlessly over some rock in a polluted pool, they are quite complex organisms. They move by sending out cellular extensions of themselves, called **pseudopodia** ("false feet"), and surround food and engulf it. Many relatives of the common *Amoeba* not only have pseudopodia, they have flagella for swimming about. Amoebas are thought to have evolved from heterotrophic flagellates, which capitalized on the ability to distort the cell surface, extend it, and withdraw it. They developed this capacity to such an extent that they have come to rely on it as a way of life.

The movements of amoebas speak once again of the basic unity of life. Amoebas use the very same proteins you use for movement: the muscle proteins actin and myosin. If actin isolated from an amoeba is mixed with myosin isolated from a mammalian muscle, a jellylike clot forms; if ATP is added to the clot, it contracts forcefully! In amoebas, the actin and myosin proteins are attached in what might seem to be a random fashion to the cell membrane. When contraction occurs in one region, a pseudopodium is slowly extended. Contraction of two pseudopodia causes them to sweep around food and engulf it (Figure 5.3a).

Some amoebas provide protection for their fragile body by encasing it in a layer of sand grains. Others secrete a hardened shell that surrounds and protects them. From one or more openings in the shell, pseudopodia poke out. This trend reaches its peak of development in the diverse group known as the **foraminiferans** (bearers of windows). They secrete a hardened case of limestone that is peppered with tiny holes. The organism lives within its beautiful shell, but it can project hundreds or thousands of threadlike pseudopodia through the holes. Because the pseudopodia are sticky, they can trap bits of phytoplankton floating in the seas even while the foraminiferan hides in its house.

When they die, the foraminiferans fall to the ocean floor. In the past, the products of their accumulated bodies were the primary source of the world's oil (petroleum) deposits. Where there are foraminiferan shells, oil is likely to be nearby.

Figure 16.10 Body shells of radiolarians, relatives of the amoeba and foraminiferans. 170×.

R. Kessel and C. Shih/Springer-Verlag

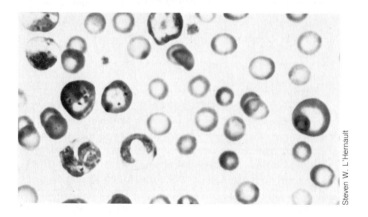

Figure 16.11 Malaria parasites in red blood cells. At this stage, each parasitic form divides repeatedly. At regular intervals, they burst out and reinfect more cells. Recurring, serious fevers accompany this cycle. When a mosquito draws blood from an infected individual, gametes of the parasite undergo sexual fusion to form zygotes, which eventually develop into spindle-shaped forms (sporozoites) within the mosquito. When the mosquito bites again, sporozoites enter the victim and begin the cycle anew.

Steven W. L'Hernault

Sporozoans Calling an organism a **sporozoan** is not so much a reference to its pedigree as to its life-style. For the sporozoans are a diverse group of tiny, single-celled organisms that are probably more closely related to several other kinds of protistans than they are to one another. They share one feature: they are all parasitic, living within the bodies of animals. As in the case of viruses, the parasitic life-style has led to a loss of nearly all features except those required to infect a host and to reproduce. Their host performs all the tasks connected with obtaining food. One group of sporozoans in particular has been prominent in human history: the members of the genus *Plasmodium,* which are the source of the disease malaria (Figure 16.11). Both a human host and a certain species of mosquito are needed if a malaria parasite is to complete its life cycle. Thus a key to eradicating malaria has been to drain and fill in the breeding grounds—swamps and stagnant ponds—of the malaria-carrying mosquitos.

Ciliates: Peak of Single-Cell Complexity Nowhere in the living world are the fantastic potentials of a single cell expressed more fully than they are in the group of protistans known as **ciliates.** Covered with thousands of cilia synchronized for swimming, armed in many cases with hundreds of deadly, poison-charged harpoonlike weapons that can be fired at prey and predator, and possessed of a voracious mouth, the ciliates prowl through or wait in ambush in woodland ponds, the intestinal contents of various organisms, and a variety of other habitats.

Perhaps the most widely occurring ciliate is *Paramecium* (Figure 16.12). Like most ciliates, *Paramecium* has a gullet, a cavity that opens to the external watery world. Rows of specialized cilia beat food particles into the gullet. Once inside, the particles become enclosed in food vacuoles, where digestion takes place. Unusable leftovers are carted off to a region known as the anal pore, which functions to eliminate wastes to the outside. Living as it does in fresh-water environments, *Paramecium* depends on contractile vacuoles (Chapter Five) for eliminating the excess water that is constantly flowing into the cell.

Predatory cell that it is, *Paramecium* is built for rapid movement. Between 10,000 and 14,000 individual cilia project like tiny oars from the cell surface. Through the mechanisms for cellular movement described earlier in Chapter Four, each orderly row of cilia beats in a coordinated way with adjacent rows (Figure 16.12). So efficient is this coordination that some species of *Paramecium* can move 5,000 micrometers through their surroundings every second! That rate of movement allows them to far outstrip their flagellated prey.

Ah, but that which works for one predatory ciliate works for others. Often *Paramecium* itself is outmaneuvered by another voracious ciliate, *Didinium,* as you saw in Figure 9.1. When a didinium engulfs a paramecium (which is

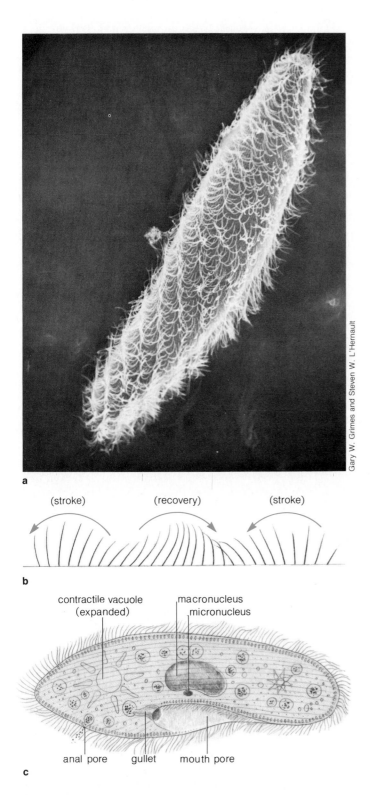

a

(stroke)   (recovery)   (stroke)

b

contractile vacuole
(expanded)

macronucleus
micronucleus

anal pore   gullet   mouth pore

c

Gary W. Grimes and Steven W. L'Hernault

Figure 16.12 (**a**) *Paramecium* on the prowl, using coordinating ciliary beating (**b**) to drive its body through its watery environment. (**c**) The sketch hints at the astonishing complexity of this single cell.

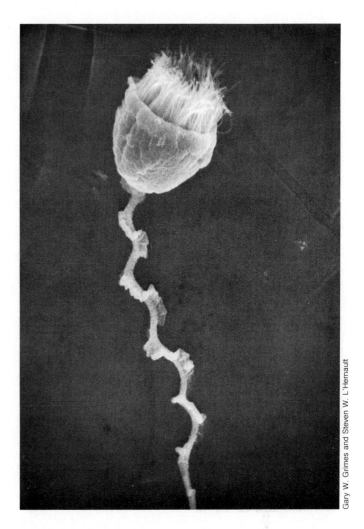

Gary W. Grimes and Steven W. L'Hernault

Figure 16.13 *Vorticella*, a sedentary ciliate.

almost as large as the predator), the didinium seems to be all mouth. Less visibly dramatic but equally stunning in the contest between predator and prey are the didinium's **trichocysts**—long contractile filaments anchored within the cell that are used to hold onto and paralyze its dinner.

Both *Paramecium* and *Didinium* are free-swimming, but other ciliates crawl about or simply stay put. *Vorticella,* one of the largest of all ciliates, normally rises stalklike from its holdfast on the floors of ponds, lakes, and streams. With utmost grace, it turns its highly efficient, cilia-rimmed mouth into the currents flowing past (Figure 16.13). *Stylonychia* (Figure 16.14) crawls along in search of prey. Although they are astonishingly diverse in form, all such single-celled ciliates are of the same sort: they are the lions of the microscopic world.

Figure 16.14 *Stylonychia*, a crawling ciliate.

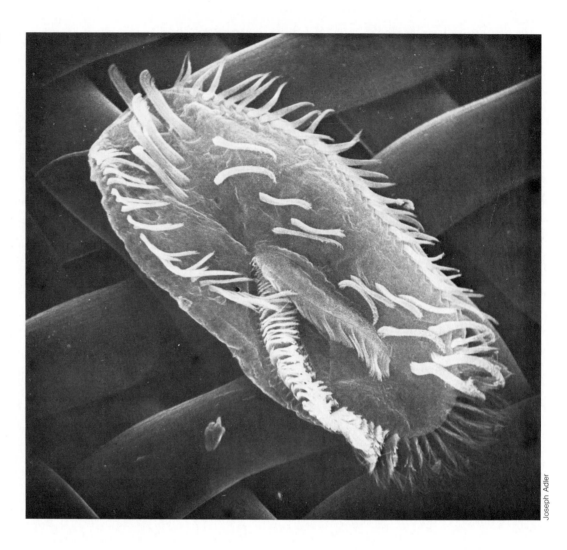

On the Road to Multicellularity

We can guess that somewhere among the ancestral lines of single-celled protistans are the long-since vanished gateways to multicellularity. The transitional forms probably arose many times in many evolutionary lines that were radiating into the world's shallow Precambrian seas. Perhaps the earliest were no more than simple colonies, designed by chance when single-celled organisms divided and the newly formed daughter cells failed to separate. Such inadvertent adherence meant increased size. To the extent that it deterred smaller predators, or improved motility, or offered resistance to strong currents, the factors causing adherence were selected for and perpetuated. Indeed, colonial forms have persisted even to the present. They are structurally diverse, and they clearly are related by way of some such ancestors to protistans of many different types.

In a colony, each cell benefits from the loose association but each acts independently; it is incapable of modifying its behavior according to what is happening to i:s neighbors. All the cells are sensitive to the same things (light, food, temperature) and all respond to change in the same way. They feed and reproduce in the same way. It is only with the division of labor that there is interdependency, where one type of cell can't exist without the other. As Figure 16.15 indicates, the transition from colonial to multicellular forms could have occurred gradually, with an ever-increasing division of labor. The potential of multicellularity has been realized in three major pathways: the plants, fungi, and animals. These remarkable developments are topics of the next two chapters.

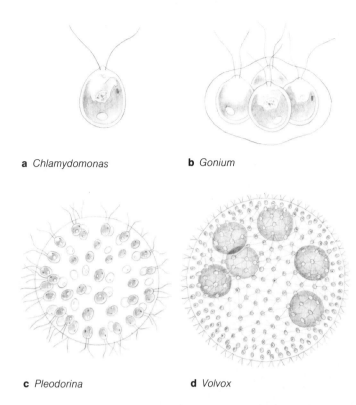

a *Chlamydomonas*　　　　**b** *Gonium*

c *Pleodorina*　　　　**d** *Volvox*

Figure 16.15 The Volvocales, a group of related organisms that suggest a possible route from unicellularity to a multicellular organism having a division of labor among cells. (**a**) *Chlamydomonas* is a free-living single cell thought to resemble the ancestor of the entire group. In the simplest colonial forms (**b**), all cells—held together in a jellylike sphere—are identical in appearance and function. In somewhat more complex species (**c**), cells at one end of the sphere are larger and retain the capacity for movement (with flagella) and for reproduction. The most complex member of the group, *Volvox* (**d**), has two distinct cell types, each incapable of existence without the other. The smaller cells around the sphere's surface are motile but incapable of reproduction. The larger, interior cells are immotile; their only function is reproduction. Each is capable of dividing to give rise to a new individual having both cell types.

Perspective

In the realm of monerans and protistans, we have evidence for one of the most significant evolutionary events. At the molecular level, these two kinds of single-celled organisms are so similar it seems likely that one evolved from the other. With the appearance of protistans, life was destined for stunning departures from the simple prokaryotic body plan. And those departures would be based largely on two things: a more complex genetic apparatus and self-contained energy centers.

Except during replication, prokaryotic DNA is a single, circular strand in the cytoplasm. In contrast, the DNA of eukaryotic protistans is separated physically from the cytoplasm by its own nuclear membrane. Moneran reproduction relies mostly on simple binary fission. Protistan reproduction relies on mitosis and, in sexually reproducing forms, on meiosis to assure the orderly parceling out of far greater stores of DNA into daughter cells. With nuclear isolation, protistans were on their way to a complex cellular division of labor. Soon after, perhaps through engulfment of bacteria and algae that managed to become symbionts instead of dinner, the eukaryotic protistans became equipped with energy-generating mitochondria and with energy-trapping chloroplasts. (Mitochondria are suspiciously like bacteria, just as chloroplasts are suspiciously like blue-green algae.) However nuclear isolation and energy control centers came about, they apparently helped trigger the extensive and rapid evolution into forms called plants, fungi, and animals.

It would be foolish, however, to portray the monerans and protistans as being somehow left behind during this magnificent surge toward complexity. By remaining structurally simple, they have survived astonishing environmental extremes. By retaining their simple strategy of reproducing quickly and in large numbers, they are able to adapt genetically in a short period to new environmental windfalls and pressures. Witness, for instance, how little time it takes for new resistant strains of bacteria to evolve following the introduction of some new antibiotic. In contrast, the body plans of plants, fungi, and animals, with all their potential for meeting many new and different challenges, are too complex for meeting others. Even as new doors were being opened for them, others were being closed behind—and that is why you won't find plants, animals, and fungi flourishing in hot springs. They cannot displace their simpler relatives in such places. In a given range of environments, monerans and protistans became masters of survival long before the appearance of the spectacularly diverse forms of multicellular life, and thus they continue to exist with them even into the present.

Recommended Readings

Jurand, A. and G. Selman. 1969. *The Anatomy of Paramecium aurelia.* New York: St. Martin. A tribute to the astonishing complexity achieved in a single cell.

Margulis, L. 1970. *Origin of Eukaryotic Cells.* New Haven, Connecticut: Yale University Press. Outstanding treatment.

Stanier, R. et al. 1970. *The Microbial World.* Third edition. Englewood Cliffs, New Jersey: Prentice-Hall.

Chapter Seventeen

Figure 17.1 Sunlight filtering through a grove of California coast redwoods.

Plants and Fungi

In turning to the patterns of plant diversification over time, we find compelling evidence for the evolution of structure and function as adaptations to new environments. Between 3.5 billion and 400 million years ago, the evolution of life appears to have been confined to the shallow seas. And for aquatic plants, at least, selective pressures must have been much the same everywhere, for the fossil record shows comparatively little diversity in plant form. But by 400 million years ago, when the continental plates collided (Chapter Fifteen), vast regions of the earth became permanently elevated above sea level. As shallow seas drained away and this new and vacant environment opened up, there must have been selection for those aquatic plants having adaptations that would make possible the transition to life on land.

In tracing the story of how the land was colonized, we will be reminded again and again of life's utter dependency on water. It is true that some aquatic plants can grow in the low wetlands bordering almost any body of water. But in all other terrestrial zones, some special means for acquiring and conserving water is essential. Even if water is available, there must be a system for conducting the life-sustaining fluid and its dissolved nutrients through the plant body. Ordinarily, aquatic plants don't need water transport mechanisms. Most grow as filaments or sheets that are only one or two cells thick. They rely on simple diffusion and active transport of materials and wastes across external cell membranes, which are in direct contact with the surrounding water. Adapting to life on land, however, has meant multicellular expansion and differentiation into roots, stems, branches, and leaves. It has been accompanied by the development of stomates and waxy leaf coverings, which control water loss and gas exchange (Chapter Eleven). Every land plant around us, no matter how unique it is in outward appearance, exists because of its ability to take in nutrient-rich water through its roots, transport water through its vascular tissues, and control water loss through its stems and leaves.

Dependency on water has also influenced the development of plant support tissues. The surrounding water supports much of the body weight of aquatic plants. Even in large, complex forms, support tissues are devoted more to keeping the plant body intact during wave action than to holding it upright. Where it is advantageous to stay near the water's surface, some plants simply rely on oil vacuoles inside their cells or on gas-filled floats for buoyancy. When the first few plants had made the transition to land, perhaps support tissues were not needed; perhaps they simply sprawled over the surface of the ground, soaking up sunlight. But as plant populations increased, the ones that held their photosynthetic structures a little higher than their prostrate neighbors captured more light and began to shade out competition. Regardless of when it happened, competition for sunlight began, and a course of events was set in motion that would eventually lead to vascular tissues for water and nutrient transport as well as strong, reinforced stems and trunks.

But even with adaptations in body form, there could not have been explosive radiation into the vacant land environment without the added potential for variation in the mode of plant reproduction. Like animals, virtually all plants undergo sexual reproduction. This means that at some point in the life cycle, meiosis occurs and is followed by formation of haploid cells which give rise to haploid gametes. The gametes then fuse to form a diploid zygote. This basic sexual mechanism seems to have originated with the ancient protistan forms from which both plants and animals were derived. But even assuming this was the case, plants and animals have long since gone their separate ways.

The haploid stage for most animals is brief, consisting of the sperm and eggs produced following meiosis (Figure 17.2a). Once male and female nuclei of these gametes fuse during sexual reproduction, a new diploid generation begins. All subsequent stages of the life cycle, up to the time when gametes are produced again, are composed of diploid cells.

In many primitive aquatic plants, however, the opposite is true: the plant body is composed exclusively of haploid cells (Figure 17.2b). In such plants, sexual reproduction is often triggered by the onset of adverse environmental conditions—dwindling water supplies, a drop in temperature heralding the approach of winter. When that happens, some or all of the cells making up the haploid parent body simply divide mitotically to produce gametes; and every two gametes that fuse form a diploid zygote. But the zygotes of most primitive water-dwelling plants never undergo mitosis. Instead they enter a *resting stage* and wait out adverse conditions. As soon as the environment becomes hospitable again, the resting cells undergo meiosis.

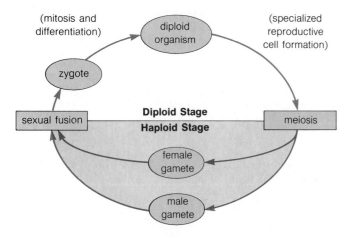

(mitosis and differentiation)

(specialized reproductive cell formation)

a *Generalized life cycle for animals*

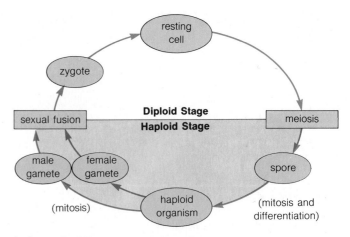

(mitosis)

(mitosis and differentiation)

b *Generalized life cycle for primitive plants*

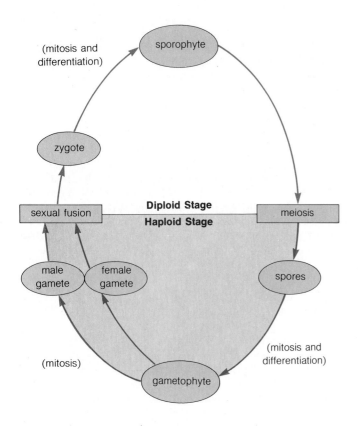

(mitosis and differentiation)

(mitosis)

(mitosis and differentiation)

c *Generalized life cycle for more advanced plants*

Figure 17.2 Generalized comparison between the sexual reproductive patterns seen in the life cycles of animals, primitive plants, and complex land plants.

This leads to formation of a number of **spores**—cells that give rise to a new haploid generation. The significant thing about all this is the resistance of the diploid stage to conditions unfavorable for vegetative growth. *For it was through the development of a complex diploid stage that some plants came to have the fundamental means for adapting to conditions on dry land.*

In moist land environments, we can still find plants that resemble ancient fossils of early pioneers. For them, mitosis and differentiation lead to a highly developed plant body during the haploid stage, and another during the diploid stage. The haploid body produces the gametes that give rise to the diploid stage; hence it's called the **gametophyte** (the "gamete-producing plant"). The diploid plant produces spores that give rise to the haploid stage; it's called the **sporophyte** ("spore-producing plant"). Gametophytes and sporophytes are usually quite different in appearance, for they are adapted to different conditions (the sporophyte is less dependent on a constant supply of free-standing water). Whenever a life cycle contains alternating haploid and diploid stages of this sort, it is known as an **alternation of generations.** We can outline this tendency toward the development of a prolonged sporophyte generation by examining representatives from a few existing plant groups.

Plants That Never Left the Water

Algae is a term that originally came into use to define simple aquatic "plants." It no longer has any formal significance in most classification schemes, however, because the phyla (divisions) lumped together under the term are now generally recognized as belonging to more than one kingdom:

a   b   c

Figure 17.3 Representatives of three algal groups. (**a**) A small red alga that grows off the coast of England. (**b**) *Ulva* (sea lettuce), a green alga. (**c**) Underwater view of kelp, one of the brown algae.

Cyanophyta (blue-green algae)   }   Monera

Chrysophyta (golden algae, diatoms)
Euglenophyta (photosynthetic flagellates)   }   Protista
Pyrrophyta (dinoflagellates)

Rhodophyta (red algae)
Phaeophyta (brown algae)   }   Plantae
Chlorophyta (green algae)

The red, brown, and green algae remain in the plant kingdom because some or all of the members of these phyla have multicellular body plans that are more complex than simple colonies or filaments. These three groups represent the peak of plant diversity in aquatic environments.

Red Algae By at least 600 million years ago, **red algae** appeared in the warm, shallow waters of the earth. Today, with the exception of a few fresh-water forms, the 4,000 or so species of this phylum are an important component of marine environments ranging from the shallow intertidal zone to the basement of tropical reefs. They can be found growing at depths of about 175 meters below the surface, depending on light penetration. Their body form ranges from single cells, to simple filaments, to sheets that fan out a meter or so (Figure 17.3). In certain aspects, red algae still resemble the prokaryotic blue-green algae from which they are thought to have been derived. They, too, contain chlorophyll *a*, various carotenoids, phycocyanin, and phycoerythrin. There are also structural similarities in the photosynthetic membranes of these two phyla. But the life cycles of these plants have no obvious parallels with the patterns of other algal groups and in some cases are poorly understood. Whatever their derivation, they complete the sexual part of their life cycle without relying on *motile* gametes: the sperm lack flagella.

Red algae range in color from true red to purple, depending on their accessory pigments. This pigment range is one of the sources of their adaptive success in deeper water. Chlorophyll, recall, functions most efficiently in light of red wavelengths. But red wavelengths do not penetrate far below the water's surface; there, available wavelengths are blue-green. The accessory pigments in red algae absorb these wavelengths and pass on some of the energy to nearby chlorophyll molecules engaged in photosynthesis. In a very real sense, much of the community structure of tropical reefs depends on the pigments in these photosynthetic producer organisms.

Brown Algae The 1,500 species of **brown algae** are found along rocky coasts, where they anchor themselves to the substrate by special structures called holdfasts. Many forms have hollow, air-filled structures (floats). Floats often occur at regular intervals along the plant body, enabling the plant to remain suspended in the water. Some brown algae grow well into the uppermost regions of the intertidal zone. During the hours when the tide is out, their thick, leathery surface helps keep the plant from drying out. Also included in this group are the giant kelp, which may grow 100 meters upward from their holdfasts on the sea floor. As strangely beautiful as kelp "forests" may be, more than one scuba diver has been in real trouble by becoming tangled in the dense, waving plants.

Kelp has specialized conducting cells in its stem that move the products of photosynthesis from plant parts near the sunlit surface waters to the parts growing below, in deeper coastal waters. Several species also thrive in the open sea. *Sargassum* floats as an immense, tangled mass through the Sargasso Sea, which lies between the Azores and the Bahamas. This brown alga helps support a community of unique marine animals.

In the brown algae, we see signs of increasing reliance on the diploid stage of plant life cycles. In simpler filamentous species, the gametophyte and the sporophyte body may be essentially the same in outward appearance. But in more developed species, the large, complex sporophyte dominates the life cycle; the gametophyte is extremely reduced in size. What selective pressures have been at work on the life cycles of brown algae? That we don't know. It is only when we turn to the green algae that we gain insight into the significance of the dominance of the sporophyte generation over the gametophyte generation.

Green Algae Although **green algae** are relatively inconspicuous in the sea, they represent an evolutionary line that is thought to have given rise to plants that made the transition to land. Today most green algae are fresh-water organisms, and some live on land. The photosynthetic system in these 7,000 species is identical with that of higher plants, consisting as it does of chlorophyll *a* and *b*, carotenoids, and xanthophylls. In all aspects of energy metabolism, they are similar to the more complex land plants. Body plans range from single motile cells to the broad multicellular "leaves" of *Ulva* (sea lettuce). But the most common body plan is a series of straight or branched filaments, each only a single cell thick.

For most green algae, the gametophyte (haploid) stage is dominant; it is the stage in which all active growth occurs. The diploid stage usually consists of a single resting cell that waits out adverse conditions. And yet, a few green algae live in regions of the sea where conditions are stable, so a resting cell has fewer advantages. These species undergo mitosis during the diploid stage, and the outcome is a sporophyte

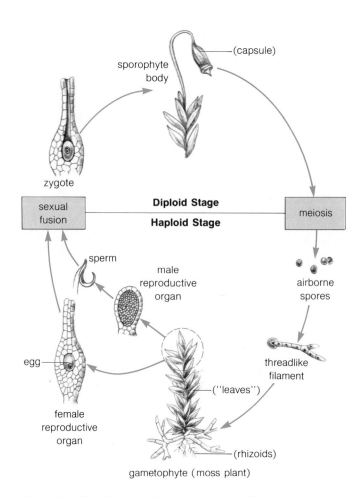

Figure 17.4 Moss life cycle. (As in the other plant life cycles, the various structures are not drawn to scale.)

body. A sporophyte generation is not a common strategy among green algae, and the sporophyte body itself is not particularly notable in terms of special adaptations. *But it does indicate that the capacity to develop a sporophyte generation might have been present in ancient green algae, and could be acted upon by natural selection when environmental conditions changed or when new environments opened up.*

Interestingly, among green algae there is a trend toward more complex sexual mechanisms. In simpler forms that rely on **isogamy** (gametes of opposite sexes are identical in appearance), the single-celled organisms themselves function as gametes. In more complex forms that rely on **oogamy,** gametes are differentiated in size and motility, much like the motile sperm and nonmotile eggs of animals. The egg remains protected in the plant body. The sperm travels to it, sometimes attracted by chemical substances the egg secretes into the water. Freed of the requirement to swim about, the egg can develop to greater size and complexity. It also can incorporate food reserves and substances that enhance the survival chances of the next generation. *Thus oogamy in ancestral plants may have been one mechanism by which increasingly complex plant forms developed that could survive on land.*

Finally, it is significant that green algae lack appreciable amounts of the specialized accessory pigments required for photosynthesis in dimly lit places. The absence of such pigments restricts these plants to shallow water or the upper layers of clear, deep water. Thus ancestral green algae were likely candidates for surviving in the shallow bogs and lagoons that opened up some 400 million years ago, when continental land masses became elevated. Through subsequent adaptations in body form, green algae have come to live not only in diverse aquatic environments but in land zones where the substrate remains moist.

With such potential for life on land, why is it that existing species of green algae have not made more than the most tentative forays away from the water? The reason is that even forms which rely on oogamy are restricted to areas where, at least from time to time, there is enough standing water around the reproductive organs to assure the sperm of a watery route to the egg. Without further modifications in their sexual life cycles, some lines of green algae simply continued to exist in those environments where existence was possible. Presuming the right kinds of modifications appeared in other lines, the stage was set for the move onto land. Let's turn now to explore what some of those modifications must have been, by examining some living plants that are largely terrestrial but still require water for sexual reproduction.

Transitional Land Plants

Mosses and liverworts, members of a group called the **bryophytes,** have not exactly overwhelmed the earth's surface, even though they have one of the most ancient lineages. But they are about 24,000 species strong. Moss plants are the most common members of the group. Although they grow on (and to some extent above) the land's surface, they have no vascular system. The plant body has no vascular roots; it has only tiny cellular threads (rhizoids) that are anchors, not organs for absorbing water. Mosses have no vascular stems; most have only long parenchyma cells into which water diffuses rather than flows through. Their "leaves" are outgrowths a few cells thick, and they don't have vascular structures, either. Moss "leaves" have a thin waxy coat and tend to curl up as they become dry, which prevents additional water loss. Moss plants also tend to grow in tight clusters, which prevents

loss of water from the soil and from individual plants. They simply can't reproduce sexually unless they live in areas where water is periodically available.

Mosses share a characteristic with the green algae. When moss spores germinate, they form a green, threadlike filament that resembles many filamentous green algae. This gametophyte filament goes on to become the familiar plant, with male and/or female reproductive organs (Figure 17.4). The male reproductive organs produce sperm. A diploid zygote forms only when sperm from a male organ is able to swim to the female organ. Both male and female reproductive organs sometimes can be found on the same plant body in some mosses; in most mosses they are likely to be found on separate plants that are still near each other. But even this short journey is impossible unless a film of free water covers the entire route. Assuming a few water drops are present, though, sperm and egg can fuse to form a zygote. The zygote then grows into a sporophyte—in this case, a spore capsule carried on a stalk growing right out of the tip of the gametophyte body (Figure 17.4). Eventually, specialized cells of the sporophyte undergo meiosis and produce more spores, which can survive unfavorable periods. In various ways, the sporophytes thus resemble one model of the earliest land plants—little more than a green body with a spore capsule on top.

Vascular Plants

About 400 million years ago, the true vascular plants—the **tracheophytes**—appeared. Within the relatively short span of 50 million years, the major lines of vascular plants had become established on land. Many of these early evolutionary lines have descendants that have survived to the present, and they provide us with examples of how these plants may have functioned in the past.

The **horsetails** are one such group. They were once highly diverse and included treelike plants about 15 meters (about 50 feet) tall. They are now restricted to a single genus *(Equisetum)* of relatively inconspicuous plants that are often found in vacant lots, along railroad tracks, and around ponds and lakes in northern regions of North America and Eurasia. The leaves, which range from tiny scales to larger structures, are arranged in whorls about the stem (Figure 17.5). This plant body is the sporophyte stage of the life cycle.

In horsetails, the spores are produced on highly modified stems. These modified stems are borne on a central stem at the apex of the plant, forming a structure called a **cone.** The spores are dispersed by wind. If they land on a suitably moist surface, they germinate to form the haploid gametophyte: a small haploid structure that has no roots, stems, or leaves. The gametophyte bears male and female sex organs. The motile sperm produced must have free

Figure 17.5 Transitional land plants. (**a**) Haircap moss plant, with the capsule and stalk of the sporophyte rising above the gametophyte plants. (**b**) Gametophyte plants of a liverwort, another representative of the nonvascular bryophytes. The umbrella-like structures house the sex organs. (**c**) A horsetail (*Equisetum*). (**d**) A lycopod (*Lycopodium*). (**e**) Underside of a spore-bearing leaf. The dark structures are *sori*, in which spores are produced. (**f**) Tropical tree fern. (All photographs National Audubon Society/PR)

water (dew, rain) to reach the female sex organ. Once fertilization has occurred, the diploid zygote develops by a series of mitotic divisions that ultimately produce the diploid horsetail plant.

The **lycopods** are another group of ancient ancestry. They were highly diverse during the Paleozoic era, and some forms were full-sized trees. The modern genus *Lycopodium* is a common representative of the group. A *Lycopodium* sporophyte has true roots and small scalelike leaves that spiral around the stems and branches. It also has cones, much like those of the horsetails. These plants are diploid. Meiosis occurs in the cones and is followed by formation of haploid spores. The spores are dispersed and germinate to form small, free-living gametophytes. For these gametophytes, free water is required for the sperm to reach the female organs. Following fertilization, the diploid zygote develops into a new sporophyte body (Figure 17.5).

The **ferns** make up another group of vascular plants with ancient origins. Although ferns are more diverse than most of the other vascular plants we have described (about 12,000 species have been identified), they were even more richly represented during times past, particularly during the Paleozoic. Ferns have an extensive root system and well-developed leaves (Figure 17.5). Most often the stem is underground (some tropical "tree ferns" are notable exceptions).

The conspicuous fern plant is the sporophyte (diploid) stage of the life cycle (Figure 17.6). Spores are dispersed through the air and germinate to form a small, heart-shaped gametophyte (a prothallus). The gametophyte plant bears the sex organs. Again, motile sperm require free water to reach the eggs. The new sporophyte generation develops from the diploid zygote and soon overshadows the small gametophyte plant, which then disintegrates.

Adaptive Importance of Seeds

Although the horsetails, lycopods, and ferns all have sporophyte stages adapted for life on land, their distribution is still limited by the vulnerable nature of the gametophyte stage that is needed to complete the sexual life cycle. The gametophytes of these plants usually lack vascular tissue. Hence they are limited to growing where water is freely available for sexual reproduction. Only the diploid sporophyte has vascular tissues. More than this, for the horsetails, lycopods, and ferns, the sperm are motile and must have free liquid water to swim to the eggs. Thus they have made it only halfway to the land. Although their sporophytes are terrestrial plant bodies, in a very real sense their gametophytes are aquatic.

But among the lycopods and some ferns are species that hint at how a new evolutionary road appeared. Unlike the other primitive vascular plants, in which every gametophyte bears both male and female reproductive organs, these plants produce two distinct kinds of spores, which differ in size. The larger spores are **megaspores** (mega-, "large"), and the smaller ones are **microspores** (micro-, "small"). Plants that produce spores of two sizes are said to be **heterosporous** (hetero-, "different"). In contrast, the horsetails, most ferns, and *Lycopodium* are said to be **homosporous** (homo-, "same") because only a single type of spore is produced. In some of the heterosporous lycopods, the condition is carried one step further: the spores produce different types of gametophyte plants. The haploid plants growing from microspores bear only *male* sex organs, whereas those growing from megaspores bear only *female* sex organs. In itself, this development hasn't proved to be especially useful for these lycopods, because ample water must still be around for gametophytes to develop and for gametes to get together. Even so, this type of heterospory appears to have been the route by which the dominant group of vascular plants—the **seed plants**—arose.

Seeds have a resistant seed coat, which minimizes water loss, and internal food reserves, which nourish the embryo during dormancy and germination (Chapter Eleven). Thus it is not only that the seed is an ideal "package" for dispersing the sporophyte; the strategy culminating in seed formation eliminates the dependence of the gametophyte stage on the availability of free water. *All seed plants are heterosporous: they produce megaspores that develop into female gametophytes and microspores that develop into male gametophytes.* But these spores are not shed from the plant to develop into free-living gametophytes, which would be somewhat vulnerable to drought. The sporophyte plant *retains* the spores; the gametophytes develop in special structures while still attached to the "parent" plant! In this way, water and food required for gametophyte development can be provided by the sporophyte plant—the stage in the life cycle that is well adapted for obtaining these

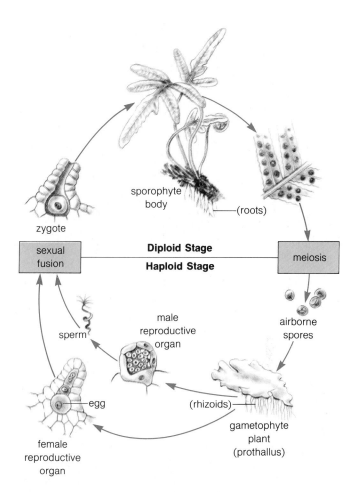

Figure 17.6 Fern life cycle, as described in the text.

resources on dry land. More than any other factor, the evolution of seed production freed vascular plants from the lowlands and other wet areas, permitting them to spread over much of the earth's surface.

Gymnosperms: First of the Seed Plants

There are four major classes of living seed plants: cycads, ginkgos, conifers, and flowering plants. The first three classes are known as **gymnosperms.** The term means "naked seed," and it refers to the fact that the seeds of these plants are carried on the surface of reproductive structures without the protection of additional tissue layers. The gymnosperms first appeared during the Paleozoic era

a

b

Figure 17.7 (**a**) Cycad. (**b**) Fernlike leaves of the maidenhair tree, *Ginkgo biloba*.

(Chapter Fifteen) and reached their peak of diversity during the Mesozoic era. The **cycads** are gymnosperms that flourished along with the dinosaurs of the Mesozoic. The few cycad species that have survived to the present are confined to semiarid regions of the tropics or to warm, temperate zones. At first glance you might mistake a cycad for a small palm tree (Figure 17.7). Despite the superficial resemblance, palms and cycads are not at all closely related, for the palms are true flowering plants. Cycads, like most gymnosperms, have massive cones. Their leaves are cov-

ered by a thick cutin layer, which retards water loss. The spiny leaf tips and sharp spines on some cycad cones are probably adaptations that minimize seed predation, for seeds are a sought-after food in areas where water (hence plant growth) is limited. In most cases, it takes several years for cycad seeds to mature. Their slow reproductive rate has probably put the cycads at a competitive disadvantage compared with other seed plants of more recent origin.

Ginkgos are even more restricted in distribution than the cycads. Only a single species (Figure 17.7) has survived to the present, despite the success of this class of gymnosperm during the Mesozoic. It seems that several thousand years ago, this attractive tree species was extensively planted in the cultivated grounds around Asian temples. But the small natural population from which these "domesticated" trees were derived must have become extinct. The near-extinction of this "living fossil" from the age of dinosaurs is puzzling, for ginkgos seem to be hardier than many trees. Especially in cities, they are planted not only because they are attractive, they seem resistant to insects, disease, even air pollution! But cultivated plants grow under unique conditions. Under natural conditions, ginkgos may be less adaptive than other species in some critical respects. Together with geological and geographic factors, such vulnerability may have restricted the range of the species, bringing it to the edge of extinction.

In terms of sheer diversity and distribution, the most significant gymnosperms are the **conifers:** cone-bearing woody trees and shrubs with needlelike or scalelike leaves. Most conifer leaves are retained for a period of several years. Because some leaves are always present on the plant body of most species, the conifers are commonly called "evergreens." The class includes such familiar plants as pine, spruce, fir, hemlock, and redwood (Figure 17.1). Spirally arranged, highly modified branch systems form the seed-scale complexes of their reproductive cones. These heterosporous plants produce woody female cones (which bear the megaspores) and smaller male cones (which bear the smaller microspores).

The pine life cycle (Figure 17.8) typifies the general adaptations of the group as a whole. We can begin with the sporophyte plant (the pine tree) to trace the parallel events that culminate in the development of the male and female gametes. The female gametes are produced on individual woody cone scales. Each cone scale bears two **sporangia** (spore-producing structures) on its upper surface. Within each sporangium is a diploid "megaspore mother cell." This cell divides by meiosis to produce four haploid megaspores. Three abort; but the remaining functional megaspore divides mitotically to form the female gametophyte. The female gametophyte is an oval mass of cells within the walls of the sporangium. The sporangium, tissue layers, and female gametophyte together constitute an **ovule.** As Figure 17.8 shows, the gametophyte has reproductive organs, each containing a single egg. Sperm cells will gain entry to the eggs through an opening in the wall surrounding the gametophyte. Gametophyte tissues secrete a sticky fluid that spreads out over the surface of the cone scale. As they do, the cone scales open up slightly in preparation for the arrival of the male gametophytes.

Even as the female gametes have been developing, parallel events have been going on in the male cones. Each male cone scale has a pair of sporangia on its lower surface. And within each of them are numerous diploid "microspore mother cells." Each of these cells undergoes meiosis to form four microspores. The microspores undergo limited mitotic divisions, culminating in the development of an immature male gametophyte: a **pollen grain.** A pine pollen grain consists of only four cells by the time it reaches its maximum development in the sporangium. At that point, the sporangial wall bursts, releasing many pollen grains into the air. Wind dispersal is enhanced by a pair of inflated bladders attached to each pollen grain. The transfer of pollen from the male to the female reproductive structures is known as pollination.

Each spring, a normal pine tree produces millions of pollen grains which float through the air, sometimes as thick yellow clouds. Some settle on the upper surface of the female cone scales, near the sporangia containing the female gametophytes. The sticky fluid on the scales traps the pollen grains and holds them in place. Now the cone scales close, thereby protecting both gametophytes during the delicate events of fertilization. As the sticky fluid dries, it contracts and slowly pulls the pollen grains toward the opening in the wall surrounding the female gametophyte.

When the pollen grain is drawn into the opening, it produces a **pollen tube:** a tubelike projection that starts to grow toward the female gametophyte (Figure 17.8). While this is happening, the pollen grain produces two sperm cells. These sperm move through the protective passageway of the pollen tube to the eggs in the female reproductive organs. As soon as one egg becomes fertilized, the other becomes nonfunctional. The stage is now set for the development of a pine seed. The sporangial wall hardens and thickens to become the seed coat; the diploid zygote divides mitotically to form several embryos, one of which survives to become the next sporophyte generation. The tissue of the female gametophyte is the food reserve for nurturing the embryo during dormancy and its eventual germination.

Conifers were the dominant land plants during the Mesozoic, and unquestionably their mechanisms of seed production, protection, and dispersal helped assure their success. Since then, their distribution has gradually been reduced for various reasons, the most significant apparently being the rapid evolution of flowering plants. Even so, conifers are still important components in many communities, especially in northern regions and at high altitudes (Chapter Twenty-Two). They also are major sources of lumber, pulp, and numerous industrial products.

a

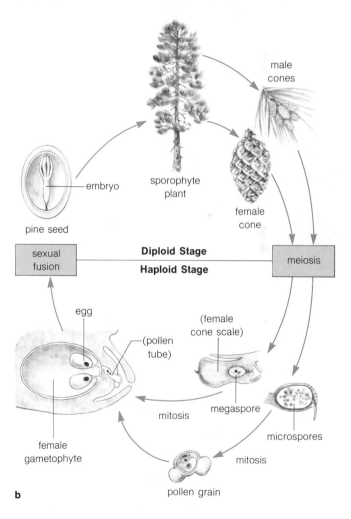

b

Figure 17.8 (**a**) Large female cone and smaller male cones from a lodgepole pine tree. (**b**) Pine life cycle, as described in the text.

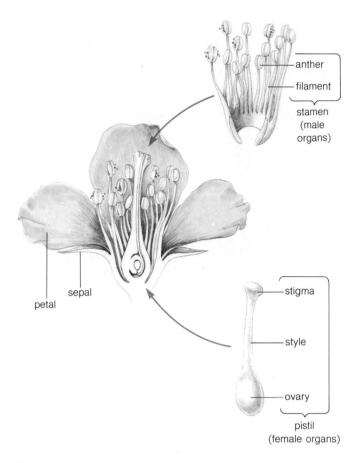

Figure 17.9 Typical organization of a flower.

Flowering Plants

The **angiosperms,** or flowering plants, are the fourth and by far the most diverse class of seed plants. Although they appeared only about 135 million years ago, they quickly rose to dominance. Today there are about 250,000 known species of flowering plants, grouped into two subclasses: monocots and dicots (Chapter Eleven). The dicots number at least 200,000 species; they are the most diverse group of flowering plants. The 50,000 or so species of monocots include such familiar plants as grasses, palms, lilies, and orchids. They have played a central role in human population growth, for our main crop plants (wheat, oats, rice, rye, barley, and corn) are monocots. All these crop plants are derived from members of the grass family.

Flowering plants have a unique reproductive system that involves the formation of flowers. Despite tremendous diversity in form, all flowers are built according to the same general plan (Figure 17.9). The parts are highly modified leaves, arranged spirally or in whorls about the floral axis. In the gymnosperms, leaves have been modified to form cone scales, which produce either microspores or megaspores. In the angiosperms, leaves have been modified to form microspore- and megaspore-producing structures as well as accessory parts that have both direct and indirect roles in sexual reproduction.

As we move upward from the base of the floral axis, we first encounter the **sepals,** the flower bud scales. Often sepals are green; in some flowers they are highly colored. They may be separate and distinct, or they may be fused together in various ways. Upward or inward from sepals are **petals:** modified leaves that are highly variable in form and color from one species to the next. Like sepals, the petals may be distinct from one another or fused together. Although sepals and petals are not themselves sexual organs, they often have a major role to play in the functioning of the reproductive system, as you will soon see. Inward from the petals are the **stamens** (male reproductive organs). Each stamen is composed of a supporting *filament* and an *anther,* a cluster of chambers in which pollen grains are produced. The female structure, known as a **pistil,** is composed of three parts. The base is the *ovary* (an expanded chamber in which megaspores are produced), a *style* (a stalk above the chamber), and a *stigma* (a special sticky surface cap on top of the style). Most pistils are compound, being made of separate units (carpels) fused together.

If we were to section the ovary, we would find one or more ovules made of two tissue layers that surround the megaspore and are attached to the ovarian wall. In some plants, hundreds or thousands of individual ovules may be attached to the inner wall of the ovary or to its internal partitions. Meiosis of the megaspore mother cell occurs *inside* an ovule to produce four haploid cells. Three usually abort, leaving a single functional megaspore that divides to form a single female gametophyte.

The flower just described is a "perfect" flower, having both stamens and one or more pistils. Some plant species have "imperfect" flowers, with only stamens or only pistils. They are often called male or female flowers. In some species, male and female flowers appear on the same plant; in other species, male and female flowers are on separate plants.

Regardless of the type of flower, the key to the sexual process is the transfer of pollen grains from the anther to the stigma. Once a pollen grain has been deposited on the stigma, it forms a pollen tube that grows down through the style to reach the ovarian chamber. Two sperm nuclei from the pollen grain break through the end of this tube and penetrate the female gametophyte. One of these nuclei fuses with the egg to form the diploid zygote. In most organisms, fusion of egg and sperm is a singular event in the sexual cycle. But in flowering plants, we have a process called **double fertilization:** in addition to the fusion of one sperm nucleus with the egg, the second sperm nucleus fuses with the two "polar nuclei" (Figure 17.10) to form a single triploid (3N) nucleus!

Following double fertilization, the ovule expands and develops into the seed. The tissue layers thicken and harden, becoming the seed coat. The fertilized egg develops into the diploid embryo. The triploid nucleus divides repeatedly, thereby forming the endosperm (Chapter Eleven). In many seeds, the endosperm is massive and functions as the food storage tissue that will be used for early growth when the seed germinates. In other plants with limited endosperm, the cotyledons act as food storage tissue. In contrast to the "naked seeds" of gymnosperms, the angiosperm seed is contained within the ovary as it develops. The ovarian wall expands to form another structure characteristic of flowering plant reproduction: the **fruit.** We tend to think of fruits as juicy edible structures, as indeed many are. But fruits are any structures that develop from the ovary and contain the ovules or developing seeds. Nuts and grains are dry fruits; tomatoes are fleshy fruits. A raspberry is an aggregate of many units that later separate; a pineapple is a cluster of multiple units that remain together at maturity. Fruits, be they edible structures, small capsules, or a variety of other forms, are adaptations for efficient seed dispersal, which helps assure successful reproduction.

Coevolution of Flowering Plants and Their Pollinators
Flower-bearing plants may be found in alpine regions, in forests, in deserts, even in aquatic settings. More than any other kind of plant, they have spread farther across the earth's surface and have become the most diverse group of all. What underlies their adaptive success? For the answer we must return to the Mesozoic, when the wind-pollinated gymnosperms were the dominant land plants. It was at this time that the flowering plants appeared and began to take hold. By the close of the Mesozoic, they had undergone expansion in numbers and kinds through a remarkable range of environments.

It happens that this period of rapid expansion was paralleled by expansion among insect groups. Although insects first appeared far earlier than flowering plants (during Silurian times), they initially did not undergo rampant diversification. But during the early Mesozoic, some beetles apparently were already feeding on pollen or other parts of some wind-pollinated plants. They must have been secondary in importance as pollinating agents, because they probably visited plants on a somewhat haphazard basis. Yet, we can readily imagine that plants which were slightly more conspicuous or tasty would be able to attract more insects, and dispersal of pollen would be more effective because of it. Hence there would be a selective advantage for plants having not only ample pollen, but special glands for secreting a sugary fluid (*nectar*) and flamboyant floral advertisements. However the interactions

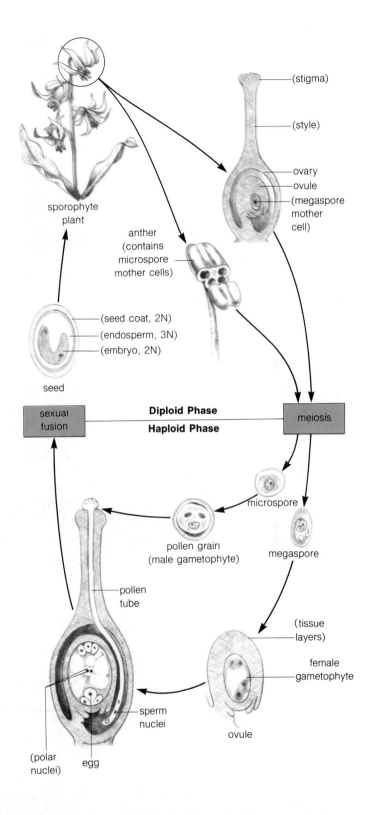

Figure 17.10 Life cycle of a typical flowering plant, as described in the text.

a

Harlo H. Hadow

b

Harlo H. Hadow

c

Harlo H. Hadow

d

R. Taggart

e

R. Taggart

f

R. Taggart

g

R. Taggart

h

Ted Schwartz

i

Edward S. Ross

j

Edward S. Ross

k

Edward S. Ross

l

m

n

o

p

q

Figure 17.11 Pollination of plants through reliance on winds, deceit, spring traps, and various exotic and luscious inducements.

(**a,b**) Reproductive structures of the saguaro cactus brush against the head of a bird pollinator, the gila woodpecker, and pollen is thus carried from blossom to blossom in return for sips of nectar. (**c,d,e,h**) Plants pollinated by bees tend to be blue or yellow and brightly colored. (**f,g**) Some orchids tend to look like females of the pollinating species (in which case the males attempt to copulate with the flower) or like rival male bees vibrating in the wind (in which case the territorial male insect attacks the flower and pollinates it). (**i,j,k**) Scotch broom is an "explosive" flower. The weight of its pollinator is needed to force open the flower, which releases the pollen-laden stamens that are positioned to strike against the underside of the pollinator. (**l**) Close-range visual guides to nectar. Fuschias (**m**), with their suspended flowers, are pollinated by hummingbirds. Wind-pollinated flowers include cattails (**n**) and grasses (**o**). (**p**) Arctic lupine, an "explosive" flower of the far north. (**q**) Stephanotis is serviced by night-flying moths; it is light-colored and astonishingly fragrant during the night. (Day-flying humming-birds are opportunistic when it comes to stephanotis, and are secondary pollinators of it.)

developed, by the dawn of the Cenozoic, diverse species of insects and flowering plants existed and were locked in interdependence.

Today the arrangement of flower parts, the patterns in which they fuse or remain distinct, their color, odor, nectar, size—all these things generally have evolved in ways that attract bees, beetles, butterflies, moths, flies, wasps, birds, even bats. Although it is true that some successful flowering plants are wind-pollinated (their flowers typically do not have nectar or perfume and tend to be green), most depend on animals to disperse their pollen. Color is an important attractant for many pollinators. Bird-pollinated flowers, for instance, tend to be red—a wavelength to which bird eyes are more sensitive. Bee-pollinated flowers tend to be blue or yellow with prominent ultraviolet components. Odors, too, lead the way and bring in pollinators from considerable distances. Not all are sweet: many flowers have what we think are foul scents that are irresistible to flies and other carrion-loving insects.

Once the pollinator has located the flower, color patterns and petal shapes guide it to the nectar—and not incidentally channel its movements in a way that ensures pollination. For example, the petals of hummingbird-pollinated flowers often form a long tube that accommodates the shape of the bill. These flowers often exclude other kinds of potential pollinators (such as heavy bumblebees). In some flowers, the ovary is off to the side of the main axis, which prevents damage from the probing beak of the hovering bird. (The ovary tends to be centered in butterfly-pollinated flowers; after all, who ever heard of a deadly butterfly proboscis?) And then there are the orchids! Some mimic the form and coloration of the pollinating insect species. The male insects attempt to mate with flower after flower, spreading pollen about as they are led ever onward by this cruel deception. Fortunately, there eventually comes along a real female, so the insects do keep trying, and the orchids do keep getting pollinated.

In tropical environments, there typically is extreme specialization between plants and pollinators. But in more rigorous climates, plants tend to be less picky about their pollinators—which points up something about the nature of the coevolutionary process between flowering plants and their cohorts. On the one hand, *the more refined the "fit" between plant and pollinator, the more efficient pollination can be.* For instance, less energy has to be channeled into producing a lot of pollen grains. On the other hand, *the more specialized the plant–pollinator relationship, the greater the likelihood that the plant may face extinction if the pollinator should happen to disappear.* And in the rigorous, low-diversity climatic zones of the world, such disappearances are a real possibility. In attempting to identify the nature of such interactions, we therefore must look to the delicate feedback relationships that may exist not only between coevolved species, but between those species and their environment. More will be said about this in Chapter Twenty-Two.

Fungi: Decomposers of a Varied Sort

Fungi, recall, are multicellular eukaryotes. Most are **saprophytic:** they secrete enzymes into their environment that break down the energy-rich remains of dead organisms and their by-products, such as leaf litter on the forest floor. A small number are **parasitic:** they obtain nutrients from organisms that are still alive. In almost all cases, there is reliance on enzyme secretion that promotes digestion outside the fungal body, and then nutrient absorption across the fungal plasma membrane. This strategy is known as extracellular digestion and absorption. It is largely because of this unique energy-acquiring strategy that many biologists now place the fungi in a separate kingdom even though many primitive fungi have some plantlike characteristics. The question of whether fungi were derived directly from prokaryotic ancestors, from protistans, or from primitive plants may never be resolved. Regardless of where they are placed in a classification scheme, the important thing to keep in mind is that the fungi represent a *different* evolutionary line of development. And the fact that we are describing them near the end of this chapter does not mean they are more advanced than land plants! Fungi have a much simpler organization.

All **true fungi,** the Eumycophyta, are much the same in basic characteristics if not in appearance. They send out cellular filaments, called **hyphae** (singular, hypha), which grow into a tangled mat (a **mycelium**) or into a complex, spore-producing body. The club fungi (basidiomycetes) are perhaps the most advanced forms: among the 25,000 species are edible mushrooms, puffballs, rusts, and smuts. Another important group are known as the sac fungi (ascomycetes), which include various forms of yeasts, cup fungi, morels, and truffles.

The common edible mushroom may be used as one example of the diverse reproductive cycles of true fungi. There is an alternation between haploid and diploid stages, with asexual spores produced during favorable times (Figure 17.12). A new generation begins when an airborne haploid spore lands on a moist substrate and germinates. The spore grows into a small, loosely organized haploid mycelium. If two compatible hyphae from these mycelia grow near each other, they may fuse without actual fusion of their nuclei. Thus a hypha with two nuclei per cell is formed. This hypha forms an extensive mycelium, destined to become a long-lived network beneath the soil. Under moist conditions, a spore-bearing structure (the mushroom) forms above ground. Inside the mushroom's reproductive structures, the pairs of nuclei derived from the two different hyphae fuse. Immediately the fused diploid nuclei divide meiotically and spores are produced. Once the spores are dispersed by winds or insects, the mushroom withers. As the spores fall on moist ground and germinate, the cycle begins again.

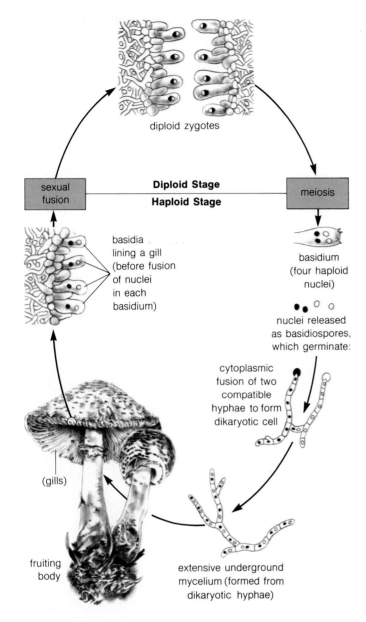

diploid zygotes

Diploid Stage

Haploid Stage

sexual fusion

meiosis

basidia lining a gill (before fusion of nuclei in each basidium)

basidium (four haploid nuclei)

nuclei released as basidiospores, which germinate:

cytoplasmic fusion of two compatible hyphae to form dikaryotic cell

(gills)

fruiting body

extensive underground mycelium (formed from dikaryotic hyphae)

Figure 17.12 Life cycle of a true fungus, using the common edible mushroom as an example.

Fungi As Symbiotic Organisms Many fungi have come to form symbiotic relationships with other organisms. Pine trees, for instance, won't grow well unless hyphae of certain fungi form dense mats around their roots. These fungi are actually parasitic, for they absorb carbohydrates from the live trees. At the same time, the thick fungal mats around the roots help the tree absorb nutrients from the soil more rapidly than it otherwise would. Or consider the **lichen:** a "composite" organism made up of a fungus and an alga

living together in total interdependence. The fungal hyphae form a dense mat above and below the algal cells. In this arrangement, the fungus is assured of photosynthetically derived nutrients, and the alga is protected by the hyphal mats from such environmental extremes as damaging light intensity or temperature, which few other organisms can tolerate. Thus about 15,000 species of lichens can do well not only in open patches of forests (they need some sunlight to survive), but in deserts, in tundra, in Antarctica, even on bare granite rocks of isolated oceanic islands.

In most communities, however, different species of parasitic and saprophytic fungi compete with other organisms and with themselves for available substrates. If we were to measure adaptive success simply on the basis of numbers and distribution, we would find it depends on several things. Usually the fungal species dominating a community are capable of rapid hyphal growth and spore germination, and the enzymes they produce are potent enough to break down substrates quickly. More than this, by-products of their metabolic activity often are poisonous to other species—and they themselves show resistance to the toxic by-products of others! We are left to conclude that poisonous mushrooms aren't "getting even" with us because humans eat mushrooms; they are interacting with one another according to patterns of competition laid down long before mushroom lovers ever stalked the forest.

And this reinforces an important point: an organism is not inherently "good" or "bad," even though it may come to have some bad consequences for us. In their natural environment, parasitic fungi have their place. Normally they attack organisms that are in some way damaged or weakened, as with age. Young, healthy plants or animals are usually resistant to fungal attack unless environmental conditions shift drastically and somehow make them vulnerable. Their resistance developed over time, as the action of natural selection weeded out less resistant strains. Even as saprophytic fungi break down the remains of dead organisms, so also do parasitic fungi break down the weakened or less resistant organisms. In this way, minerals and other materials are returned to the water or soil, where they can be picked up and recycled through the community.

Late Blight What happens when we fail to recognize the interactions among plants and fungi in the community? Consider the potato, a plant native to the cool, dry regions of the Peruvian Andes. This plant was under cultivation almost 2,000 years ago. A parasitic fungus that causes the disease **late blight** in potatoes (and tomatoes) was also native to this region, but it posed no serious threat to the crops because environmental conditions kept it in check. The potato was introduced to Europe in the sixteenth century, and it became a major food crop. In Ireland it became *the* major food crop. During the growing season, Ireland happens to have cool, moist nights and warm,

Roger K. Burnard

All other photographs Victor Duran

Figure 17.13 A few representative fungi. (**a**) A coral fungus. (**b**) Scarlet cup fungi. (**c**) Bird's nest fungi. (**d**) Shelf fungi, growing outward from a tree trunk. (**e**) Common edible mushrooms. (**f**) Edible morels, prized by gourmets.

humid days—which is quite unlike the climate of the Andes. In this setting, and with uninterrupted fields of host plants, the fungus went wild. Once a plant was stricken with late blight, its vines rotted within fourteen days. Between 1845 and 1860, the years of the great potato famine, a third of the population of Ireland starved to death, or died in the outbreak of typhoid fever that followed as a secondary effect, or fled the country.

Black Stem Wheat Rust Even though we are beginning to appreciate the complex relationships that exist between organisms and their environment, it is not always a simple thing to come to terms with that fact. Consider, for example, that two-thirds of the dry bean crops and almost all of the pea crops in the United States consist of only two varieties each—which means these staple crops are extremely vulnerable to parasitic fungal attacks. Consider that **black stem wheat rust** claims an appalling amount of annual wheat crops—one of our most basic of all food sources. The fungus causing the disease has coevolved with two hosts: the wheat plant and the barberry bush. And it goes through not one but *five* spore stages during the year! It is a never-ending battle to develop increasingly resistant wheat strains, because each time a new wheat strain appears in the fields, the rust may undergo mutations and develop into a form able to attack the resistant variety. For a time it was thought that eradication of barberry plants (on which sexual reproduction of the fungus takes place) would interrupt the life cycle. It appears that enough mutations can produce new fungal strains even in the absence of sexual reproduction. Finding a way out of this precarious situation will not be easy. In economic terms, it is far more efficient to plant, tend, and mechanically harvest fields of the same kind of crop than it is to have diversity among crops. And efficiency *is* a crucial part of our survival equation. As long as the human population continues on its course of explosive growth (Chapter Twenty-Two), it is a race against time to produce the most food in the fastest way possible. We are, as a consequence, extremely vulnerable in the worst possible way: in our total dependence on monocrop agriculture. Again, we must continue to meet existing demands with existing technology. *But simultaneously, we must begin to design agricultural systems that will put us in balance, not in constant battle, with communities of organisms in the natural world.*

Perspective

How often, if we think about them at all, do we think of plants as little more than greenery in the scenery, and of fungi as little more than curious and/or terrible intruders into human affairs? Yet, without these organisms, without their invasion of the land, we could scarcely have even made it onto the evolutionary stage. Until they came to cloak the earth, inch by inch, continent by continent, there could be no other life forms of any sort on the vacant land. Their evolution, while not smacking of the melodrama of, say, the rise and fall of the dinosaurs, was remarkable nevertheless when we stop to consider the odds.

Plants are immobile. They cannot crawl, leap, run, or fly. Beginning with ancestral forms that could not have been more complex than simple green algae, a long series of chance mutations apparently led to plants having a few seemingly unimpressive traits: a few threadlike anchors to sink into the muddy margins of the sea, a tiny chloroplast-filled stalk, a few scalelike leaves to spread out stiffly beneath the sun. And yet, within those anchors, stalks, and scales, hollow tubes eventually appeared as forerunners of highly adaptive transport systems through what we now call roots, stems, and leaves. The diploid stage of aquatic life cycles proved to be an adaptation of remarkable potential. Resistant as it was to adverse conditions, the diploid sporophyte stage gradually became more and more pronounced in the move to higher and drier land. Among land plants, the gametophytes came to be housed in the parent sporophyte body which, through its transport system, could nurture them with water and dissolved nutrients through adverse periods. Reproductive structures became attuned to seasons, to times of rain and winds. And most astonishing of all, some plants coevolved with insects and other animals. The life cycles of these plants came to depend on animals to carry pollen grains to the female reproductive organs and to disperse seeds; simultaneously, the life cycles of animal pollinators came to depend on the plants for food. Over time, the fungi became part of life on land, as they have been in water—decomposers of the first rank, recyclers of life-giving nutrients for the animals and plants with which they are linked in the web of life. Like plants, most fungi rely on winds and insects for dispersal. Thus, without means of their own for motility, plants and fungi have moved over the barren plains, to the mountains, to land everywhere. And wherever they have come to rest, they have become foundations for new communities of life.

Recommended Readings

Andrews, H. 1947. *Ancient Plants and the World They Lived In.* Ithaca, New York: Comstock. Well-written, well-illustrated account of plants through the ages.

Bold, H. 1976. *The Plant Kingdom.* Fourth edition. Englewood Cliffs, New Jersey: Prentice-Hall. Excellent survey of the plant kingdom, with good illustrations.

Jensen, W. and F. Salisbury. 1972. *Botany: An Ecological Approach.* Belmont, California: Wadsworth. More advanced reading, but excellent discussions and illustrations of ecological aspects of plant life.

Chapter Eighteen

Figure 18.1 Will the real animals please stand up?

Evolution of Animals

It is the last scene of an epic Western film; the hero has finally tracked down the rustlers, saved the longhorns, fought off a mountain lion with his bare hands, narrowly escaped the lunge of a rattlesnake, and found the rancher's daughter (who had been kidnapped and hastily abandoned in the desert) by spotting the vultures circling in the sky above her. Now, with his faithful dog beside him, he waves goodbye to the girl, mounts his restless stallion, splashes across a shallow stream, and rides off into the golden west.

When you think back over the kinds of stories on which we are raised, it's easy to see how we acquire a stereotyped picture of what "animals" are. For instance, how many kinds of animals would you say are mentioned in the pared-down saga opening this chapter? Seven, right? But did you happen to notice that all the animals mentioned are *vertebrates*—animals with backbones? The saga is just a little lopsided: more than 99 percent of all the kinds of animals in the world are backboneless *invertebrates.* Undoubtedly just out of camera range were all manner of worms, grasshoppers, spiders, scorpions, ticks, centipedes, millipedes, water fleas, mosquitoes, flies, beetles, butterflies, moths, aphids, snails, bees, and various other spineless things. Long after we finished counting all the kinds and numbers of vertebrates alive today, we would still be counting up the kinds and numbers of a staggering array of invertebrates. Only by beginning with a few examples of these less conspicuous animals can we gain insight into how all animals evolved—into what they all have in common and how diversity arose among different groups. Only by putting aside our culturally shaped preconceptions can we piece together a picture of the evolutionary progression that led to organisms as complex as horses, dogs, snakes, and longhorns, and ourselves.

A Few Thoughts on "Colonial Animals"

Somewhere between 2 billion and 1 billion years ago, the first animals arose. They left behind nothing of themselves other than a few petrified feces, and a few tantalizing tracks and burrows frozen in ancient rocks. At the dawn of the Cambrian they were leaving behind hard parts, so we know something of what they looked like by then—but by then they were already well-developed multicellular animals, or **metazoans**. Thus we can assume that long before 700 million years ago, there must have been any number of experiments in multicellularity, most likely of the sort described in Chapter Sixteen.

One such experiment is thought to have led to the **sponges**—diverse, abundant, but relatively simple animals found in shallow seas throughout the world. In one sense, they are like a multicellular vase with holes in the sides. Seawater flows in through the holes (pores) and then out through the opening at the top. The inside is lined with flagellated *collar cells*, which trap and ingest microscopic organisms carried in on the water flowing through. These collar cells link the sponges to far simpler relatives (Figure 18.2). Certain flagellated protistans also sport a delicate, sticky collar, which collects food particles that the flagellum beats up around it. There are also several colonial forms of collared flagellates that are functionally independent but are held together in a jellylike matrix. Yet sponges clearly are a step above these colonial forms in organization, for they have four distinct cell types that are incapable of independent existence. Besides collar cells, sponges have thin cells lining the outside surface, thick cells lining the pores, and various amoebalike cells creeping about in the body wall. To these amoeboids fall the tasks of food catering (passing on some food from the collar cells to the other cell types) and reproduction.

Sponges do resemble animals—they are multicellular and they ingest other organisms. Yet communication between cells and integration of activities are truly primitive. Water flows into a central cavity, and it flows out; if microorganisms flow in with it, they are captured when they brush against the sticky collar cells. The cells lining the pores show only simple responsiveness to simple changes in internal conditions (they contract and shut off the inward flow of water if it contains certain noxious substances). *Thus sponges differ from all other animals in three important ways: they have no nerve cells, they have no specialized muscles, and they have no gut.*

Does this put sponges outside the animal kingdom? Perhaps not. Sponges possess certain proteins found in all other animals—but not in protistans, plants, or fungi. More than this, cells resembling sponge collar cells have been found throughout the animal kingdom, even in humans. It

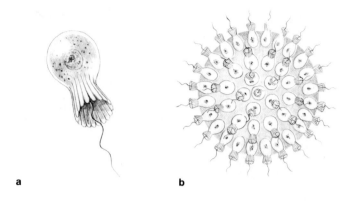

a

b

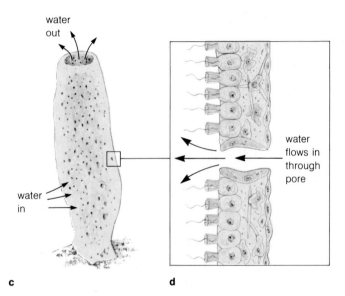

water
out

water
in

c

water
flows in
through
pore

d

Figure 18.2 Sponges and some simpler relatives. Certain protistans called collared flagellates (**a**) resemble sponge collar cells. Several colonial organisms are also made up of collared flagellates embedded in a jellylike matrix (**b**). A sponge (**c**) is multicellular (**d**), but there is no real coordination among its four cell types (collared cells, amoeboid cells, pore cells, and thin outer cells).

is possible that sponges are an extremely ancient offshoot of the main line of animal evolution, but true animals nevertheless. Whatever their ancestry, their body plan leaves little room for further developments. Diversification has largely meant increased folds in the body wall, making ever more twisted channels for water flow and thereby permitting ever more efficient filtration for food. For the sponges it has been enough: for about a billion years, they have survived in their own quiet way in various marine and some freshwater environments.

Radial Body Plans for Movement and Integration

If not in the sponges, in what forms do we find the simplest systems for intracellular communication and integration? Again we must peer into the shallow waters of the earth, this time to observe the **coelenterates**—the jellyfish, sea anemones, and their relatives. At first glance the anemones and the jellyfish seem quite different. But if we reduce their body plans to the basics (Figure 18.3), it becomes clear that a jellyfish resembles an upside-down anemone, and that an anemone resembles an upside-down jellyfish. Both are made up of two cell layers separated by a jellylike layer. (In a jellyfish the inner layer is quite thick; hence its name.) The main difference is that anemones attach themselves to the sea bottom, and jellyfish are free-floating.

Regardless of whether they float or sit, coelenterates face similar challenges and meet them in similar ways. For them, food or danger are likely to appear not on the water's surface or on the bottom but anywhere in the waters in between. Thus the systems they need to sense and respond to the environment are arranged radially about a central axis. For example, tentacles bearing harpoonlike weapons radiate outward from the mouth in all directions. Body plans of this sort are said to show **radial symmetry**.

Coelenterates are the simplest organisms having true nerves. Nerve cells in the outer and inner body layers branch out through the body, making contact with one another and with contractile cells. But they are more of a **nerve net** than a nervous system of precise, channeled response (Figure 18.4). For all other animals, a nerve conducts impulses in one direction; for the simplest coelenterates, a stimulus on one branch of a nerve cell generates an impulse that runs out along all other branches—which in turn can send impulses of their own along the same branches. Such multidirectional, diffuse message-sending is appropriate for organisms that must be prepared to respond in any direction to change.

The response system is a network of T-shaped cells, which are filled with contractile proteins just like those in human muscle cells. In one body layer, contractile cells work like longitudinal muscles: when they contract, the body becomes shorter and fatter. In the other body layer, contractile cells work like circular muscles: when they contract, the body becomes longer and skinnier. (Figure 18.1 shows the results of such movements in sea anemones.) Thus the two networks of contractile cells operate as **antagonistic muscles**. It is the same pattern that earthworms depend on for movement. And it is the same pattern that animals ranging from worms to humans use to churn food and pass it along in their gut.

Finally, coelenterates have one of the most primitive guts known. It's little more than a large chamber, lined with glands that secrete digestive enzymes. Newly captured food

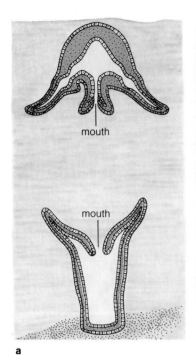

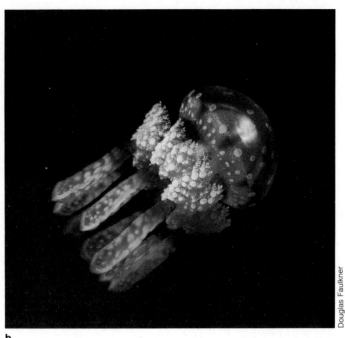

b

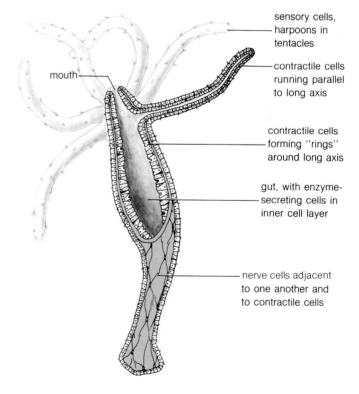

Figure 18.3 Coelenterate body plans. (**a**) All mature coelenterates are shaped either like a vase (a *polyp*) or like a bell (a *medusa*). All body parts in either a polyp or a medusa radiate away from the central body axis (running from mouth to base), and all parts are generally symmetrical (equivalent to one another) even though some variations do occur on the basic plan. (**b**) A mature jellyfish. Compare this coelenterate with the sea anemones shown in Figure 18.1.

is stuffed into the chamber, where initial breakdown occurs. Cells of the gut wall ingest and digest the resulting particles, and then the soluble products diffuse to the rest of the body's cells. Undigested residues are simply expelled through the mouth. In all the nearly 10,000 species of coelenterates, there is no circulatory system for transporting nutrients absorbed from the gut. *But coelenterates have a true gut nevertheless—along with rudimentary integration of neural and muscular activity.*

Flatworm Brains, Bilateral Symmetry, and Hints of Other Things To Come

So far we have a model for the evolutionary beginnings of integration, but where do we go from here? A clue comes from the embryonic development of modern coelenterates. Characteristically, coelenterate offspring don't look at all like their parents. Each is a **planula**: a larval form made of an undifferentiated cell mass enclosed in an outer layer of richly ciliated cells. At first it uses its little cilia as feet to swim or to crawl on the sea floor. Then it settles down on one end, a mouth forms at the other, and a gutlike cavity opens up. Tentacles develop around the mouth; it has become a **polyp** (Figure 18.3).

Suppose, in the past, chance mutations led to a form of planula that ignored the developmental signal to stop

Figure 18.4 The simple nerve, muscle, and digestive systems of coelenterates, as illustrated here by *Hydra*.

sensory cells, harpoons in tentacles

contractile cells running parallel to long axis

contractile cells forming "rings" around long axis

gut, with enzyme-secreting cells in inner cell layer

nerve cells adjacent to one another and to contractile cells

mouth

brain, and light-sensitive eyespots

branching gut

circular and longitudinal muscles under outer cell layer

"mouth"

nerve cord

a

b   "mouth"   c

Figure 18.5 Body plan of the forward-crawling, bilaterally symmetrical flatworms, shown (**a**) sliced lengthwise, (**b**) sliced crosswise, and (**c**) in real life.

crawling about, or in which the signal was never given. Suppose, further, its mouth opened up not on one end but on the underside. Such a mutant would bear striking resemblance to an existing group of animals: the **flatworms**. It would at once have a selective advantage. At its feet—*at its mouth*—would be an untapped energy source: living micro-organisms and the remains of others that had died and settled to the sea floor. An organic feast! Now selection would go to work on the diffuse, radially symmetrical nervous system, for food and danger would be encountered more by the end going first. Sensory organs would gradually become concentrated at the "head" end, for the forms that could sense and turn toward food (and away from danger) more quickly than others would tend to leave more offspring.

We can speculate that the shift to a creeping life style meant further changes in body plan. Just as there would be a selective advantage in having a head end different from the tail end, so would there be an advantage in having the vulnerable, exposed top side different from the underside. At the same time, there would be little pressure for the right half of the body to become much different from the left half. After all, predator and prey would more than likely be encountered on both sides, not just one. So the forward-crawlers would be under pressure also to be **bilaterally symmetrical**: the left half of the body would remain approximately equivalent to the right half (Figure 18.5). *A shift from radial symmetry to bilateral symmetry could have led, in some evolutionary lines, to paired nerves and muscles, paired appendages such as legs, and paired sensory organs such as eyes and ears.*

Even with these adaptations, the flatworms never have had the potential to increase much in size. With their body plan, all cells must remain close to the gut to obtain nutrients and dispose of metabolic wastes. In larger forms, the gut has become more and more branched to accommodate more and more cells. Some even have simple tubes for picking up dissolved wastes from internal cells and expelling them outside the body, much like the far more complex kidneys do for other animals. But the real limiting factor has been the need for gas exchange. Oxygen enters and carbon dioxide leaves the cells by simple diffusion—so any cell too far from the surface will suffocate. That is why even the flatworms that reach a length of 60 centimeters (about 2 feet) remain thin, almost leaflike.

More than this, if you can envision a freeway at rush hour, with cars entering and leaving on the same ramp, you have an idea of how inefficient it is to have only one opening present in the flatworm gut. What goes in but cannot be digested must go out the same way. Newly ingested food is traveling from the mouth to the far branches of the gut, even as undigested residues are traveling back to the mouth. As long as two-way traffic is necessary, various regions of the gut cannot undergo specialization. And as long as the opening is near the center of the body (which at least cuts the distance that food and residues must travel), the mouth can't be positioned forward, near the sense organs that detect the presence of food in the first place. Hence the selective advantage of the **anus**—the second opening to the gut. The development of the anus could not have been that difficult to achieve. It probably arose many times among ancient flatworm groups, and apparently it led to several evolutionary lines of further development—the ribbon worms, nematodes, nemerteans, and annelids. Thus the ancestral flatworms are thought to occupy a position of central importance in the evolution of all the more complex animals. *For with one-way traffic through the gut, regional specialization for efficient food processing was now possible.*

Despite their overall simplicity, flatworms have not in any sense been "left behind." In fact, they are so successful it would be difficult to find many shallow aquatic settings where they don't flourish. Many parasitic flatworm species—the **flukes** and **tapeworms**—have come to live in the moist digestive tract of complex animals. Adult parasitic flatworms lack external cilia. These parasites and the larger free-living flatworms must rely on muscles for creeping or for undulating motion. Accompanying this muscular activity has been greater nervous activity—and more nerve cells concentrated in the brain. Thus the flatworm brain hints at interesting possibilities, even though flatworms are so well adapted to their environments that further possibilities have not yet been explored.

Such finely tuned adaptations speak of other supposedly "simple" animals as well. We are, by the nature of our interests, unfolding a story of increased complexity in organ systems and their integration that ultimately assured our own evolution. But it's important not to lose sight of the many living things that do quite well without complexity. Consider another kind of worm, the **nematode**. There are perhaps *half a million* kinds of nematodes in the world, in more settings than any other animal group occupies. Countless species live as parasites inside plants and animals. Others are free-living forms ranging from deserts to polar snows to hot springs to the ocean depths. And are they abundant! A single rotting apple may house 100,000 of them; an acre of rich farm soil may contain up to 10,000,000,000 nematodes in its top few inches.

It isn't that nematode "complexity" assures their success; they are little more than a tube-within-a-tube. Although they do have an anus and a one-way gut, usually the only specialized region is a muscular, horny-toothed, or plated zone behind the mouth. Reproductive cells are crammed in between the gut and the body wall, where they function at a rapid pace to produce more nematodes. The nervous system is scarcely more complex than a flatworm's, and with only a few longitudinal muscles, nematodes move about only in an awkward thrashing way.

Figure 18.6 Body plan of a nematode.

Their secret, it seems, is the way they can resist so many environmental insults. Nematodes are covered with a remarkably tough cuticle. They can survive astonishing levels of acidity or alkalinity, terrible temperatures, and noxious compounds. Sometimes they live for hours in fixatives and preservatives that would instantly kill other animals being prepared for microscopic examination. One kind even thrives and reproduces in the digestive juices of the carnivorous pitcher plant. Nematodes are incredibly resistant to suffocation. If oxygen is present, they use it; if it is not, they switch to anaerobic pathways. We should be so blessed, especially during smog alerts. In fact, if we continue to pollute our environment in every way imaginable, we can be sure the nematodes will be among the last to suffer.

Two Main Lines of Divergence: Protostomes and Deuterostomes

As successful as nematodes have been, they appear to be an evolutionary side branch in terms of potential for diversification. In contrast, two other groups derived from ancestral flatworms have been remarkably successful in this respect. They differ from each other in the way the second opening to the gut is formed during embryonic development. In the simpler **protostomes**, the first opening formed becomes the mouth; the anus forms later. The group includes the annelids (such as earthworms), arthropods (such as insects), and molluscs (such as snails and squid). In the simpler **deuterostomes**, the first opening becomes the anus and the

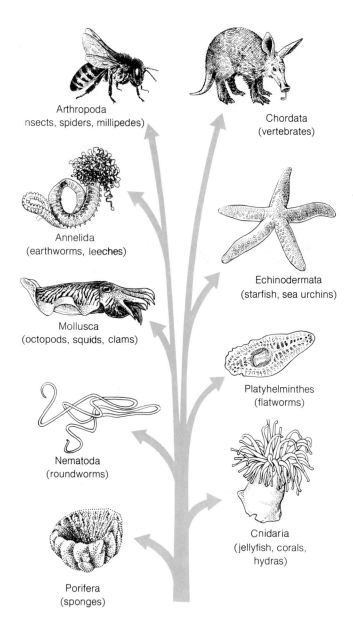

Arthropoda
(insects, spiders, millipedes)

Chordata
(vertebrates)

Annelida
(earthworms, leeches)

Echinodermata
(starfish, sea urchins)

Mollusca
(octopods, squids, clams)

Platyhelminthes
(flatworms)

Nematoda
(roundworms)

Cnidaria
(jellyfish, corals,
hydras)

Porifera
(sponges)

Figure 18.7 Phylogenetic tree of animal evolution.

second, the mouth. Deuterostomes include the echinoderms (such as sea stars) and all vertebrates—from fish to amphibians, reptiles, birds, and mammals.

In itself, the distinction between protostomes and deuterostomes is trivial, but it helps us identify two major lines of animal evolution that diverged long ago (Figure 18.7). Although many of the structural adaptations of these

two lines appear to be strikingly similar, they are evidence of parallel solutions to common challenges, not evidence of some recent common ancestry. Thus, even though both beetles and reptiles evolved on land, they produced different sorts of legs. Even though butterflies and birds took to the air, they did so with different sorts of wings.

Profound Adaptations in Worms Called the Annelids

The Coelom: Cradle for Internal Development One of the lines that diverged from ancient flatworms came to differ from them in some important ways. These were the **annelids**, the true segmented worms. They were probably the first major group of protostomes to appear. They had a long, straight gut, with a mouth and an anus. And between the gut and body wall, they had a fluid-filled space, the **coelom** (Figure 18.8), divided into segmented compartments.

Without a coelom, the ancestral flatworms could not have a vigorously active gut or complex internal organs. Between their gut and outer body surface, they were solid with cells—so whenever the gut churned to digest food, whenever body muscles contracted for movement, all cells were subject to physical stress. When organs became suspended in a coelom, the digestive system could get down to the serious business of churning and grinding up a variety of foodstuffs, without tearing the body apart in the process. *But a coelom not only insulated various internal organs from the stresses of body movement; it also permitted them to be bathed in liquid, through which nutrients could diffuse inward and waste products could diffuse out.*

Circulatory Systems Had the coelomic fluid remained the only medium for exchanging materials, annelids might have had limited success as a group, for distribution of materials *within* the wall of the gut or outer body would depend on the slow process of diffusion. But early in annelid evolution, a **circulatory system** appeared: through epithelial–muscular tubes, fluid containing nutrients and wastes traveled from the body's surface to its inner regions, then back again. In this system, the circulating fluid (blood) provided a means by which materials could be efficiently exchanged between internal and external environments—*which provided the potential for larger and more massive bodies.*

In the annelid system, there is no true heart. Blood is kept circulating in one direction because of forceful contractions in certain muscular vessels, and because of one-way valves in some of the larger vessels (Figure 18.8). Smaller vessels leading away from the big longitudinal vessels carry blood to cells of the gut, nerve cord, and body

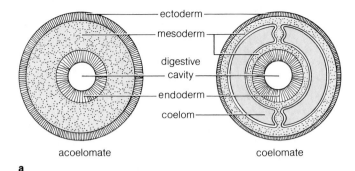

ectoderm
mesoderm
digestive cavity
endoderm
coelom

acoelomate   coelomate

a

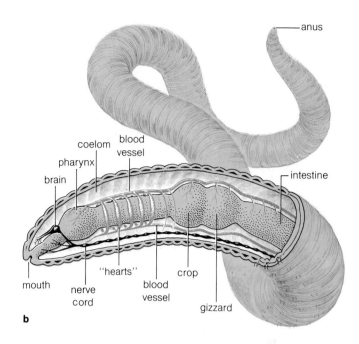

anus
coelom
blood vessel
pharynx
brain
intestine
mouth
nerve cord
"hearts"
blood vessel
crop
gizzard

b

Figure 18.8 (**a**) Comparison between acoelomate and coelomate body plans, shown in cross-section. In acoelomates such as flatworms, mesoderm forms a solid matrix between the outside layer and the digestive cavity. Coelomates such as earthworms have a central fluid-filled, epithelium-lined cavity in which various internal organs are suspended.

(**b**) The segmented body plan of the common earthworm, showing its true digestive, nervous, and circulatory systems suspended in the coelom.

(**c**) Except for the first three head units and the last tail unit, each body segment contains two coiled excretory organs called *nephridia* (singular, *nephridium*). The earthworm nephridium may resemble the ancestral structure from which the vertebrate kidney gradually evolved. Each is composed of a tubule system—around which fine capillaries are intertwined—and a storage bladder. One end leads from the coelomic cavity of the body segment in front of it, and the other opens to the outside. As body fluids pass through this system, water and solute levels as well as metabolic wastes are regulated, with some materials probably being selectively reabsorbed by the capillaries, much as they are in the vertebrate kidney. (After Gardiner, *Biology of Invertebrates,* McGraw-Hill, 1972)

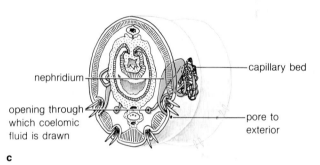

nephridium
capillary bed
opening through which coelomic fluid is drawn
pore to exterior

c

wall. In the blood of some annelids, we see hemoglobin. Hemoglobin dramatically increases the blood's oxygen-carrying capacity, which increases the capacity for cellular respiration (Chapter Six). *And with its high energy yield, cellular respiration means greater potential for muscular activity.*

Body Segments and the Fine Art of Walking In annelids, we also have the first signs of **segmentation**: each repeated body segment has its own flexible wall (a cuticle) surrounding its own watertight coelomic chamber. The chambers may be repeated over and over, thereby forming a long, large body. And various segments, especially those near the head, may be modified in ways that make the body increasingly adapted to specific environments (Figure 18.8).

Each flexible segment has its own set of paired antagonistic muscles and a set of bristles. Waves of alternating

contractions and expansions of circular and longitudinal muscles pass through the series of segments. With these waves, bristles on the contracted segments grip the ground, which keeps the body from slipping back (Figure 18.9). Thus the earthworm advances with far more coordination than the thrashing nematodes. More than this, in many annelids the bristles are embedded in fleshy, paddle-shaped lobes (*parapodia*) projecting from both sides of the body wall. Figure 18.9 shows one example of this development. In some annelids, a third set of muscles, which run obliquely to the others, extends out to the lobes. And by alternating contractions of oblique muscles, the lobes can be swung up, forward, down, and backward. *Thus, in the annelids, we see for the first time paired appendages and a mode of walking forward.*

We also see a new level of sophistication in nervous systems. Without precise control over the activation of circular, longitudinal, and oblique muscles, there would be

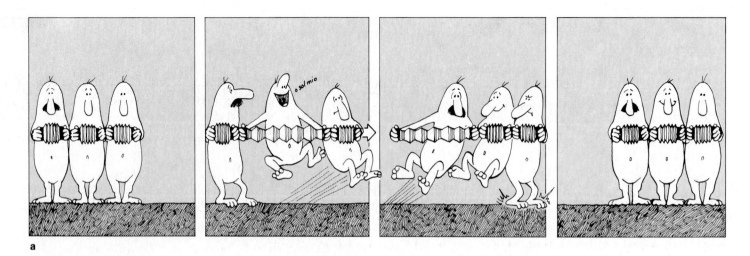

Figure 18.9 (**a**) Principle of forward movement in the earthworm, based on alternating expansion and contraction of body segments through the use of antagonistic muscles, and on anchoring the contracted sections with good, strong grips on the ground. (**b**) Puzzling *Peripatus*, with features suggesting evolutionary links to annelids *and* arthropods.

anarchy among segments and the body would get nowhere. But a double nerve cord travels the length of the body, coordinating all movements. In each segment, the cords are swollen with cells controlling local activity. And the brain, located in the head, serves as a center for input from eyes and other sense organs.

Hinged Suits of Armor for the Arthropods

The cuticle covering each annelid body segment is thin, strong, and flexible. It is well adapted for a life-style that depends on being able to bend each segment. But through chance mutation, one or more lines began that had an increasingly hardened cuticle. Because its bearers were more protected from predators, they had a selective advantage. We see this pattern of cuticle thickening in their descendants, the **arthropods.** Each body segment is covered with a rigid "shell." Yet the cuticle between certain segments remains pliable, thereby providing "hinges" for movement.

Only the most minor rearrangement of muscles would have been needed to accompany the change in segmentation. As in many annelids, some long muscles could run continuously between adjacent segments. Thus, contraction of a single longitudinal muscle on one side of the body would bend the body in that direction. Through contractions of appropriate muscles, the body could be bent up,

down, or to the right or left at the pliable boundaries between segments. More than this, lateral appendages could become segmented into hard sections, with flexible regions in-between spanned by oblique muscles. These modifications would be enough to give rise to the jointed leg, which would be capable of far more movements than the simpler, soft appendages from which they were derived. (The word arthropod means "jointed leg.")

The arthropod body plan, with its unique hardened cuticle on the outside (an **exoskeleton**) and its network of muscles connecting the segments, obviously demands a more sophisticated nervous system of control. Thus it is no surprise that, although the arthropod nervous system closely resembles that of annelids, the brain is often larger and more highly developed.

In the seas, the group of arthropods known as crustaceans (crabs, for instance, and lobsters) have come to be well represented. But it is on land that the arthropod potential for variation is most pronounced. We can assume the hardened exoskeleton of arthropods was first selected for as an effective means of defense, then capitalized on for locomotion on the floors of ancient seas. But it also turned out to be an adaptation that would be of considerable survival value when some of these animals moved onto land. *For the arthropod hardened exoskeleton was not only a superb barrier to water loss; it was adaptive also in the sense of providing support for a body deprived of water's buoyancy.*

The first land arthropods were the arachnids. Their descendants (spiders, scorpions, ticks, and mites) are still with us. They were followed by centipedes and millipedes.

b

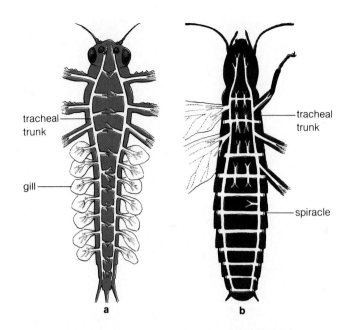

Figure 18.10 Tissue adaptations allowing for gas exchange in arthropods. (**a**) Many aquatic insects have closed tracheal systems, which extend into lateral body wall extensions that serve as gills. (**b**) In land arthropods, tracheal systems open to the outside through tubes in each abdominal and thoracic segment. (After Gardiner, *Biology of Invertebrates*, McGraw-Hill, 1972)

And from centipede-like ancestors, the insects evolved. The transition to land required still another adaptation. Like many other animals in the seas, the larger marine arthropods extract oxygen from the water by using **gills:** thin tissue flaps that protrude into the water and are richly supplied with blood vessels. In ancestors of land arachnids, gills gradually came to be enclosed in a chamber below the cuticle, where body fluids could keep them moist. This chamber for gas exchange was the world's first **lung.** In some arthropod stocks, another means of breathing developed. A series of branching tubes, called **tracheae,** evolved as routes for air flow in all directions through the body. These fine tubes are stiffened with chitin, which keeps them from collapsing under the pressure of the body's weight. *The success of arthropods on land depended not only on the jointed exoskeleton but on a series of tissue adaptations that led to a system for taking in oxygen.*

It is among the insects that tracheal breathing reaches its peak of development. The tracheal tubes branch into ever smaller tubules that terminate next to interior cells; each cell in the insect body has its own supply of fresh air! As a result, insect cells are capable of the highest metabolic rates known—rates that underlie the development of the powerful, tiny muscles required for insect flight. With the potential for flight, insects began their explosive radiation into almost all environments. A distance that might take a spider a lifetime to traverse could be spanned in minutes.

An insect wing is not a modified leg, as it is in birds. Rather, it has developed through modifications of a flap of the body wall, and no muscles thread through it. In some insects, muscles are attached only to the base of the wing and pull on it, leverlike, to make it pivot around the point of attachment. In many other insects, the wing is hinged to body wall plates so that when tension distorts the body wall, the wing flaps. In both cases, the wing moves in a figure-eight pattern that not only lifts the insect but propels it forward as fast as 48 kilometers (30 miles) an hour! The rate ranges from about 4 beats per second for a butterfly to 1,000 beats per second for a mosquito. Insect flight is accomplished with a tiny system of nervous control that is amazing for its capacity to do so much with so little. Each cell has a rigidly defined role, so there is no room left for learned behavior. Insect flight, as well as the rest of insect behavior, is with few exceptions preprogrammed (Chapters Thirteen and Twenty).

Flexible Flaps for the Molluscs

The ancestral annelids gave rise not only to the arthropods, they or their predecessors gave rise to still another successful group: the **molluscs,** an estimated 50,000–100,000 spe-

a — Douglas Faulkner

b — Douglas Faulkner

c — Lynn M. Stone/National Audubon Society/PR

d — Roger K. Burnard

e — Roger K. Burnard

cies of clams, oysters, abalones, limpets, snails, slugs, squid, octopus, and their diverse kin. *If the arthropod plan was based on hardening the body wall, then the molluscan plan was based on keeping it flexible and emphasizing its musculature.* The surface in contact with the sea floor became thicker and more muscular, until it became a large "foot" for creeping about. On the backside, a flap of body wall—the **mantle**—was formed. Molluscs have come to differ from one another in the ways

the mantle has been put to use. In several molluscan lines, the cells of the mantle secrete substances which, together with various mineral inclusions, come to form a rock-hard shell.

In chitons, limpets, and snails, the shell is a protective shield from all but the most persistent, clever, and forceful predators. But there is a price to pay for protection: the shell is so heavy that it slows movement literally to "a snail's

f

g

h

Figure 18.11 Marine arthropods: (**a**) a banded coral shrimp with delicate sensory structures, (**b**) two spiny lobsters vaguely reminiscent of Las Vegas chorus lines, and (**c**) a male fiddler crab, displaying his oversized claw as a warning to intruders.

Land arthropods: (**d**) a stunning *Cecropia* moth, (**e**) a many-footed millipede, (**f**) a grasshopper with powerful, chevron-striped hind legs, (**g**) two dragonflies mating, and (**h**) a spider on moss.

All photographs this page Roger K. Burnard

pace." Other molluscan lines, most notably the squid and octopus, have opted for speed rather than protection. Over time, their shell has been reduced to a tiny internal "cuttlebone" or has been left behind altogether. Concurrently, the mantle has come to be used for movement. It has been modified into a muscular tube that can be contracted with great force, thereby emitting a powerful stream of water in one direction and propelling the body in the other.

Thus freed from the weight of a shell, and equipped with a jet-propulsive mantle, the squid has become the most magnificent swimmer of the invertebrate world. Some species are among the largest marine animals and are preyed upon by giant sperm whales. Squid more than 18 meters (60 feet) long have been observed. Large sucker scars on whale bodies suggest there may be squid of even more monstrous dimensions in the deep oceans.

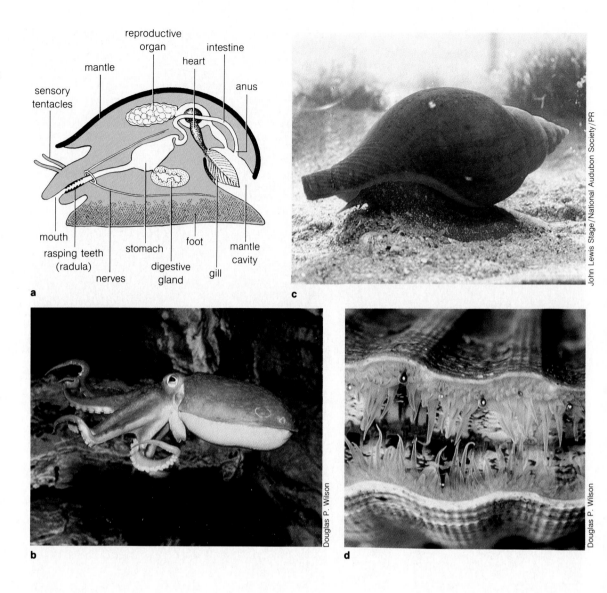

Figure 18.12 Representative molluscs. (**a**) Generalized body plan for molluscs. All have a flexible mantle, and many have an external shell. None has an internal skeleton. In some forms such as snails, movement occurs by wavelike contractions beginning in the back of the foot and continuing forward. In other forms, such as squid, water jets are expelled forcefully from the mantle cavity, which propels the animal in the direction opposite the jet stream. (**b**) Octopus, a cephalopod whose relatives include the chambered nautilus (Figure 1.4f) and the giant (40-foot-long) squid. The nervous system and sensory organs (eyes especially) are highly developed in these fast-swimming predators. (**c**) Whelk, a marine gastropod whose relatives include land snails and slugs. (**d**) Feeding structures and eyes along the shell margin of a scallop, whose relatives include mussels, oysters, and clams.

Figure labels: sensory tentacles, mantle, reproductive organ, heart, intestine, anus, mouth, rasping teeth (radula), nerves, stomach, digestive gland, foot, gill, mantle cavity

Both squid and octopus have amazingly well-developed eyes. And their nervous systems represent the peak of development in the invertebrate world. The massive brain of these animals approaches those of mammals in size and complexity. As in mammals, their brain includes a cortical region capable of storing information about experiences, which can be used to modify subsequent behavior. Their complex behavior patterns are not rigidly determined by inheritance but are capable of being modified considerably through learning and memory. When compared with vertebrates, the squid and octopus are a clear example of **parallel evolution:** because of similar evolutionary pressures, the entirely separate vertebrate and cephalopod lines developed similar structures, having similar functions.

Everything But the Kitchen Sink for the Chordates

Compared with protostomes, the branch of the animal kingdom in which we reside is far less exuberant in its diversity. Only a small number of deuterostomes have managed to survive to the present. Of those, only two groups are significant. They are the **echinoderms** (sea stars, sea urchins, and related forms) and the **chordates** (including fish, amphibians, reptiles, birds, and mammals).

The first deuterostomes may not have been much different from the most primitive flatworms. It seems likely they used cilia for swimming about. With a one-way gut

All photographs Douglas Faulkner

sieve plate
(through which
sea water
filters in)

water-vascular
system

extended

retracted
tube foot

a

b

c

d

e

Figure 18.13 Representative echinoderms ("spiny-skins"), marine animals that may be distantly related to chordates. (**a**) Brittle star, (**b**) sea urchin, (**d**) crinoid, and (**e**) cobalt sea star. Almost all echinoderms have an endoskeleton of hard plates, which often bear spiny projections (hence the name). All go through a free-swimming, bilaterally symmetrical larval stage. But adult forms generally show radial symmetry. (Their radial symmetry may be a later evolutionary development, arising when free-swimming ancestral forms gradually became adapted to life on the sea bottom.) Their unique way of moving about is based on constant circulation of sea water through a water-vascular system of canals and tube feet (**c**). Each tube foot leads from short branches of radial canals. One end is a muscular bulb, the other is suckered. Contraction of the bulb wall forces water into the tube part, making it rigid and extended. Coordinated extension and retraction of many suckered tube feet (combined with directed contraction and relaxation of muscles in the tube foot walls themselves) enable the echinoderms to move body parts in specific directions, and to become attached to rocks, gravel—and prey.

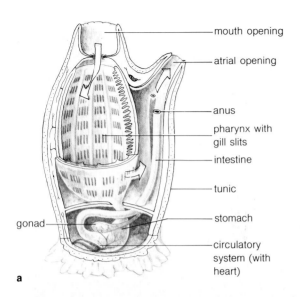

mouth opening

atrial opening

anus

pharynx with
gill slits

intestine

tunic

gonad

stomach

circulatory
system (with
heart)

a

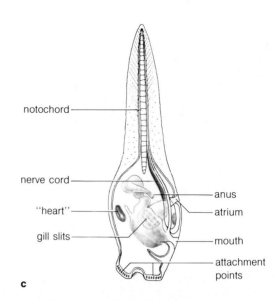

notochord

nerve cord

"heart"

gill slits

anus

atrium

mouth

attachment
points

c

b

Douglas P. Wilson

Figure 18.14 Cutaway view (**a**) and photograph (**b**) of adult tunicates, or "sea squirts." The larval form is shown
in (**c**), although it is not drawn to the same scale. (Drawings after W. D. Russell-Hunter)

and a swimming life style, these organisms would have been on the way to the development of bilateral symmetry, and to the concentration of sensory structures and nerves about the mouth. But in at least two deuterostome lines, further development of the head was arrested when they moved down an evolutionary road to a sedentary life-style. During the Precambrian, when predatory animals were appearing in greater size and numbers, these deuterostomes developed heavy, defensive armor of mineralized plates embedded in the body wall. But heavy armor not only helped keep them from being eaten, it forced them to settle to the bottom. These were the first echinoderms (the name means "spiny-skinned"). Ever since, echinoderms either have been living attached to or have been slowly lumbering over the ocean floor, much like miniature Sherman tanks. For them, bilateral symmetry no longer offers selective advantage; and a pattern of radial symmetry like that of the bottom-dwelling coelenterates has emerged. But extensive modification associated with a sedentary life is not restricted to echinoderms. Similar processes apparently took place among other primitive deuterostomes, and among the most primitive of all living chordates: the tunicates.

From Notochords to Backbones As tiny larvae, **tunicates** look and swim like tadpoles. They are bilaterally symmetrical, with a front end and a tail end (Figure 18.14). They have a nervous system with a nerve cord running the length of the body. Running down the center of the tail is a support structure, a rod of stiffened tissue called the **notochord.** Muscles attached to the notochord contract rhythmically to bend the tail and propel the larva through the water. *As simple as the notochord appears to be, it represents a profound evolutionary step: it was the forerunner to the chordate's internal skeleton, or* **endoskeleton.**

For the tunicates, the promise of the notochord has not been fulfilled. After a period of swimming, the larvae settle down and become attached, at the front end, to the sea floor. There they undergo drastic metamorphosis. The tail, the notochord, and most of the brain are lost by the time metamorphosis is complete and feeding begins. The adult animal also becomes encased in a tough outer "tunic" (hence the name).

Tunicates practice **filter feeding,** whereby water drawn in through the mouth is passed out through a series of pores known as **gill slits** (Figure 18.14). Cilia lining the gill slits create a water current, and a sheet of mucus captures food suspended in the water, which is then passed into a gut for digestion. It is an efficient system, as long as plenty of food floats past. A critical moment in the life of a tunicate is settlement and attachment of the larva. Everything depends on the larva finding a suitable site. If the larva becomes attached to a spot unfavorable in any respect, it has blown its one chance. The adult form can't move on. Thus sedentary organisms of all sorts do one or both of two

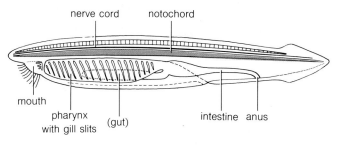

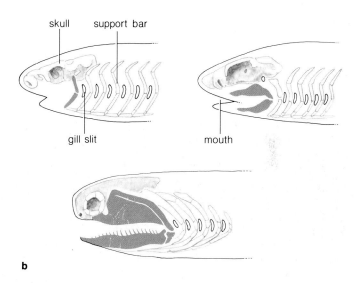

Figure 18.15 (**a**) Cutaway view of *Amphioxus*. (**b**) Suggested evolution of gill-support structures, as found in jawless fishes, into the hinged vertebrate jaw. (After J. Young, *The Life of Vertebrates*)

things: (1) they produce many offspring, so that at least some survive by attaching to suitable sites; and (2) they produce larval forms with sensory structures that improve chances for detecting favorable sites.

Some zoologists see our ancestral beginnings in primitive tunicates. They readily picture ancient seas in which great numbers of larval chordates swam about to locate living sites. They picture further that there may have been great variation in how fast the larvae became attached to a substrate and metamorphosed into sedentary adults. Rare individuals may have failed to make the change altogether, thus retaining the option of escaping from a bad spot. Today, living in sea-floor sediments throughout the world are a few animals of this sort. They are the lancelets, or *Amphioxus* (Figure 18.15a).

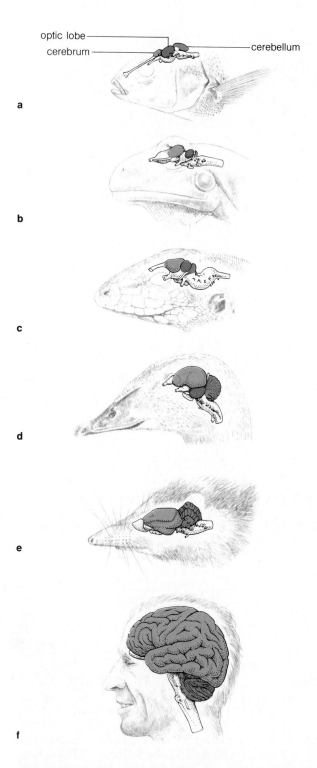

optic lobe
cerebrum
cerebellum

a

b

c

d

e

f

Figure 18.16 Trend toward increased neural complexity among five classes of vertebrates: (**a**) bony fishes, (**b**) amphibians, (**c**) reptiles, (**d**) birds, and mammals such as shrews (**e**) and humans (**f**). Compare these sketches with Figures 13.9 and 13.12, which detail the human brain and nervous systems. (After A. Romer, *The Vertebrate Body*)

a

b

Roger K. Burnard

Roger K. Burnard

Figure 18.17 (**a**) A green tree frog, representing the amphibian line of evolutionary development, and (**b**) a Florida alligator, representing the reptilian line.

Adult lancelets are fishlike; they retain a bilaterally symmetrical body and the capacity to swim about. And they keep the notochord. On locating what appears to be a suitable feeding site, they burrow into the sediments, leaving only the mouth exposed. Then they draw in water and extract food particles from it in much the same way as adult tunicates do. But if food doesn't float past, they can swim away in search of it.

Now consider what would happen if this trend went one step further. It is possible that the swimming forms sampled the sea floor as they moved about. And in drawing in the organic matter (microorganisms, wastes, and scraps) through their feeding currents, they would have hit upon a rich, underexploited resource. By retaining the larval form through adulthood, a roaming, searching life-style would be possible—which would give a selective advantage to those organisms tending toward advanced sensory structures and nerve specialization at the head end. At that point, the capacity to swim effectively through tides and currents would be more important. But better swimming abilities would require stronger muscles—and stronger muscles would be useless without stronger body parts to pull against. *Free-swimming chordates which, by chance mutation, came to have their notochord strengthened while remaining flexible would have a selective advantage—and it is through just such a development that vertebrates could have emerged.*

The first vertebrates were the **jawless fishes** in which the reinforced notochord was replaced with a series of hard bones, called **vertebrae.** These bones were arranged in a vertical column to form a backbone. These vertebrae had three unique features. *First,* the shape provided firm attachment sites for muscles running in several directions from the backbone to the body wall. *Second,* they fit next to each other with smooth joints, which allowed the backbone to flex and bend. *Third,* because they were hollow in the center, they served as a protective shield for the nervous system. Through chance mutation and selection, the brain could become protected by various hard structures which eventually became a **skull.** And gradually, additions and extensions to the vertebral column and skull could develop. For as soon as the capacity to make bone had appeared, natural selection could go to work on individual organisms that varied in the number, size, shape, and position of their bones. For example, once a series of bones had encased the brain, a piece of bone could develop beneath the mouth, becoming attached to the muscles controlling mouth movements. This feature is characteristic of the **true fishes:** vertebrate animals with jaws capable of biting, chewing, and grinding food, with the means to tear off chunks of almost everything edible in sight.

The development of formidable jaws surely was an evolutionary force of the first rank. As the number of these voracious fishes increased, there must have been selective advantage in being able to discern food in the distance, which could be pursued and consumed. Hence the origin of sensitive eyes, of a "nose" for detecting distant odors, of pressure sensors for detecting movements in the water. All such sensory structures were integrated with a brain and spinal cord of impressive complexity. *The trend toward increasingly specialized sensory detectors and neural integration among the true fishes prevailed through descendant forms that moved onto land—through amphibians, reptiles, birds, and mammals.* Figure 18.16 outlines this trend.

Or consider the more fundamental implications of an internal skeleton. The arthropod skeleton, recall, is a hardened external shell, so it is a direct physical constraint on growth. The only way arthropods can outgrow their shell is to shed it, which they do in a series of periodic **molts.** The skeleton splits and the animal crawls outside. Underneath is the complete outline of a new skeleton, which will become hardened by deposition of chitin. The arthropod typically expands greatly before hardening is completed. But in the meantime, it is vulnerable to predators and, without a hard skeleton for muscles to pull against, is somewhat clumsy. Needless to say, the number of molts is kept to a minimum, as is increased growth.

The arthropod skeleton imposes another constraint. As any animal doubles in length, the weight of the skeleton must be increased at least eightfold, for it must also become thicker as well in order to support the added body weight. That is why, thriller movies to the contrary, you will never

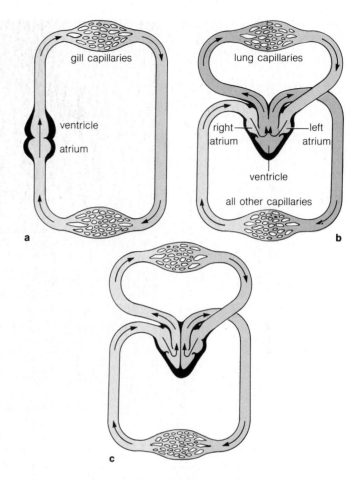

Figure 18.18 Generalized sketches of circulatory systems for (**a**) fishes, (**b**) amphibians, and (**c**) birds and mammals.

have to worry about being attacked by a 10-ton ant. It is only because of the potential of *internal* skeletons that the world has seen dinosaurs, elephants, and whales. The slender bones of the vertebrate endoskeleton contribute far less than exoskeletons to total body weight, even while providing strength.

Lungs and a Many-Chambered Heart Among primitive filter-feeding chordates, gills were devices for feeding and for respiration. Simple diffusion of oxygen into the body tissues, supplemented by some absorption through the gills, was enough for these animals. But with the increasing activity of the jawless fishes came the need for more muscles—and muscle cells had to be supplied with more oxygen. Thus, gills became more important in respiration. Blood pumped through the thin tissue flaps picked up oxygen from the water, then transported it to the rest of the

Figure 18.19 Feathers and fur—two distinct adaptations for temperature regulation in variable environments.

body. With the development of the jaw and the evolutionary lines leading away from filter feeding, the gills came to be devoted entirely to the task of oxygenating the blood. They still do this in all true fishes.

But gills must remain moist in order to function. Because they are located more or less externally, they quickly become ineffective when exposed to air. When the first vertebrates began moving onto land during the Devonian, their internal swim bladder was used to extract oxygen from the air. (Chapter Fifteen). Thus, in the moist and thin-walled air bladder, the air slowly gave up its oxygen to the surrounding blood and tissues. The vertebrate lung had appeared.

But what to do with the oxygen so received? First in amphibians and then in permanent land dwellers called reptiles, there must have been strong selective pressures for developing more efficient means of circulating the blood. For instance, even holding the body off the ground requires greater muscle activity, hence more oxygen. The forerunner of the amphibian and reptilian heart, as seen in fishes, was a rather inefficient pump having only two chambers: an atrium for receiving blood from the body, and a ventricle for sending it back (Figure 18.18). The blood left under high

pressure and moved rapidly because of the heart's contraction. But as it passed through the fine capillary mesh in the gills, it slowed down considerably. It was freshly supplied with oxygen as it left the gills, but it traveled sluggishly through blood vessels to the brain, muscles, and other tissues, and even more sluggishly on the return trip to the heart. Apparently the rate was enough for life in water; it was not enough for life on land.

In the earliest amphibians, a second atrium developed. It received oxygen-rich blood directly from the lungs, then passed it on to the ventricle. At the same time, the other atrium was receiving oxygen-poor blood from the rest of the body—and that blood, too, was being dumped into the ventricle. In that chamber, both kinds of blood were mixed. Some went back to the lungs through a network of blood vessels; some went back to the rest of the body through a separate network. In a way, the amphibian heart was more efficient than its predecessor. Blood flow from the heart was more rapid because it was being pumped directly, under high pressure, to both the lungs and the other body tissues. Of course, the mixing-up process meant half the blood reaching the lungs had just come from there; and half the blood sent to the body from the heart had failed to travel to the lungs (Figure 18.18). But again, it was fast enough for the slow-moving amphibians and reptiles.

A simple modification to this heart structure eventually appeared in two separate lines of reptiles, one of which was destined to give rise to birds and the other to mammals. For them, a partition now divided the ventricle into two chambers. One atrium received oxygen-depleted blood from all over the body and sent it to one ventricle, which pumped it to the lungs. The other atrium received oxygen-rich blood from the lungs and sent it to the second ventricle, which pumped it to the rest of the body. *What began as a single circulatory system among primitive chordates became, in birds and mammals, a pair of circulatory systems operating in synchrony.* Does it sound familiar? It should: it's the pattern of human blood circulation described earlier in Chapter Twelve.

On the Potential of Fur and Feather It was when the world climate began to grow cooler and drier during the Cenozoic (Chapter Fifteen) that the birds and mammals rose to dominance. Both these groups share one feature that undoubtedly has contributed to their continuing success: they have specialized adaptations that help them maintain a *constant internal temperature* even as the external temperature rises and falls. In almost all cases, life cannot exist outside a narrow temperature range: from about 0°C (the freezing point of water) to about 50°C (the temperature at which the bonds holding proteins in their three-dimensional shape break apart). This doesn't normally present much of a problem for life in the seas, in which temperatures generally don't vary by much more than 10 degrees over the year. Which is fortunate for fish: the same richly vesseled gills that take in dissolved oxygen from the water just as readily give up the heat of metabolic activity to it! Most fish simply have no internal mechanisms for coping with sudden changes in temperature. External temperatures on land are more variable. Some land-dwelling animals are more vulnerable to temperature extremes than others. Reptiles are much like fish in this respect. Although they are able to dissipate excess internal heat by evaporation, they must respond behaviorally to external energy sources to maintain high enough internal temperatures when outside temperatures drop. Thus a desert lizard spends much of the day moving in and out of the shade and changing its position with respect to the sun. If its body temperature rises above or falls below its range of tolerance, the lizard dies.

Birds and mammals have greatly extended their environmental temperature tolerance range through both internal and external adaptations. They both have special internal controls for increasing or decreasing metabolic activity. They also have various devices at the body's surface for increasing or decreasing heat loss. For example, complex controls governing the blood vessels in such exposed body parts as ears and feet can regulate blood flow, which influences how much body heat is lost to the

a

b

Figure 18.20 Representatives from two orders of mammals: marsupials and monotremes. (**a**) Kangaroo. For marsupials, there is a short embryonic development period in the uterus, then further development in the mother's abdominal pouch on the outside. (**b**) Duck-billed platypus. As with other female mammals, the female platypus secretes milk for offspring—but embryonic development proceeds in eggs outside the mother's body. Such forms are probably descendants of an early offshoot of the mammalian line.

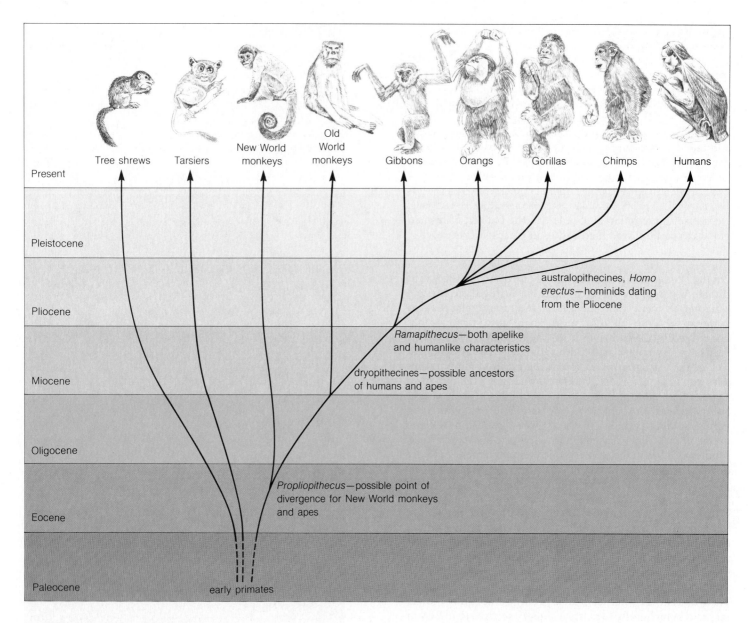

Figure 18.21 Family tree for primates.

Labels in figure:

Present · Pleistocene · Pliocene · Miocene · Oligocene · Eocene · Paleocene

Tree shrews · Tarsiers · New World monkeys · Old World monkeys · Gibbons · Orangs · Gorillas · Chimps · Humans

australopithecines, *Homo erectus*—hominids dating from the Pliocene

Ramapithecus—both apelike and humanlike characteristics

dryopithecines—possible ancestors of humans and apes

Propliopithecus—possible point of divergence for New World monkeys and apes

early primates

surrounding air. In most land environments, air temperature is usually below the temperature at which many living cells function best (37°C to 41°C), so heat must be generated from within. Thus birds and mammals must take in food on a fairly constant basis in order to maintain a high rate of metabolism. Beyond this, birds rely on the insulative properties of feathers; and mammals, of fur. With such adaptations, these two major animals groups have survived environmental extremes and have radiated into almost all the diverse climatic zones on earth.

The Human Heritage

In these few pages, we have spanned a tremendous range of evolutionary history. We have pieced together a picture of the flow of adaptations leading to nervous systems and brains; to bilateral symmetry and segmentation, foretelling of paddles, fins, and legs; to specializations in head, trunk, and appendages. We have brought other adaptations into the picture: internal circulatory systems for supplying ever larger multicellular bodies with nutrients even while removing metabolic wastes; backbones for strength and flexibility; internal temperature control to match the changing tests of the external environment. In doing so, we have watched the sprouting of the first metazoans and their subsequent growth over 2 billion years into the mammalian stem, one of many stems on the animal phylogenetic tree. It is here that we exist, one branch among a diverse array of other mammalian forms that have in common all these structural and functional antecedents.

With apes, monkeys, and prosimians, we share further adaptations that make us uniquely **primate.** We have a large cerebral cortex, highly integrated systems of muscles and nerves, and flexible shoulder joints. We have forward-directed eyes astonishingly good at discerning color, shape, and movement in a three-dimensional field; we have fingers and toes adapted for grasping rather than running. We have teeth of a varied sort, capable not only of tearing in the manner of jawed fishes but of piercing, crushing, and grinding. We give birth not to large litters but generally to only one offspring at a time.

When and where did these adaptations develop? Here the evolutionary story picks up with a small, perhaps insect-eating mammal that lived in tropical forests at the dawn of the Cenozoic, some 65 million years ago. It surely evolved from the ratlike mammals scurrying over the Mesozoic landscape. But its move to life in the trees called for grasping digits and flattened fingernails rather than pointed, ratlike claws. It called for greater hand-eye coordination and forward-directed eyes for stereoscopic vision, hence better depth perception. Climbing and, much later, reaching for and hanging onto overhead branches were adaptations that led all primate progeny to view the world from an upright position. Long before the striding, two-legged walkers called humans appeared, natural selection was testing out the adaptations of tree-dwelling primates that would make possible the eventual long-range migrations of the human species.

These tree-dwelling primates got around. By the Miocene, apelike forms known as the **dryopithecines** were well represented in parts of Africa, Europe, and Asia. Some seem to qualify as the stock from which modern humans and apes eventually arose. If anything, the dryopithecines were transitional forms. They were capable of climbing about in the trees, yet also capable of walking about on the ground—explorers, perhaps, ever ready to seek the protection of the branches above. Interestingly, tree-dwelling apes alive today are typically plant eaters, gentle in disposition and hiding in the forest canopy rather than venturing into the open. But apes living in open grasslands are of a different sort. They eat plants but they also eat meat—and they are far more aggressive. (For instance, a chimpanzee that is threatened or on the attack is quite capable of throwing stones and making a general ruckus.) In the dryopithecines, we have a possible antecedent for both.

Geologic upheavals of the late Cenozoic may well have been the environmental trigger for the sudden adaptive radiation of many primate lines out of the forests. By that time, Old and New World primates had already diverged and were going their separate ways even as the continents of Africa and South America were going their separate ways (Chapter Fifteen). But to the east in Africa, repeated crustal fracturing was creating the Great Rift Valley (Chapter One), flanking it with volcanic mountain ranges and carving it up with lakes, swamps, and rivers. Within this immense valley, a mosaic of geographically isolated environments emerged—cloud forests on wet mountain slopes, tropical forests, savannas, open plains, deserts. And if we know anything at all about environmental diversity, it is that it is paralleled by diversity in life forms. The microcosms of the Great Rift Valley surely fostered distinct adaptations in the small, scattered bands of early apes.

The fossil record in East Africa does in fact point to a very rapid evolution of apelike forms during this period. Isolation in one region beset by alternating wet and dry climates apparently led to the chimpanzees. Their half-arboreal, half-terrestrial way of life may well reflect the changing environments to which their ancestors had to adapt and readapt. Isolation in another region of continually warm, humid forests apparently led to the evolution of the gorilla. And in regions of increasing aridity, we have fossil signs of primates in the human family—the first **hominids.** They may have been already adapted to foraging on the ground for seeds, fruits, nuts, and the remains of small animals when they were forced to move into open grasslands. Were they already standing on two legs? Possibly. There would be a strong selective advantage bestowed on those forms which could stride about, with eyes on the horizons not only for potential food but for the carnivores—leopards, lions, tigers, hyenas, jackals—that preyed on them. More than this, an upright stance would free the hands for defense if one were caught out in the open. The first hominids never abandoned their foraging habits. But through use of hand-held objects—protoweapons, prototools—they also developed the capacity to hunt. Hominids had become both predator and prey.

The earliest indisputable hominids were the **australo-pithecines,** the "southern ape-men" that apparently emerged during the Pliocene (Figure 18.21). These forms at first had an average brain size no larger than a chimpanzee's. Somewhere around 2 million years ago, there were also members of the genus **Homo**—of which we are the only living representatives. Their brain size was intermediate between that of most australopithecines and **Homo erectus**—a later form with jaws and teeth not all that different from modern hominids. During the Pleistocene, at least two species seem to have been coexisting in the same regions. They could not have done so and remained distinct species unless they were not competing for the same things. One, derived from the early australopithecines, was a scavenger and gatherer, perhaps of seeds from wild grasses and trees. The other, *Homo erectus,* was becoming specialized as a tool-using food gatherer—and hunter. Perhaps it was that the smaller sized and smaller brained species came to be recognized as potential food by the hunters. Whatever the case, by 1.2 million years ago the scavengers had become extinct.

Homo erectus populations evolved past them, and not only in the physical sense. Their repertoire of cultural adaptations included clothing made of animal pelts and the use of fire as well as various types of weapons. With such adaptations they fanned out from Africa into Europe and Asia. All the while, their brain size became increasingly more modern as survival came to depend more on intelligence and the capacity to accumulate and transmit experience. The **Neanderthalers,** a distinct hominid population appearing about 100,000 years ago in Europe and Asia, were one of our closest relatives. They may have developed in the relative isolation of Europe, cut off from the rest of hominid evolution by the Pleistocene glaciations. They developed a distinctive and considerably complex culture. Their cranial capacity was indistinguishable from our own. Yet, about 30,000 years ago they seem to have disappeared. Did they simply merge with the overall human population when the retreat of the glaciers again permitted migrations, or were they overwhelmed in an early show of our capacity for genocide? We do not know. But by 30,000 years ago, there was one remaining hominid species—**Homo sapiens,** man the reasoner, self-proclaimed man the wise.

Perspective

Many thousands of years ago, as human bands radiated across the continents, they began their increasing manipulations to bring both themselves and their environment closer to their visions of a better life. As visions evolved, so did these dispersed human groups evolve in culturally divergent ways. Even while geographic isolation led to different adaptations in superficial physical traits, so also did it lead to the divergence of habits, opinions, adornment, songs, words, foods, and myths. All these evolving forms of distinct behavior might be called the beginnings of cultural speciation. But biological speciation did not follow in its wake. Physical barriers proved permeable, gene flow between populations continued despite cultural barriers, and we have remained one.

Too often we view culture as an inviolate gift separating ourselves and others—with "others" meaning not only members of regionally distinct human groups but all nonhuman forms of life. What is it about culture that makes us think it comes mysteriously into being on its own? Culture is no more and no less than a culmination of the long evolutionary processes we have been describing throughout this book. It is an expression of a trend toward behavioral diversity and versatility that appears to have begun in the Precambrian seas. Culture did not come suddenly into being with *Homo sapiens.* In the most basic sense, it is simply an ability to accumulate experience, to modify present and future behavior in terms of the past, to learn from the environment and anticipate it. Since the development of the first nervous systems, animals of many sorts have been traveling the route toward increasing reliance on learned behavior. You will be reading about some of these animals in the next unit. For them, each increment in the capacity for learning has proved adaptive and has been selected for. Long before the appearance of humans, other animals developed the capacity for local traditions, so to speak—for behaviors learned by direct observation and imitation, then transmitted by repetitive behavior in overlapping generations of social groups. Such traditions are, in a sense, protocultural.

We have evolved far beyond that point, of course, for we have developed the capacity to communicate with symbols. We combine and recombine thoughts in a dream-like interplay of images in our head. Some we translate into artifacts as lasting as the pyramids, others as fleeting as words uttered into the wind. It is a remarkable degree of behavioral evolution we share with no other life form—but it is, in the final analysis, an outgrowth of structures and processes we share with others traversing the long road of animal evolution. It is a magnificent heritage.

Recommended Readings

Campbell, B. 1974. *Human Evolution: An Introduction to Man's Adaptations.* Second edition. Chicago: Aldine. Marvelous comparative treatment of hominid form and function; well-written and certainly provocative.

Romer, A. and T. Parsons. 1977. *The Vertebrate Body.* Fifth edition. New York: Saunders. About the best book around on the subject.

Russell-Hunter, W. 1968. *A Biology of Lower Invertebrates.* New York: Macmillan. An excellent little paperback covering biological unity and diversity from coelenterates to molluscs.

———. 1969. *A Biology of Higher Invertebrates.* New York: Macmillan. Same skilled writer, this time exploring the physiology and adaptations of animals ranging from annelids to sea squirts.

Unit V

Life in a Changing Environment

Chapter Nineteen

Figure 19.1 European storks on the wing to Africa—destination of two billion birds of many sorts during their annual migratory journeys.

Roger K. Burnard

Individual Behavior in Space and Time

During their individual journeys through time, all animals begin with a similar set of challenges. Each must find food and other vital resources on a fairly regular basis. At the same time, of course, each must avoid becoming food for some other animal. And if it is to leave offspring, each must be responsive to certain internal and external cues, acting in concert, that will help assure reproductive success. Food, danger, potential rivals and potential mates—the directed responses an animal makes when confronted with such environmental stimuli are called forms of **behavior.**

But "behavior" is not all that easy to define. For one thing, the range of behavioral responses is simply astonishing. A response may be as simple as orienting the body relative to a source of light energy; it may be as complex as the seasonal migrations that take vast populations of birds from one continent to another and back. For another thing, there is a range of variation *within* the same category of response! The intensity of a given response, how rapidly it is carried out, and how long it lasts depend on the nature of external cues (which varies) and on an animal's internal state (which also varies). However, we can say this:

In the broadest sense, there are two aspects of behavior: one genetic, and the other learned.

No matter what the genetic program or the learned components may be—no matter how they may interact to produce intricate behavior patterns—the behaviors that endure are the ones that are adaptive.

Consider what this means. As a result of natural selection, the behaviors characteristic of each species have been finely tuned for life in a certain predictable setting. By virtue of birth, an animal normally finds itself in the kind of environment—pond, woodland, desert, shore—where its species has evolved. Thus it comes into the world with structural adaptations and a genetic predisposition toward certain behaviors that will enable it to survive there, just as generations of its kind have survived before it. In its search for food, the golden eagle of the American Southwest takes to the skies. It has a built-in structural and behavioral capacity to stay aloft silently, hour after hour, by gliding on outstretched wings; it has sensitive eyes that scan the sparse scrublands below. And structures and behaviors merge in the stunning power dive that sends it plummeting unde-

tected toward its prey. But say that same eagle is captured and released in the Amazonian rain forest. Wouldn't the dense tropical foliage hide potential prey from its view? And wouldn't the tangled foliage make a mess of the power dive? Most likely that eagle would perish, for adaptations to a certain range of environmental conditions have been stamped into the structural and behavioral fibers of its species, as they are generally for all species. The eagle, like every animal, has its own destiny in space.

Every animal also has its own destiny in time. An underlying genetic program carries it from development through maturation, through journeys of long or short duration, through feeding and resting and reproductive cycles characteristic of its species. Even before birth, its life is linked to the rhythms of day and night, sun and moon, the seasons, the years, to all the environmental fluctuations that mark its species' perception of the passage of time.

But the behavioral patterns dictated by the genetic program are not necessarily so rigid that they cannot change! Individuals of the species may well have the capacity to respond in novel ways to the unexpected. With varying degrees of success, behaviors may be matched to experiences in ways that are *not* anticipated by the genetic program but instead are new expressions of its potential. If such responses work, they may become embedded in an animal's memory—nongenetic adaptations, to be sure, but just as capable of promoting individual survival. Thus in this chapter, we will be looking not only at the genetic basis for individual behavior. We will also be looking at various ways in which behavior may be modified by experience.

The Genetic Basis of Behavior

There are, in the natural world, many patterns of movement that predictably run their course once some stimulus sets them in motion. They are built-in behaviors of all members of the species; they are *innate*, or genetically determined. For instance, a newborn chick confronted with small particles on the ground will spontaneously peck at them, even if it has never observed another chicken performing this decidedly chickenlike behavior. A newborn duckling raised in isolation and then shown mud for the first time in its life will spontaneously sift the mud through its beak, which is a ducklike feeding behavior. Innate motor responses that are linked to relatively simple environmental stimuli are called **fixed action patterns.** The stimuli that trigger a fixed action pattern, or even some part of it, are called **releasers.**

Assuming there is a genetic basis for the details of fixed action patterns, is there also some aspect of gene expression that influences *which* stimuli will serve as releasers for those patterns? Consider the common fruit fly, *Drosophila.* Reproduction does not occur automatically any time a male and female *Drosophila* happen to meet. The male first must perform an intricate courtship dance, which acts as a releaser for a fixed action pattern in a receptive female. Then the female assumes a certain position that releases a fixed action pattern in the male, a pattern required for successful copulation. Now, different populations of this species have slight variations in mating behavior. A female normally responds most readily to the dance of a male from her own population, not to the dance of a male from a more distant population. When they mate, the female donates a set of genes whose structural and functional expressions somehow affect her preference in the mating dance. And the male donates a set of genes whose structural and functional expressions represent the details of the dance she preferred. The female's selective mating probably increases survival chances for the next generation. Why? Because her mate belongs to a population that has been molded, generation after generation, to a certain environment. Thus her preference for him enhances the likelihood that their offspring will receive precisely the collection of genetic traits that are most adaptive for the place in which the offspring are destined to live.

We can make another important point about the genetic basis of behavior. *As with all heritable features, innate behavior may be modified through gene mutation.* The variant behaviors promoting survival and reproduction tend to be favored—and the ones that don't tend to be eliminated. *Drosophila* provides us with a dramatic example of the way natural selection may act to eliminate maladaptive behavior. A recently discovered mutation in one strain of *Drosophila* has been given the colorful and descriptive name "fruity." A male that is heterozygous for this mutation behaves the same as wild-type males. But a male that is homozygous for the fruity mutation displays bizarre mating behavior: he ignores females and goes into a frenzied courtship dance for every male he meets. In this case, the fixed action pattern is normal, but a genetic mutation has modified the way stimuli release the fixed action pattern! Because the fruity male is doomed to die without leaving offspring, the mutant behavior is quickly lost from the population.

Not many mutations have such obvious and direct effects on reproductive success. But virtually every modification of behavior will have some long-term effect on the probability of the individual reaching reproductive age. Those individuals who reproduce determine the genetic composition of the next generation. Thus, in each inter-

Figure 19.2 A male finch fluffing his feathers not because it is cold outside but because a receptive female finch is on the perch.

breeding population, there is continual selection for those genetically controlled behavior patterns that enhance reproductive success in a specific environment.

Over time, the forces of natural selection modify innate behaviors, sometimes restricting them and other times extending the circumstances under which they are released. For instance, a single behavior pattern may come to be released by more than one environmental stimulus. Feather fluffing in birds is an example. This innate behavior has its origins in a physiological advantage: feather fluffing helps the bird keep its body temperature constant as the outside temperature drops. But somewhere along the line, feather fluffing has undergone divergence in function. Not only is it used to retain body heat, male birds of many species use it in both courtship and territorial displays (Figure 19.2). By fluffing their feathers, the males create a larger silhouette. We can well imagine that the first bird to be genetically programmed to fluff its feathers on sight of another member of its species was somewhat more effective in turning off a potential rival and turning on a prospective mate. If that were true, he would have increased the probability that his genes—including those governing the expression of his novel behavior—would be passed on to the next generation.

This example emphasizes an important point about the evolution of behavior. It is not the animals themselves that consciously bring about innate behavioral diversity. Rather,

---

Diversity of innate behavioral responses evolves through the action of natural selection; it is based on the increased reproductive success of individuals whose genetic make-up caused the behavior to be released by additional stimuli.

Control and Modification of Fixed Action Patterns

Given the fantastic range of responses, we cannot help being impressed with one of the paramount features of behavior in many species—its *plasticity*, the capacity to be modified to meet new environmental conditions. Plasticity of response is adaptive to changing times and circumstances. Even in supposedly simple animals, the same stimulus doesn't always trigger the same response. The response may be modified by secondary factors and by the recent past history of the individual. For example, substances found in the flesh of all animals will, when squirted in the water near a hungry sea anemone, elicit a vigorous feeding reaction. But the same substances squirted near a well-fed anemone elicit an avoidance reaction. The internal state of the animal influences its responsiveness to external stimuli. The animal somehow "stores" events of the recent past and accordingly modifies its responses to new stimuli. This physiological capacity to store information is the basis for memory and learning (Chapter Thirteen).

There has been a tendency to assign somewhat mystical names to the internal states regulating such innate responses as feeding behavior. Thus we hear about "motivations" of animals, or the "drives" that compel them to eat, mate, and fight. For the most part, such names are misleading in that they imply a kind of psychic control, which certainly hasn't been demonstrated. What has been demonstrated repeatedly is this:

Innate responses are under the control of specific brain centers and endocrine glands, which are themselves responsive to various physical and chemical aspects of the internal and external environments.

The "hunger drive" has been studied in laboratory rats. If a region on the lower surface of a rat's hypothalamus is destroyed surgically, the rat overeats so much that it becomes outrageously fat. This region must be a "satiety center" controlling the rat's hunger. Normally, a high glucose level in the blood triggers the satiety response because glucose molecules bind to receptors in that region. But when the satiety center is not functional, feeding is not turned off as it usually would be when enough food is eaten to meet normal body needs for glucose. This satiety response is highly specific to glucose. In one experiment, glucose that had been slightly modified was injected into a rat. It could still bind to cells in the rat's satiety center, but the modification was enough to prevent the triggering effect, and the rat would not stop eating.

Other brain centers are known to regulate such behaviors as sex and aggression. Even if an animal observes a potential mate (or rival), it will not respond with sexual behavior (or with aggression) if stimulation of the corresponding brain center is artificially blocked. Yet artificial stimulation of the appropriate brain center has little effect unless the animal does observe a potential mate (or rival)! Analysis of such brain centers is far from complete. But it strongly suggests that "motivation" will become understandable as interactions between the neural and hormonal activities that regulate an animal's responsiveness to external signals.

The activities of brain centers regulating innate behaviors may themselves be a product of natural selection. We can readily envision how different species may have come to display characteristically different patterns of activity. For instance, how often have you spent a day at a zoo and noticed that large cats—lions, leopards, jaguars, pumas—usually are lethargic, but that most of the wild dogs—wolves, cape dogs, coyotes—usually are pacing back and forth for hours on end? What ecological and evolutionary factors might have given rise to this innate behavioral difference between the two types of predatory carnivores?

Learned Behavior

In many species, some innate behaviors show little capacity to be modified. As long as a fixed action pattern contributes to adaptability, selection tends to favor individuals in which that behavior is rigidly locked into the genetic program. But say an individual has the genetic potential to modify a fixed action pattern. If the modification makes it more successful than its relatives at finding food, escaping from predators, or producing offspring, then its plasticity of response will be selected for.

If such modifications are more than one-time responses to a new cluster of stimuli—*if there is an enduring potential for adapting future responses as a result of past experience*—then the ability is called **learning.** In a constant or highly predictable environment, rigid fixed action patterns are adaptive for many animals; hence those patterns generally tend to persist. But in a changing and unpredictable environment, the capacity for behavioral plasticity and learning tends to be fostered.

Learned behaviors range from simple modifications of fixed action patterns to the complex insight learning of some primates, especially humans. However, much of learned behavior is still a puzzle. One problem is that learning has generally been studied not in the natural environment but in experimental situations. Because the natural setting contains many variables operating at once, it is not likely that the "pure" learning responses observed during experiments exist so crisply in the wild. We will survey some aspects of learning here, but it's useful to keep in mind that there are many elaborations on these simple processes.

Figure 19.3 Two kinds of large, predatory carnivores characterized by two kinds of activity patterns during the day. How might natural selection have shaped the difference in these innate behaviors?

The classic work of Ivan Pavlov, a Russian physiologist who was interested in the secretion of digestive juices, grew into one of the first controlled studies of learning. Pavlov noted that his laboratory dogs salivated immediately after he placed a meat extract on their tongue. This he interpreted to be a simple reflex response (Chapter Thirteen). But then he found that if he rang a bell just before giving the dogs the extract, the dogs began salivating at the sound of the bell alone. Pavlov called this new response a "conditioned reflex," for the dogs had come to associate the sound of the bell (a conditioned stimulus) with the taste of food (the reinforcing stimulus). This behavior, whereby a connection is made between a new stimulus and a familiar one, is a form of **associative learning.**

Why is associative learning adaptive? If an animal learns to anticipate certain events, it can respond more rapidly to them when they do occur. If the sound of an airplane engine always precedes shotgun blasts from the hunter who picks off animals from the air for the fun of it, the arctic wolf that learns to respond quickly to the sound of an approaching airplane may have a better chance of surviving the mindless slaughter than its less adaptive relatives.

Another form of associative learning is "instrumental conditioning." Here, a reinforcing stimulus (reward or punishment) appears *after* a behavioral response is given. The animal learns by trial and error. Earthworms show learned behavior in tests with a simple T-maze (a maze with a base and two arms shaped like a "T"). The earthworms enter the maze at the base, and if they turn down one of its arms, they encounter an irritating stimulus (say, an electric shock). If they turn down the other arm, they encounter a moist, darkened chamber. After many trials and enough shocks, they generally learn which turn leads to the more hospitable ending.

In the forms of learning just described, the behavior persists as long as the reinforcement persists. But if the reinforcing stimulus is withdrawn (say, if the bell is rung again and again without being followed by the taste of food), the learned behavior soon becomes extinguished. This, too, is a learning process; it is called **extinction.**

An interesting finding of maze studies is that some animals show a **latent learning** ability. A rat will learn its way through a complicated maze more quickly if it's first given a chance to poke about, even without any reward when it happens to find the way out. Such exploratory behavior is popularly called "curiosity." But what is actually causing the rat to explore? And without reward, why does it remember what it discovers? We can speculate that because "curiosity" leads to learning about the physical nature of the environment, it must be adaptive. Such learning might include knowing about a good hiding place, which might well mean the difference between life and death, given an unexpected encounter with a hungry predator. Thus there would be gradual selection for latent learning ability. (Of course, exploratory behavior may also leave an animal dangerously vulnerable, and it will be selected for *only* if the rest of the behavioral potential of the individual allows it to cope with what it happens to come across while poking about!)

It is only in some primates that **insight learning** has been adequately demonstrated. With this behavior, alternative responses to a situation are first evaluated mentally. The evaluation may lead to a sudden connection of separate bits of knowledge, which tells the animal what the appropriate response might be. Thus insight learning is a "trial-and-error" learning process that goes on in the brain, a putting together of accumulated experiences to try out in new situations.

As we look at examples of learned behaviors in this chapter and the next, keep in mind that our separation of learned from innate behaviors has been done to simplify the discussion. In the natural world, a behavior pattern may have an intricate mix of innate *and* learned components, and it often is impossible to make the kinds of distinctions we have here.

Nature Versus Nurture: Is Innate Behavior Learned?

As biologists in the first half of this century explored one form of learning after another, they began to wonder if *all* behavior is learned. Maybe even the simplest patterns of movement were not innate but instead were the result of trial-and-error learning by the embryo as it develops in the egg or womb. In one of the first tests of this notion, frog embryos were raised in water containing an anesthetic, which deprived them of all sensory stimulation and prevented all of the normal squirming motions a tadpole makes as it develops within an egg. When the tadpoles hatched, the anesthetic was removed. And the tadpoles swam away and began feeding! Their fixed action patterns were exactly like those of normal tadpoles. Similar studies of other species confirm that some forms of behavior are indeed innate.

As another example, a newly hatched chick always flaps its wings together but always moves its legs one after the other. To find the basis for the different movements in these two sets of limbs, Viktor Hamburger and his coworkers operated on a two-day-old chick embryo. They transplanted a piece of neural tube from the chest to the abdominal region, and vice versa. When nerves formed and nerve processes grew out from the neural tube many days later, nerves that normally would have developed in the legs actually developed in the wings, and vice versa. At hatching time, the chick was incapable of normal movement. Its legs moved forward and backward at the same time, hence walking was impossible. And its wings could not be flapped together at the same time. Clearly the patterns of limb movement are rigidly fixed in potential nerve cells long before nerves or limbs form. *When* limbs are to be moved is partly subject to learning; *how* they move is not.

Selective Responses to Environmental Stimuli

No matter where we are, there are myriad chemical, auditory, tactile, and visual cues in the environment even though they are not all obvious to us. Worm-eating snakes can smell an earthworm at some distance; robins can hear an earthworm under the ground. We can do neither. Nor should it be otherwise: *To survive, different animals must respond only to a limited number of stimuli in the total environment.*

If we or any other animal were able to tune in on all environmental stimuli at once, we would waste time and energy filtering stimuli that have no adaptive value for us. Besides, an animal with too many alternatives at its disposal is more likely to suffer from sensory overload. The more sights, sounds, touches, and smells bombarding its brain, the less decisive its responses will be. A predator is not likely to wait for prey to stop and figure out which stimulus is the most significant (the fangs, you dummy, the fangs!). Certain stimuli herald food or shelter, danger or safety, predator or mate, but they must be perceived selectively.

The degree of development of an animal's sensory receptors corresponds to the requirement for selective perception. Natural selection is constantly testing for variations in perceptive abilities that might enhance survival, even as it by-passes or eliminates those that are not. For instance, most birds have little or no sense of smell. They live in an open-air environment in which odors disperse quickly, so there is little or no selective advantage to an acute sense of smell. On the other hand, birds have complex visual receptors for locating food from a distance, through three-dimensional space. Thus, in approaching the behavioral responses of any animal, we must keep its sensory filtering systems in mind, for they largely dictate what behaviors we will see.

Tactile Stimuli Tactile stimuli are among the signals that animals receive from the external world. For instance, embedded in the skin of most animals are sensory cells keyed to touch, pressure, and movement. One tactile stimulus that commonly evokes feeding behavior is direct contact between predator and prey. Another is the vibration that travels from a web-entangled and struggling insect to the receptors on the feet of a spider lurking in anticipation of the meal. Tactile stimuli are important in locating shelter. The hermit crab will crawl up to an empty snail shell and explore it with its claws to decide if it is a proper house. Once satisfied, the crab deserts its old shell and takes up residence in the new one. The dances of honeybees on the vertical comb of a hive (Chapter Twenty) normally are performed in the dark, so tactile signals must be at least one way that dance messages are communicated to other

Figure 19.4 Tactile stimuli as an important means of communication between mother and infant baboon. As you will see in the next chapter, much of baboon society depends on the fine art of body language.

Timothy Ransom

members of the hive. Among birds and mammals, tactile stimuli help maintain social bonds. Especially among primates, tactile interactions seem to promote group cohesion (Figure 19.4).

Chemical Stimuli All animals have **chemoreceptors**—sensory cells that directly or indirectly transform chemical stimuli impinging on them into nerve impulses. Two main kinds of chemoreceptors, those of taste and smell, are found throughout the animal kingdom, especially among insects and most vertebrates. They are of adaptive advantage in finding food, in avoiding nonfood, and in avoiding becoming food. Some are also used in locating and courting a mate.

In the marine environment, for instance, sharks are notorious for the way chemoreceptors lead them to slight traces of blood in the water. In the air, molecules of chemical substances attract flies to flesh and moths to flowers. On land, an ant extends its stinger intermittently to lay down a chemical trail that leads other members of its nest to a food source. When the food dwindles, the chemical message quickly vanishes because ants cease to reinforce it, and the search goes on elsewhere. Chemicals used in this way, as behavioral releasers for some target organism, are called **pheromones.**

In many insect species, pheromones function as powerful sex attractants. Even though they usually go undetected by members of other species, they are practically irresistible to members of the species producing them. Because they have the distinct advantage of being so highly selective in their action and so innocuous as far as other species are concerned, pheromones are finding some use as lures in pest detection and control programs. They are used to detect whether a certain pest (such as the Oriental fruit fly) is beginning to invade a region (such as the citrus belt of Southern California, where it could cause heavy damage to citrus crops). Pheromone traps are set out in strategic places, such as around airports and shipping ports through which overseas pests may inadvertently be introduced. If invaders are detected before they become widespread, effective pesticides can be applied exactly where needed rather than having the entire region blanketed with chemicals. Such programs are aimed at minimizing the massive ecological disruption that occurs with indiscriminate application of chemicals such as DDT and malathion (Chapter Twenty-Two).

One problem is that many naturally occurring pheromones are difficult to isolate from insects in large enough amounts to permit extensive use. Chemists have been working to develop synthetic pheromones that appeal to such insects as the crop-destroying Oriental fruit fly and the gypsy moth. Work is proceeding slowly in many cases, though, for the life cycles and behaviors of many insect pests are not yet understood in detail—and pheromones usually appeal to the target insect during only one stage of its development. But application of synthetic attractants as they are developed, in combination with ecologically sound practices (restricted chemical applications, crop rotation, diversified crops), does show promise.

Auditory Stimuli Sounds are also important in the way many animals respond to the environment and communicate with one another. In the sea, certain fishes grunt and chatter; the humpback whale emits eerie "songs" that travel

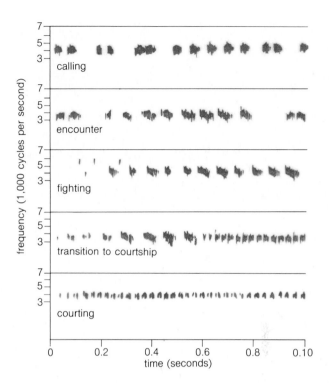

Figure 19.5 Repertoire of sounds of the cricket *Telogryllus commodus.* Because insects are tone deaf, the signals are distinguished on the basis of volume and rate of emission. (From the work of Richard Alexander, University of Michigan)

hundreds of kilometers through the deep. On land, the use of sounds is richly varied, but always in adaptive ways. Consider the diversity in the duration, pitch, intensity, and complexity of bird songs. Even with their variation, all these songs fall within a common range of frequency. If we assume the use of sound in communication is an adaptive behavior, then we should be able to find some factors in the environment that have led to its evolution. The presence of a predator in the environment is likely to elicit a series of loud, short calls of varying frequency. The calls serve as warnings, and it is difficult for the predator to locate their point of origin. In forests, high-frequency sounds are restricted because they are scattered by trees and dense foliage; they seem to be used among members of the same species for locating one another over short distances. Volume, too, depends on circumstance. Loud sounds can attract predators. They also can intimidate potential rivals of the same species and can be used as a beacon for guiding lost offspring back to their parents. Bird song is one of the most complex of all forms of auditory communication, especially the identification songs of male birds as they stake out territories during the spring.

Figure 19.6 Fish-eating bat of Central America. Here, a captive bat was trained to trawl in a tub of water for submerged bits of fish. It could not detect a totally submerged tidbit. But when fine wires projecting above the water's surface were used to mark the food's location, the bat quickly learned to find it in the dark. Such experiments suggest the fish-eating bat hears echoes from parts of fish that protrude above water. The original pulse and its returning echo must diminish so much in intensity at the water–air interface that underwater parts can't be detected.

Not all auditory stimuli are vocal. Arthropods such as crickets create distinctive mating songs by rubbing frictional body parts against resonant body parts. This behavior is called **stridulation.** Cicadas vibrate membranes to produce sounds. Male gorillas thump their hands against their chest to make unmistakable warning signals. Objects in the environment, too, are used to create auditory signals: beavers slap their tail on the surface of a pond, and woodpeckers communicate by drumming away on dead trees.

One of the most dramatic uses of sound is **echolocation:** the generation of sound waves as a way of orienting toward or away from objects. Bats, porpoises, and some birds and insects echolocate. Donald Griffin proved that bats, which feed at night, orient themselves spatially by emitting high-frequency sounds (which our ears can't detect) and then listening to the echoes of those sounds off objects in their surroundings. The bats could avoid flying into vertical wires strung a foot apart in a darkened room—but they could not do so when their ears were plugged. Again we find that the perceptions of other animals don't necessarily correspond to our own perceptions of the world.

Visual Stimuli Of all external stimuli, the most important include those sensed by **photoreceptors.** It's probably no accident that the narrow portion of the electromagnetic spectrum animals can perceive as light (Figure 5.5) is the same portion that readily penetrates water—the environment from which all life is thought to have evolved. Through the properties of light, photoreceptors detect not only color and brightness but information about shapes, space, depth, and movement.

In Chapter Five you read that plants are selective in the wavelengths of light they use. Similarly, photoreceptors of animals respond differently to different wavelengths. In dimly lit environments, at dusk or at night, some animals have the capacity to perceive light from the red (longer wavelength) portion of the spectrum. Most insects can respond to light ranging from ultraviolet to red-orange, but many are most sensitive to green or blue-green light (probably because that's the color associated with their food). Honeybees can perceive ultraviolet light, which we cannot do. By photographing white flowers in ultraviolet light, it's possible to show that many actually have intricate visual patterns for attracting bees.

As you will read in the next chapter, visual stimuli function in the intraspecific behavior of many birds and mammals—in courtship, territorial, aggressive, and cooperative behavior, and in the raising of young. Natural selection, it would seem, has been at work to develop elaborate visual displays of the sort typified by the male peacock. This bird has a truly long, cumbersome tail whose only function seems to be to dazzle the female with vibrant colors and patterns. Vivid examples are also found in predator-prey relationships. Capitalizing on the properties of light, predatory animals have developed complex visual systems for capturing light reflections that tell of form, position, and movement of prey. And because natural selection does not operate in a vacuum, the prey in their turn have acquired devious and equally complex ways of avoiding detection. For instance, with **cryptic coloration** and **cryptic behavior,** some animals take on not only the appearance but the "behavior" of other objects in their environment (Figure 19.7).

Behavior Through Time

So far, we have looked at primarily the stereotyped behaviors of an animal as they occur at any given moment. But if we are to understand behavior, we must also look at behavior through time. As you will now read, even fixed action patterns vary because of past events and future events that are anticipated by the development program of the species.

Critical Periods in Behavioral Development For many animals, certain behaviors can be learned only during certain brief stages of development. Because the time for learning these behaviors is so limited, they are called **critical**

Figure 19.7 Cryptic coloration and cryptic behavior. (**a**) To confuse the bird predator, one katydid species has become so adapted to its background that its members look like a leaf—even down to blemishes and chewed-up parts! (**b**) Caterpillars of some moth species tend to look like bird droppings because of their coloration and the body positions they assume. (**c**) By lurking motionless against its like-colored backdrop, then lunging at the appropriate moment, the yellow crab spider secures dinner. (**d**) What bird? With the approach of a potential predator, the American bittern stretches its reed-colored body and thrusts its beak upward—and even sways gently. like the surrounding reeds in a soft wind.

a   b

(vertical credits: Nina Leen in Animal Behavior, *Life Nature Library; F. Schutz)*

Figure 19.8 (**a**) Human imprinting objects. No one can tell these goslings that Konrad Lorenz is not Mother Goose. (**b**) A sexually imprinted rooster wading out to meet the objects of his affections. During a critical period of the rooster's life, he was exposed to a mallard duck. Although sexual behavior patterns were not yet developing during that period, the imprinting object became fixed in the rooster's mind for life. Then, with the maturation of sexual behaviors, the rooster sought out ducks, forsaking birds of his own kind, and lending further support to the finding that imprinting may be one of the reasons why birds of a feather do flock together.

periods in development. Consider the process by which newly hatched ducklings come to recognize members of their species. If they are exposed to a large moving object shortly after hatching, they tend to show an enduring preference for it. Normally, of course, the object is the mother duck. But experiments show these birds will also attach themselves to artificial models of various sizes, shapes, and colors—even to human models (Figure 19.8). This preferential behavior toward an object presented during the critical period is called **imprinting.**

Effects of Stimulus Deprivation on Human Development Humans appear to have an important learning period during their first two years of life. It had been known for some time that stimulus deprivation during early postnatal life leads to severe, irreversible learning deficiencies in laboratory animals. But it was not until Rick Heber's prolonged Milwaukee Project that stimulus enrichment during the first two years of human life was found to be crucial. Early enrichment programs have more beneficial effects than such extended programs as Operation Headstart, which normally have been initiated during later periods of development.

In this study, a test population of socially underprivileged children was divided into two matched groups. The children were raised in their own homes. But Heber and his researchers visited the children in one group for intensive, individual stimulation from birth to age two. During the visits, they held the children, showed them illustrated books, sang to them, and so forth. (The researchers had no contact with the other group.) When this sensory enrichment program was introduced early enough, the improvement was dramatic. The average "IQ" for children in the test group soared some twenty to thirty points above the control group by the end of preschool. And the greatest difference did not even show up until the children had become teenagers. Around puberty, when the nervous system undergoes final maturation, the differences in learning ability and achievement between the two groups was indisputable. It is postulated that during early postnatal life, connections are made in the brain that do not become functional until much later. But in the absence of rich and varied stimuli during the early learning period, these connections simply are not made.

Effects of Learning on Developmental Programs If all components of a fixed action pattern show up in all members of a species, does it mean none of those components is learned during development? Not necessarily. The general structure of the chaffinch song, which has three parts, is imprinted on male chicks when they are a few weeks old as they hear adult male chaffinches singing. But they don't join the chorus until the next spring, about eleven months later, when they set out to establish territories of their own. During this critical spring period, the young birds develop the song by imitating other males in the neighborhood. Although the general song structure is similar from bird to bird, local "dialects" exist. And these are the versions that are learned and will be sung year after year, no matter how many versions are heard after that.

Similarly, the manner in which some birds react to the edge of a cliff varies with experience. Normally, newly hatched gull chicks withdraw from the edge of small cliffs near the nest, a response with clear survival value. As they mature, however, they readily sit by the edge and don't even hesitate to jump off. But this behavior is not age·dependent. In one experiment, gull chicks were reared on flat ground, and when they were older they were released near a small artificial cliff. Their response was the same as that of newly hatched birds. Only with experience was there a change in their willingness to approach the cliff's edge, let alone jump from it.

Cyclic Behavior Patterns During at least one stage of development, all animals move about in response to the environment—in search of food, protection, a mate, a place to give birth, a place to die. These movements may be individual interactions with the environment, with nothing more than a short-term response to changing conditions. They also may be intentional movements of individuals or populations over great distances. We could give many examples of these movements through time. But here we will focus on perhaps the most dramatic of all: the **migration,** the cyclic surging of whole populations away from one region toward another that beckons with compelling force—and then a returning to the place from which the journey began.

Migration is a predictable behavioral aspect of many species. Crustaceans of the open ocean heed a recurring call to enter water of different depths. Insect populations may show migratory patterns, but the life span of most insects is so short that they may not be able, as individuals, to complete the round trip. Most likely their offspring are the ones to journey home, thereby giving the illusion of cyclic migration. The monarch butterfly (Figure 19.10) is a notable exception. Certainly many fishes show migratory behavior. Each year, newly hatched salmon leave the headwaters of their mountain home for the rich feeding grounds of the open seas, sometimes thousands of kilometers away. They live for several years in the open seas, feeding and developing to sexual maturity. Then, in the spring of their last year, they leave the seas and return to their birthplace, where they will reproduce and die.

Hormones trigger their quest, and production of those hormones is stimulated by the lengthening days of spring. Once migration is set in motion, chemoreceptors function in

Figure 19.9 Oh, the advantages of recognizing cliffs, especially when you are born and raised on the edge of one.

Figure 19.10 Monarch butterflies at a recently discovered wintering grounds in Mexico. A long-term study with thousands of tagged individuals proved populations of these insects do indeed make the immense round trip from North America each year.

Edward S. Ross

Birds have a high rate of metabolic activity. Because they burn up energy quickly, they must always have a reliable food source. Few environments are so productive that they yield abundant food all year long. With the onset of winter, for example, food supplies dwindle in the north temperate regions, where many migratory birds reproduce. Even though birds show effective adaptive behaviors to changing temperatures, the simple fact remains that the food must be there for birds to survive. That requirement alone would be enough of a selective factor over time to compensate for the dangers of their ambitious undertakings—flights over open oceans, through fog and storms, with predators along the way. But with the return of spring, they are rewarded for making the return trip by finding breeding grounds rich in new plant growth and relatively free of predators.

Generally speaking, *annual migrations involve not only environmental events but physiological events that depend on the function of endocrine glands, especially the pituitary.* The long journeys demand high energy expenditure. Accordingly, there is a change in physiology that consists of an accumulation of energy reserves (fat) and the ability to use it rapidly. And just before the onset of migration, molting occurs: new, stronger feathers replace old ones. There is also a hormonal link between migration and reproduction, as Albert Wolfson demonstrated in his studies of the western junco and the white-crowned sparrow. Interwoven with that hormonal activity is at least one environmental factor: the cyclic changes in the length of daylight. Such changes are called **photoperiodism.** Wolfson exposed birds to artificial periods of long and short days. When exposed to a short-day cycle, the birds lost weight and their gonads regressed. But on a long-day cycle, the opposite occurred. It is now thought that such environmental cues as increasing day length somehow affect the pituitary through the intervention of the pineal gland, which is light-sensitive. Wolfson's cytological studies support this idea. He found that during the breeding season, the Golgi complex of pituitary cells enlarges and is involved in intense glandular secretions. And under artificial light conditions simulating the lengthening of days, the Golgi complex responds in the same way.

Once their migratory journeys begin, birds use external cues and internal mechanisms to guide themselves to their destination. It may be that latent learning plays a part in how birds find their way about. In their **homing behavior,** birds learn to recognize familiar landmarks in the area around the nest. Homing behavior is limited for nonmigratory birds; few can find their way back if they are taken very far away from their home range. For migratory birds, the range can be astonishing. The literature is filled with such examples as the attempt to remove albatrosses from their home on Midway Island in the South Pacific Ocean. Because the birds were interfering with military planes landing and taking off there, they were transported to potential new homes as far away as North America. One

the salmon's unerring ability to return from its journey, following a trail of odors specific to its home stream and never forgotten. Other environmental cues—water currents, temperature gradients, sounds of waterfalls and rapids— may also function in salmon migration. But how they function remains largely unexplained.

Among bird populations, migration takes on impressive dimensions. Many bird species make an annual round trip between winter feeding grounds and their breeding range. Birds have the advantage of flight, which means they have a built-in physiological capacity to traverse vast regions quickly. Their physiology might help explain why they undertake their immense migrations in the first place.

albatross released in Washington State refused to be displaced by the United States Armed Forces. It returned to Midway in a little over ten days, having averaged 317 miles a day for the 3,200-mile journey—a journey never before undertaken and most of it occurring away from any landmarks, over open ocean!

Underlying short-range and long-range homing behavior is a form of **piloting**—an orientation toward home by topographic or meteorological cues that a bird seeks out through either random or systematic explorations. These might be river systems, forests, or landforms such as mountains or coastlines. They might be prevailing winds, air turbulence over various landforms, or even the cyclic shifts in the temperature and humidity of air masses throughout the year. Perhaps one of the clearest examples of environmental cues comes from Gustav Kramer's studies of caged starlings. He proved these day-flying birds use the sun's position as a compass. Throughout most of the year, his caged starlings fluttered randomly. But in the spring and autumn, their flutterings became oriented in the direction characteristic of migration for their species. Even when the sides of the cage were covered, so that landmarks could not be seen, the starlings still oriented in the right direction—as long as the cage remained open to the sky. When the cage was entirely covered, their flutterings became random again. Kramer then built a cover for the cage that had shuttered windows on its sides. Next to each window was a mirror, which could be moved to shift the angle of incoming sunlight. When the mirrors were shifted so that the light entering the cage was deflected 90 degrees, the starlings changed the orientation of their flutterings according to where the sun *appeared* to be, rather than where it really was.

But these studies raise a perplexing question. If starlings navigate by the sun's exact position, how do they take into account the variations in the sun's position in the sky—not only with the seasons but with the time of day? Because starlings migrate for at least six hours during a day, they somehow must compensate for the sun's shifting position. It seems they have some internal mechanism—a "clock," so to speak. A **biological clock** enables animals to correlate rhythmic changes in the environment, and to change their behavioral response accordingly. Evidence for biological clocks is fairly widespread throughout the animal kingdom. Again, there seems to be a connection between the endocrinological system, light stimuli, and migratory behavior—between some internal clocks and the external world. These and other parts of the migratory puzzle aren't easy to sort out. Like so much of animal behavior, migration is replete with exceptions to the general rule. Variations on the innate disposition to migrate arise from physiological differences among species, from different timings of migration, from varying environmental conditions, even from variations among individual birds. So far, *all that can be said is that the pituitary is the key to several aspects of cyclic migration, but it is a master key that fits an almost infinite number of locks.*

Perspective

In this chapter, we touched on the links between the physiology of individual animals and the external environment—on the union of stimuli and the sensory systems filtering them, the internal systems keyed to them. The directed responses to these stimuli are called forms of behavior. The examples we used covered both the genetic and the learned aspects of behavior. Two major points have run through the discussion. First, all behavioral responses have their basis in the way an animal perceives its environment: in the limitations and the potential of its physiology. Second, all the diverse ways of perceiving and responding to external cues have a common source. Each provides a selective advantage that in some way helps an animal escape predation, find food and mates, or survive changing environmental conditions. The cues may be chemical, tactile, auditory, or visual. They may also be endocrinological. But in all cases, cues are heeded in a *selective* way that enhances the individual's probability of surviving and reproducing.

In the next chapter, we will show how these stimuli and responses come together into the rich tapestry of behavioral displays. These displays are elaborate and often ritualized signals that animals use to communicate information. They herald a new level of organization in nature—the extension of the individual "need to survive" to the *cooperation* of individuals in the environment.

Recommended Readings

Animal Behavior, Life Nature Library. 1972. New York: Time-Life Books. Entertaining, well-illustrated introduction to behavior.

Dethier, V. and E. Stellar. 1970. *Animal Behavior.* Englewood Cliffs, New Jersey: Prentice-Hall. A good short book taking a neurological and evolutionary approach to behavior.

Dorst, J. (translator C. Sherman). 1962. *The Migration of Birds.* Boston: Houghton Mifflin. An excellent book, with descriptions of migrations that occur all over the world.

Heber, R. and H. Garber. 1975. "The Milwaukee Project." Chapter 19 in *The Exceptional Infant,* vol. 3: *Assessment and Intervention* (B. Friedlander, G. Sterritt, and G. Kirk, eds.). New York: Brunner/Mazel. Extraordinary study of the interplay between intelligence and environment.

Marler, P. and W. Hamilton. 1966. *Mechanisms of Animal Behavior.* New York: Wiley. More advanced reading, but one of the best treatments of the subject.

Chapter Twenty

''It's M//////////////////////////////////NE!''

The Social Basis for Population Structure

Each year, as biological alarm clocks go off more or less in inevitable synchrony, birds erupt in seasonal songs. Trills, screeches, whistles, crescendos, hoots—in this manner they announce their presence on strategic branches, fenceposts, treetops, chimneys, telephone poles. Through their individual ruckuses they are proclaiming title to food and space. But regardless of variations in number, kind, and volume of notes, all of these songs are meant to carry messages to prospective mates and prospective challengers. In their most fundamental sense, they mean the patterns of distribution for each feathered population are once more being renewed, tested, and redefined.

In turning to the manner in which animals are distributed in space, we must look not only to their interactions with the environment or with predators and prey. We must look also to their interactions with **conspecifics**—individuals of the same species. Underlying all such interactions are two opposing mechanisms. On the one hand, there are **dispersive mechanisms** for spacing individuals far enough apart to provide each one with enough resources for survival. On the other hand, there are also **cohesive mechanisms** for enabling individuals to come close enough together to mate (at the very least) and often for collective food gathering, mutual protection, and the rearing of young. The outcome of these dispersive–cohesive interactions among conspecifics is known as **social behavior.**

In this chapter, we will look briefly at the nature of social groups—how they can be shaped by individual distance and territoriality (dispersive behaviors) and by social bonds (cohesive behaviors). Along the way, we will look at some of the communicative displays that incorporate chemical, tactile, auditory, and visual stimuli in ways that are significant to the social group.

Selective Advantages of Simple Social Groups

If you have ever watched cars bunch up to a standstill on a freeway during rush-hour, you already know that not all groupings are social. In natural settings, too, simple aggregations can form that are nothing more than a response to some condition in the local environment. For example, if there is a moist and rotting log in a forest, wood lice tend to collect there—not because they thrive on one another's company but because they thrive in moist and rotting logs. As each individual wanders into the log, it begins to feed and its rate of movement slows to a virtual standstill. Hence it would be there even if there weren't another wood louse in the log.

Consider, by way of contrast, the true social groupings of emperor penguins in Antarctica. During the continuous night of winter, winds there can reach 140 kilometers (90 miles) per hour, and temperatures can plummet to −60°C. The penguins face these icy blasts in huddles of thousands of individuals. Their use of one another's presence to help break up the force of the winds and to keep warm enough to survive is a large-scale social behavior. But their social behaviors include more than group responses to the environment. They also include a true division of labor between parents. Parent penguins incubate their eggs during the long winter months, for their offspring must have time enough to grow and develop before entering the surrounding seas during the short summer, when food is abundant. Immediately after laying their eggs, all the females shuffle off to sea, which can be as much as 160 kilometers (100 miles) away. There they feed themselves during the entire incubation period. Meanwhile, back on the ice, each male is left holding an egg on his feet. They make do with what is available: feet. There are no grasses and twigs in Antarctica for building other sorts of nests. The males do more: they incubate the eggs by tucking them up into a warm body fold. But how do the males swim after food in the sea with an egg tucked under the tummy? They don't. They simply stop eating. The males live off their fat reserves until the females return, some two months later, to feed the chicks. Only then do they stagger off to find food.

As unique as the emperor penguin social group seems to be, it shares some basic features with social units found throughout the animal kingdom. For one thing, **pair formation** (in which a male and female join together to mate and perhaps share such tasks as food getting, nest building, brooding, and defense) is a common social unit. In some species, pairs last just through copulation or through the reproductive season. In other species they last a lifetime. Among fish such as the seahorse, it is the male alone that raises the young; among hummingbirds, wolf spiders, polar bears, squirrels, and deer, it is the female alone. But in many cases, both parents stay together to raise their young to

a

b

Figure 20.2 (**a**) A portrait of male and female wolf, paired in the cooperative venture of raising the young.
(**b**) Greeting ceremony typical of the wolf pack. Guess which one is the dominant male.

Figure 20.3 Defensive formation of a family grouping of musk oxen, on sighting their natural predator, the wolf.

maturity. Then the social unit may be composed of one or more parents and one or more sets of offspring; they are **family groupings.** Among species such as beaver, there may be as many as three sets of offspring living with the parents at any one time, which creates a rather large and heterogeneous family group. Among species such as wolves and lions, the social units are derived from family groups but the organization is a bit more flexible. Even as some offspring leave, young adults from other family units join up to form a relatively stable hunting unit.

Coherence of a family group bestows certain advantages on the young and thereby promotes their survival. The young are protected from predators and environmental extremes until they are mature enough to fend for themselves. More than this, they have their parents as models in learning where and how to obtain food and to avoid danger. Among certain migratory birds, adults know where to find a local feeding source along the route that a young, first-time traveler with the flock might miss. In effect, the prolonged association of adults and offspring extends the amount of information that can be transmitted to offspring. *Thus, in addition to inheriting innate behavior patterns, the young of a family group acquire a set of learned behavior patterns of proven adaptive value.*

Evolution of More Complex Social Groups

We can assume that under certain circumstances, the stable family grouping has offered enough advantages so that larger and more complex social units have evolved. Baboons, for instance, live and travel in large units (troops). All the females born into the troop stay on as part of the family, bearing young of their own when they reach maturity. But none of the young males stays on: eventually they all leave to join other troops. Sometimes the young males succeed in mating one or more times with the females before they depart, but often they do not. Sooner or later, the only adult males remaining to sire offspring are the ones that migrated in from other troops. This kind of structure offers both social stability (by the continued presence of females) and genetic diversity (by the continuing migrations of males).

The larger social units also include schools, herds, and flocks. They vary considerably in size, structure, and permanency. But they are alike in that they all extend selective advantages to the coherent family group. One of the most important of these advantages is a greater degree of protection from predators. For example, a herd of musk oxen (Figure 20.3) will cluster into a defensive formation around offspring at the sight of their natural predator, the wolf. Gazelles of the African savanna will thunder away in an undulating mass as lions launch their attack.

How did such defensive formations evolve in the first place? Usually, larger social units form in open environments—the arctic tundra, savanna, plains, open seas—where food resources are adequate for supporting large populations but where there simply aren't many places to hide. For lack of any other hiding place, it seems that cover is sought among other animals being preyed upon. Faced with such numbers, predators must attempt to separate potential prey from the periphery of the group. Falcons first make mock attacks against a flock of small birds, trying to separate one flying on the fringes of the group before attacking it (and thereby avoiding a possible mid-air collision). In the seas, the schooling behavior of one fish may be a way of taking cover by putting other fish between it and the predator. (Even fish that don't normally travel in schools will clump together in a tank in which there are no hiding places.) By virtue of sheer numbers, the individual apparently attains a measure of protection it would not have on its own. The potential predator usually is restricted in such cases to pursuit and capture of the very young, the weak, the old, or the injured individual that can't keep up with the group.

Both the individual members of a social group and the group as a whole are constantly being tested for their capacity to survive and reproduce in a given environment. As you will read in Chapter Twenty-One, built-in control mechanisms govern the actual size that any given group can maintain. *But within those broader environmental constraints, only those individuals having behavior patterns consistent with survival of the group are successful in their own survival and reproduction.* As a result, extremely complex group-oriented behavior patterns have tended to evolve.

In evolutionary theory, the trend toward complexity in social groups shows similarities to the trend that led to multicellular organisms. First there were casual associations of individuals of common descent. Such associations proved significant in the struggle to survive. They were fostered by natural selection, and the bonds between individuals grew stronger. But it was only with the division of labor that the complex kind of animal group called a **society** could have formed. It's true there is some division of labor within many smaller animal groupings. In a baboon troop, for instance, some individuals act as sentinels on the periphery of the main unit. There they must detect and alert the troop to impending danger. Other baboons occupy the center of the troop and concentrate on caring for the young. But it is among the social insects such as bees, ants, and termites that division of labor reaches the level of a complex society (Figure 20.4).

Insect societies show such specializations and interdependence of individuals that they take on the character of "superorganisms." Just as a single cell of a multicellular organism can't survive and reproduce on its own, neither can a single social insect survive and reproduce on its own. In each case, only one insect of the group becomes the

a

b

c

d

Figure 20.4 Life in an insect society. (**a**) The queen bee and her court. Attracted by pheromones, attendants constantly lick her body. The circle of attendants changes continually, which ensures that, through food transfer, the queen's pheromones are passed through the hive. Thus her presence is communicated to all bees in the colony.
(**b**) Transfer of food from bee to bee. Considerable antennae tapping during food transfer helps the bees orient and communicate.

(**c**) Bee dance. Forager bees returning from the field communicate the location of pollen and/or nectar by performing a dance for potential recruits. Together, the angle of the "wag-tail" part of the dance on the comb, its duration, and the sounds that the dancer emits communicate the food's location. Floral odors on the dancer's body also help the recruits find the right flowers once they arrive at the location indicated by the dance. Foragers carry pollen in "baskets" located on their hind legs. One pollen pellet is visible on this forager's leg.

(**d**) Guard bees. Worker guard bees assume a typical stance at the colony entrance. With front legs uplifted and antennae outstretched, they are alert and ready to "examine" all approaching bees.

(**e**) The queen, the single most important member of the colony. She is the only egg layer. During active brooding, she lays about 1,200 eggs each day. There is only one queen for each colony; normally she lives for about two years.

(**f**) The drone. The male member of the colony has no sting. His only function in life is to mate with a virgin queen. The matings take place during flight, with seven to ten drones mating with a single queen. Each mating takes but a few seconds, after which the drones fall dead to the ground. About 300 to 500 drones are found in a colony. Those which do not mate with the queen live about one to two months.

(**g**) The worker. The workers do virtually all the labor in the colony.

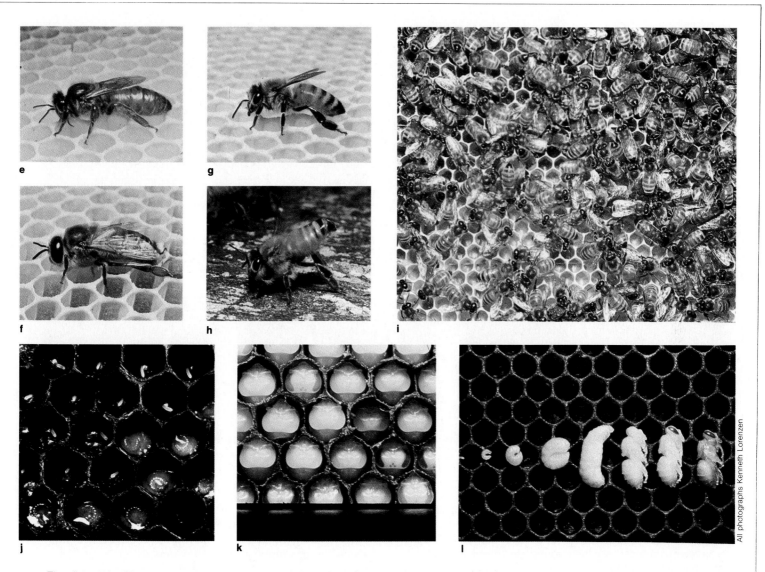

They forage, feed larvae, guard the colony, construct honeycomb, and clean and maintain the nest. In the event that the queen is suddenly lost, young worker larvae can be selectively reared into new queens. Between 30,000 and 50,000 worker bees are present in a colony. They live about six weeks in the spring and summer, and can survive about four months in an overwintering colony.

(**h**) Scent-fanning. As air is fanned, it passes over the exposed scent gland of a worker bee. Pheromones released from this gland help other bees orient to the colony entrance or to the queen during swarming.

(**i**) Worker bees on new honeycomb. Worker bees secrete wax, which is used in constructing honeycomb—the site of pollen and honey storage, and brood rearing.

(**j**) Eggs and young larvae. Like other insects, honey bees pass through several life stages prior to becoming adults. These cells have been exposed to show one- and two-day old eggs and larvae. The larvae are floating on royal jelly, a food secreted by worker nurse bees.

(**k**) Pupae. When cell cappings are removed, the uniformly aligned heads of worker pupae become visible.

(**l**) Developmental stages of the worker honey bee. The small egg is visible on the left. After three days, the larva emerges and grows rapidly for about six days. The larva then stretches out in its cell and becomes a ''pre-pupa'' for two days. The next nine days are spent as a pupa. As the pupa matures, eye pigments are the first to darken, followed by pigments in the rest of the body. After twenty-one days, the adult emerges from its cell, ready to take part in bee society.

Figure 20.5 (**a**) A stunning show of individual distance among a group of porpoise, fishing through the surf. (**b**) On shore, individual distance among gulls.

a

b

queen, the reproductive "organ" for all. But the queen depends totally on the other, nonreproductive members of the society for food, protection, and construction and maintenance of the nest in which young are raised. These nonreproductive members have unique structural and behavioral modifications.

For instance, some members of termite societies have structures and behaviors adapted for constructing and repairing the nest. Others, the "soldiers," have a long proboscis that secretes a sticky substance, which is used in defense of the colony. In some insect societies, the division of labor corresponds not only to structural traits but to the developmental stage. In the first ten days of its life, a worker honeybee (Figure 20.4) devotes itself to cleaning honeycomb. Only toward the end of this stage does it begin making short exploratory flights. In the next ten days it not only cleans the hive, it engages in construction activities and transports nectar being brought in by foragers to storage cells. From that point on—from about twenty-one days after birth until the day it dies—it joins the force of foragers collecting the pollen and nectar that sustain the society.

Such social groups are noted for their efficiency. When some members become specialized in food gathering, others in defense, and others in reproduction, there is little wasted energy in the group as a whole.

In highly integrated insect societies, natural selection must act through the cooperative behavior of all the individuals. Suppose a queen bee has genes that enable her to produce large numbers of highly efficient foragers. In itself, the potential advantage for the hive would be wasted unless she also produces enough worker bees that will automatically sink their stingers into invaders of the hive, even though their defensive behavior means certain death (their entrails are ripped out in the process). If the queen does not give rise to a diverse array of cooperative members, she herself probably won't survive to pass on her genes. Only if behaviors mesh into coordinated activity will they be perpetuated through the queen's offspring. Again we can draw an analogy with the cells of a multicellular organism. Skin and bone cells, for instance, can't function in the reproduction of a new individual. But the genes determining the adaptive characteristics of skin (toughness) and bone (strength) reside in and are perpetuated by the reproductive cells—which they protect and support.

Insect societies are the most extreme expression of cooperative behavior, for they represent total suppression of the individual. Far more common is some combination of competition and cooperation between individuals in a social group. This dichotomy is expressed in many ways, and helps determine both the spacing of animals within groups and the spacing of groups within the environment.

Signs of Dispersive Mechanisms: Individual Distance and Territories

Typically, individual animals tend to arrange themselves at some generally predictable distance from one another. This **individual distance** is defined by the balance between cohesive and dispersive forces acting on each individual. It's not a rigidly defined space. It varies with unexpected events, such as chance encounters with another individual. It also varies with internal rhythms and external rhythms keyed (for example) to the changing seasons. Between late summer and fall, a song sparrow generally couldn't care less about the presence of conspecifics. But come spring and the breeding season, sparrows are most intense about keeping one another at what they consider a suitable distance.

Aggression and Dominance The most unambiguous way to maintain individual distance is to show **aggressive behavior,** a readiness to do injury to another individual or at least threaten to do so. Interestingly, we call such behavior "aggression" when it's directed at conspecifics, "feeding" when it's directed at prey, and "defense" when it's directed at predators. The same mechanical expressions—pulling, pecking, beating, and so forth—necessarily underlie all three behaviors because an animal has, by virtue of its physical makeup, only certain expressions available to it. Only the adaptive significance of the expression differs in each case.

Aggression to establish individual distance is an expression of competition within a group. Initially there is quite a bit of fighting, which results in *dominance* by one individual and *submission* by others. Such encounters work to separate the stronger animals by a greater distance from the rest of the group. More than this, they work to establish a ranking of its members—a **social hierarchy** of individual distance laden with social meaning. All members of the group with any survival sense at all come to recognize that meaning. As a result, actual fighting dwindles. The mere threat of aggression from top-ranking animals becomes enough to elicit submissive behavior from conspecifics and to maintain social order. *Because less energy is wasted on aggression, more is available for the business of survival. At the same*

Randall Eaton

Figure 20.6 Ritualized threat behavior of a dominant male cheetah. Notice the exaggerated ruff of fur, the ears pulled back flat against the skull, and oh yes, the fangs.

time, selection favors the dominant animal, whose competitive show of strength assures it will get the choicest food and its pick of potential mates.

A dominant animal advertises its higher status in formal displays, or **ritualized behavior.** Such displays are exaggerations of ordinary functional movements, and they are clearly distinct signals to conspecifics. Thus there is no mistaking a dominant wolf or dog, which carries its head and tail erect and walks with a stiff, formal gait. And there is no mistaking the dominant cheetah (Figure 20.6).

But what becomes of the weaker of the two combatants? Submission finds expression in two ways. In **avoidance behavior,** the submissive individual stays out of the way of the victor. But this maneuver is not necessarily as

a

b

Figure 20.7 Appeasement behavior among baboons. In (**a**), notice the assured position of the dominant animal—and the abject stare and groveling posture of the subordinate one, who is intent on making little conciliatory smacking noises with its lips. In (**b**), a young male presents his genitals to a dominant female; he already is aware of appeasement behaviors that help shape baboon social structure.

simple as it sounds. If all must use the same feeding grounds, the same paths, the same watering holes, and sometimes the same females, then the dominant and submissive individuals could be bumping into each other all the time, and the sight of the loser could provoke another attack. So the loser may learn the dominant animal's routine movements—and schedule its own activities at the watering hole, the feeding grounds, or on the paths to avoid confrontation.

Ah, but animals are not automatons, routines are not inflexible, and some confrontations are inevitable despite the best-laid schemes. This is especially true if the group lives in a relatively confined area. In such encounters, the submissive animal may resort to **appeasement behavior** as a further show of deference. No uplifted tails for the losers: instead they use exaggerated displays of submission. For wolves and dogs, appeasement displays include tucking the tail between the hind legs and presenting oneself before the dominant animal in a generally miserable, abject cower.

And if those displays don't work, they can always resort to exposing the vulnerable throat area and/or genitals to the dominant animal. Such appeasement behaviors are highly ritualized among baboons (Figure 20.7).

The interesting thing is that, in the natural environment, a dominant animal rarely takes advantage of such total vulnerability. Only under the stresses produced by overcrowding in experimental situations (Chapter Twenty-One) may the dominant animal carry its aggressive behavior through to the kill. *We are left to conclude that in the natural setting, selection must favor the behaviors helping to keep the group together rather than the ones that would promote mutual destruction.* Indeed, there is probably selection against overly aggressive animals. They can frighten away potential mates, perhaps kill their own offspring (hence eliminate their own gene complement), and drive away other members of the population that afford, by their very clustering, mutual protection and assistance. A male baboon, if left alone as a result of its excessive aggression, is much more likely to fall victim to a leopard—which takes its prey from the fringes of the troop or takes those baboons isolated from the troop by their own antisocial behavior.

Territorial Behavior So far, a dominance hierarchy has been described as being relative to each animal's perception of the individual distance between itself and its conspecifics. But assume environmental conditions are such that the animals are kept constantly on the move. Then individual distance is a portable thing, something each animal carries around with it. Such is the case for the baboon troops of the savanna, where scarcity of food and watering holes sends these animals foraging over a broad area. At the same time, the threat of predators keeps the troop from spreading too far apart, so they move as a unit.

But what happens when animals take up permanent residence somewhere? Then expressions of dominance become attached to a more-or-less geographically definable area—a **territory**—which may be aggressively defended. The nature and size of any territory is determined by many variables—topography, the availability of resources, the number of predators, the number of conspecifics and how they are dispersed, even the time of year. Thus it's difficult to make generalizations about territorial behavior. Given its potential variability, natural selection has simply run rampant with such behavior.

For example, in the breeding season a mockingbird becomes highly aggressive in its defense of a certain area, delineated perhaps by treetops, hedgerows, and open fields. From the trees it can scan its entire territory and readily observe any intrusions by other mockingbirds. It sings loudly to announce its presence and its identity, thereby alerting potential invaders that if they approach they will be observed—and will face a fight. How large an area can it carve out? *For any territorial animal, how much space it can defend*

and still have enough energy left for other essential activities partly defines the upper limits on its domain.

In contrast, ground-dwelling animals establish territories and they patrol them—but they can't patrol everywhere at once and they can't see all conspecifics in the territory at once. Often their territories are not even inside a fixed boundary but instead are networks of paths and places where activities occur. Consequently, some territories overlap extensively in space. Thus the scent markings of wolves and dogs are not only olfactory signals of territorial dominance, they also may be traffic signals for maintaining spatial exclusion. Paul Leyhausen suspects that "a fresh mark means 'section closed,' an older mark means 'you may proceed with caution,' and a very old mark means 'go on, but before you use this please put your own mark so that the next one knows exactly what to do.' " In this way, competing animals avoid direct confrontation with one another.

When territories are first established, there usually are aggressive encounters of some sort. Rarely are they fights to the death. Once an animal wins a battle, it accepts submissive behavior of the loser or permits the loser to flee. Such fights are most vigorous near the center of what an animal claims as its territory, as has been shown in the case of the male stickleback fish. As the male moves farther and farther away from the territorial center, the attacks on intruding conspecifics become less and less vigorous. Edward Wilson thus defined territory as "a space in which one animal or group generally dominates others which become dominant elsewhere."

And yet, having a defined living area does not necessarily imply aggression. The whiptail wallabies live in distinct groups. Each group has its own restricted living area. But the areas occupied by such groups are not defended—one group simply avoids neighboring groups. Areas occupied by one group or individual to the exclusion of conspecifics, but which are not aggressively defended, are called **home ranges.** The home ranges of adjacent wallaby groups often overlap. Where such overlapping occurs, members of different groups feed peacefully together. Beyond the normal expression of hierarchical dominance (in this case established to determine access to females), a male invading the home range of another group encounters no aggression whatsoever.

Signs of Cohesive Mechanisms: Courtship and Raising the Young

Courtship ~~Always~~ Usually Conquers All Many behaviors go into dispersing conspecifics and maintaining individual distance, even when that distance reaches the dimensions of territoriality. Thus you might suspect that an aggressive animal would encounter a few problems when it decides to let another animal—most notably a mate—into its private

a

b

c

d

Figure 20.8 Courtship behavior among the albatross. (**a**) The male spreads his wings as part of a complex court-ship ritual that also includes show of a puffed-up chest (**b**). Bill-touching (**c,d**) represents the breakdown of individual distance barriers and the onset of a year-long pair bonding and the raising of young.

domain. And you would be right. Before mating can take place, the responses of male and female must coincide. The dovetailing of interests occurs not only because of hormonal factors, it occurs also because of specialized behavior patterns called **courtship displays** (Figure 20.8). *The presence of a potential mate may simultaneously give rise to sexual interest and fight–flight reactions; through courtship displays, these opposing behaviors are resolved.*

The male stickleback fish, for instance, can't quite seem to make up his mind about what to do. This highly aggressive fish spends a good part of the time vigorously defending his territory, which centers on the nest he builds. This territory he defends from males whose bellies turn red during the breeding season. The red belly acts as a signal for mutual recognition among males of the species, and it triggers aggression. Ah, but the red belly also tends to attract female sticklebacks. When a female entering the territory is embellished with a certain signal of her own—a silver belly swollen with eggs—then the male proceeds with an astonishing combination of courtship and aggression. It is normal for territorial invaders looking for a fight to turn upward, thereby displaying their belly and their intention. The female, too, turns upward, but her swollen belly deters rather than provokes the male's attack. He (usually) does not attack her and turns away to lead her to the nest, and the female follows. Then the male, who no longer sees the belly stimulus, turns back to attack her; she counters by displaying her belly; and so on until their zigzagging brings them to the nest. When the female peers into it, the male shoves her in and then trembles against her until she discharges her eggs. With that, her swollen belly disappears, aggression once more reigns supreme, and the male drives her away. He then proceeds to fertilize the eggs and, later, to protect the developing young.

At least among stickleback, the mate is allowed to escape. Among many species of spiders, males face the titillating possibility of becoming a meal as well as a mate for the larger female. The male's presence in her domain may just as easily evoke her breathtakingly eclectic predatory behavior—which may be one of the reasons why males are a good deal smaller and (hopefully) less conspicuous to her. Understandably, he approaches the dangerous female with elaborate courtship displays and the utmost caution. Sometimes he remains a safe distance away from her and waves conspicuously marked appendages for hours; only if he manages to put her in a trance with these hypnotic movements can he get away with it. One male garden spider announces himself by plucking in a species-specific way at a strand of the female's web. If there is no return signal—or if the web trembles under the pattering of eight rapidly approaching feet—dispersive behavior takes hold, he drops the strand, and runs like the devil. If, however, she responds with a gentle tug on the strand, he proceeds with his mission, tugging seductively on the strand until he is close enough to mate.

Mutual Recognition of Parents and Progeny With all the aggression surrounding the stickleback's defense of its nest-centered territory, it may come as no surprise that it is an equally attentive parent, up to a point. It is singleminded in the way it protects the eggs from rival sticklebacks (who like to eat them) and in the way it constantly fans fresh water over the eggs to supply them with oxygen. But come the moment of birth, the male abandons the newly hatched offspring and swims into other territories, with an eye to some other stickleback's yolky treasure.

Delayed aggression is even more pronounced among mouth-breeding fish. One species of catfish not only holds up to fifty fertilized eggs in its mouth until they hatch (which in itself is no small feat: each egg is about the size of a marble, and the catfish itself is no more than about 2 feet long). It also carries the newly hatched offspring in its mouth to protect them from predators, spitting them out only to let them forage. Although its parental care is commendable, it is offered only until the young grow to a certain length. That length signals the end of the male's cohesive behavior. Should one then return after being spit out for the last time, it is looked upon as perfectly yummy fair game.

Among almost all mammals, bonds between parents and offspring are established by the chemical signals in licking behavior, which takes place during a short period following birth. In fact, in some experiments in which the young were isolated before this behavior could occur, the mother showed no recognition of her own when they were brought back to her. Mother goats will drive such kids away when they try to nurse.

Visual releasers are vital in establishing and maintaining cohesive bonds between parents and offspring of many species, especially birds. Among birds that are born blind, weak, and featherless, practically the only motor response of which they are capable is to open their mouth in a wide gape (Figure 20.9). But it's enough: gaping is a strong visual signal that elicits parental feeding responses. Soon insects or seeds or fruit or meat are shoved into little gullets at an amazing rate. The record may be held by a parent bird that made 800 food-bearing trips to its nest in a single day. Nestlings often can eat their own weight in one day, which can lead to a fiftyfold increase in body size in a matter of weeks.

The young seem capable of trying to hang onto a good thing when they see it. Even when they are first beginning to forage with their parents, they will still try gaping and quivering their wings, demanding the feeding response even when there is food all about them. The parents eventually reach the point of exhaustion, physically and motivationally, so that even gaping stimuli are no longer recognized. Before the coming of the first winter, the young will be dispersed to fend for themselves, either alone or within the hierarchy of a flock, until the cycle of reproduction and the cohesive behavior it implies begins again.

Figure 20.9 Under the attentive eye of their father, and pressing against their mother's body, which helps break the cold offshore wind, these Caspian tern chicks convey the bonds between protector and protected.

Geoff Moon/Bruce Coleman Inc.

Perspective

If we assume natural selection is the basis for evolutionary change, then behavioral as well as structural traits that are the most adaptive, that provide the edge in competition for resources, will be selected for. Behavioral competition among individuals often takes the form of aggressive behavior. On the surface, such behavior would seem to preclude the formation of social groups. For social groups are based, to varying degrees, on mutual cooperation. Aggression is dispersive, cooperation is cohesive. But even though competition and cooperation are opposing behaviors, they are not necessarily mutually exclusive. Both offer advantages that are competitive in terms of survival, and both may be subject to natural selection. Within a given social structure, competition can give dominant individuals first access to resources—food, shelter, space, mates. But within that same structure, cooperation leads to such advantages as the protection afforded by numbers—protection not only for submissive members but for dominant ones as well. It offers the potential for the division of labor, and the possibility for prolonging learning through the developmental period for offspring, which is exceedingly adaptive in variable environments.

In many animal groups, dominance moves past the preservation of individual distance to geographic territories. In other animal groups, dominance maintains individual distances but does not get translated into territorial aggression. More than this, the intensity of aggression and territoriality are not constants. They vary with time, with environmental disruption, with changes in the numbers and distribution of conspecifics. They are only two of many factors—some known, many not yet identified—that shape and balance the behaviors of individuals and groups. Finally, sociality as well as aggression appears to have an innate basis in animal behavior.

How do these generalizations apply to human social behavior? We are, after all, products of our biological heritage. Our flesh is like the flesh of all other mammals; our blood flows when we are injured, as their blood flows. We, too, must carve out a share of life's resources to survive. Based largely on a sense of this kind of unity with the nonhuman world, many writers—both trained scientists and popularizers—have attempted to decipher what is the inherent, "natural" behavior of humans. One such writer has drawn the conclusion that humans are born murderous, aggressive, and warlike in defense of territory. Others have attempted to use highly regimented, socially stratified insect societies as models for human behavior. Both approaches are based on the notion of **biological determinism:** they have in common a tendency to overestimate the role of heredity in determining behavior and status in human societies. At the outset, these approaches often are rooted in observable facts. But as they develop, they bypass the essence of human nature.

No one can deny that the behavior of ants, for example, is encoded in their genes. Given the extremely limited space the brain occupies in the ant body, it could not be otherwise. Either ants will be preprogrammed to behave in a manner that promotes the welfare of the colony into which they are born or they will perish. But to attribute to humanity the same level of biological determinism is a mistake, for it totally ignores the unique contribution of culture. Culture in its broadest sense is the accumulation of the experience of past generations, which can be used to modify present behavior. The *capacity* for culture is determined genetically, for our DNA encodes the blueprint for constructing our storehouse of behavioral potential, the human brain. But that brain happens to be the most complex brain in the living world. It allows us to store the wisdom and folly of past ages—and learn from them. *Thus we have the capacity to move beyond the innate behavioral potential encoded in our DNA.*

In the final analysis, does it truly make much difference whether the first humans were gentle seed-eaters or marauding bands preying on one another? For it is here and now that we have both the capacity for competition and the capacity for cooperation. It is here and now that we have the capacity to choose a path of aggression or a path of coexistence. It is true that competition for resources has been a driving force of evolution. But over time, in species after species, the *unit* of competition has shifted as cooperative behavior leads to ever larger social groups.

Among all animals, there are variously defined "in-groups" and "out-groups" that form the basis for their competitive behavior. Among humans, too, there are various definitions of who is "in" and who is "out." But these two terms are relative; they are not constants. For example, there are times when we compete with one another as Northerners and Southerners, as blacks and whites—and then unite as Americans in competition with, say, the Russians. Yet, in rare times of crisis—for instance, in the tragic aftermath of a devastating earthquake in a densely populated country—we remember our common humanity. Then, the cooperation we have achieved within the parochial confines of our regional "in-group" flows out to embrace those of us in pain in the distance. Such moments illuminate our capacity for self-recognition.

If there has been any progress at all in human cultural evolution, it has been in our formation of ever more inclusive in-groups and in our attempts to extend our circle of trust. We still have a long way to go, for we all know these circles are as fragile as our shifting priorities. With this in mind, someone joked that we might try populating the planet Mars with imaginary enemies in order to achieve worldwide solidarity, a universal circle of trust against a newly defined out-group. But perhaps in place of this imaginary external threat, we might come to unite before very real ones common to all of us. For we face not only the threat of weapons capable of destroying much of the world, but the threat of self-intoxication with our own technology as it carries the human population further and further away from ecological stability. Overcoming these threats, as you will read in the last three chapters, will begin with collective consciousness of the human population as one social group, with our territorial responsibility the entire world.

Recommended Readings

Crook, J. 1970. *Social Behavior in Birds and Mammals.* New York: Academic Press. Broad summary of intraspecific social structure as adaptations to the environment.

Esser, A. 1971. *Behavior and Environment.* New York: Plenum. Excellent coverage of social behavior and social space.

Klopfer, P. and J. Hailman. 1974. *An Introduction to Animal Behavior.* Second edition. Englewood Cliffs, New Jersey: Prentice-Hall. Covers historical development of basic concepts in animal behavior.

Lorenz, K. (translator M. Wilson). 1952. *King Solomon's Ring.* New York: Thomas Crowell. Delightful animal lore.

Thorpe, W. 1956. *Learning and Instinct in Animals.* Cambridge, Massachusetts: Harvard University Press. Important survey of natural behavior in all major animal phyla.

Chapter Twenty-One

Norman Myers/Bruce Coleman Inc.

The optimist proclaims that we live in the best of all possible worlds; and the pessimist fears this is true. — James Branch Cabell

Populations in Space and Time

Suppose this year the federal government passes legislation limiting the size of your family to three children; no more under any circumstances. Suppose, further, the new law specifies that each father must be sterilized after the birth of the third child—and that if he refuses, he will be jailed for a few years and sterilized anyway, with or without his consent. *It could never happen here,* you might be thinking, and not only because of the furor it would raise over the violation of individual rights. Decades ago, Americans began to suspect that the quantity of life has a pronounced effect on its quality. Since the baby booms following the return of overseas troops after the world wars—an era when a family of twelve was considered a great achievement—our concept of family planning has been slowly changing. As a nation we are now moving in the direction of **zero population growth,** in which the birth rate over the long term equals the death rate.

But you are among the fortunate few in the world who have the luxury of a choice. You belong to a population that presently has enough food and enough money to buy whatever other resources it is lacking. You enjoy superior standards of hygiene and medical care. And all these things mean you don't need a lot of children to help you survive and perpetuate the family line.

Most of the people in the world don't have the luxury of making such "rational" choices. Consider the conflict going on today in India between individual rights to reproduction and the rights of the overall population. Each year 9 million Indians die—but 22 million are born. To a population now standing at well over 600 million, *13 million new members are added annually.* There are more people in this one country than there are in North and South America combined! Yet India does not have enough food for its population now, nor does it have enough money to buy more food and other resources. Most of the population does not have adequate medical care, and living conditions are for the most part appalling. Each *week* 100,000 more people enter the job market, looking for jobs that are nonexistent.

Because of these circumstances, the Indian government has been supporting population control programs for more than two decades. But the programs have not succeeded. For one thing, 70 percent of the people are illiterate and can be educated only by word of mouth. And sustained population control programs are hampered because 80 percent live in remote agricultural villages that are not easy to reach.

Finally, Indian villagers don't see the meaning of total population growth; they see only that many children die of disease, of starvation. They believe that only by having more children will they increase their own personal chance for survival. Without large families, they say, who will help a father tend fields? Who will go to the cities in the hope of earning money to send back home? How can a father otherwise know he will be survived by at least one son, who must, by Hindu tradition, conduct the last rites to assure that the soul of his dead father will rest in peace? Clearly there is a serious discrepancy between the way villagers see their world and what their world has actually come to be. Because of this discrepancy, twenty-five years from now the population may reach *1 billion*—which would wipe out all of India's efforts to increase crop production and to break the cycle of poverty, disease, and starvation that plagues it. That is why, in desperation, their federal government passed legislation in 1976 calling for compulsory sterilization. For them, the hypothetical example opening this chapter briefly became reality, until public outrage became so great that the law was rescinded.

Population control, in short, is an issue complicated by the fact that the conditions under which others must make their decision to limit family size are enormously different from the conditions under which we have made ours. We live in the best of existing worlds; others live in the worst. The chasm between us is an unfortunate legacy of the nineteenth century. That was an age of rampant colonization and control by Europeans and, to a lesser extent, Americans over the vast natural resources of undeveloped countries. It was an age when trade agreements were established, generally by force, that benefited powerful nations at the expense of everyone else. For centuries the pattern has remained one of taking raw materials from these countries and giving them mostly advice in return. Even now, as industrialized nations send 2.5 million tons of low-grade protein (such as wheat) to undeveloped nations, they take out 3.5 million tons of high-grade protein (such as fish and shrimp). It is true native sellers of such goods are being paid more than their own people can pay—but it is also true they are not being paid a fair market price. Of all goods and services produced in the entire world during a given year, the United States alone takes at least 30 percent of it; the share of all developing countries *combined* is less than 10 percent. Yet the United States is paying no more per capita for its raw materials than it was paying at the beginning of this century. *Perhaps, then, advice on population control will become more palatable when there is more of a balance to the flow of resources on which <u>any</u> population depends—a more secure basis for individual life that will make population control seem the rational approach that it is.*

In this chapter, you will be reading about the biological principles underlying the growth and survival of populations. These principles can tell you a great deal about the human condition—where it is now, where it is heading. But this chapter can only hint at the complex reasons why there is resistance in many parts of the world to accepting what these principles mean. For that reason, perhaps you will want to supplement the biological perspective with reading on your own into the histories of East and West. For in those histories, in their convergence, lie clues to what must be undone and what must be built anew before we can hope to achieve some semblance of long term, dynamic stability for the human species.

Population Size

Birth, Death, and Exponential Growth A population, recall, is a group of interbreeding individuals separated (geographically or behaviorally) from others of their species. To define its character, we can begin with **population size** (its number of individuals) and **population density** (the way its individuals are distributed through their environment). Population size and density are not static things. They tend to fluctuate because individuals come and go—by birth, death, and inclination—all the time. But how do we pin down the character of a population if it is more or less changing all the time? We can look first to the way it grows in size. **Population growth** is the difference between the birth rate and death rate, plus or minus any inward or outward migration. As a first approximation, we can begin with this principle of population growth:

Any population that is not restricted in some way will grow in size at an increasingly accelerated rate.

A convenient way to express the growth rate of a given population is known as its **doubling time** (how long it takes to double in size). Some bacteria can double their number every thirty minutes; humans are now doubling their number every thirty-five years. Regardless of the unit of time, in the absence of control factors the pattern is always the same. What starts out as a gradual increase in numbers turns into explosively accelerated increases—a pattern of **exponential growth.** All populations, from those of amoebas to humans to whales, grow exponentially under ideal conditions no matter what the reproductive

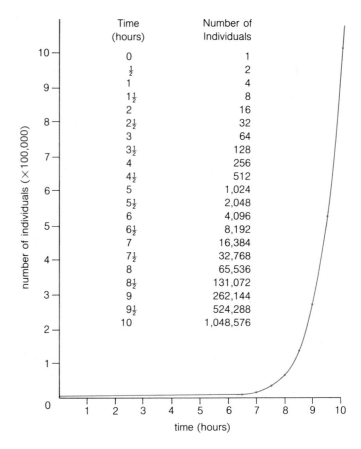

Time (hours)	Number of Individuals
0	1
$\frac{1}{2}$	2
1	4
$1\frac{1}{2}$	8
2	16
$2\frac{1}{2}$	32
3	64
$3\frac{1}{2}$	128
4	256
$4\frac{1}{2}$	512
5	1,024
$5\frac{1}{2}$	2,048
6	4,096
$6\frac{1}{2}$	8,192
7	16,384
$7\frac{1}{2}$	32,768
8	65,536
$8\frac{1}{2}$	131,072
9	262,144
$9\frac{1}{2}$	524,288
10	1,048,576

Figure 21.2 Exponential growth of a population of bacteria. Division occurs every half hour.

rate of their species. The reason for exponential growth is simple. As new individuals are added, *they enlarge the potential reproductive base.* The next generation enlarges it further, and so on. For a while, many new individuals can be born and nothing seems to be getting out of hand. But then an ominous thing starts to happen. With successive doubling periods, the size of the population actually begins to skyrocket.

We can see how this happens by putting a single bacterium in a culture flask with a rich supply of nutrients. In half an hour the bacterium divides in two; half an hour later, the two divide into four. Assuming no cells die between divisions, every half hour the number doubles. But the larger the population base becomes, the worse things get. After only $9\frac{1}{2}$ hours (nineteen doublings), the population will be over 500,000—and by 10 hours (twenty doublings), it will simply soar past 1,000,000! When we plot the course of such exponential growth, we end up with a **J-shaped curve** (Figure 21.2).

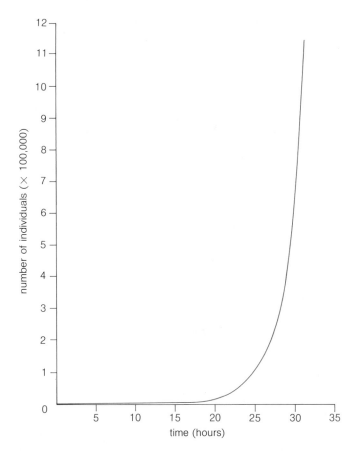

Figure 21.3 Exponential growth of a population of bacteria in which division occurs every half hour, but in which 25 percent of the individuals die between divisions. Although deaths slow things down a bit, in themselves they are not enough to stop the explosive pattern of exponential growth.

For a slowly reproducing species such as a whale, it may take a hundred years for population size to double. Even so, after only twenty doublings (2,000 years), there would be a million whales derived from each original pair. Although the doubling time varies from one species to the next, fantastic population sizes are relatively short-term prospects when the rate of growth is measured on the evolutionary time scale.

Ah, you might say, but a whale doesn't live to be 2,000 years old; even as some are born, others are dying—which surely must keep population size under control. We can see whether the death rate is a significant control factor by starting over with our bacterium in its nutrient-rich culture flask. But this time let's assume 25 percent of the population dies between each doubling time. The death rate does slow things down a bit, in that it takes almost 2 hours instead of half an hour to double population size.

But as Figure 21.3 shows, only the time scale has changed—we still have a J-shaped curve! It's just that it now takes 30 hours instead of 10 to arrive at a million bacteria. Thus we have another principle governing population growth:

As long as the birth rate remains even slightly above the death rate, any population grows exponentially.

Carrying Capacity of the Environment Obviously, you *know* there must be controls of some sort on population size that prevent berserk expressions of exponential growth. For example, chances are you know that when you take a walk through a forest you are not going to be trampled to death by a billion rabbits. Something about a stable natural system such as a forest keeps its populations in check; somehow birth rates are balanced with death rates. But any natural system is bound to have complex interactions going on among its diverse populations. So let's go back to that (by now exhausted) bacterium in its culture flask, where we know we can control the variables. First we will feed it a balanced diet of glucose, minerals, and nitrogen, then we will allow it to reproduce for many generations. Initially the bacterial population goes through an exponential growth phase. Next, population growth tapers off, only to give way to a **plateau phase** in which population size remains relatively stable. Then the population begins to decline—suddenly at first, followed by a more gradual pace until the entire population is dead. In Figure 21.4, curve *a* depicts this characteristic rise and fall in numbers.

What caused the growth curve to be patterned in this way? For these bacteria, glucose meant food and energy—but the culture dish held only so much glucose. We can deduce that, as the population expanded faster and faster, the glucose had to be getting used up faster and faster, too. When glucose supplies began dwindling, so did the basis for exponential growth. Hence we have another principle governing population growth:

When any essential resource falls in short supply, it becomes a limiting factor on population growth.

So we try again—only this time we keep adding glucose to the culture dish. And now the growth curve takes on the shape of curve *b* in Figure 21.4. Notice how this curve shoots up higher than curve *a* but then falls more rapidly. Although the extra glucose encouraged further growth, there were no extra supplies of nitrogen and minerals. Different resources had now become the limiting factors. To keep our bacterial population growing explo-

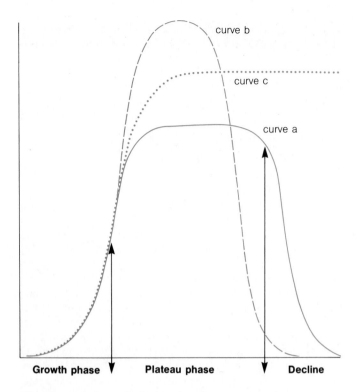

Growth phase ↓ **Plateau phase** ↓ **Decline**

Figure 21.4 Effect of carrying capacity of the environment on population growth curves, as described in the text.

sively, we are obviously going to have to feed it more nitrogen and minerals, too.

But even with all the nutrients it needs, the population not only declines, it crashes! What went wrong? As the bacteria multiplied, so did their metabolic by-products— which drastically changed the nature of the environment. So many waste products were given off that the bacteria actually poisoned themselves to death. Only if the toxic medium were removed every so often from the culture dish and replaced with a fresh medium would things stabilize. Only then would we end up with an **S-shaped curve:** an exponential growth phase that leads into a stable plateau phase (curve *c* in Figure 21.4).

These experiments have helped us identify another principle controlling population growth:

A population of any species can tolerate only a certain range of environmental conditions.

If some environmental factors exceed this **tolerance range** for a population, death follows. Tolerance ranges limit the growth of a population in a given place and they limit its

geographic distribution. Thus resource availability and tolerance to prevailing environmental conditions interact to define where population growth must level off. They define the **carrying capacity** of the environment: the maximum size at which a stable population can be maintained. This point is emphasized in Figure 21.5. There is no escaping the limits imposed by the carrying capacity of the environment. For all species, it is like a brake on potentially runaway growth.

Maintaining Population Stability

When a natural population is said to be stabilized, that doesn't mean its size is frozen at some level. It means the population size typically fluctuates *within a predictable range*, neither dropping to very low levels nor exceeding certain upper limits. It is a dynamic kind of stability, with feedback mechanisms constantly regulating the birth and death rates.

There are two ways of looking at these mechanisms. We can see how they relate to population density. We can also see which ones are built into the species and which are dictated by the environment.

Density-Dependent and Density-Independent Factors Certain control mechanisms come into play whenever populations change in size; they are **density-dependent factors.** When population size increases, various pressures arise that serve to cut it back. When population size decreases, these pressures ease up, which encourages renewed growth. For instance, if population size increases, the amount of food available for each individual declines. The decline in food may lead to impaired health, which may affect the reproductive rate. But if population size falls, the survivors have more food available, their health may improve, and their reproductive rate may rise—so population growth resumes. Food availability, then, is a density-dependent factor.

Other control mechanisms are at work regardless of how large or small a population may be; they are **density-independent factors.** You could, for instance, boil a culture flask full of bacteria and the entire population would drop dead no matter how many or how few there were. Such catastrophic events are not confined to culture flasks. They await any natural population whenever environmental variables soar above or plummet below the tolerance ranges characteristic of the species.

Population size of many desert insects is controlled in this way. For them, the brief spring rains break the dormancy of many thousands of insect eggs. In a short time, insects emerge, they feed, they mate, they lay eggs all over the place. They cannot dally, for with the onset of the

Figure 21.5 How many bananaquit birds in the West Indies? As first perceived by Thomas Malthus, an English clergyman and economist, *food supplies increase at an arithmetic rate, but populations increase at an exponential rate.* Without controls, every population—from bacterial to bananaquit to human—has the potential to outgrow its food supply and ultimately to face widespread starvation. The availability of food is one of *the* major factors determining the carrying capacity of the environment for a given species.

 — caption note in margin: Thase Daniel

prolonged dry season, the cuticle covering their body dries out in the intense heat. The populations then die simultaneously and totally. Explode and crash, explode and crash—only by quickly re-creating the thread of dormant, heat-resistant eggs between rainy seasons can these species survive (Figure 21.6).

It is not often, however, that we can clearly isolate density-independent factors from density-dependent ones. They usually interact to such an extent that it's difficult to know where one ends and another begins. For instance, whether a rabbit population tolerates a sudden freeze depends on whether its members have enough food and burrows. Availability of food and desirable burrows are density-dependent factors—yet here they interact with a density-independent change in temperature.

To help shed more light on how interactions such as these control population size, let's take a closer look at both their intrinsic and extrinsic nature, as determined by observations of experimentally controlled populations and naturally occurring ones.

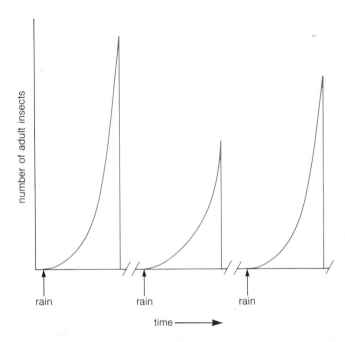

Figure 21.6 Fluctuating population levels of adult insects in a desert environment. Periodic rains trigger the hatching of dormant eggs, and the population level rises rapidly—only to fall rapidly as dry conditions return and bring about the near-simultaneous deaths of all adults.

Intrinsic Limiting Factors We can define **intrinsic limiting factors** as built-in features of the individual; they relate to gross physiology, metabolism, structure, and behavior. Perhaps one of the most intriguing demonstrations of how intrinsic factors operate comes from ecologist John Calhoun's studies of extreme overcrowding among Norway rats. His experiments suggest that changes in the endocrinological system of these animals accompany increases in population density; and that these intrinsic changes in turn influence population distribution and reproductive behavior.

For Calhoun's experiments the rats, which are normally territorial, were confined in pens. They were given plenty of food and protection against their normal predators, and disease was kept to a minimum. As population density increased under these artificial conditions, normal territorial behavior began breaking down, only to be replaced by behaviors of a bizarre sort. Some rats became cannibals even though there was plenty of other food. Some became frenzied, sexually deviant, or pathologically withdrawn. Underlying these changes was an increase in the size of adrenal glands, probably because of the steadily increasing aggressive encounters as more and more rats were forced to interact in the restricted space of the pens. Sometimes

dominant males would go on a rampage, attacking not only subordinate males but females and offspring. Individuals low on the social ladder would eat and move about only when the other rats were asleep.

At the same time, there was an overall reduction in the size of glands other than the adrenals. The reduction was most pronounced in glands concerned with reproductive hormones. Miscarriage was common; so was death of pregnant females during delivery. Among the most stressed groups, only 4 percent of the offspring survived. Maternal behavior deteriorated as females first became careless in building nests, then began simply to heap up strips of paper as makeshift nests, then finally delivered their offspring directly on the sawdust of the open pen. If they encountered another rat while they were toting their infants about, they would drop them and forget all about them. The infants were helpless before the cannibalistic adults.

In studies of experimentally induced overcrowding among mice, the social structure that prevailed was based on varying levels of withdrawal. Mice at the low end of the ladder showed chronic withdrawal symptoms punctuated by short bursts of violence. One would chew on another, which would passively submit to being chewed. Mice at the top of the ladder were healthy and sleek—but absolutely passive. They slept; when presented with food, they ate; when presented with water, they drank. But they never fought, they never went out of their way to find food, and they never copulated—they seemed no longer interested in reproduction. Eventually the survivors of those stressful conditions were placed in a spacious, uncrowded pen. But they could only huddle desperately together. They seemed absolutely terrified of the opportunity of less crowded distribution.

(Similar results have been noted in studies of a number of social mammals. It has even been suggested that crumbling social structures and other signs of aberrant behavior in overcrowded human societies may be partly related to the same kind of intrinsic limiting factors. As tempting as the similarities may be, the data available do not conclusively indicate the findings may also apply to overcrowded human populations.)

In natural settings, too, intrinsic limiting factors affect how populations come to be distributed in space. Territorial behavior in natural communities is under intrinsic controls that are tied to the carrying capacity of the environment. So, to a large extent, is migratory behavior. Migration is an intrinsic response to extrinsic limitations of a local region. For instance, elk herds in North America migrate on a daily basis and a seasonal basis. During daylight hours, they bed down in the forest or do some light feeding beneath the trees; at night they move into open meadows to do most of their eating. But even with ample food sources, the carrying capacity of a meadow is low because it doesn't provide ample protection against predators. In contrast, the forest

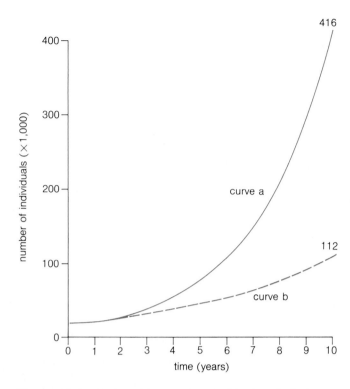

Figure 21.7 Comparison of the growth rates of two hypothetical populations that differ only in the timing of reproduction. In both cases, each individual lives five years, and each breeding pair produces four offspring. In *curve a*, reproduction occurs at age two. *Curve b* shows what happens when reproduction is delayed until age four. Delayed reproduction has some impact on population growth over the short term. But in the long term, *curve b* would also show exponential growth.

some time afterward, they may continue to coexist with several new generations. Such overlapping may contribute to higher population levels. If reproduction occurs later in life or if the post-reproductive life span is short, there is less overlapping and there may be lower population levels. Reproductive timing has become adapted to the carrying capacity of the environment for different species. Some species are long-lived and start reproducing late in the life cycle. Some go through a long period of embryonic development and give birth to only a few individuals. Perhaps the fullest expression of delayed reproduction is found among species such as Pacific salmon. Their members grow to maturity, then die right after reproducing. Because generation overlap is eliminated, a higher rate of reproduction is possible without adversely affecting the carrying capacity of the environment. More will be said about delayed reproduction in a later section on human populations, for it is sometimes considered to be a possible way to control our own rate of growth.

Extrinsic Limiting Factors We can define **extrinsic limiting factors** controlling population growth and distribution as external conditions—not only temperature, rainfall, nutrients, and such, but the presence or absence of other species. Interactions among species are especially important, for they help define *where* a species will actually be found in its potential range of distribution. Perhaps the most dramatic extrinsic check is the **predation** of one kind of species on another. The interactions between the Canadian lynx and the snowshoe hare are a classic example of the feedback mechanisms at work in a predator–prey system. Although evidence for this set of interactions is secondhand (Figure 21.8), it has been collected every year for more than a century. And even allowing for unknown variables, certain patterns undeniably are there. The lynx and hare population sizes fluctuate together in cycles, with the lynx population lagging slightly behind that of the hares. When the number of hares is low, there is a period of rapid population growth. The number of lynx, which is also low, rises right after the growing number of hares—then both populations crash. The cycle repeats itself about every decade.

An intrinsic factor (the explosive reproductive cycle of the hares) triggers the cycle. An extrinsic factor (the scarcity of lynx at the start of each cycle) permits rapid fulfillment of that potential. With more hares running around, the lynx are better fed; more lynx survive, and more reproduce. But the intensified predation, along with diminishing food and perhaps with disease, exerts more pressure on the hare population, which soon plummets. The lynx population has only one way to go: down. And once their numbers have also declined, the cycle can start again. In this example, predation works to help control the size of *both* populations as they interact with one another.

provides ample protection but not enough food. The daily movement of the herd creates a composite environment of forest and meadow, with a higher carrying capacity than either area alone can offer. Under the same kind of pressures, the herds have come to migrate between lowland meadows during winter and high mountain meadows from late spring to early fall. Changing weather in the high country gradually drives them back to the lowlands before the onset of winter. But the thing that keeps them from living all year long in the lowlands is the limited productivity there. Only because the herds show an intrinsic tendency to migrate is vegetation in both places given time for renewed growth—which supports a larger elk population over the long term.

One more intrinsic limiting factor worth mentioning is the **timing of reproduction** in individual life cycles. If reproduction occurs early in life, and if the parents live for

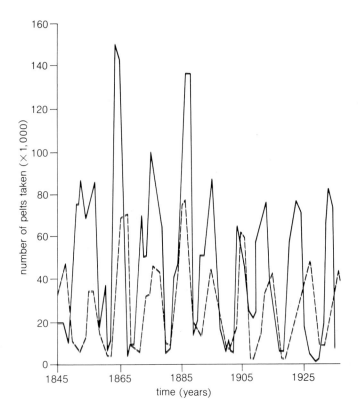

Figure 21.8 Correlation between the lynx population (dotted line) and the snowshoe hare population (solid line) in Canada over a ninety-year period. These long-term data are derived not from field observations but from counts of the pelts that trappers sold to Hudson Bay Co. The curves are taken to be a general index of the way predation can control the populations of both predator and prey. (Data from D. MacLulich, University of Toronto Studies, Biology Series 43, 1937)

This figure is a good test of how willing you are to accept conclusions without questioning their scientific basis. (Remember the discussion of scientific methods in Chapter Two?) For example, what other extrinsic and intrinsic control factors could have been influencing the population levels? Were there also fluctuations in climate over this time span? Were some winters more rigorous, thereby imposing a higher death rate on one or both populations? Although this is called a simple predator–prey system, weren't the hares preying on the vegetation, which may have been overbrowsed in some years but not in others? What about owls, martins, and foxes—which also prey on hares? What if some years there were fewer trappers because of such very real variables as Indian uprisings? What if some years there was more lynx trapping than hare trapping, or vice versa? And in looking closely at the curves, can you really conclude there is an unvarying correlation between them, or is there evidence here and there of a random drifting apart—which might indicate other factors at work?

Human Population Growth

So far, we have touched on a few intrinsic and extrinsic controls to show how they can regulate the growth of natural populations. Thus you can see how populations can grow explosively under certain combinations of short-term variables, only to decline in the long term as controls reassert themselves. Let's see how this pattern applies to the growth of the human population.

Doubling Time for the Human Population In 1975 the human population reached 4 billion. Even if this number represented a stabilized population level, we would still have to contend with monumental problems. In any given year, between 5 and 20 *million* people now die of starvation or malnutrition-related diseases (Chapter Five). The industrialized nations have only a fraction of the world's population, yet they use almost all the world's resources—most of which are nonrenewable. And even though so much is being channeled to so few, the wealthiest of nations are facing chronic shortages of energy and materials along with the poorest. Thus even the self-chosen few are facing restricted economic growth. There are so many people in the world, even the most humanitarian effort to share all resources equally would raise the average worldwide standard of living only slightly—even as the standard of living for the industrialized nations plummeted.

At present, there is no realistic prospect of advancing the quality of life for all. The future is more grim. If anything, the by-products of existence are causing the overall quality of life to deteriorate. As you will read in the last two chapters, our self-inflicted pollutants carry the same danger that the self-inflicted pollutants carried for those bacteria described earlier. In the expansionist mentality of the nineteenth century, the world was our oyster. Today the world is our culture flask, and no great experimenter in the sky is going to take away our wastes and renew resources for us.

What makes future prospects especially chilling is that our population is by no means static. We are a rapidly growing population of 4 billion! For every thirteen individuals who die, thirty-three are born. That translates into an average growth rate of 2 percent a year. If we take the average human life span to be seventy years, we can calculate how long it takes to double world population. The average life expectancy divided by our annual growth rate gives us our population's **doubling time:**

$$\text{Doubling time} = \frac{\text{life expectancy}}{\text{annual growth rate}} = \frac{70 \text{ years}}{2\%} = 35 \text{ years}$$

That means by the year 2010—well within the lifetime of most of us reading this book—world population may soar to 8 billion!

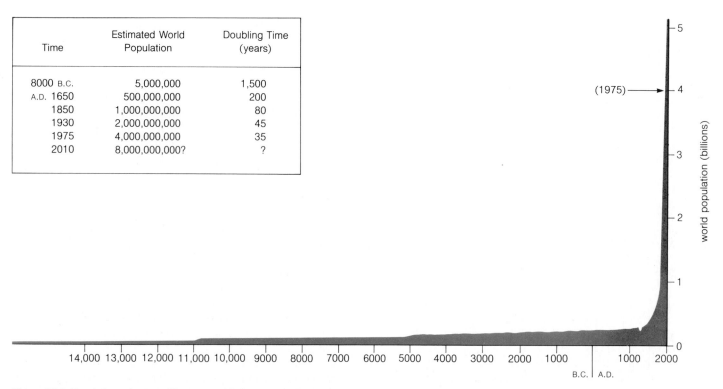

Time	Estimated World Population	Doubling Time (years)
8000 B.C.	5,000,000	1,500
A.D. 1650	500,000,000	200
1850	1,000,000,000	80
1930	2,000,000,000	45
1975	4,000,000,000	35
2010	8,000,000,000?	?

Figure 21.9 The J-shaped curve of human population growth. The slight dip just before the period between 1347 and 1351 shows the time when 75 million people died in Europe as a result of the bubonic plague, or Black Death—a virulent disease spread by fleas that thrived on the massive rat populations in cities. At current growth rates, it would take only thirteen months to replace 75 million individuals.

With the most intense effort, we just *might* be able to double food production over the next thirty-five years to keep pace with growth. But we would succeed in doing little more than maintaining marginal living conditions for most of the world. Under such "ideal" conditions, deaths from starvation might *only* be 10 to 40 million a year! For a while, it would be like the Red Queen's garden in Lewis Carroll's *Through the Looking Glass,* where one is forced to run as fast as one can to remain in the same place. But then what happens when resources other than food become limiting factors? What happens in the year 2045, when the population has doubled again to 16 billion? Can you brush this picture aside as being too far in the future to warrant your concern? *It is no farther removed from you than your own sons and daughters; that world is their legacy.*

Where We Began Sidestepping Controls How did we get into this mess in the first place? For most of our existence as a species, our population has been growing slowly. But in the past century, there has been an astounding acceleration of growth (Figure 21.9). We have no reason to assume there

are special rules governing human population size and its rate of growth. Instead, we must assume the human population is controlled by factors much like those controlling populations of other large animals having relatively long life spans. If that is true, there are only three possible reasons why our long-term growth rate accelerated:

1. We must have developed the capacity to expand steadily into new environments.

2. The carrying capacity of the environment was increased in some way.

3. A series of limiting factors was removed so that more of the available resources could be exploited.

Let's consider the first possibility. We know that by 50,000 years ago, the human species had migrated over much of the world. For most animal species, such extensive radiation into new environments is a much slower process. Not only is there competition with species that may already occupy an area, there must be time to modify tolerance ranges in response to pressures from new environmental conditions. With the human species, environmental constraints were bypassed not by genetic mutation but by the application of learning and memory—how to build fires, assemble shelters, create clothing and tools, plan a community hunt. Learned experiences were not confined to individuals but raced like the wind through one human group after another because of language—our ability for cultural communication. It took millions of years for certain animal groups to evolve the ability to fly. Through our intelligence and through cultural communication, it would take less than seven decades from the time we first ventured into the air until we landed on the moon.

What about the second possibility? Since the human species first appeared, there have been several profound shifts in climate. Climate is a major influence on the amounts and kinds of vegetation that will grow in a region, hence on the numbers and kinds of animals that vegetation supports. As you read in Chapter Eighteen, a general trend toward long-term drought may have forced our ancestors from their dwindling forest homes and into the more productive grasslands. And then, the general warming of the earth that began 11,000 years ago apparently triggered the shift from the hunting way of life to agriculture—from risky, demanding moves after the game herds to a settled, more dependable basis for existence in more favorable environments. Even in its simplest form, the agricultural management of food supplies bypassed one of the most basic limits on the carrying capacity of the environment. Populations could expand to new limits. And with each cultural innovation—irrigation, metallurgy, social stratification to provide a labor base, the development of fertilizers and pesticides—the limits were expanded and were met again with a resurgence of growth. Thus, with the domestication of plants and animals, and the development of agriculture, the carrying capacity of the environment has risen abruptly for human populations.

And what about the third possibility—the removal of certain limiting factors from the environment? The potential for growth that was inherent in the development of agriculture began to be fully realized with the suppression of contagious diseases. Until about 300 years ago, contagious diseases kept the death rate high enough to counteract the growing birth rate. For contagious diseases spread like wildfire through crowded settlements and cities; they are density-dependent factors. Without proper hygiene, without sewage disposal methods, and plagued with such disease carriers as fleas and rats, population levels increased only slowly over the long term. But plumbing and sewage treatment methods did appear. Bacteria and viruses were recognized as disease agents. And vaccines, antitoxins, and drugs such as antibiotics were developed. Thus one after another major limiting factor on human population growth has been largely pushed aside. Smallpox, plague, diphtheria, cholera, measles, malaria—many diseases have been brought under control in the developed countries. Con-

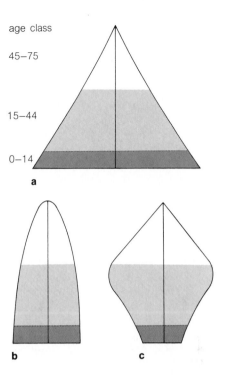

age class

45–75

15–44

0–14

a

b   c

Figure 21.10 Age structure diagrams for (**a**) a rapidly expanding population, (**b**) a slowly expanding population, and (**c**) a declining population.

currently, medical technology has been exported in a humanitarian effort to control disease in the developing countries as well. With old age and starvation the only remaining checks, population growth has been skyrocketing ever since.

We are faced with two options. Either we make a global effort to limit our numbers, so that our population stabilizes according to the carrying capacity of the environment, or we passively wait until the environment does it for us. Our choices and the time we have to make them are far more limited than we would like to think. At our best, we are creatures of compassion, and any move to deliberately cause deaths would mean losing something of what being "human" is about. Even deciding to do nothing at all is to deny part of what we are. At the same time, because we are on a course to alleviate suffering and premature death, our medical advances will continue to lower the death rate. Thus the most humane, reasonable option—short of biding time until the crash—is to reduce birth rates as dramatically as we have lowered death rates.

Age Structure and Fertility Rates Two important factors influence just how much we can expect to slow down the birth rate. The first has to do with **age structure:** how individuals are distributed at each age level for a population (Figure 21.10). By placing individuals of a population

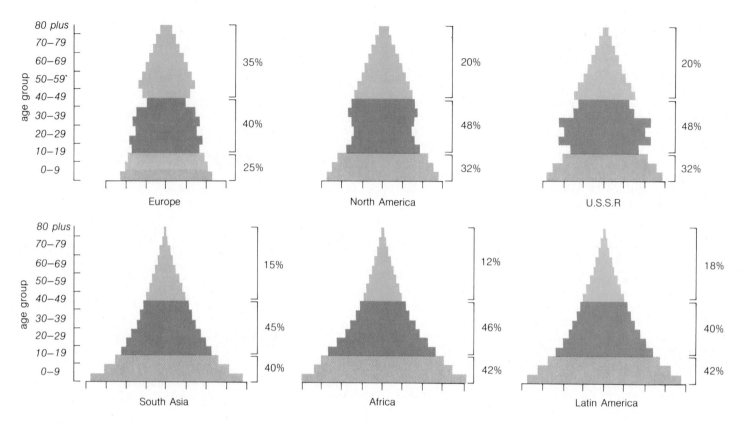

Figure 21.11 Age structure diagrams for the world's major geographic regions.

into three categories—those before, during, and after reproductive age—we can create diagrams that readily show us the prospects for population growth. The average range of childbearing age for women is 15 to 44. The age structure pyramid for a rapidly growing population has a broad base, filled not only with reproductive-age men and women but with a large number of children who will move into that category during the next 15 years. As Figures 21.11 and 21.12 suggest, *more than a third of the world population now falls in the broad pre-reproductive base.* Thus we have an idea of the magnitude of the effort that will be needed to control birth rates on a global scale—and of the urgency surrounding birth control implementation.

One other factor influencing the short-term picture of population growth is the **fertility rate,** or how many infants are born to each woman during her reproductive age. Frequently the general fertility rate is calculated on the basis of number of births each year per 1,000 women between ages 15 and 44. Today the average number of children in a family is 2.6 in industrialized countries, and 5.7 in developing countries. But even assuming we can achieve *and maintain* a world average of 2.5 children per family—which is the fertility rate that is said to be needed to bring us to

zero population growth—it will be 70 to 100 years before our population stops growing. Why? Because there is an immense number of existing children yet to move into the reproductive age category! And even then, when and if the world population does level off, there will be a staggering number of people on earth. Even if a replacement level of 2.5 children per family were achieved by the year 1980, world population would continue to grow until it leveled off, in the year 2070, to about *6.3 billion.* Obviously, it is going to take planning and implementation of birth control programs over the next seven decades to achieve population control. At present even 5- and 10-year plans are virtually nonexistent.

Birth Control Methods The best we can hope for in the immediate future is to slow the growth rate and buy some time so that we *can* implement more effective measures. One simple way to slow things down would be to encourage **delayed reproduction**—childbearing in the early thirties as opposed to the midteens or early twenties. In Ireland, women customarily marry later in life; in China, the government has raised the age at which marriage is allowed.

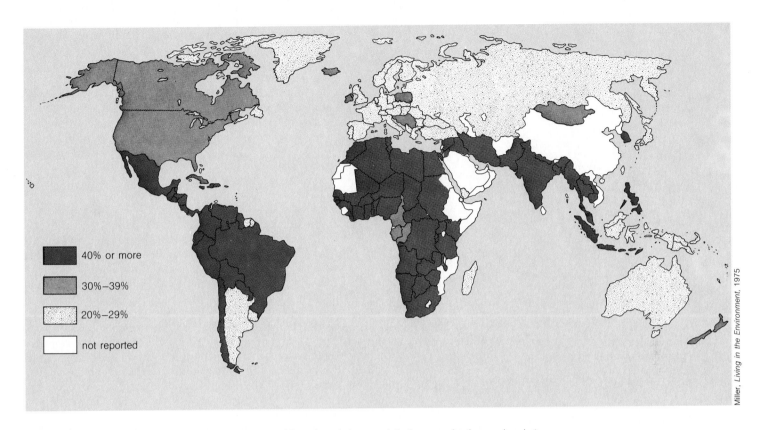

40% or more

30%–39%

20%–29%

not reported

Figure 21.12 Percent of world population under the age 15, and ready to move into the reproductive age bracket.

Such cultural constraints tend to lower the number of children in each family. Figure 21.7 showed how delayed reproduction can slow down population growth rates. But remember that as long as birth rates even slightly exceed death rates, populations will experience exponential growth. Thus short-term strategies ultimately must give way to long-term birth control measures.

And here we come to the most gnawing problem of all. At present, there simply is no completely effective, completely safe, low-cost way to control births. Instead there are various methods that more or less reliably prevent conception, and more drastic surgical ones that are used to abort the developing embryo. A related problem is the lack of enough educational programs concerning birth control. Each year in the United States we still have about 100,000 shotgun marriages, about 200,000 unwed teenage mothers, and *1,000,000 abortions.* This is the legacy of our confusion over what to do about a sexual revolution that has swept through our society down to the primary school level—but that has not been accompanied by adequate sex education programs. On the one hand, many parents promote boy–girl relationships for younger and younger ages. On the other hand, they close their eyes to the possible outcome in terms of premarital intercourse and unplanned pregnancy. Birth control advice to teenagers is often condensed to a terse, "Don't do it, but if you do, be careful!" In this lack of resolution lies part of the controversy over what birth control represents. For most of the world, birth control is a question of utmost urgency, a matter of sheer survival; for sexually liberated Americans, it is often touted as a matter of convenience. More than 500 million years of sexual evolution have gone into assuring a powerful motivation to engage in sex and thereby reproduce. Overlaid on those 500 million years are a few centuries of moral sanctions that demand suppression of the compelling sex drive—and yet simultaneously demand that we go forth and multiply. We have not managed to suppress sexuality, but we *have* managed to multiply far beyond any previous imaginings. It seems we ought to start thinking about reconciling our biological past with the need for a stabilized cultural present. Given the alternatives, we don't seem to have much choice about it.

Once conception has occurred, the only form of birth control is **abortion,** in which the implanted embryo is

Table 21.1 Comparative Effectiveness of Available Birth Control Methods*

Method	Theoretical Effectiveness (Percent)	Actual Effectiveness (Percent)
Totally effective		
Abortion	100	100
Sterilization		
Tubal ligation (female)	99.96	99.96
Vasectomy (male)	99.85	99.85
Highly effective		
Oral contraceptive (Pill)	100	99.9
IUD plus spermicide	99	98
IUD		
Newer Copper-T	99	98
Older loops	98.3	97
Condom (good brand) plus spermicide	99.9	95
Effective		
Diaphragm plus spermicide	99	90
Spermicide—vaginal foam alone	98	90
Condom (good brand) alone	99	85
Diaphragm alone	98	85
Moderately effective		
Rhythm, temperature method	95	80
Spermicides—cream, jelly	90	75
Relatively ineffective		
Condom (cheap brand)	85	70
Withdrawal	85	70
Rhythm, nontemperature method	90	Variable, but normally below 60
Unreliable		
Douche	—	10

*Percentages are based on the annual number of pregnancies per 100 women of reproductive age who are active sexually.

dislodged and removed from the uterus. (In miscarriages, the embryo is expelled spontaneously.) Until recently, abortions were generally forbidden by law in the United States, unless the pregnancy endangered the mother's life. Supreme Court rulings in this past decade have held that the State does not have the power to regulate abortions in the first trimester. The outcome has been legalization of abortion in this country and others. Moving the large number of backroom operations to modern medical facilities at least reduces the frequency of dangerous, traumatic,

and often fatal attempts to abort embryos, either by pregnant women themselves or by quacks. Newer methods have made it relatively rapid, painless, and free of complications if performed during the first trimester. Abortions in the second and third trimesters will probably remain extremely controversial unless the mother's life is clearly threatened. For both medical and cultural reasons, however, it is generally agreed that the preferred route to birth control is not through abortion but through control of conception in the first place.

The most effective method of preventing conception in the first place in complete **abstention:** no sexual intercourse whatsoever. It is also silly to expect people to follow it. A modified form of abstention is the **rhythm method,** in which intercourse is avoided during the woman's fertile period. The fertile period, recall, begins a few days before and ends a few days after ovulation (the release of an egg from an ovary). It is determined either by keeping records of the length of a woman's menstrual cycle or by taking her temperature every morning when she wakes up. (Just before the fertile period, there is a one-half to one degree rise in body temperature.) But ovulation can be irregular, and miscalculations are frequent. The method *is* inexpensive (it costs nothing after you buy the thermometer) and it doesn't require fittings and periodic check-ups by a doctor. But its practitioners do run a 40- to 50-percent risk of becoming pregnant.

Withdrawal, the removal of the penis from the vagina prior to ejaculation, is a truly ancient contraceptive method dating at least from Biblical times (not because overpopulation was seen to be a problem then, but because our ancestors didn't always live by the rules, either). It requires extraordinary will power. And even if the mind manages to conquer the body, the method may fail anyway: the fluid released from the penis just before ejaculation may contain viable sperm cells.

Other methods are based on using physical or chemical barriers to prevent sperm cells from entering the uterus and moving up the ovarian ducts. **Condoms** are thin, tight-fitting sheaths of rubber or animal skin that are worn over the penis during intercourse. They are about 85 to 95 percent reliable (Table 21.1), and they do help prevent venereal disease. But condoms can only be put on over an erect penis, which calls for an interruption of activities at a time when rational behavior somehow seems not as interesting as immediate fulfillment. Also, they have been known to tear and leak, which renders them useless. A **diaphragm** is a flexible, dome-shaped disk used with a spermicidal foam or jelly. It is placed over the opening of the cervix just before intercourse. A diaphragm is relatively effective if it has been fitted by a doctor, and if it is inserted correctly with each use. **Spermicidal foam** or **spermicidal jelly** is packaged in an applicator and is emptied into the vagina just before intercourse. It is toxic to sperm cells, yet it is not always reliable when used without another device

Table 21.2 Procedures for Some Birth Control Methods*

Method	How It Works
Abortion (by qualified physician):	
Dilation and curretage (D & C) during first 12 weeks of pregnancy	Surgical widening of cervix; scraping of placenta from walls of uterus
Vacuum aspiration during first 12 weeks of pregnancy	Cervix dilated and placenta removed by vacuum pump
Intraamniotic injection after 16 weeks of pregnancy	Needle inserted through abdominal wall into uterine cavity; small amount of amniotic fluid withdrawn and replaced with salt solution; labor induced within 25 hours
Sterilization:	
Tubal ligation (female)	Surgical procedure for cutting and tying ovarian ducts
Vasectomy (male)	Tiny incision in scrotum to sever tubes that carry sperm; small section is removed and the tubes are retied; in a new procedure designed to be reversible, a metallic-plated valve is inserted that can be switched on or off with a magnet
Oral contraceptive (Pill)	Combinations of synthetic female hormones (estrogen and progesterone) taken once a day to inhibit release of egg from ovary
Intrauterine device (IUD)	Small S- or T-shaped plastic or metal device inserted through the vagina into the uterus
Diaphragm with spermicide	Flexible disk inserted into vagina to fit over cervix before intercourse; mechanically blocks sperm; spermicide kills sperm

*See Table 21.1 for comparison of the effectiveness of these methods. After G. T. Miller, Jr. *Living in the Environment.* (Belmont, California: Wadsworth, 1975.)

such as a diaphragm or a condom. The practice of **douching,** or rinsing out the vagina with a chemical immediately after intercourse, is almost useless. Sperm cells can move past the cervix and out of reach of the douche within ninety seconds after ejaculation. And no panicky flight to the medicine chest and frenzied rinsing is that rapid.

Still other contraceptive methods are based on hormonal control of the reproductive cycle. Most widely used is **the Pill**—an oral contraceptive of synthetic estrogens and progesterones. These synthetic hormones substitute for hormones normally produced by the ovary during the menstrual cycle. They suppress the release of gonadotropins from the pituitary and thereby prevent the cyclic maturation and release of eggs (Chapter Fourteen). Birth control pills are prescription drugs. Formulations vary and should be selected to match the individual patient's needs. That's why it isn't wise for a woman to borrow the Pill from someone else.

If the woman doesn't forget to take her daily dosage, the Pill is one of the most reliable contraceptives available. There is no interruption of the sexual act, and the program is easy to follow. Often the Pill corrects erratic menstrual cycles and decreases associated cramping. Even so, the Pill is not without potential side effects. In the first month or so of use, it may cause nausea, weight gain and swelling, and minor headaches. Its continued use may lead to blood clotting in the veins of a small number of women (3 out of 10,000) predisposed to this disorder. There have been some cases of elevated blood pressure and abnormalities in fat metabolism (which may be linked to a growing number of gall bladder disorders among women). Proponents of the Pill argue that for most women, the known risks associated with using it are far lower than the risks associated with pregnancy.

Newer, estrogen-free formulations now eliminate some of the side effects. The potent "morning-after Pill" now becoming available eliminates pregnancy after intercourse. There are certain hazards connected with this drug, however, and it can't be used on a regular basis. An inexpensive form of the Pill—strips of edible paper treated with oral contraceptives—is now being tested on a large scale. It may extend use of the Pill to developing countries for mass birth control programs.

A birth control method that once seemed most promising for use in developing countries is the **intrauterine device,** or **IUD.** A doctor must insert this small plastic or metal device into the uterus. With this foreign object in the uterus, a fertilized egg often cannot become implanted in the uterine wall. In cases where implantation does occur, the IUD stimulates processes that cause the embryo to be dislodged. The IUD is relatively inexpensive, and once it is inserted there is no need to give it much further thought unless it is accidentally expelled from the uterus. Especially with the newer designs, such as the Copper-T, expulsions are rare.

But the IUDs can give rise to other complications. Typically, there is a marked increase in menstrual discharge. Usually the increase is merely inconvenient, but in a few cases there are increased possibilities of anemia or hemorrhaging. Some research also indicates a greater probability of uterine disease. In such cases, supervision by a doctor is necessary; in all cases, a checkup every six months is advised so the woman can verify that the device is in place and functioning properly. Thus, even though the low cost of the IUD makes it seem ideal for mass birth control in developing countries, its potential benefits must be weighed against potential complications—especially where ready access to a doctor, let alone medical facilities, are not readily available to women.

Hormonal control of male fertility is a more difficult matter. The hormonal methods for suppressing female fertility are based on the cyclic nature of the woman's reproductive system. But sperm production in males is not cyclic, and it is under a more diffuse kind of hormonal control. Various medications have been developed, but they are still experimental. It will probably be several years before they become available, and their effectiveness is in doubt.

Surgical sterilization methods hold the greatest promise for completely reliable birth control. In male **vasectomy,** a tiny incision is made in the scrotum so that the seminal ducts (which transport sperm cells) can be severed and tied off. The simple operation can be performed in twenty minutes in a doctor's office, with only a local anesthetic. So far there is no firm evidence that it disrupts the male hormone system, and surveys suggest there is no noticeable difference in sexual activity. The only major objection to vasectomies is that they are generally irreversible. Some recent surgical advances promise to improve the possibility of reversing the operation, if that should be desired. For females, the most common sterilization method is **tubal ligation,** in which the ovarian ducts are cut and tied off. Because tubal ligation is more complex than vasectomy, it usually is performed in a hospital and requires about a two-day stay after surgery.

There is little doubt that surgical sterilization is one of the safest, most effective, and most convenient birth control methods. It is a permanent method right now, however, so it is best considered by couples or individuals who have made a mature commitment to limit their family size. Many communities maintain family planning services that can be contacted for assistance in evaluating the available birth control options.

Perspective

In this chapter, we began with the premise that all populations have the potential for exponential growth—growth at an ever-accelerating pace to explosively high levels. We proceeded to analyze the nature of intrinsic and extrinsic limiting factors that keep populations in check and thereby assure stability according to the imperatives of the environment. The biological implications of population instability are, as we have seen, enormous—but so are the social implications of achieving stability, of achieving and maintaining zero population growth.

We can say with some certainty that the human population has already outgrown many of its resources. It has postponed the consequences only by means of a tremendous manipulation of such nonrenewable resources as fossil fuels. But suppose we finally determine what our population size must be in order to live within the carrying capacity of the environment. What will be the repercussions for human society?

Will we continue to enforce a situation whereby most of the world's population can expect only quiet or not-so-quiet desperation, or will we initiate programs to distribute available resources on a more equitable basis? What economic strategy will prevail—a minimum standard of living for the largest number possible, or a smaller population with a higher standard of living for all? The achievement of zero population growth is a biological necessity, but it surely will give rise to a whole new set of problems.

For instance, modern economic systems are based on constant growth, on uninterrupted consumption of goods and services. The growth rate of the gross national product is the standard measure of the nation's economic health. An economic system in which consumption is suddenly lowered would require a totally new approach to government, to employment, to resource distribution, to individual life-styles and expectations. There presently are no plans for easing some new system in as old systems are eased out. And consider the fact that most members of an actively growing population fall in younger age brackets. Under conditions of constant growth, the age distribution means there is a large work force that is capable of supporting older, nonproductive individuals with various welfare programs, such as social security, low-cost housing, and health care. But with zero population growth, far more individuals will fall in the older age brackets. How, then, can essential goods and services be provided for nonproductive members if the productive ones are asked to carry a greater and greater share of the burden? These are not abstract questions. Put them to yourself: How much are you willing to bear for the sake of your parents, your grandparents? How much will your children be willing to bear for you? We clearly have arrived at a major turning point, not only in our biological evolution but in our social evolution as a species. The decisions awaiting us are among the most difficult we will ever have to make, but we must come to recognize that they *must* be made, and soon.

All species face limits to growth. Natural selection has given rise to a host of control mechanisms that work to contain the potentially explosive reproduction of all life forms. In one sense we have proved ourselves different from the rest, for our unique ability to undergo cultural evolution has allowed us to postpone the action of most of these control mechanisms. But the key word here is *postpone*. No amount of cultural intervention can hold back the ultimate check of limited resources. We have repealed a number of the smaller laws of nature, but in the process we have become more vulnerable to those laws which cannot be repealed.

Visions of the future are many, but all too many foretell of famine, disease, and war as an overextended population readjusts itself to available resources. Even now, the industrialized world arms itself and talks of war if imports of what it considers to be essential resources are restricted. If you doubt this, think back on the meaning of talk at the local, regional, and national levels of invading the Near East in the spring of 1974, in the wake of an oil embargo. At the same time, developing nations dream of using the weight of sheer numbers to wrest away the resources they, too, so desperately need. All nations of the world are being drawn inexorably into the conflict. For there no longer is such a thing as a "regional conflict" over resources in the world we have made for ourselves. With the tangled, interdependent threads of trade and alliance reaching out from the nineteenth century, the specter has become one of another world war.

An individual facing these issues inevitably recoils, for what can a single person do to change the momentum of a species? But remember that cultural as well as biological evolution acts through *individuals* in a population. The technology that has made our greatest achievements possible has also originated with individuals—from the vision of the first deliberately shaped tool to a vision of a species at peace with the environment, and with itself. Whether productive human societies continue depends largely on as yet undefined social and economic innovations. But these innovations will emerge only when the collection of individuals making up society recognizes the new set of imperatives. If you renounce your responsibility to use resources wisely, if you choose not to consider the urgent need for population control, then we surely must fail. For how can others be expected to sacrifice more than you yourself are willing to sacrifice? *The momentum of our species, be it in the direction of catastrophe or survival, has become the sum total of our commitments as individuals.*

We have come far enough to realize, perhaps for the first time, that what has been considered "good" in the past is not necessarily consistent with survival. It is not merely survival but survival with dignity and purpose that constitutes the greatest long-term "good" for our species. Thus we must now call on an ability we share with no other life form—an ability to make a choice regarding limited resources and reproduction of our kind. We are fast approaching real limits to growth, and decisions can no longer be left to the next generation. It is not likely that we will in our time be able to conceive of and implement new social patterns. But if we don't make a beginning now, the luxury of choice may be lost. Facing our self-inflicted dilemma will mark the end to the childhood of our species. But with enough commitment, it may yet come to represent the finest application of the unique talents with which we have been endowed.

Recommended Readings

American Academy of Arts and Sciences. 1973. "The No-Growth Society." *Daedalus*, 102:4. See especially D. Kingsley's analysis of zero population growth.

Ehrlich, P. and A. Erhlich. 1972. *Population, Resources, Environment.* Second edition. San Francisco: Freeman. Sobering look at human ecology. See especially the chapters on population dynamics.

Elliott, R. et al. 1970. "US Population Growth and Family Planning: A Review of the Literature." *Family Planning Perspectives* (October), vol. 2. Balanced analysis of the literature.

Polgar, S. 1972. "Population History and Population Policies From an Anthropological Perspective." *Current Anthropology*, 13:2, pp. 203–241. Analyzes often-ignored cultural barriers to programs for population control.

Population Reference Bureau. 1972. "Population Statistics: What Do They Mean?" *Population Profile* (March). One of the best introductions to demographic terms and concepts.

Silverman, A. and A. Silverman. 1971. *The Case Against Having Children.* New York: David McKay. Looks at traditional motherhood myths, explores alternative roles for women.

Chapter Twenty-Two

Figure 22.1 One consequence of human population growth—the replacement of stable, mature ecosystems with vast and vulnerable monocrops.

Stability and Change in the Biosphere

Tropical reef, savanna, tundra—before you begin this chapter, turn back to the photographs in Chapter One. You may be surprised to discover how much more you know about life in these and all places, for you have learned something about the meaning of its unity and diversity. You know that from the time of their common molecular and cellular beginning, all organisms have had to interact with the physical environment and with one another in acquiring energy and material resources. You know these interactions have led to diversity in form and behavior. **Ecology** encompasses all these interactions. *Thus it encompasses, directly or indirectly, all you have read in this book.*

In its attempt to integrate all the parts into the whole of life, ecology is perhaps the most ambitious of all undertakings. Consider what it must mean to follow the consequences of just one event through the levels of biological organization. Say that a mutation in the DNA of a disease-causing virus makes its target host cell more vulnerable. Greater vulnerability at the level of cells can lead to the collapse of tissues, of an organ; the collapse of one organ can cause all other systems to break down; individuals can die. If the new viral strain is potent enough, a population may be brought perilously close to extinction. And the dwindling numbers of one population may have major effects on the entire **community**—the interacting association of organisms in a common environment.

The effects of such ecological disruptions are something like ripples radiating outward from a stone that has been tossed into a pond. Much of ecology is concerned with identifying the consequences of those ripples at some specific level of biological organization. But in a broader sense, ecology is also concerned with the stability of the whole pond and its surroundings—with the **ecosystem,** the sum total of all the interactions linking organisms in a community with one another and their environments.

Of course, this sweeping perspective is one reason why ecologists in general find it difficult to remain passive bystanders as they observe not only the pebbles but the rocks and boulders of human activities being thrown into one pond after another, until all that's left is rubble. But in actively seeking to explain why it is important to maintain the integrity of natural ecosystems, ecologists have found they must first ask why people are disrupting them in the first place.

Consider, by way of introduction, an ecosystem on the northern edge of the Kalahari Desert. Here, in the nation of Botswana, the longest river in southern Africa ends in an immense fresh-water delta. The delta supports lush grasses, brush, and groves, which in turn sustain one of the most magnificent arrays of wildlife imaginable—exotic birds, crocodiles, hippopotamuses, buffaloes, elands, kudus, wildebeests, elephants, lions, and many more. The animals, in their turn, are host to bloodsucking mosquitoes and tsetse flies. These insects transmit the dread sleeping sickness to humans and cattle. Hence, by the very fact of their presence in the delta, they have effectively kept humans and cattle out.

But it happens that cattle raising is Botswana's economic mainstay. Cattle need water and grazing land, which are becoming increasingly scarce. Pressure is on to open up the delta for the cattle industry by eradicating the tsetse fly and the mosquito. That means clearing the woodlands, burning off the brush—and destroying the wildlife. Why would people do such a thing? For profit, of course, but also to feed other people. Anyway, how can we demand that they *not* do such a thing? Would Californians so readily turn over the fields and vineyards of their fertile inland valleys to quails and rabbits, coyotes and hawks? Would Texans so readily set aside their open range for the preservation of zebra, kudu, wildebeest, not to mention the occasional lion? Would Nebraskans so readily donate their fields of waving grain to support human populations in Africa, so that the African wildlife can be left alone? The questions we ask are relative to what part of the world we happen to live in. And there simply are no easy answers at the regional level. If there are answers at all, they will come only with recognition of the whole earth as one ecosystem—as one **biosphere,** with mutual cooperation for the long-term stability of the whole.

Community, ecosystem, biosphere—these are the interconnected topics of this chapter. First to be explored are principles governing community structure. You will see how these principles affect the interactions among organisms in ways that tend to stabilize energy and material resources within the space of a given natural environment. You will also see how these principles operate as communities change with time. Finally, you may come to see that in working with instead of against these principles, we may at last achieve the long-term stability that is potentially ours to share with the world of life.

Community Structure

The Niche Concept In attempting to identify the structure of a natural community, ecologists often begin with the niche concept. A **niche** is a functional description of a species' role in the community; it is an expression of the range of all variables that influence whether a species has the resources it needs and whether it can carry out all the activities needed to live and reproduce. What energy and materials does it demand, and in what amounts? How much water does it need? What is the range of temperature, wind, shade, and sunlight it can tolerate? What is the extent of its **habitat**—the physical environment, with its characteristic array of organisms, in which the species lives and reproduces? What other species does it depend upon or compete with, and in what numbers? What other organisms are predators or parasites on it? Answers to such questions describe the niche of a given species, so let's take a look at what some of the answers can be.

Symbiotic Relationships First of all, what does it really mean to define a niche partly in terms of the dependency of one species on others? After all, today no species can live for long in isolation from others. But some species happen to be more demanding than others about the kinds of organisms with which they interact. Sometimes one or both members of two species come to rely on the presence of the other for survival. Such a relationship is called **symbiotic** (the word simply means "living together"). Not all symbiotic relationships are the same. They differ in degree of dependency, in how exclusive the attachments are, and in the extent to which one species is helped or harmed by the presence of the other.

The weakest symbiotic attachment between species is **commensalism.** Here, one species (the "guest") simply lives better in the presence of another species (the "host"). The host species does not garner distinct benefits from the presence of the guest species, but neither does it suffer serious harm. For instance, robins and fruit flies are both commensal with humans. It's not that they find it impossible to live without us. It's just that, for them, the carrying capacity of the environment is usually *highest* in places where humans live. There are concentrated resources such as shade trees, earthworm-rich lawns, and berries (which robins exploit) and overripe fruit (which fruit flies exploit). We may find ourselves enjoying the territorial song of a male robin, or we may be mildly disturbed as fruit flies swarm persistently over a fruit bowl. But most of the time we aren't even aware of the presence or absence of either species. Such is the weak involvement in most commensal attachments.

In **mutualism,** the bonds between members of two species are stronger than they are in commensalism, for positive benefits flow in both directions. Most of these attachments probably begin in commensalism and gradually become transformed into mutual dependence. Stunning examples are found among flowering plants and their insect pollinators. As you read in Chapter Seventeen, a long evolutionary history lies behind this form of symbiosis. The pollinator depends on the plant for food, even as the plant depends on the insect for reproductive success. Even so, their mutual bonds are not always exclusive. If honeybees find clover blooming profusely, they concentrate on collecting food from clover. When the blossoms of that plant species fade, they turn to honeysuckle, apple blossoms, or whatever other flowers are available to them. Similarly, if honeybees are not around, then bumblebees, butterflies, or moths may pollinate the clover.

As time goes on, mutualism may become more and more exclusive. Two species can become structurally or behaviorally coadapted to each other because of some selective advantages. The yucca moth (Figure 22.2) has come to obtain its pollen from one plant only; even its larval stage dines only on yucca seeds. The yucca plant depends exclusively on this one moth pollinator. Hence the moth's private energy source, available throughout its life cycle, helps assure reproductive success. At the same time, the moth helps assure reproductive success for the plant: its pollen is carried exactly where it must go instead of being randomly spread about by a less picky pollinator that visits different plant species.

Whenever selective advantages of this sort lead to specialized structures or behaviors that make each partner totally dependent on the other, the two species have entered an **obligate relationship.**

Perhaps the most extreme form of symbiosis is the one-way relationship called **parasitism,** in which one species benefits at the expense of the other. A parasitic relationship probably begins in commensalism. But one member of the guest species may undergo random genetic change that permits it to exploit the energy reserves of its host. Then it gains a survival advantage. Suddenly it takes less energy to get energy, which has no small bearing on survival and reproduction. Because the population to which the variant form belongs gradually gains a competitive edge, there is selection for the parasitic life-style. Compared with their nonparasitic relatives, parasites tend to be structurally simple. Once they have tapped the energy stores of their host, they are no longer under selective pressure to retain their own energy-procurement systems. (Thus some species of mistletoe, a parasitic plant that lives by sinking its roots into live oak trees and tapping the phloem tissue of its host, may lack chlorophyll.) Of course, in every stable parasitic relationship, the benefits gained by the parasite must stay within certain limits. A parasite that could extract all available energy from its host would kill the host—which would mark the end of the parasitic species' preferred niche in the community.

a

b

c

d

All photographs Harlo H. Hadow

Figure 22.2 Coevolution in the high desert of Colorado. There are several species of succulents called yucca plants (**a**), but each has coevolved exclusively with only one kind of yucca moth species (**b**).

The adult stage of the moth life cycle coincides with the blossoming of yucca flowers. Using mouth parts that have become modified for the task (**c**), the female moth gathers up the somewhat sticky pollen and rolls it into a ball. Then she flies to another flower and, after piercing the ovary wall, lays her eggs among the ovules. She crawls out the style and shoves the ball of pollen into the opening of the stigma.

When the larvae emerge (**d**), they devour about half the seeds of the yucca plant and then gnaw their way out of the ovary to continue the life cycle. The seeds remaining are enough to give rise to a new yucca generation.

So refined is this coevolved dependency that the moth and larva can obtain food from no other plant, and the flower can be pollinated by no other agent.

The Concept of Competitive Exclusion Symbiotic relationships are a major consideration in defining the niche of a species. Still another important consideration is the competition that may exist between them. If two species have some requirement or activity in common, their niches will overlap. The greater the overlap, the greater the possible competition between them. They may compete for some resource that limits reproductive success—say, a certain food, a nesting place, a scarce mineral in the soil, or sunlight. When that happens, one of the two species may experience **competitive exclusion:** a gradual displacement from the site. The reason one gives way to the other is that no two species are exactly the same in their activities or their ability to acquire resources. In some small or large way, one will have the advantage. The other will be forced to modify its niche or it will perish. As a result, *no two species in a community occupy the same niche indefinitely.*

Of course, there are ways of minimizing niche overlap other than total exclusion. For instance, water is a limiting factor in a desert community. So some species might dominate a waterhole during different times of the day, others might dominate during the night. One species might dominate during a prolonged dry season, others during a brief rainy season. *In any number of ways, niches in a community might be partitioned in time as well as in space in ways that minimize friction between competing species.*

But balancing the tendency to minimize overlap is a tendency for all species to expand niches in search of resources. Thus a continual state of tension underlies natural communities. *It is a constructive tension, for it means each species is constantly being tested in terms of its ability to maintain or expand its niche, to fill in gaps that may exist in the current community structure, or to compete more effectively than other species as conditions change.* The overall effect is a fine-tuning of all species to their environment and to one another's presence, so that optimum use is made of available resources.

Food Webs and Energy Flow The niche organization of a community is structured around **trophic levels:** an interlocked array of species through which materials and energy flow (Figure 22.3). The **producers,** mainly photosynthetic organisms, are responsible for the community's net primary productivity (which is, recall, the photosynthetically produced food left after the producers have used what they need). Other organisms in the community are **consumers** of one sort or another. The primary consumers are *herbivores;* they feed directly on photosynthetic organisms. The secondary consumers include *carnivores,* which eat herbivores (as well as other carnivores); *scavengers,* which eat organic refuse or carrion; and *parasites,* which thrive on the tissues of living hosts. Finally, there are **decomposers:** fungi and bacteria that break down organic debris and thereby help recycle vital materials to the producers.

This general sequence of who eats whom is sometimes called a "food chain." But this term implies a simple linear relationship that is seldom seen in real communities. Imagine a fisherman dropping his net in the sea to catch some fish that are feeding near the surface on suspended algae. Come lunchtime, he fries up some of his catch, demonstrating his carnivorous tendencies. Should he later lose his footing on the deck and fall into (whoops!) shark-infested waters, the "chain" might be portrayed:

$$algae \longrightarrow fish \longrightarrow fisherman \longrightarrow shark$$

But clearly this would be an oversimplification of the feeding patterns going on in the community; it would exclude any number of alternative routes for energy flow. Most likely, protistans and microscopic animals (foraminiferans, sea stars, copepods, shrimp, sea urchin larvae, and molluscs) were also grazing on the algae. Squid and assorted large fish might have been munching through the school of fish. The fisherman might have accompanied his fish fillets with potatoes and apple pie; he might even have washed everything down with too much wine (hence his sloppy footing). Thus he would shift back and forth between carnivore and herbivore—and would even relish the waste products (alcohol) of decomposers. If he had harpooned a shark instead of falling prey to it (and assuming he was not a picky eater), he might have ended up as top carnivore. Thus

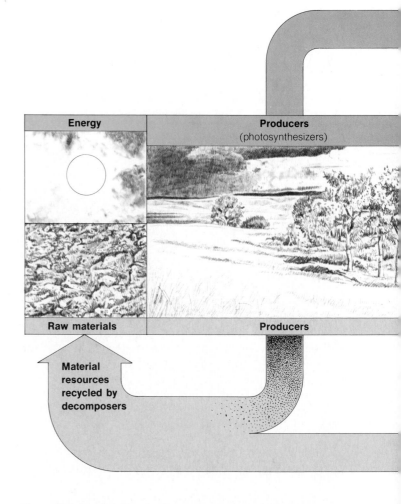

Figure 22.3 Simplified version of trophic levels in a natural community. Notice that raw materials may be recycled, but that with each transfer to a new trophic level, some energy is lost (hence the diminishing size of energy flow arrows). Sunlight provides the only energy input for the entire community.

the pattern of "who eats whom" in a community is better expressed as a **food web** rather than a food chain.

An important aspect of energy conversions through a food web is that at each higher trophic level, there is considerably less energy to be passed on. Because of built-in conversion losses, usually less than 10 percent of the energy transferred to a new trophic level remains as potential energy. Thus, as you read in Chapter Six, *a growing animal must take in 10 kilocalories of photosynthetically derived energy to produce every 1 kilocalorie of stored energy (or potential food energy for predators).* As a first approximation, we could say it took 1,000 kilocalories of energy stored in the algae to produce 100 new kilocalories stored in the fish, which

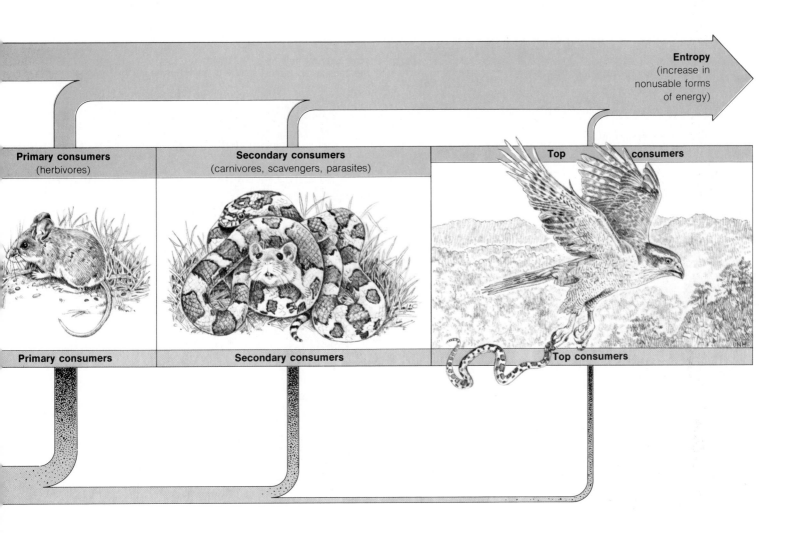

Primary consumers
(herbivores)

Secondary consumers
(carnivores, scavengers, parasites)

Top consumers

Entropy
(increase in
nonusable forms
of energy)

Primary consumers

Secondary consumers

Top consumers

produced 10 new kilocalories in the fisherman—which yielded only 1 new kilocalorie of stored energy in the shark.

But there is still another important factor that influences this picture of energy conversions through a food web. Energy storage becomes more and more inefficient over time, because considerable energy must be spent to maintain the body once it has stopped growing. Assume the fisherman is forty years old. Assume also he took in an average of 2,700 kilocalories a day ever since he was a teenager. That comes to about a million kilocalories a year. Although his food intake was a lot less during early development, his total consumption probably exceeded 30 million kilocalories. But once he reached adult size and

weight, every kilocalorie he consumed was used up in maintaining all his body's cells, and in providing them with energy for work. When he fell into the sea, he had retained only about 250,000 kilocalories—or less than 1 percent of the total number he had consumed during his lifetime! *The greater the time interval between energy conversions from one trophic level to the next, the lower the efficiency of the transfer—above and beyond the built-in 10-to-1 conversion loss.*

Energy Budgets for the Community Is it possible to determine how much energy flows not through a single individual but through an entire community in a given time

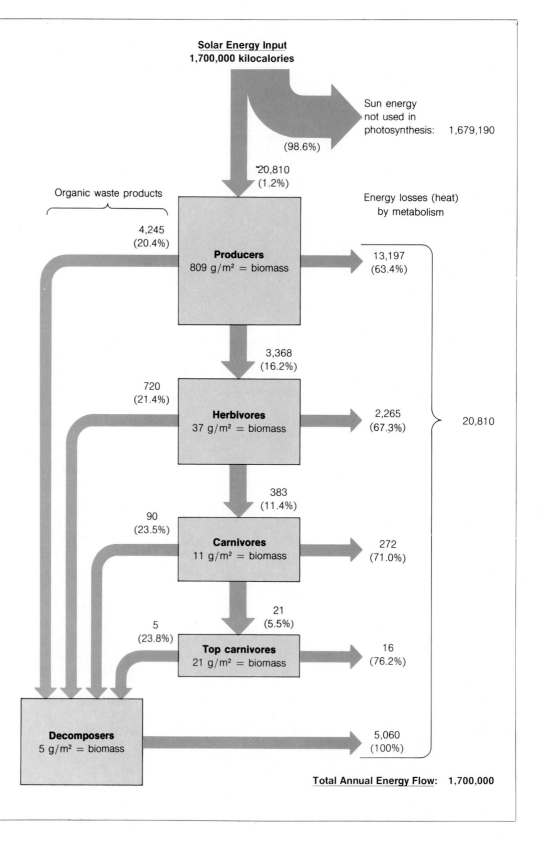

Figure 22.4 Annual energy flow (measured in kilocalories per square meter per year) for an aquatic ecosystem in Silver Springs, Florida.

The producers are mostly green aquatic plants. The carnivores are insects and small fish, and the top carnivores are larger fish. The energy source (sunlight) is available all year long.

Only 1.2 percent of the incoming solar energy is actually trapped in photosynthesis to generate new plant biomass. And more than 63 percent of the photosynthetic products are metabolized by the plants themselves to meet their own energy needs. Only 16 percent is harvested by herbivores, and the remainder is eventually decomposed by bacteria and fungi. Similarly, most of the herbivore energy is expended in metabolism and goes into the decomposer system; only 11.4 percent is consumed by carnivores. Once again, the carnivores burn up most of the energy they take in and only 5.5 percent is passed on to top carnivores. The decomposers cycle and recycle all the biomass received from all other trophic levels. Eventually all of the 5,060 kilocalories will appear as heat produced during metabolism. (Decomposers, too, are eventually decomposed.)

This diagram has been deliberately oversimplified. No community is completely isolated from all others. Organisms and materials are constantly dropping into the springs. And there is a slow but steady loss of other organisms and materials that flow outward in the stream that leaves the community. Over time, these inflows and outflows balance one another. (After H. T. Odum, "Trophic Structure and Productivity of Silver Springs," *Ecological Monographs*, 27:55—112, 1957. Copyright 1957 by the Ecological Society of America)

Solar Energy Input
1,700,000 kilocalories

Sun energy not used in photosynthesis: 1,679,190

(98.6%)

20,810
(1.2%)

Organic waste products

Energy losses (heat) by metabolism

4,245
(20.4%)

Producers
809 g/m² = biomass

13,197
(63.4%)

3,368
(16.2%)

720
(21.4%)

Herbivores
37 g/m² = biomass

2,265
(67.3%)

20,810

383
(11.4%)

90
(23.5%)

Carnivores
11 g/m² = biomass

272
(71.0%)

21
(5.5%)

5
(23.8%)

Top carnivores
21 g/m² = biomass

16
(76.2%)

Decomposers
5 g/m² = biomass

5,060
(100%)

Total Annual Energy Flow: 1,700,000

period? More than this, is it possible to find out how the energy is divided up among different trophic levels? Let's consider a few ways of approaching these questions. In one model, energy flow and apportionment is depicted as a pyramid. The shape of the pyramid roughly corresponds to actual counts made of all the organisms in a community. For example, we might find the following relationship in a bluegrass field:

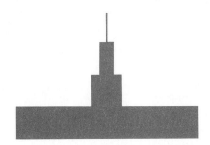

3 top carnivores
(birds, moles)

354,904 carnivores
(spiders, ants, beetles)

708,624 herbivores
(invertebrates)

5,842,424 producers
(grasses, weeds)

Aside from the monumental patience needed to count the organisms and prepare the pyramid, the model itself is not very instructive. For one thing, it defines the community structure at only one brief moment—and the structure may change cyclically or permanently over time. For another thing, the size of the organisms being counted in each trophic level can distort the picture. A similar count in a redwood forest would yield a small number of large producers (the trees), which still manage to support a large number of herbivores and carnivores (insects). And one deer would be counted as a single herbivore, as would a single insect—even though a deer eats far more than an insect does.

We can overcome a few of these problems if we weigh individuals in each trophic level instead of counting them, thereby coming up with their *biomass* (the total weight of all organisms in a given category). It's better than a simple head count. But it still has its shortcomings. For instance, in communities where producers are tiny but grow and reproduce rapidly (algae of marine communities do this), the biomass pyramid may be upside-down! The consumer biomass at any instant would actually exceed the producer biomass, because algae are consumed about as fast as they reproduce. It's just that the survivors, few as they may be, are reproducing at an astonishing rate.

Such variations can be accounted for if we calculate the *total biomass* each species produces in a year's time. This approach brings us closer to understanding energy flow through a community, but there is one last problem. Not all organisms of the same biomass have the same impact on the community. A warm-blooded mouse, for instance, is much more active and consumes far more food that a cold-blooded lizard of the same size and life expectancy. Thus our picture of energy flow will be complete only if we work out the **energy budget** for all its individuals. We have to

determine how much energy they take in, how much they burn up in metabolism, how much they excrete as waste products, and how much they store in their body. It is a difficult thing to do, but it has been done in some cases (Figure 22.4). And such studies have revealed that underlying all communities—even those of highly diverse types—is a general principle:

In no way can a natural community survive indefinitely if it expends more energy than it takes in.

A community can't spend any more than the energy contained in its annual primary productivity. It may underspend, in that it may not use up all the primary production available. (The excess might be stored in the wood of trees, for instance, or in the leaf litter piled up on the forest floor.) But in no way can a community routinely overspend its energy budget. A community that expends more organic compounds to produce energy than are added by processes such as photosynthesis will inevitably dwindle away.

This simple principle reveals one of the reasons for the energy crisis that the human population is experiencing. For millions of years, the global community of life has been accumulating a reserve of fixed carbon in the form of "fossil fuels," such as coal and oil. But the rate at which we are now dipping into the global bank account so exceeds the rate at which new stores are being added that the biosphere is not only running in the red, it will soon be almost completely depleted of such resources (Chapter Twenty-Three).

Does Diversity Mean Stability? Through interlocking food webs, a natural community cycles materials and energy. In these cycles, recall, consumers and decomposers assure that the essential nutrients removed from the soil and water by one generation of producer organisms are ultimately returned to nourish a new generation. Hence we have the great carbon and nitrogen cycles (Chapters Three and Five), as well as many more, which help assure the community's overall stability.

But constantly testing the stability of natural communities is change—change in the type and number of individuals, change in the environment through floods, fires, or pollution and the sprawl of human activities. The effect of such disturbances on community structure raises two questions. First, will the community be more likely to endure if it has a simple array of species in its trophic levels, or a diverse array? Second, will it be more likely to endure if it has food webs that are rigidly fixed, or that cross-connect to provide alternative sources of materials and energy as conditions change? These questions are being actively debated at this time, for there is not complete agreement among ecologists on the answers. But most believe that the greatest stability is found at some intermediate level of complexity. *It appears that both very simple and very complex communities are less likely to recover from disturbances than a community having some intermediate level of diversity.*

A simple community such as that found in the arctic tundra is vulnerable even to slight disruptions, because few alternatives are open to the populations making up its trophic levels. At the other extreme, the trophic levels of a highly complex community are partitioned in delicate interdependence of one sort or another. For instance, the basis for the complexity of a temperate forest is the array of trees that capture sunlight energy. By the time sunlight travels from the top canopy of leaves down through the intermediate layers, less than 5 or 10 percent may reach the forest floor. All these stratified layers of stored energy are the basis of food webs, including insects, birds, mice, and squirrels. On the forest floor are larger herbivores such as moose and deer, which use the trees not so much for food as for cover against predators. The predators (such as wolf, lynx, and mountain lion) use the forest for shelter against such climatic extremes as blizzards. The dense foliage stabilizes the local microclimate by cutting down winds, acting as a buffer against fluctuating temperatures, and holding back moisture that otherwise would escape from the community through evaporation. The forest trees also enrich the surface soil with nutrients mined from the depths by their roots, and then later dropped in leaves. Extensive tree roots prevent soil erosion. Although such a complex forest community tends to be tolerant of many minor disruptions, it can't survive a severe fire, clear-cutting by lumberjacks, or the introduction of a new leaf-stripping insect. For such major disruptions destroy the trees that are the *energy foundation* for the entire community structure.

Succession: Change in Communities With Time

From Pioneers to Climax Communities Glacier Bay in Alaska is remarkable for more than its spectacular scenery (Figure 22.5). Nowhere else in the world are glaciers retreating so rapidly, and nowhere else have changes in the newly deglaciated areas been documented so carefully for so long a time. A comparison of maps from 1794 onward shows the ice has been retreating at rates ranging from 3 meters a year at the sides of the glacier to a phenomenal 600 meters a year at its tip over bays and inlets. It is here that biologists have had one of the best and most prolonged opportunities to observe the processes by which vacant land first becomes populated with life forms and goes on to become a stable, complex community.

As the glacier retreats, the constant flow of melt water tends to deplete the soil of mineral nutrients, especially nitrogen. Thus the newly exposed shoreline usually is

Figure 22.5 Ecological succession in the Glacier Bay region of Alaska. McBride Glacier is about a mile back from Muir Inlet. Early visitors include gulls and other birds (**a**). The early pioneer stage includes the flowering plant *Dryas* (**b,c**), horsetails, and fireweed (**d**). The soil in (**e**) was covered with ice less than 10 years ago. Within 20 years, young alders, cottonwood, and willows appear in drainage channels (**f**). Within 50 years, (**g**) Sitka spruce has appeared. By 80 years, mature alders are being crowded out by cottonwood and spruce (**h**). In an area deglaciated more than 130 years, a dense Sitka spruce, western hemlock forest forms the climax community (**i**).

McBride Glacier

f

g

h

i

All photographs Roger K. Burnard

unable to support plant life. But soon the water near the edge is populated with nitrogen-fixing blue-green algae. As algae die, they provide a natural fertilizer for the soil. Soon horsetails, sedges, cotton grass, and other plants begin to appear along the margins of the land. They are soon accompanied by mosquitoes, diving beetles, and water striders. Sandpipers, gulls, rock ptarmigan, and other birds begin visiting the site. All these organisms are pioneers, adapted to moving quickly into new, barren areas. By their very presence, however, they begin to alter the environment, for organic wastes from birds accumulate and raise the soil temperature, change its acidity, and add various nutrients to it. Meanwhile, the first short plants offer some protection for young plants of other species—and in this way they set the stage for their own rapid displacement. New organisms move in, competing more effectively than the pioneers for niche space and eventually crowding them out. Over the years, deciduous trees slowly take hold— alder, then mature willows and poplars. A few evergreen spruce trees germinate in the shade of the deciduous trees. In eighty years they will come to dominate the landscape. And decades later, they will give way to giant hemlocks. Through all these stages, different species of microorganisms, plants, fungi, and animals make their entrance and then gradually depart as the nature of the community changes. *And all the while, the total biomass slowly increases, offering more possibilities for partitioning of niches.* Eventually the site is dominated by a **climax community:** an array of species locked together in such an efficient use of materials and energy that it is stable and self-perpetuating.

What you have just read is a summary of the process of **succession**—a gradual, sequential replacement of communities, each somewhat more diverse than the one before, until the stable climax stage is reached. The process may take hundreds of years; sometimes it takes thousands.

Succession may also occur when an established community is disrupted in whole or in part. Consider what happens when part of a spruce forest is clear-cut for its lumber. Reestablishing the community in the exposed area is not simply a matter of seeds from nearby spruce trees drifting over and germinating. *Most dominant plant species in a climax community cannot grow or develop unless a certain integrated community structure already exists.* Normally, spruce seeds will germinate only in the litter of a forest floor, and they develop into saplings only if an existing forest canopy shades them. So they can't repopulate the stripped land right away. Instead, pioneer species tolerant of the exposed conditions become established during the first few growing seasons. In this case the first arrivals are annual weeds, which produce abundant seeds, have highly efficient ways of dispersing them, and are capable of germinating in full sunlight. Perennial weeds and grasses follow them into the area. Year after year the perennials send out more roots and shoots, increasing in size and gradually becoming dominant in the competition for sunlight and resources. As fallen leaves, dead plants, and other debris accumulate, the decomposers begin to thrive. Only when a rich layer of decomposing debris has accumulated do the trees characteristic of more mature communities start to germinate. Only then can the shrubs and trees of the climax stage go on to become reestablished once more. Such partial disruptions are one of the reasons why you can see more than one type of community in the same general region. They are passing through different stages of succession.

Much of what we consider to be permanent features of the landscape are only transitional stages on the way to a climax community. For instance, most of the lakes of North America formed when the glaciers of the last ice age began retreating, about 11,000 years ago. Succession for all these lakes, large and small, began at that time. And for all of them, the successional pattern has been unfolding in this way: Streams and rivers leading into them steadily deposit sediments, and aquatic communities begin producing organic debris. As the lakes become shallower because of the deposits, different plant species become established in the water or around the shallow lake margins. This brings in different consumers and more organic debris. Gradually the deep, clear lake waters become so enriched with plant and animal by-products that they become bogs, then swamps, until finally the basins are completely filled. Once that happens, the climax communities characteristic of the lake regions become established.

The filling in of a lake basin with organic and inorganic debris is a form of **eutrophication.** Normally it takes a few centuries for a small lake to fill in, and more than 10,000 years for a large one. By enriching lake waters with fertilizer run-off, with sewage, and with industrial wastes (thereby accelerating the growth of encroaching plants), we can bring succession to completion within our lifetime. But dead lakes rapidly on their way to becoming bogs are not necessarily desirable from aesthetic, recreational, or economic standpoints. Therefore, efforts are being made to slow down eutrophication of the 100,000 or so lakes in the United States. (Substantial progress is being made in reversing the eutrophication of Lake Erie. It will take ten times as long to clean up Lake Michigan.)

The Other Side of Smokey the Bear So far, you may have formed the opinion that all disturbances to natural communities are bad. But some communities *depend* on intermittent disruptions for maintaining stability. For example, in the Sierra Nevada of California are isolated sequoia groves, some of which contain giant trees that are more than 4,000 years old. Many of these groves are protected as part of national or state park systems. Among other things, protection has traditionally meant minimizing the incidence of fires—not only accidental fires from campsites and discarded cigarettes but also natural fires touched off by lightning. The fire prevention and suppression program has been highly successful. The only problem is that fires at regular intervals are essential for the long-term stability of the sequoia community!

Sequoia seeds can germinate only on bare mineral soil; if there is extensive litter on the forest floor, there won't be any new sequoias. Modest fires eliminate the litter. They also eliminate other shrubs and trees that compete with the sequoias—yet they do not damage the sequoias themselves. The reason mature sequoias resist fire is that they have bark as thick as your arm is long. The bark burns poorly and insulates the trees against heat damage. But when small, periodic fires are prevented, litter piles up. Other species appear that are fire-susceptible. Even though they don't actually displace the sequoias, the sequoias are no longer reproducing. The litter and undergrowth represent so much potential fuel that fires are hotter than they otherwise would be—hot enough to damage the giants. *The point is, without an understanding of the natural interactions maintaining a community, efforts to preserve them may have the opposite effect.* (Smokey the Bear has since been "retired." His simplistic campaign is being replaced by a more enlightened program of public education based on ecological principles.)

Working With Instead of Against Succession Net productivity is high in pioneer communities, for the absence of complex food webs means fewer energy conversion losses, and the time interval between conversions that do occur is brief. But as pioneer communities are replaced with ever more diverse communities, the amount of untapped biomass reaches a plateau. At the climax stage, incoming usable forms of energy are balanced with outgoing nonusable forms. Stability is achieved, but there is very little energy left over—which is why climax communities can't sustain large human populations. For us, the alternative is crops: low-diversity "pioneer" communities that are high in productivity but low in stability. They give the illusion of stability only because of massive imports of energy, fertilizers, pesticides, and (usually) water.

Obviously our dependence on crops isn't going to vanish. But in some way, we must integrate agricultural communities with the surrounding natural communities. The ecologist Eugene Odum has advocated an intermingling of communities of different ecological ages. Forest communities, for instance, are valuable for more than lumber. They are production sites and holding stations for water and minerals that gradually wash down to replenish surrounding valleys and plains. Strip the forests and we strip the surrounding lands of vital resources. Although in the short term there may be lower production levels if forests are preserved, in the long term there will be less vulnerability to insect predators and diminishing resources. *A diverse array of crops, complemented by a diverse array of estuaries, forests, and streams, would be a compromise that offers some stability for <u>both</u> agricultural and natural communities.*

Ecological Backlash: The Case for Compromise

It is not only for the sake of our crops that we must come to terms with natural communities. Too often we are finding that our disruptions of natural systems have disastrous consequences not only for our crops but for our economy, even our lives. Four diverse examples will make the point:

DDT and Thee For much of human history, we have been at war with insects that destroy crops and transmit diseases. The organic compound DDT, which sends insects into convulsions, paralysis, and death, has been instrumental in bringing many of the worst offenders (such as malaria-transmitting mosquitoes) more or less under control. But DDT is a stable compound that cannot readily be broken down. It can persist in the environment for as long as fifteen years. Because of its stability, it is a prime candidate for **biological magnification**—the increasing concentration of a nondegradable substance as it moves up through trophic levels. Why does it become so concentrated? Recall there is about a 10-to-1 conversion loss at each energy transfer from one trophic level to the next in a food web. That's the same as saying it takes 10,000 pounds of algae to produce 1,000 pounds of animal plankton that will produce 100 pounds of small fish. These small fish, when consumed by larger fish, will produce 10 pounds of food, which in turn will produce 1 pound of, say, brown pelican. But DDT is nondegradable and insoluble in water—and it dissolves in fat. Thus it tends to accumulate in the fat and oil reserves of the body. It is then passed on to the next species in the food web. Thus the DDT present in all that biomass ends up in the top consumer organism. The consequences have been unfortunate for the brown pelican, and they have also been unfortunate for us.

Consider what happened back in 1955, when the World Health Organization stepped in with a DDT spraying program to eliminate malaria from the island of Borneo, now a state of Indonesia. That step was not taken lightly. Nine out of ten people there were infected with this terrible disease, which is epidemic proportions by anybody's standards. The program worked, insofar as the mosquitoes transmitting malaria were brought almost entirely under control. But DDT happens to be what is known as a broad-spectrum insecticide; it kills nontarget as well as target species. Sure enough, the mosquitoes had company: flies and cockroaches that made a nuisance of themselves in the thatch-roofed houses on the island fell dead to the floor. At first there was much applause. But then the small lizards that also lived in the houses and preyed on flies and cockroaches found themselves presented with a veritable feast. Feast they did—and they died, too. And so did the house cats that preyed on the lizards. With the house cats dead, the rat population of Borneo was rid of its natural

Joseph R. Jehl, Jr.

Figure 22.6 Biological magnification through the food web ending with the brown pelican. The concentration of DDT had the effect of softening the shells of brown pelican eggs, and for a time the populations of these birds dwindled. With the recent ban on DDT, the brown pelicans are making their comeback.

predator, and rats were soon overrunning the island. Now, the fleas on the rats were carriers of still another disease, called the sylvatic plague, which can be transmitted to humans. Fortunately, the threat of this new epidemic was averted in time; someone got the inspired idea to parachute DDT-free cats into the remote parts of the island. But on top of everything else, some of the people of Borneo found themselves sitting under caved-in roofs. The thatch in their roofs was made of a certain kind of leaf that happened to be the food source of a certain kind of caterpillar. The DDT didn't affect the caterpillar but it killed the wasps that were its natural predator. When the predator population collapsed, so did the roofs.

On the Importance of Being a Hippo Hippopotamuses are short-legged, barrel-shaped mammals with practically hairless skin and a bulging mouth. They live in herds in African rivers and swamps, where they feed on aquatic plants and on plants lining the shore. Sometimes they also leave the water and munch on nearby crops. Sometimes (although rarely) they attack small boats. And for that they have been shot. In fact, in many regions they have been slaughtered in a campaign to clear the waterways.

But now, where hippos have been wiped out, the waterways are burgeoning with plant growth and filling up with silt. It seems the hippos kept the plant populations in check and, in their habit of digging out wallowing holes, they kept the silt from piling up. The now-shallow waterways have become ideal breeding grounds for the snail that is the intermediate host for the blood fluke—whose primary hosts are humans (Figure 22.7). About 200 million people now suffer from **schistosomiasis,** the ravages of the blood flukes. It is one of the most rampant infectious diseases in the world. And its effects are dreadful: stomach cramps, deterioration of vital organs, chronic exhaustion that permits little more than a few hours of activity a day. Sometimes death immediately follows infection. But more typically, the victims can expect only years of pain before their weakened condition makes them susceptible to a killer disease. "The Hippo's Revenge" might well be the title of a most informative documentary film.

The Incredible Spreading Imports In the 1880s, the water hyacinth from South America was put on display for the New Orleans Cotton Exposition. Flower fanciers from Florida and Louisiana carried home clippings of the blue-flowered plant and set them out for ornamental display in ponds and streams. Unchecked by natural predators and nourished by the nutrient-rich waters, the fast-growing hyacinths rapidly displaced many desirable native plants and choked off the ponds and streams. Then they went to work on rivers and canals. They are still there, and they are still bringing river traffic to a halt.

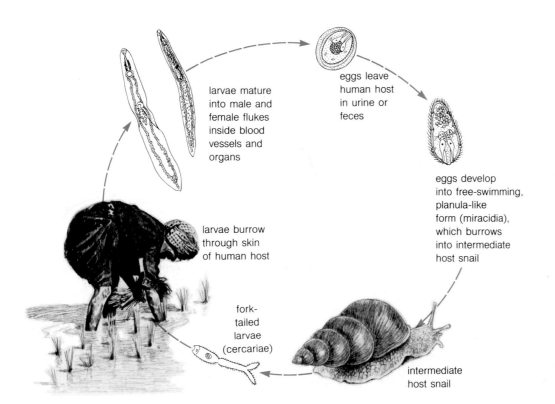

larvae mature
into male and
female flukes
inside blood
vessels and
organs

eggs leave
human host
in urine or
feces

Figure 22.7 Life cycle of a schistosome, a fluke thriving in tropical Africa and Southeast Asia. Adult flukes live in human blood vessels, where they perpetually copulate. The female fluke of some host species prefers to lay eggs in the veins around the host's bladder. Spines on the eggs pierce the vein walls, which creates local hemorrhaging. The blood carries the eggs into the bladder, then out from the host through urine. The eggs hatch into ciliated larvae, which swim about until they find and bore into a water snail. The snail acts as an intermediate host. Feeding on the snail's tissues, the larvae reproduce asexually in enormous numbers. Eventually a larval form (the cercaria) leaves the snail and swims about until it encounters a human (who is working, for example, in a rice paddy). It bores into its new host's skin and from there travels the blood vessels, to begin the cycle anew.

eggs develop
into free-swimming,
planula-like
form (miracidia),
which burrows
into intermediate
host snail

larvae burrow
through skin
of human host

fork-
tailed
larvae
(cercariae)

intermediate
host snail

Species introduction into established communities doesn't always have disastrous effects. Honeybees, mosquitofish, ring-necked pheasant—all have been absorbed into and have become part of community structures. Most of our own food—from apples, cabbage, wheat, oranges, to cattle and chicken—are the progeny of imports from other countries. But as Table 22.1 makes clear, we can't always say that nothing ever goes wrong, or that natural communities always recover.

Aswān and Other Fables Once there was a country that desperately needed food and energy for its growing population. It happened that one of the most magnificent rivers in the world flowed through this country. Each year the river deposited tons of mineral-rich silt on its fertile floodplain before it reached the sea. Why not dam the river, said the country's leaders, and use the water to irrigate more land, control the annual spring flooding of the river, and provide hydroelectric power all at the same time? The result of this modern-day fairy tale is known as the billion-dollar Aswān High Dam of Egypt, and not all Egyptians are living happily ever after.

For one thing, as water backed up behind the dam, almost 100,000 Egyptians had to choose between giving up

their family homes or being submerged along with ancient and priceless temples that were part of Egypt's cultural heritage. But there have been far more devastating results. Now that the Nile River floodplain is deprived of its annual enrichment with silt, artificial fertilizer has to be trucked in at a cost of 100 million dollars a year—a cost carried by the subsistence farmers who make, on the average, less than a hundred dollars a year each. Furthermore, now there is nothing to wash away the previous year's salt build-up in the soil. And with silt deposits no longer compensating for erosion, the fertile river delta is shrinking—and an alarming part of what remains has completely dried up. Restoring the delta with pumps, drains, and wells may cost more than the cost of the dam itself.

Ironically, evaporation as well as bottom seepage from the new lake filling in behind the dam is so great that the lake basin may never fill up to predicted levels. So nobody can live around the lake because nobody knows for sure where the shoreline will be. More seriously, there is less water to go around than there was before. And even though some 700,000 new acres (about 1.6 million hectares) have been opened up for agriculture, the population outgrew the potential food increase even before the dam was finished. At the same time, with the nutrient-rich flow of the Nile turned off, another major food source—the sardines,

Table 22.1 Effects of Introducing a Few Species Into the United States

Species Introduced	Origin	Mode of Introduction	Outcome
Water hyacinth	South America	Intentionally introduced (1884)	Clogged waterways; shading out of other vegetation
Dutch elm disease The fungus *Cerastomella ulmi* (the disease agent)	Europe	Accidentally imported on infected elm timber used for veneers (1930)	Destruction of millions of elms; great disruption of forest ecology
Bark beetle (the disease carrier)		Accidentally imported on unbarked elm timber (1909)	
Chestnut blight fungus	Asia	Accidentally imported on nursery plants (1900)	Destruction of nearly all eastern American chestnuts; disruption of forest ecology
Argentine fire ant	Argentina	In coffee shipments from Brazil? (1891)	Crop damage; destruction of native ant communities
Camphor scale insect	Japan	Accidentally imported on nursery stock (1920s)	Damage to nearly 200 species of plants in Louisiana, Texas, and Alabama
Japanese beetle	Japan	Accidentally imported on irises or azaleas (1911)	Defoliation of more than 250 species of trees and other plants, including commercially important species such as citrus
Carp	Germany	Intentionally released (1887)	Displacement of native fish; uprooting of water plants with loss of water fowl populations
Sea lamprey	North Atlantic Ocean	Through Welland Canal (1829)	Destruction of lake trout, lake whitefish, and suckers in Great Lakes
European starling	Europe	Released intentionally in New York City (1890)	Competition with native songbirds; crop damage; transmission of swine diseases; airport runway interference; noisy and messy in large flocks
House sparrow	England	Released intentionally (1853)	Crop damage; displacement of native songbirds; transmission of some diseases
European wild boar	Russia	Intentionally imported (1912); escaped captivity	Destruction of habitat by rooting; crop damage
Nutria (large rodent)	Argentina	Intentionally imported (1940); escaped captivity	Alteration of marsh ecology; damage to earth dams and levees; crop destruction

From David W. Ehrenfeld, *Biological Conservation.* Copyright © 1970 by Holt, Rinehart, and Winston. Adapted by permission of Holt, Rinehart, and Winston.

shrimp, lobster, and mackerel that flourished in the enriched waters off the delta—has declined catastrophically. Worse yet, the lake and the irrigation networks have so accelerated the spread of blood flukes that half the Egyptian populace are now carriers of schistosomiasis. In irrigated areas, where eight out of ten humans live, women can expect to live only to age twenty-seven, men to age twenty-five.

Perhaps in the long run there will be real benefits from this project. Its defenders are quick to point out that even though there isn't enough food to go around now, without the dam and its irrigation system there would be even less. The hydroelectric output should stimulate industrial development. Stocking the immense lake with fish should provide a new source of protein, and perhaps (assuming the blood flukes are controlled) it will even attract tourists. Regardless of the position taken on the dam's ultimate value, regardless of how well-intentioned the planning, development, and international support may have been, there is no defense for one important fact: warnings from ecologists *before* the project began were totally ignored.

The integrated structure and stability of existing communities must be considered <u>in advance</u> in order to determine what the impact will be of any deliberate disturbance—and we must come to recognize how vital it is to seek ways to minimize that disturbance.

Ecosystems in Space: The World's Biomes

It's one thing to isolate some of the characteristics of community structure. It's something else again when we turn to the structure of an **ecosystem**—to all the communities and environmental conditions in a large geographic region. Each part of the earth's surface is in some ways unique, with its own combination of mountains and lowlands, soil and rock, ponds or rivers or ocean currents, high elevations or low, its own nutrients, its own amount and patterns of light and moisture, its own pronounced or slight seasonal changes. And this means each of the world's communities is in some ways unique. But if we put aside the overlay of differences, we find that ecosystems can be clustered into a few basic types, or **biomes,** on the basis of their dominant array of primary producers. These arrays represent stable adaptations to prevailing environmental conditions, such as the interaction of climate and topography for a certain land mass.

When we seek to identify ecosystem structure, we must look first to the major primary producers on which it is based.

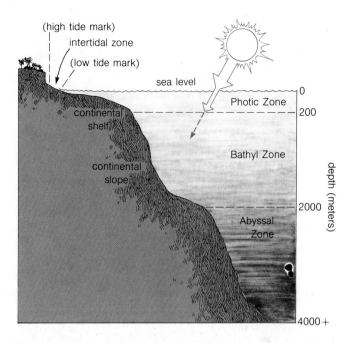

Figure 22.8 Zones of biological activity in the ocean. (The horizontal scale is enormously contracted here.)

Primary Productivity in the Seas The distribution of producer organisms in the vast province of the seas is governed by such variables as light, temperature, salinity, and available nutrients. All marine ecosystems—from the shallow **intertidal zone** between high and low tide marks, to the **abyssal zone** some 2,000 meters ($1\frac{1}{2}$ miles) beneath the ocean's surface—begin with the photosynthetic organisms in the upper 61 meters (about 200 feet) of water. Only in this **photic zone** (Figure 22.8) is the light intense enough to drive photosynthesis. Temperature and salinity in the deep ocean do not vary much, mainly because major currents tend to circulate water masses on a global scale. Hence ecosystems at great depths take on much the same character no matter what their geographic location. But at the surface, temperatures can range from 30°C (86°F) in the tropics to −2°C (28°F) at the poles. Salinity at the surface ranges from high concentrations in such tropical waters as the Red Sea to low concentrations in polar regions (because of melting freshwater ice).

In addition to these environmental variables, *we must consider the way materials are cycled back to the producers in order to understand why marine ecosystems are distributed as they are.* In shallow nearshore waters and coral reefs, decay occurs within the photic zone, so recycling of resources can proceed in much the same way that it does on land. Over open oceans, wastes and the bodies of dead organisms drift down to the bottom. Once there, this material can't decay quickly, because of the low temperatures. Even when it is

finally broken down, it is far removed from the narrow zone of photosynthetic organisms that could thrive on it. This means essential minerals such as nitrates and phosphates are constantly being removed from the surface waters. That's why the primary productivity and species diversity of open-water ecosystems are generally low compared to estuaries, nearshore waters, and tropical reefs.

Sometimes strong vertical currents, or **upwellings,** transport mineral-rich water from deeper zones to the surface. Upwellings commonly are created by offshore winds that drive warm surface water away from a coast, which allows colder water from below to move up and replace it. The productive ecosystems off the coasts of California, Peru, and Mauritania exist because of such upwellings. Similarly, the Antarctic Seas, cold as they are, have strong circulating currents moving up nutrients, which makes them highly productive regions.

Estuaries, coastal waters, upwellings—all these regions are being commercially fished at or above levels that are dangerous to their stability. More than this, as we turn to the seas to find more oil, gas, and minerals, the underpinnings of this stability may be weakened even more. We don't know for sure what the ultimate impact of our increased activities will be on this greatest of biomes, the world's oceans. But from what we know of our impact on natural communities in general, the prospects are not encouraging.

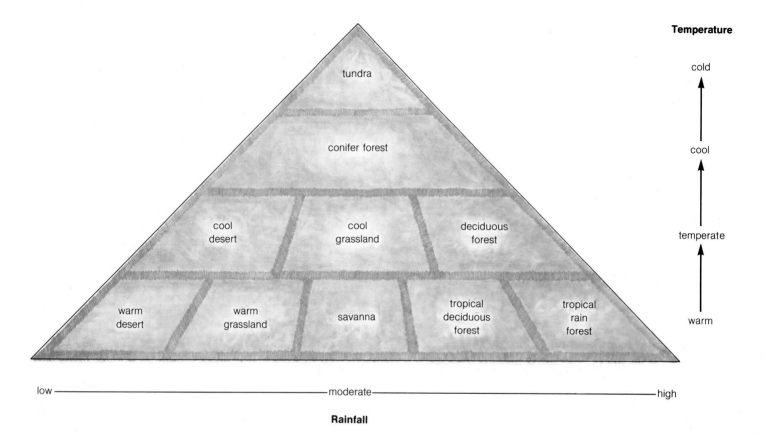

Temperature

cold

cool

temperate

warm

tundra

conifer forest

cool desert

cool grassland

deciduous forest

warm desert

warm grassland

savanna

tropical deciduous forest

tropical rain forest

low —————————————— moderate ————————————— high

Rainfall

Figure 22.9 Relation between temperature and rainfall, just two of many factors that help dictate the distribution of biomes on the earth's surface. Several major kinds of biomes are shown here. They are arranged as you might find them in traveling from the poles (top of the pyramid) to the tropics (its base), or in descending from the top of a high mountain. The colder the temperature, the less effect rainfall has on the type of primary producers at the site. But as temperatures warm up, the amount of rainfall becomes more important in determining what can grow in a given region.

Primary Productivity on Land On land, the distribution of biomes is mainly influenced not only by the amount of sunlight but also by temperature and rainfall (Figure 22.9). Moderating these basic limiting factors are other variables. For instance, the amount of incoming sunlight varies with latitude (distance from the equator), with slope exposure, with the seasons, with recurring cloud covers or clear skies. Cloud covers and clear skies depend on such factors as mountain barriers, land masses, and wind and ocean currents. These factors in turn influence air temperature. In coastal zones, moist air and fog tend to hold in heat energy, so there is less temperature variation from day to night and from one season to the next. Inland desert zones, with their clear skies and sparse rainfall, show more extreme temperature variations from day to night. The land itself—its elevation, its mineral content, whether its soil holds mois-

ture or encourages run-off, whether it is rolling or flat or mountainous—helps dictate what kinds of life will be found in what places.

Given such variables, ecosystems vary considerably even in the same general region. Yet it's still possible to characterize vast tracts of land as being generally of one type of biome or another (Figure 22.10). From Chapter One, you are already acquainted with two major types—the **tundra,** with its primary producers limited to grasses, mosses, lichens, and a few dwarfed woody plants such as willows; and the **savanna,** with its primary producers of grasses and scattered trees adapted to prolonged dry spells. Here we'll focus on five more types of biomes (grassland, desert, tropical rain forest, deciduous forest, and coniferous forest) to illustrate the primary producers of different regions.

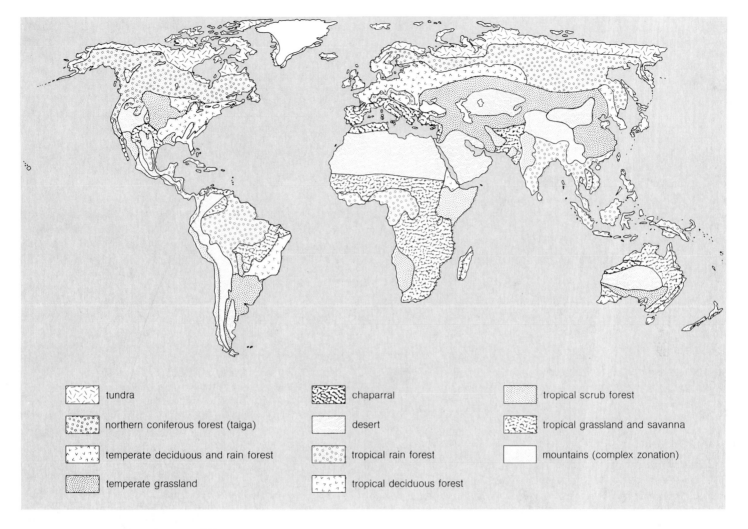

Figure 22.10 The world's major types of biomes.

Legend:
- tundra
- northern coniferous forest (taiga)
- temperate deciduous and rain forest
- temperate grassland
- chaparral
- desert
- tropical rain forest
- tropical deciduous forest
- tropical scrub forest
- tropical grassland and savanna
- mountains (complex zonation)

Grassland In parts of the United States, Russia, and Argentina, the land is flat, temperatures are moderate, and rainfall is limited by mountains that bar most of the storms moving in from the sea. Even though there is not enough rain to support forests, there is enough to support rolling seas of grass (Figure 22.11). At one time, **grasslands** extended westward from the Mississippi River region to the Rocky Mountains, from Canada through Texas. Where rain was sparse, short grasses were the dominant primary producers. Where there was a little more rainfall, grasses taller than the buffalo living there stretched out to the horizons.

Today, undisturbed grasslands are rare. For century after century, decaying leaves and litter had enriched the soil of tallgrass regions, making them inevitable targets for conversion into vast fields of corn and wheat. The shortgrass regions did not have enough rain to support these crops, but they seemed ideal for imported cattle. Enter beef cattle, exit the buffalo (the natural dominant herbivore of the system), and exit the Indian—one of the few modern human groups ever to become an integral part of a stable ecosystem. (The Indian's way of life has long since vanished, but the buffalo may yet make a comeback. With the cost of supplementary cattle feed skyrocketing, the buffalo is looking more and more attractive because of its ability to subsist entirely on the grasses with which it coevolved.) But early cattle operations led to overgrazing, and attempts to farm shortgrass regions led to erosion, on an immense scale,

a

that crippled the primary productivity of much of this land. John Steinbeck's *Grapes of Wrath* and James Michener's *Centennial* are two novels that speak eloquently of this disruption and its consequences.

But natural events, too, can bring about large-scale change. In 1976 Reid Bryson, director of the University of Wisconsin's Institute for Environmental Studies, issued a sobering report on what may soon happen to the world's major grassland-turned-agricultural biomes. Long-term studies suggest the world is entering a period of adverse climatic conditions that may last for at least four decades, perhaps for centuries. If the climate becomes anything like that of the past century instead of the unusual warm climate

we have experienced over the past five decades, the impact on the world's human population will be staggering. Overall, food production in the United States will not suffer significantly because there will be more rain, not less, in the northern half of the country. But crop production in Canada, China, India, and the Kazakhstan wheat fields in the USSR may decline catastrophically. Hence a major shift in climate can mean only one thing: famine and starvation for much of the world. Poor nations will have their population problem "solved" for them in a most inhumane way. And it is not too difficult to guess what the response will be of powerful nations whose populations are desperate for food.

b

Figure 22.11 The grassland biome: (**a**) tallgrass prairie, as it often appears in the vast flat interior of continents, and (**b**) rolling shortgrass prairie with mountain barriers to rain in the distance.

Figure 22.12 Warm, dry desert biome near Tucson, Arizona.

Desert On almost every continent, you can find **desert** biomes: arid regions that support little primary production. About 5 percent of the world's deserts are extremely torrid during the day. But because there is little moisture in the air, heat loss is tremendous during the night, which can be extremely cold. Few kinds of organisms can survive these extremes. The Sahara Desert of Africa and Death Valley in the United States are such deserts. About 100 meters below all desert regions is a zone of sediments and rocks saturated with water (groundwater). In the few spots where these water-bearing rocks break through the surface at low elevations, springs and oases form that often support small, comparatively lush pockets of plant and animal life.

Where cold winters are followed by prolonged drought, you find *cool deserts*. Here, primary production is limited to the springtime growth of grasses, herblike plants, and shrubs able to withstand frost followed by arid conditions. Cool deserts in the western United States support sagebrush communities.

In *warm deserts* (Figure 22.12), the temperature typically ranges from mild to hot, and the primary producers vary widely from one region to the next. Even so, there is remarkable similarity in the kinds of plant and animal adaptations to these temperatures. Plants such as the cactus, for instance, remain metabolically active during the long, dry spells between rains. They can do so because they have

Harlo H. Hadow

no leaves, which means they don't lose much water through transpiration (their stems carry on photosynthesis). When the brief spring rains arrive, these plants use extensive root systems just below the soil surface to soak up what they can. With water being a major limiting factor in warm deserts, cacti have evolved various devices, such as formidable spines or poisonous sap, which discourage predation on the plant body.

During the day, a warm desert appears devoid of life. But where there are producer organisms, so are there consumers. During the summer day, the heat exceeds their range of tolerance, so consumers are found in burrows, under rocks, wherever they can escape the relentless sun. It is during the cool hours of dusk and dawn, and to some extent the middle of the night, that the desert comes to life (and death, depending on who finds and eats whom).

Ironically, it is precisely because of the relentless sun that the desert may be vulnerable to ecological disruption. Hard on the heels of water lines are human populations, out for retirement or fun in the sun. On weekends and holidays, the deserts of Southern California face human hordes on that symbol of American freedom, the motorbike, tearing up erosion-resistant grasses and surface soil, chasing down and running over rabbits or coyotes in a hip Western version of fox-and-hounds. And heaven help the tarantula that wanders into Sun City.

a

Rain Forest When warm temperatures combine with abundant rainfall, the **tropical rain forest** emerges. It is a dense forest of stratified communities, dominated by trees spreading their canopy far above the forest floor. Figure 22.13a shows a tropical rain forest of the highlands of Trinidad.

At different levels beneath the tall canopy in such forests are plant species adapted to ever diminishing amounts of sun. On the forest floor, few plants are effective at photosynthesis, although any break in the canopy encourages exuberant growth of tree saplings and other plants. Competition for light is intense, but the abundant rains make possible adaptations that never would be seen under drier conditions. Vines of all sorts clamber up toward the sun. Mosses, orchids, lichens, and bromeliads (plants related to the pineapple) grow on tree branches. They obtain minerals from the falling leaves and debris, and from the wastes of animals that live high up in the canopy. The food webs sustained by all these producer plants are astonishing. Entire communities of insects, spiders, and amphibians can live, breed, and die in the small pools of

water that collect in the leaves of plants growing high above the forest floor. Many insects, birds, and monkeys spend most of their lives at a single level in the canopy. (For instance, from the time of birth until two months later, spider monkeys hang on for dear life to mother, who keeps her arms free for swinging through the trees and who rarely drops to the ground.) In such places, the kinds of organisms living in or on a single tree often exceed the kinds of organisms living in an entire forest to the north!

With all this diversity, however, a tropical rain forest is one of the worst places to grow crops. Despite all the producers, there is practically no organic debris. Decomposers quickly break down organic debris and return useful nutrients to the soil. There, layers of roots take up the nutrients for rapid conversion into new biomass. Thus, when the forest is cleared of plants and animals for agriculture, most of the nutrients are permanently cleared off with them.

Even with **slash-and-burn agriculture** (in which the forest biomass is reduced to ashes, then tilled into the soil), most of the nutrients are soon washed away because of the

Dona Hutchins

b

Figure 22.13 Tropical rain forest biomes, showing aerial plants and ever-present vines in the canopy (**a**), and the dense, lush growth on the forest floor (**b**).

Figure 22.14 The changing character of a temperate deciduous forest biome in spring, autumn, and winter.

heavy rains. After a few years, cleared fields become infertile and usually are abandoned—and the forest is left to heal itself through the extremely slow process of succession.

Lumbering operations, too, are removing mahogany, teak, rosewood, and other mature tree species that will be long in returning, for they can grow only as part of an integrated forest system. Mahogany, teak, and rosewood furniture and paneling are not essential for human survival. As essential as they seem for status in human communities,

they are far more essential to the community structure of established tropical rain forests.

Deciduous Forest The natural ecosystems of Southeast Asia resemble those of Central America. In these **tropical deciduous forests,** most plants lose their leaves during part of the year as an adaptation to the dry seasons that alternate with the wet monsoons. It happens that small shifts in the

jet stream patterns of air circulating far above the earth's surface occur naturally, and they determine whether the monsoons will fall, say, in India or in the oceans to the south. The natural ecosystems of such areas as India have evolved over thousands of years, and organisms have been selected that can survive years of drought. They are little affected by these short-term variations. But the human population is particularly dense in Southeast Asia, and the agricultural systems providing them with food depend utterly on the timely arrival of the monsoons—a dependence that often leads to shortages and famine when vulnerable crops wither and die.

Throughout much of eastern North America are **temperate deciduous forests,** where the dominant trees drop their leaves not in response to dry seasons but as a response to winter cold (Figure 22.14). Few undisturbed ecosystems of this type still exist. Depending on patterns of human settlement, they were cleared or cut over, often several

Timothy Ransom

a

Figure 22.15 Two distinct coniferous forest biomes: (**a**) the moist, temperate coastal forest of the Pacific Northwest in summer, and (**b**) the subalpine forest of Yosemite beneath the first snows of winter.

times. Today the forests of the Southern Appalachian Mountains are the best example of this disappearing biome. In precolonial times, herbivores included squirrel, rabbit, and deer, which were prey for bobcat, wolf, bear, and mountain lion. Today, because of the human population in the area, the natural predators have been largely exterminated on the grounds that they are dangerous and don't belong there. Populations of large herbivores such as deer are kept in check through controlled hunting seasons—although unlike natural population controls, hunters often seem more interested in taking the healthiest animals instead of the old and weak.

Coniferous Forest Below the tundra line in northern regions is an extensive zone known as the **northern coniferous forest,** or **taiga** (Figure 22.15). Although winters are prolonged and extremely cold, with a fairly constant snow cover, the summer season is warm enough to promote the dense growth of trees. Evergreen conifers (spruce, fir, pine, and hemlock) have the competitive edge, for they are able to carry out photosynthesis just as soon as temperatures rise above freezing. They also have thick bark and heavily waxed, needlelike leaves, which enable them to withstand extreme temperatures and to conserve water during both winter cold and drought. During winter, animals drifting

b

Ansel Adams

south to escape the blizzards of the tundra find shelter here.

Food webs built around the conifers include insects that feed on the bark, buds, cones, leaves, and tender shoots of the trees. Decomposers are not as prolific here as they are in deciduous forests, for the resin-rich needles of conifers decompose much more slowly. The carnivores feeding on insects include birds of all sorts, whose niches are carved out in zones ranging from ground level to the uppermost canopy. Climbing about on the branches and trunks are seed-eating squirrels and mice. Larger herbivores, such as hare, beaver, moose, and deer, do not use the conifers as their primary food source, even though they are sometimes

mistakenly included as part of conifer food webs. Instead they forage on low-growing vegetation and deciduous trees that spring up around the region's many lakes, streams, and marshes ("taiga" is Russian for "swamp forest"). They use the forest primarily as protection against predators.

Is it this same recognition of sanctuary that draws us to a forest? It's interesting that, at a time when our grasslands were first being tilled under, a conservation movement began to save the forests. In fact, the first bureau of conservation was the United States Forest Service. Partly as a result of its initial efforts, a third of our country is still heavily forested; of that, 27 percent is preserved in national, state, and local parks. The Forest Service manages all wildlife, watersheds, recreation, and lumbering in the national forest system. Part of its program is based on **tree farming,** the replacement of climax communities with monolithic stands of fast-growing softwood trees. There is growing concern that simplified tree farms are replacing too much of the stable, 30- to 200-year-old climax communities either because of or in spite of Forest Service management. As simplification is permitted to occur, the system becomes more vulnerable and the trees must be protected like other crops, through widespread use of insecticides and fertilizers. There is also reliance on clear-cutting to keep up with demands for wood products (the average American directly or indirectly uses approximately 560 pounds of wood products such as paper each year). This practice leaves areas scarred and vulnerable for decades. Yet 90 percent of our national forests are potentially open to clear-cutting. This is another area where our short-term demands eventually must be tempered by the need for long-term ecological stability.

Is the City an Ecosystem?

Ocean, savanna, tundra, grassland, desert, forests of different character—*as diverse as all the world's biomes are, as simple or complex as their dominant communities and ecosystems may be, they have in common a self-contained, self-perpetuating stability.* They are essentially self-contained arrays of consumers and decomposers that have achieved dynamic balance with the producers of the site by recycling nutrients even as they themselves are nourished.

For the past 11,000 years, a new kind of ecosystem has been evolving that knows few geographic or climatic boundaries. It is based on human communities that are increasingly overlapping and merging together, first into cities and then into sweeping urban areas called **megalopolises** (Figure 22.16). An urban region contains not only human consumers and their pets, but also such scavengers as rats, cockroaches, pigeons, starlings, and sparrows. An urban center interacts with the land, the forests, the rivers, even the local climate. It *is* an ecosystem. But it is not self-contained, and it is not stable.

The preceding photographs of a few biomes were presented not merely to make you gasp in wonder over the glories of natural ecosystems. And it was not out of indifference to the arrays of consumers and decomposers of each ecosystem that we excluded photographs of a squirrel or wolf here, a rabbit or mushroom there. Rather, the photographs are meant to convey that *the foundation of any biome is its characteristic array of producer organisms.* This is where all webs of life begin; without them, life ends.

Urban centers such as cities have no producers. They have only consumers and, compared to the human biomass,

Figure 22.16 The city as ecosystem.

they have a negligible number of decomposers. The environmental conditions are much the same for a city as they are for the surrounding biome. But soil is paved over, plants are bulldozed under, and water as well as other resources are diverted for direct and indirect human use. *A city endures as an ecosystem only as long as it imports energy and materials from someplace else.*

Repercussions of an Unbalanced Energy Budget The need to import materials and energy into an urban center puts amplified pressures on ecosystems far outside its boundaries. A typical United States urban center having a population of 1 million requires basic daily imports of 625,000 tons of water, 9,500 tons of fossil fuels, and 2,000 tons of food, not to mention material resources for the construction and industry needed to sustain human habitats. Immense tracts of land outside the urban sphere must be converted to agriculture. Vast transportation networks must be created to ship in agricultural products. Entire regions are mined over for metals of all sorts. Water and oil are brought in through urban tentacles that span the globe. The tapping of worldwide resources creates imbalances not only for natural

Table 22.2 A Few Characteristics of Urban Ecosystems

Benefits

Employment
Mass manufacturing and production
Excitement and stimulation of ideas
Entertainment

Internal Instabilities

Reliance on imported energy
Reliance on imported food
Reliance on imported water
Reliance on imported materials
Air polluted with heat and chemicals
Water polluted with sewage and industrial wastes
Accumulation of solid wastes
Noise pollution
Congested transportation pathways
Poverty
Potential for disease epidemics
Ethnic tensions
Lack of recreation and open space
Crowded habitats and concurrent individual stress
Crime, drug abuse

External Urban-Derived Instabilities

Simplification of natural ecosystems
Pesticides and agricultural wastes
Disturbances to all stable ecosystems
Surface mining

After Miller, *Living in the Environment.*

communities but for other human communities. Protein-rich fish and fishmeal from such protein-starved places as South America are used to feed between 70 million and 100 million city-bred cats and dogs in the United States. Shrimp from protein-starved India finds its way to urban dining rooms in the form of "appetizers." What does the city offer in return? Jobs and manufacturing (essentially to sustain the artificial environment), entertainment, a stimulating environment are some of the answers given. Table 22.2 lists a few more. The diversion of available world energy resources into cities is mostly one-way, an energy flow that leads not to stability but to amplified demands of the increasing urban population.

Everything Has To Go Somewhere—Pollution Through Nonrecycled Resources In all natural ecosystems, no by-products of existence accumulate because they are recycled. What we call **pollutants** are by-products of our existence that are not recycled through the webs of life. When they accumulate to high levels, they usually have harmful effects,

or at least perplexing ones. How, for example, can one urban center containing 1 million people absorb its 2,000 tons of daily trash—bottles, cans, newspapers, toilet paper, plastic spoons and forks, paper containers, "disposable" diapers, "disposable" razors, plastic bags and wrappers, cellophane and aluminum wrappers, cartons, heavy-duty lawn litter bags, letters, envelopes, ad infinitum? How can 500,000 tons of daily human sewage and industrial wastes—lead, chromium, mercury, sulfur—be recycled through the community life? Which community members really want to recycle 950 tons of daily air pollutants in their lungs? Where to put 500 tons of dog feces and 700,000 gallons of dog urine that daily find their way not to sewage treatment facilities but to open storm drains that flow directly into lakes, rivers, streams, and oceans?

The answer is that we cannot recycle all these things. We can reduce, collect, concentrate, bury, and burn wastes, or spread them out through other ecosystems, but we can never be completely rid of pollution. For the past four decades, for example, 350 million gallons of sewage from metropolitan New York and New Jersey have been dumped each day off the shore of Long Island where, being out of sight, it was easy to put out of mind. Then, in the summer of 1970, a dead sea of sewage unaccountably began moving back toward the land, bringing with it the specter of hepatitis, encephalitis, and other terrible diseases to the megalopolis that created it.

But the point of this chapter is not to list examples of abuse after abuse created by the artificial urban ecosystem. They do exist, they are being studied in an attempt to bring them to controllable levels, and it's important for you to know about them, for it is your life as well as all life that is being affected by them. However, our focus here has not been on the symptoms but on the fundamental source of the problem. We have presented biological models of energy flow and materials reuse that work; we have raised the idea that *existing* human-designed models inherently cannot work, because they have strayed too far from the biological models. Where do we go from here? The last chapter of the book presents some alternatives.

Perspective

Molecules, cells, tissues, organs, organ systems, whole organisms, populations, communities, ecosystems, the biosphere. This is the architectural stuff of life—a sequence that reflects a progressive complexity in the adaptations of many living systems that have appeared over the past 3.5 billion years. We are relative latecomers to this immense biological building program. Yet, in the comparatively short span of 50,000 years, we have been attempting to force our own blueprints for living upon it. We are now restructuring the stuff of life at all levels—from recombining DNA molecules of different species to changing the nature of the thin, life-giving atmosphere that envelops the entire earth.

Of course, it would be presumptuous of us to think we are the only organisms that have ever changed the nature of living systems. As you read earlier, evolutionary theory suggests that the activities of photosynthetic organisms irrevocably changed the course of biological evolution, and that the appearance of predatory organisms meant simple chemical evolution would come under the dominion of biological competition. Thus biological systems have been evolving for 3.5 billion years through natural selection and through the correlated prodding of geologic unrest. Change is nothing new to this program. What *is* new is the accelerated, potentially devastating change being brought about by our cultural evolution.

For most of evolutionary history, natural ecosystems have not been characterized by sudden, cataclysmic change. Barring outside disruptions, they have been and continue to be governed by feedback control mechanisms that keep these systems in balance even while allowing room for change. For instance, potentially runaway growth of the populations in an ecosystem is kept in check by the carrying capacity of the environment. But feedback control mechanisms can come into play only when deviations already exist in the system itself. The system has already changed before the controls work to return it to dynamic stability.

Feedback control is not enough for human-designed ecosystems. Our designs are far too vulnerable because of their utter dependency on a one-way flow of materials and energy, which gives the illusion of a stable basis for explosive population growth and resource consumption. A sudden shortage of food, a shortage of raw materials—such deviations in the system can come too fast for us to correct. And they can have too great an impact for us to know whether the imbalance will be irreversible.

What about feedforward control mechanisms? Many organisms have what amounts to early warning systems. Skin receptors tell of the outside air growing colder, which sends messages through the nervous system to the metabolic apparatus for raising internal body temperature before the body itself becomes dangerously chilled. With feedforward control, corrective measures can begin before a change in the external environment has significantly altered the system itself.

But even feedforward controls are not enough for human ecosystems, for they go into operation only when change is already under way. Consider, by way of analogy, the DEW line—the Distant Early Warning System, our nation's sensory receptor for intercontinental ballistic missiles that may be launched against us. By the time it detects what it is designed to detect, it may be far too late, not only for humans but for the rest of the biosphere.

We have, as a species, moved beyond the stability afforded by feedback and feedforward controls over natural systems. It would be naïve to assume we can ever reverse who we are at this point in evolutionary time, to de-evolve ourselves culturally and biologically into becoming less complex in the hope of achieving more stability. *But there is no reason to assume we cannot achieve stability by moving forward.* For we have available to us a third kind of control system, and it is uniquely our own. We have the capacity to anticipate events *before* they happen. We are not locked into responding only after change has begun. We have the capacity for anticipating our future—it is the essence of our visions of utopia or of nightmarish hell. Thus we all have the capacity for adapting to a future which we can partly shape. We can, for example, learn to live with less; far from being a return to primitive simplicity, it would be one of the most complex and intelligent behaviors of which we are capable.

But having that capacity and exercising it are not the same thing. Our ecosystems are already on dangerous grounds because we have not yet mobilized ourselves as a species to work toward stability. Our survival depends on predicting possible futures, on designing and constructing ecosystems that are in harmony not only with what we define as basic human values but with the biological models available to us. Human values *can* change; our expectations can and must be adapted to biological reality. *For the principles of materials and energy flow—which govern the growth and survival of all systems of life—do not change.* It is our biological and cultural imperative that we come to terms at last with these principles, and with what will be the long-term contribution of the human species to the unity and diversity of life.

Recommended Readings

Darnell, R. 1973. *Ecology and Man.* Dubuque, Iowa: W. C. Brown. Excellent introduction to principles of ecology.

Emmel, T. 1973. *An Introduction to Ecology and Population Biology.* New York: Norton. Outstanding text.

Flanagan, D. (editor). 1970. *The Biosphere.* San Francisco: Freeman. Collection of articles from *Scientific American.* Includes discussions of energy flow, materials recycling, and the effect of human activities on the biosphere.

Miller, G., Jr. 1975. *Living in the Environment: Concepts, Problems, and Alternatives.* Belmont, California: Wadsworth. Excellent source book for anyone interested in exploring details of our impact on the environment.

Odum, E. 1971. *Fundamentals of Ecology.* Philadelphia: W. B. Saunders. This book, written by a prominent ecologist, is a classic in the literature.

Smith, R. 1976. *The Ecology of Man.* Second edition. Harper & Row. Outstanding collection of articles on ecology.

Chapter Twenty-Three

Bob Campbell

Not to decide is to decide. — Harvey Cox

Ecology and the Future:
Not To Decide Is To Decide

Once there was a bullfrog who found a lovely shallow pond, and he staked it out for his own. There he sat, ribbeting away, trying to attract a female frog to this really wonderful place. Well, as it turned out the bullfrog had laid claim to a frying pan full of water. Beneath the frying pan was a campstove turned on low. Slowly the water began to heat up—so slowly that the bullfrog didn't even notice. Gradually the water got warmer, but it was such a gradual thing and the frog was so busy ribbeting his little heart out that he still didn't notice. Soon there was no way *not* to notice. But by then it was so hot his muscles no longer responded to signals from his nervous system. And so it got hotter, his croaks got weaker, until he dropped dead. The moral of this story is not (as you might suspect) that you can't build a marriage on the promise of sex alone. The moral is that it is possible to live in the midst of a gradual trend toward unlivable conditions and not even be aware of what is going on.

Like the bullfrog, we generally are so busy concentrating on so many other pressing matters that we are indifferent to ominous changes going on in the world around us. Where is indifference leading us? More than this, can ecological awareness really take us in another direction? Now that we have looked at some basic ecological principles, and have some idea of energy flow and materials recycling through natural ecosystems, let's see what we have as some of our alternatives.

On the Nature of Pollution

Solid Wastes Lisa is entering college in the fall. On the outskirts of Del Mar, the town where she lives, a tiny plot of land has finally been set aside as a recycling center for newspapers, glass, and aluminum cans. Last week Lisa decided to take part in the recycling program. There was a brief period of adjustment: all the members of her family had to get used to stacking up the newspapers in the garage, to putting glass jars and bottles in one trashcan and aluminum cans in another. By the week's end, she was mildly surprised at how much had accumulated; somehow she had assumed it would take months to stack up. It took only a few minutes to run everything down to the recycling center. But along the way she became aware, for the first time, of just how many trashcans were lined up along the streets, waiting to be carted out of sight, even though the recycling center had been open for several months. Two or three trashcans sat in front of each house. There are about 5,000 homes in town, she mused. That meant about 12,000 trashcans a week had to be emptied somewhere. In the surrounding metropolitan area, there are about 200,000 homes. Could it be—400,000 to 600,000 trashcans *each week?*

After Lisa deposited everything in the collecting bins, she decided to take a run out to the county "sanitary landfill station" where all the nonrecycled solid waste ends up. She was stunned by what she saw. One trash-filled truck after another sped through the gate, antlike in their line to the dumping ground. Bulldozers were hastily shoving mounds of all manner of refuse down the sides of what had once been chaparral-covered canyons. What, thought Lisa, will happen when these canyons are filled? To the east are the rolling hills of Rancho Santa Fe; there are too many moneyed people there to let the bulldozers in. To the south and north, hundreds of tract homes are being built—even on top of the filled-in canyons as fast as the bulldozers level off the site. To the west, the ocean. *Surely not the ocean!* This same thing must be happening all over the nation, she thought. What, then, could her isolated commitment possibly mean? How could one small action help turn such a tide of solid waste pollution? Somebody ought to start a local campaign for public awareness, she mused. They should call it "Take Someone You Love to the County Dump." But with college starting soon, how could she get involved? Somebody else would have to do it.

Which "somebody" is going to tackle the 4.5 billion metric tons of solid wastes we dump, burn, or bury each year in the United States? Who is going to decide where the landfills go next? Conversely, is recycling itself a workable alternative? Recycling is part of the answer, but it can't be a hit-or-miss thing. For instance, in her solitary drive to the recycling center, Lisa used up some gasoline—a nonrenewable energy source which ultimately must be figured as part of the energy cost of the program. Although the energy cost of recycling is lower than it would be to extract and use new raw materials, it still can't be ignored—especially when you multiply the energy cost by all the individuals driving separately to the center.

What it is going to take, in the long run, is a change in our basic living habits. We have been brought up to follow

Table 23.1 Energy Needed to Produce or Recycle a Few Raw Materials

| | Pounds of Coal Needed To Produce One Pound of Metal | |
Resource	From virgin ore	From recycled material
Aluminum	6.09	0.17 to 0.26
Steel	1.11	0.22
Copper	1.98	0.11

Data from Oak Ridge National Laboratory.

Table 23.2 Comparison of Three Programs for Handling Solid Wastes

Item	Throwaway System	Recycling System	Ecologically Based System
Glass bottles	Dump or bury	Grind, remelt; remanufacture; convert to building materials	Ban all nonreturnable bottles and reuse (not remelt and recycle) bottles
Bimetallic "tin" cans	Dump or bury	Sort with magnets; remelt	Limit or ban production; use returnable bottles
Metal objects	Dump or bury	Sort; remelt	Sort, remelt; but tax items lasting less than ten years
Aluminum cans	Dump or bury	Sort; remelt	Limit or ban production; use returnable bottles
Paper	Dump, burn, or bury	Incinerate to generate heat	Compost or recycle; tax all throwaway items; establish national standards to eliminate overpackaging
Plastics	Dump, burn, or bury	Incinerate to generate heat or electricity	Limit production; use returnable glass bottles; tax frivolous throwaway items and packaging
Garden and food wastes	Dump, burn, or bury	Incinerate to generate heat or electricity	Compost; return to soil as fertilizer or use as animal feed

From G. T. Miller, Jr., *Chemistry: A Contemporary Approach* (Belmont, California: Wadsworth, 1976).

the patterns of a "throwaway" culture: use it once, discard it, buy another. For instance, between 50 and 65 percent of urban wastes are paper products—of which only 19 percent is now being recycled. If we were to recycle merely half the paper being thrown away each year, we would do more than conserve trees. For the *energy* it takes to produce an equivalent amount of new paper could be diverted to provide electricity to about 10 million homes! Or consider that about 60 billion beverage containers are sold each year in the United States. About 50 billion are nonreturnable cans and bottles, too many of which are discarded in public places. These containers represent three-fourths of the litter that must be picked up along highways—a time-consuming, energy-draining activity that costs thousands of barrels of petroleum (not to mention hundreds of millions of tax dollars) each year.

In the early decades of this century, before it became "easier" and "cheaper" to throw things away, beverage containers were used over and over again, sometimes as often as fifty times. The same system could be made to work again. Even now, some states are passing laws requiring deposits on all beverage cans and bottles. The refundable deposit has helped check the waves of litter that typically washed over Yosemite National Park each year. Oregon, Vermont, Maine, and Michigan have had similar success with their programs.

Adherence to ecological principles of materials reuse is long overdue. A transition from a "throwaway" life style to one based on conservation and reuse is already economically feasible, and we have most of the technology needed to do it. The question becomes whether we are willing to commit ourselves to crossing the threshold. We can, for instance, bring consumer pressure to bear on manufacturers by refusing to buy goods that are lavishly wrapped, excessively boxed, and designed for one-time use. We can ask our local post office how to go about turning off the daily flow of junk mail, a flow that represents an astounding waste of paper, time, and energy—and higher mail delivery rates for everyone. We can work to see that our local city and county governments plan to develop large-scale resource recovery centers of the sort described in broad outline in Figure 23.2. In such systems, existing dumps and landfills would be viewed as urban "mines." Nearly one-quarter of our past and present solid wastes might be potentially recoverable from these mines. When will such conservation and recovery programs start? They won't start without sustained commitment from Lisa. And they won't start without sustained commitment from you.

Water Pollution Adán is not happy about the housing development scheduled to be built right across the freeway from the small Arizona town in which he lives. Forty thousand homes constructed over the next five years. He saw sketches of the development in the local newspaper; the area is destined to become a high-density suburb. What's worse, there is no way existing public facilities—schools, roads, power plants, waste disposal systems—can absorb the demands of the increased population. Even putting

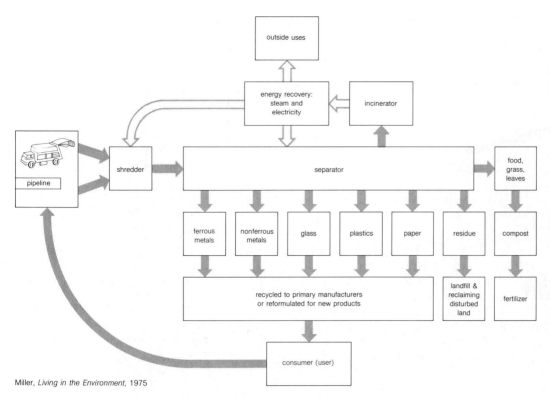

Figure 23.2 Generalized resource recovery system. Solid wastes are treated as urban "ore." Bulk items are sorted out, and the remainder is shredded and separated. Electromagnets could be used to extract steel and iron; air blowers could send plastic and paper to different recovery chambers. Mechanical screening, flotation, and centrifugal hurling could sort out metals, glass, and garbage.

Once sorted, metals could be returned to mills, smelters, and foundries; glass to various glass processing plants. Organic matter could go to compost centers, later to be used as fertilizers or soil conditioners, perhaps as fuel or animal feed. Wood could be used as fuel. Paper could be recycled or used as fuel. Even residues from incinerated materials (including particles removed from smoke to limit air pollution) could be processed into building or road material (such as bricks).

Miller, *Living in the Environment*, 1975

aside the question of who will pay for expansion of facilities, where will the by-products of the increased population go? Solid waste disposal is only one aspect of the problem. What about something as basic as waste water treatment? The present facilities are operating to capacity, and there simply isn't enough available water in this part of the country to process more raw sewage before releasing it into the few existing waterways. At the same time, more and more people are being attracted to Arizona because of its magnificent climate. Adán remembered the pointed remark of a real estate agent when he questioned the wisdom of permitting such massive development before these issues were resolved: "Did you want to keep people out *before* or *after* you moved in yourself?" Yet, with real limits on natural resources as basic as water, isn't it insane to pretend otherwise? Can more water be brought in? Even assuming someone will pay for it, where will the water come from?

For the first time in his life, Adán was facing tangible evidence that there *are* limits to growth. True, there is a tremendous amount of water in the world, locked into one vast closed system called the **hydrologic cycle.** For as short a time as it takes for water vapor to form over the oceans and fall on the continents as rain, for as long a time as it takes for water from lakes and rivers and glaciers to flow again to the oceans, there is the promise of movement through the

biosphere. Still, the promise is greater for some than for others: three out of every four humans alive today do not have enough water or, if they do, their supplies are in some way contaminated. Aside from the unequal distribution of water throughout the world, most of it (95 percent) is in the oceans and can't be used for human consumption or agriculture; it's too salty. Of the remainder, nearly 90 percent is locked in glaciers or in polar ice caps. Most of the rest is deep in the ground or suspended in the atmosphere. In fact, for every 1 million gallons of water in the world, only about 6 gallons are readily available in a usable form!

Pure water is one of our most precious and most rapidly disappearing resources. It is not that water is leaving the hydrologic cycle. Rather, it is that more and more is becoming polluted. **Water pollution** refers to a condition in which a body of water has been altered chemically or physically in such a way that renders it useless for a given purpose. We have, in effect, tapped into the hydrologic cycle and are using it directly or indirectly as a dumping ground for by-products of human existence. Thus water becomes unfit to drink (even to swim in) because it is polluted with human sewage and animal wastes, which in turn can encourage the growth of large populations of disease-causing bacteria and viruses. Through agricultural runoff, water becomes polluted with sediments from land

Direct Personal Use

*160 gallons, or 8 percent
of your daily use. Examples:*

Bath *30–40 gallons*
Shower *5 gallons/minute*
Shaving *3 gallons, water running*
Washing clothes *20–30 gallons*
Preparing and cooking food *8 gallons*
Washing dishes *10 gallons*
Housecleaning *8 gallons*
Flushing toilet *3 gallons/flush*
Leaking toilet *35 gallons/day*
Watering lawn *80 gallons/8,000 square feet*

Appalling Little Fact

Total daily faucet and toilet
leaks in New York City alone
200,000,000 gallons

Indirect Agricultural Use

*600 gallons, or 33 percent of
your daily use. Examples:*

One egg *40 gallons*
One ear corn *80 gallons*
One loaf bread *150 gallons*
One gallon whiskey *230 gallons*
Five pounds flour *375 gallons*
One pound beef *2,500 gallons*

Indirect Industrial Use

*1,040 gallons, or 59 percent of
your daily use. Examples:*

Cooling water for electric power plant *720 gallons*
Producing consumer goods *320 gallons*
Sunday paper *280 gallons*
One pound synthetic rubber *300 gallons*
One pound aluminum *1,000 gallons*
One pound steel *35 gallons*
One gallon gasoline *7–25 gallons*
One automobile *100,000 gallons*

Miller, *Living in the Environment*, 1975

Figure 23.3 Average daily use of water by an average American: 1,800 gallons, compared with 12 gallons for an average person in developing nations.

erosion, with substances such as pesticides and herbicides, and plant nutrients (such as nitrate and phosphate from fertilizers). Through industrial activity and power-generating plants, water becomes polluted with chemicals, radioactive material, and excess heat (thermal pollution).

Table 23.3 highlights some of the ways to start reversing the trend toward unmanageable water pollution. But to give you an idea of how difficult the task will be, let's take a closer look at the problems inherent in **waste-water treatment** methods used today. To begin with, about half the waste water in the United States is not being treated at all. The other half is more or less purified and restored while it is channeled through primary, secondary, and/or tertiary treatment centers before it is discharged into waterways and seas. In **primary treatment,** mechanical screens and sedimentation tanks are used to force coarse suspended solids out of the water to become a sludge. In a process called flocculation, chemicals such as aluminum sulfate are added to accelerate the sedimentation process. About 30 percent of the waste water in the United States goes only through primary treatment before the liquid (the *effluent*) is discharged. Although it is treated with chlorine to kill disease-causing microorganisms in the effluent, chlorine alone is not really enough to get rid of them completely. **Secondary treatment** depends on microbial action to degrade the sludge. Either the sludge is sprayed and trickled through a large bed of exposed gravel in which assorted monerans and protistans live, or the sludge is aerated with pure oxygen to promote microbial activity. As it happens, the microorganisms on which these treatment processes depend are vulnerable to pollution themselves: toxic substances that enter the treatment center can destroy them. Then treatment activities must be shut down until new microbial populations are established. Because levels of pollutants are continuing to rise, the search is on to find chemicals that can degrade sludge. Such chemicals will permit us to be less dependent on microorganisms for handling waste water.

Although secondary treatment is an improvement over primary treatment alone, a host of substances still remains in the effluent: oxygen-demanding wastes, suspended solids, nitrates, phosphates, and dissolved salts such as heavy metals, pesticides, and radioactive isotopes. The water discharged from secondary treatment plants can be used in irrigation, industry, and (if it has been held for a period in settling lagoons), it is often dumped into whatever local body of water may be available.

Table 23.3 Major Water Pollutants: Sources, Effects, and Possible Controls

Pollutant	Main Sources	Effects	Possible Controls
Organic oxygen-demanding wastes	Human sewage, animal wastes, decaying plant life, industrial wastes (from oil refineries, paper mills, food processing, etc.)	*Overload depletes dissolved oxygen in water; animal life destroyed or migrates away; plant life destroyed*	Provide secondary and tertiary waste-water treatment; minimize agricultural runoff
Plant nutrients	Agricultural runoff, detergents, industrial wastes, inadequate waste-water treatment	*Algal blooms and excessive aquatic plant growth upset ecological balances; eutrophication*	Agricultural runoff too widespread, diffuse for adequate control
Pathogenic bacteria and viruses	Presence of sewage and animal wastes in water	*Outbreaks of such diseases as typhoid, infectious hepatitis*	Provide secondary and tertiary waste-water treatment; minimize agricultural runoff
Inorganic chemicals and minerals	Mining, manufacturing, irrigation, oil fields	*Alters acidity, basicity, or salinity; also renders water toxic*	Remove through waste-water treatment; stop pollutants from entering water supply at source
Synthetic organic chemicals (plastics, pesticides, detergents, etc.)	At least 10,000 agricultural, manufacturing, and consumer uses	*Many are not biodegradable; chemical interactions in environment are poorly understood. Some create noxious odors and tastes; others clearly poisonous*	Prospects are poor at present. Push for biodegradable materials; prevent entry into water supply at source
Fossil fuels (oil particularly)	Two-thirds from machinery, automobile wastes; pipeline breaks; offshore blowouts and seepage; supertanker accidents, spills, and wrecks; heating; transportation; industry; agriculture	*Varies with location, duration, and type of fossil fuel; potential disruption of ecosystems; economic, recreational, and esthetic damage to coasts*	Strictly regulate oil drilling, transportation, and storage; collect and reprocess oil and grease from service stations and industry; develop means to contain spills
Sediments	Natural erosion, poor soil conservation practices in agriculture, mining, construction	*Major source of pollution (700 times more tonnage than solid sewage discharge), fills in waterways, reduces shellfish and fish populations, reduces ability of water to assimilate oxygen-demanding wastes*	Put already existing soil conservation practices to use

Only with **tertiary treatment** is there ecologically adequate reduction of the pollutants remaining in the effluent. The process involves advanced methods of precipitation of suspended solids and phosphate compounds, adsorption of dissolved organic compounds, reverse osmosis, stripping of ammonia to remove nitrogen from it, and disinfecting the water through chlorination or ultrasonic energy vibrations. It is largely in the experimental stage and it is expensive; hence tertiary treatment is rarely used.

What all this means is that present treatment centers are not properly treating existing levels of waste pollution. What, then, is going on? A typical pattern, repeated thousands of times along our waterways, is this: water for

drinking is removed *upstream* from a town, and wastes from industry and sewage treatment are discharged *downstream*. It takes no great leap of the imagination to see that pollution intensifies as rivers flow toward the ocean. In Louisiana, where the waters drained from the Central United States flow toward the Gulf of Mexico, pollution levels are considered high enough to be a real threat to public health. Water destined for drinking does get treated to remove disease-causing bacteria—but the treatment doesn't remove poisonous heavy metals, such as mercury, which are dumped into the waterways by thousands of factories upstream. More than this, it is at each river's end—the estuary bridge to the sea—that most of the animal life of the sea is born. It is here, in the region we depend most upon for the continued productivity of the world's oceans, that water pollutants are becoming most concentrated.

We must move beyond the mentality of treating an ever-growing mountain of sludge and then quickly dumping it into the environment. We must come to see the folly of dumping industrial wastes into the waters on which we and all life depend. For instance, instead of sending barges of sludge to sea (Chapter Twenty-Two), we could search for ways to convert it safely to fertilizers or compost. The methane gas that arises from anaerobic activities of micro-organisms during secondary treatment could become a fuel source. A few progressive industries and cities have initiated tertiary treatment, and the water they discharge into waterways is actually more pure than the water they originally take out. Tertiary treatment, again, is not without cost. But can we afford *not* to pay the price?

Adán was just beginning to explore these kinds of problems and alternatives when he went to a city council meeting in which the housing development was to be discussed. Except for the council members, the city planner, and lawyers representing the builders, he was the only citizen who showed up. When he tried to voice his concern, the lawyers breezily quoted statistics, planning documents, letters. The council moved on to other things. Did Adán just imagine he was being brushed aside? As he left the council chambers he thought angrily: "I'm no lawyer, and I haven't got the time or the background to present the other side of the picture. And I don't even know how to get together with anybody else who does."

Air Pollution Denise was about to throw up. The flight from Dallas to Chicago was so crowded that she had to sit in the "Smoking Allowed" section of the plane. When the man sitting next to her lit up a cheap cigar, when the stenchy blue smoke began to cling to her nostrils and even her clothing, she suddenly compared herself to a condemned woman about to be gassed to death. The flight attendant asked the man to put out the cigar (only cigarette smoking was allowed), but the stench remained. Denise barely made it to Chicago. Taking deep breaths of the cold night air, she walked to the parking lot where she had left her old car. She quickly started up the engine and drove away, cursing all cigarette and cigar smokers for their outrageous insensitivity. Clouds of exhaust billowed from the tailpipe all the way home, although Denise wasn't aware of it. When she got home she turned on all the lights and turned up the furnace, but she didn't stop to think about the fuel burning going on at the region's power plant, which was belching all manner of pollutant-filled smoke into the air as it worked to provide business, industry, and homes everywhere with energy.

In a way, the finite space inside the airplane cabin is something like the finite space of our atmosphere. Denise was acutely aware of the tobacco smoke because it had reached high concentrations in a relatively short time. But the same kind of thing is happening to the air around us. In fact, if we were to compare the earth with an apple from the supermarket, the atmosphere would be no thicker than the layer of shiny wax applied to it. *Yet into this thin, finite layer of air we dump more than 700,000 metric tons of pollutants each day in the United States alone!*

If we define **air pollution** as air that contains substances in concentrations high enough to damage living and nonliving things, then we are already in trouble. It isn't just that the pollutants can make the air smell, or cut down visibility, or discolor buildings. Air pollutants can *corrode* buildings. They can ruin oranges, wilt lettuce, stunt the growth of peaches and corn, and damage leaves on conifers hundreds of miles away from their source. They also can cause humans to suffer everything from headaches and burning eyes to lung cancer, bronchitis, and emphysema (Chapter Thirteen).

What are the major sources of air pollution? Tobacco smoking is a localized source, affecting the smokers themselves and, to some extent, nonsmokers who are stuck in confined quarters with them. The pollutants that affect life everywhere are by-products of fossil-fuel burning in transportation vehicles, home heating units, power plants, and the countless furnaces of industry. Among the most obvious are the **particulates** (solid particles or liquid droplets suspended in the air), which include dust, smoke, ashes, soot, asbestos, oil, and bits of poisonous heavy metals such as lead. Among the most dangerous air pollutants are the **oxides of sulfur.** They attack marble, metals, mortar, rubber, and plastic; they cause extensive crop loss near large cities, especially in Southern California, New York, New Jersey, Pennsylvania, Connecticut, and Delaware. At times they become so concentrated in the moist air of downtown St. Louis, they form droplets of sulfuric acid that dissolve holes in stockings of women who step outside office buildings at lunchtime! At best, sulfur oxides can make you cough and wheeze; at their worst they can figure in asthma, bronchitis, and emphysema. When rains wash them into rivers, lakes, and streams, they can destroy entire aquatic communities.

In terms of sheer tonnage, **carbon monoxide** from cars and trucks is the most common air pollutant. Because it can combine firmly with hemoglobin and displace oxygen, it can cut down the supply of oxygen being transported throughout the body. The results? The heart and lungs are forced to work harder to accomplish gas exchange, which does not bode well for people with heart or respiratory disorders. In the prolonged, concentrated fumes of rush-hour traffic, commuters may be subject to headaches, nausea, stomach cramps, and impaired coordination and vision. **Oxides of nitrogen,** other by-products of transportation, also reduce the blood's oxygen-carrying capacity and contribute to similar disorders.

It is in air pollution that we have an example of how environmental interactions can amplify problems. Short-term air pollution depends largely on the weather and topography. Los Angeles, for instance, is in a basin ringed with mountains that is open to the moist marine air from the west. Often a layer of dense, cool air gets trapped beneath a layer of warm air (something called a **thermal inversion**). When that happens, pollutants can't diffuse into the higher atmosphere, and they accumulate to dangerous levels right above the city.

Thermal inversions are most deadly in large urban areas. Here, car exhaust contains nitric oxide that reacts with oxygen in the air to produce nitrogen dioxide. When hydrocarbons and nitrogen dioxide are exposed to sunlight, they become converted to noxious substances collectively called **photochemical smog.** It is featured in Los Angeles, Salt Lake City, and Denver; it is brown, it smells, and it is hazardous to living things. There is another kind of urban air pollution, called **industrial smog,** in which fossil fuel burning gives off particulates and sulfur oxides (produced by oxidation of sulfur contaminants in coal and oil). It appears over Chicago, New York, Baltimore, Birmingham, and Pittsburgh; it is gray, sooty, and potentially lethal. Industrial smog was the source of London's 1952 air pollution disaster, in which 4,000 people died.

Denise can complain to the airlines in an attempt to ban smoking on airplanes. But where can she direct her complaints when it comes to air pollution from transportation and industry? Believing that we suddenly are going to stop relying on fossil fuel burning is about as silly as expecting us to return to a reliance on the horse (which, with all the millions of horses that would be needed, undoubtedly would create a pollution problem of another sort). Denise could keep the thermostat on her furnace turned down five degrees. She could also push for improved mass transit instead of depending on her private automobile. But is she really ready to moderate her personal comfort in such ways? Is she really willing to be a little inconvenienced by waiting around for a bus late at night at the air terminal, after a long flight home from Dallas? What about you?

Some Energy Options

Paralleling the J-shaped curve of human population growth is a steep rise in energy consumption. It is not only that more population means more demands on available resources. An individual can survive on about 2,000 kilocalories of energy a day. But in the highly industrialized United States, the average individual directly or indirectly uses more than 200,000 kilocalories a day. In the United States, energy means not only survival, but survival draped in a bizarre array of nonessential creature comforts. For instance, in one of the most temperate of all coastal climates,

a major university constructed seven- and eight-story buildings with narrow, sealed windows. With this design, a comfortable temperature is maintained not by relying on prevailing ocean breezes but on massive cooling systems—and by relying on massive heating systems instead of aligning the buildings to take advantage of the almost year-around warm sunlight! In support of this kind of life-style, more than half the energy (consumed in the form of nonrenewable fossil fuels) is given off in the form of entropy (wasted heat) and polluting by-products. *Thus the energy problem is not only a result of overpopulation; it is also a result of overconsumption and waste.*

What are our energy options? In exploring our alternatives, we can't escape the realities of the laws of energy; we can't get something for nothing, and we can't ever break even. The **net energy** available from any source is really what is left over after we have subtracted the energy it takes to locate, extract, ship, store, and deliver energy to the consumer. The cost of conventional nuclear energy must also include supplemental input of fossil fuels, the cost to mine and process uranium fuels, the cost of transporting and storing nuclear wastes. Harnessing the seemingly free energy from the sun must be measured also in terms of the energy it will take to build and maintain solar collectors as well as energy storage and transmission systems. When you hear talk of how much energy is available from different sources, keep in mind that there is often an enormous difference between the *total* amount and the *net* amount available.

Fossil Fuels Three hundred million years ago, the trees of immense coastal forests spread their leaves, netlike, into the energy streaming from the sun. They grew, they decayed; they were buried. For hundreds of millions of years, photosynthetic algae thrived in the oceans and served as food for billions of protistans, the remains of which accumulated in layers on the ocean floor. Thus the sun helped give form and substance to living treasures that would come to be buried and transformed, under the sediments and ooze and compression of time, into **fossil fuels:** coal, oil, and natural gas. For many generations, fossil fuels have been a basis for human population growth. But in merely the past three decades, we have used up more fossil fuel than in all our preceding history combined! If we continue to expand our use of oil and natural gas at the rate we have been using them for the past fifty years, these resources may be almost completely exhausted in the United States by the year 2000 and in the world by 2010 (Table 23.4). Even with stringent conservation efforts, we will be running out of these fuels early in the next century. In addition to the petroleum reserves listed in Table 23.4, there are in the states of Colorado, Utah, and Wyoming vast **oil shale** deposits that probably contain more oil than the entire Middle East. Oil shale is buried rock that contains kerogen (a hydrocarbon

Table 23.4 US Energy Reserves and Expected Times of Depletion

Resource	Quads*	Year
Petroleum	930	About 1990
Natural gas	790	About 2010
Uranium**		
Known reserves	610	About 2000
Most optimistic prediction	2,600	About 2040
Coal	14,000	?

*One quad equals one quadrillion BTUs (standard energy units).

**Uranium reserves are expressed on the basis of the amount of energy produced in existing reactors. If used in breeder reactors (which have not yet been shown to be feasible), the same uranium reserves would supply about sixty times as much energy as shown.

compound). The trouble is, by the time the kerogen is collected, concentrated, heated, and converted to shale oil, it may produce only a slight net energy yield (possibly even an energy deficit) at a tremendous environmental cost. The land would be disfigured, water and air pollution would rise, and there would be less surface water to go around in regions already facing water shortages. One of the main by-products of the extraction process is a known cancer-causing agent (benzopyrene). And it would be produced by the ton, even though there is no known way to use it or dispose of it safely. More than this, oil shale processing produces 12 percent more solid waste than the space the original rock formation occupied. Where do the leftovers go? Some have suggested that controlled atomic blasts deep in the rock formations might distill the kerogen in place. But nobody has yet figured out what six underground atomic blasts a day (the number needed to supply merely 10 percent of our current demands for energy) would mean in terms of not only ecological but geological stability of the surrounding lands.

Perhaps the most impressive figure in Table 23.4 is the one given for the coal reserves in the United States. These reserves represent fully one-fourth of the world's coal supply. In principle, they are enough to meet all the world's energy needs for at least several centuries. The problem here is that coal-burning has been the largest single source of air pollution in industrialized nations (which is one of the reasons why worldwide use of petroleum has been escalating). Much of the world's coal reserves is in the form of low-quality, high-sulfur material—which means high levels of sulfur oxides. A recent study at Stanford University suggests that for every new coal-burning power plant of modest size, built anywhere near a city, we can expect an average of seventy more deaths a year from the effects of sulfur oxides. If expensive (and energy-consuming) devices are installed to lower the pollutant levels at the source of emission (smokestacks), deaths and damage from pollution

might eventually be reduced by 90 percent. Locating the power plants far from the cities will further decrease human deaths due to coal burning. But it would be at the expense of crop damage.

To keep pace with our current energy demands, we almost certainly must burn more coal over the next two to four decades. But the consequences will extend far beyond air pollution. Even now, more than *1 billion* dollars a year are being paid in benefits to coal miners (and their families) afflicted with black-lung disease. Modern air-quality standards in mines minimize but do not eliminate this dangerous consequence of breathing coal dust. And mine tragedies—explosions, collapsed mine shafts, poisonous gases—are still frequent. For such reasons, pressure is on to permit widespread strip mining. Some of our coal reserves are close enough to the surface to be gouged out of the earth. But how many millions of acres are we willing to have permanently scarred and rendered useless for agriculture in order to maintain our present life-style? As Table 23.5 indicates, there are other options for use of fossil fuels, but they are at best limited options, with moderate-to-serious consequences for the environment.

Nuclear Energy As Hiroshima burned in 1945, the world recoiled in horror from the destructive potential of nuclear energy. But optimism soon replaced the horror as nuclear energy became publicized as an instrument of progress. That was the beginning of Operation Plowshare—a massive effort to harness the atom for peacetime use. Now, three decades later, more than sixty nuclear-powered, electricity-generating stations dot the American landscape. They produce 8 percent of the nation's electrical energy. Yet today, plans to extend our reliance on nuclear energy have been canceled or delayed. The reason? Serious questions have been raised about the cost, efficiency, environmental impact, and safety of nuclear energy.

One part of the problem with costs has to do with construction. Nuclear plants must safely contain radioactive fuel that gives off extreme amounts of heat energy. At the outset, it was clear that nuclear power plants would be more expensive to build than coal-burning plants, but the thinking was that its less expensive fuel costs would more than offset high construction costs. In the past decade, however, construction costs have soared faster than inflation. This means as time goes on, there must be greater and greater fuel savings to make nuclear power economically feasible. A second part of the cost problem relates to the fuel itself. How expensive is nuclear fuel, how efficiently is it used, and how large are the reserves? The total cost of uranium fuel used to power existing reactors is impossible to determine because of hidden costs of government subsidies to the nuclear fuel enrichment program. On the basis of costs now being paid by the power companies, nuclear-generated electricity should be considerably less expensive than coal if

Table 23.5 Evaluation of Energy Options for the United States

Option	Short Term (Present to 1985)	Estimated Availability* Intermediate Term (1985 to 2000)	Long Term (2000 to 2020)	Estimated Net Energy	Potential Environmental Impact**
Conservation	Fair	Good	Good	Very high	Decreases impact of other sources
Natural gas	Good (with imports)	Fair (with imports)	Poor	High but decreasing***	Low
Oil					
Conventional	Good (with imports)	Fair (with imports)	Poor	High but decreasing***	Moderate
Shale	Poor	Moderate to good?	Moderate to good?	Probably very low	Serious
Tar sands	Poor	Moderate? (imports only)	Good? (imports only)	Probably very low	Moderate
Coal					
Conventional	Good	Good	Good	High but decreasing***	Very serious
Gasification (conversion to synthetic natural gas)	Poor	Good?	Good?	Moderate to low	Very serious
Liquification (conversion to synthetic oil)	Very poor	Poor to moderate?	Good?	Moderate to low	Serious
Wastes					
Direct burning	Poor to fair	Fair to poor	Fair	Moderate (space heating) to low (electricity)	Fairly low
Conversion to oil	Poor	Fair to poor	Fair	Moderate to low	Low to moderate
Hydroelectric	Poor	Poor	Very poor	High	Low to moderate
Tidal	Very poor	Very poor	Very poor	Unknown (moderate?)	Low
Nuclear					
Conventional fission	Poor	Good	Good to poor	Probably very low	Very serious
Breeder fission	None	None to low	Good?	Probably low	Extremely serious
Fusion	Poor	Moderate to low?	Moderate to low	Unknown (could be low)	Unknown (probably moderate to low)
Geothermal	Poor	Moderate to low?	Moderate to low	Unknown	Moderate to low
Solar	Poor (except for space and water heating)	Low to moderate?	Moderate to high?	Unknown	Very low
Wind	Poor	Poor to moderate?	Moderate to high?	Unknown	Very low
Hydrogen	Negligible	Poor	Unknown****	Unknown (probably moderate to low)	Unknown****
Fuel cells	Negligible	Poor	Unknown****	Unknown (probably moderate to low)	Unknown****

*Based on estimated supply as a fraction of total energy use and on technological and economic feasibility.

**If stringent safety and environmental controls are not required and enforced.

***As high-grade deposits decrease, more and more energy must be used to mine and process lower grade deposits, thus decreasing net energy.

****Depends on whether an essentially infinite source of electricity (such as solar, fusion, wind, or breeder) is available to convert water to hydrogen and oxygen gas by electrolysis or direct heating. Impact will vary depending on the source of electricity.

the nuclear plants are operating (as conventional coal-burning plants do) at 80 percent capacity. Through 1976, however, nuclear plants were operating at an average of 59 percent capacity; 41 percent of the time, they were shut down because of technical problems. Because of their high construction cost, the less electricity a nuclear generator produces in a year, the more expensive that electricity must be. In short, nuclear-generated electricity is still somewhat cheaper than coal-generated electricity, but it is by no means the bargain it was predicted to be.

What about efficiency? How much *net* energy does the nuclear power program produce? Once again the answer is difficult to find, for the government does not release information about the nuclear fuel enrichment program (because a portion of nuclear fuel is diverted to military purposes). Current estimates are that nuclear enrichment (the processing needed to convert uranium to a form that can be used as fuel) takes nearly 3 percent of the electricity used in the United States—even though nuclear power plants themselves produce only 8 percent of the electricity. Aside from the costs of enrichment of nuclear fuel, there are mining, refining, transportation, and generator construction costs. All these things require a large portion of the energy that nuclear fuel will produce in the form of electricity—which means the net energy yield will probably be low.

But the major concern is for safety. What about release of radioactive by-products during normal operation? Or release of radioactive materials as a result of malfunctioning? Or escape of radioactive material from stored wastes? The amount of radioactivity escaping from a nuclear plant during normal operation almost certainly is far less of a threat to someone living right across the fence than the pollutants escaping from a coal-burning plant pose for *its* neighbors. But when it comes to the question of accidents, the picture is far from clear. It isn't that a nuclear plant is a potential bomb about to go off. The fuel of existing nuclear plants can't explode like a bomb under *any* circumstances. But there is potential danger of a **meltdown.** As nuclear fuel breaks down, it releases enormous amounts of heat. In the most common type of nuclear energy plant, the heat normally is removed by water that circulates under very high pressure over the nuclear fuel. The water so heated is used to produce steam, which is used to drive the turbines that generate electricity. Should a leak develop in the water-circulating system, the water level around the fuel might drop rapidly, and the temperature of the nuclear fuel would rise rapidly until it exceeded its melting point. The fuel would then melt and pour to the floor of the generator. There, it would contact the remaining water and instantly convert it to steam. Depending on the amount of molten fuel and the amount of water, the steam so suddenly produced might be enough to blow the system apart, spewing radioactive fuel to the exterior. If the force of the blast were great enough, it conceivably might even damage the concrete building in which the nuclear reactor is located and permit radioactive material to escape. If that were to happen, it could cause many deaths and extensive environmental damage for miles around. How likely is such an event to occur? The question is basically unanswerable. It's asking how likely is something to happen that has never happened before. All nuclear reactors have secondary cooling systems that are supposed to flood the reactor with water at once if the initial cooling water is lost. A one-minute loss of water may be enough to start a meltdown. But such systems have never been put to the test. The first facility to test an emergency cooling system is now being built, but it will be many more years before it is operational. By then, about a hundred reactors will already be in use, relying on the untested system.

One comprehensive study suggests there is only one chance in a million reactor-years of an accident severe enough to kill 70 people, cause acute radiation sickness in 170 more, and do nearly 3 billion dollars worth of property damage. There is only one chance in a billion reactor-years of an accident severe enough to kill more than 2,000 people at once, cause radiation poisoning in another 5,000 or 6,000, and cause 6 billion dollars worth of property damage. These estimates are lower than the better-known probabilities (for example) that a major dam will break, killing many more people and inflicting much more damage. But are the assumptions valid, and are the risks that low? Many would suggest the risks have been underestimated by a hundred-fold. Even if that is true, the risks would still seem to be far less than they are with many alternatives we seem to be willing to live with in modern technological society.

What of nuclear wastes? Are they a threat? Nuclear fuel can't be burned to harmless ashes, like coal. After about four years, the fuel elements of a reactor are "spent." At that point, it still contains about a third of the useful uranium fuel, but it also contains hundreds of new isotopes of various elements that have been produced during the operation of the reactor. Many of these isotopes could make the reactor nonfunctional. One of the products is an isotope of plutonium which, like the leftover uranium, could be used as fuel for other reactors if it were removed and purified. (At the moment, the United States does not have a plant capable of recovering usable materials from radioactive wastes.)

Altogether, the wastes are an enormously radioactive, extremely dangerous collection of nasty materials. As they undergo radioactive decay, they produce tremendous heat. Thus they are immediately plunged into specially constructed water-filled pools at the power plant, and there they are stored for several months. The water cools the wastes and keeps radioactive materials from escaping. During this period, the level of radioactivity and the amount of heat being generated decrease by as much as 99.9999 percent as the short-lived radioactive isotopes break down. But even at the end of this time, the remaining levels of radioactivity due to long-lived isotopes are extremely lethal. The wastes must then be transported in special equipment to centralized facilities. There they must be held in temporary storage for another five years, at least. But the decay rates of some of the remaining isotopes mean they must be kept out of the environment for *a quarter of a million years!*

Since the beginning of the nuclear age, no radioactive wastes have been put into permanent underground storage in the United States. But such storage will soon be initiated. The nuclear wastes will be mixed with molten glass. The glass will be cast into rods a foot in diameter and ten feet long. (One year's wastes from one nuclear power plant may produce ten such rods.) Then the rods will be encased in sealed steel cylinders. Ultimately these cylinders will be inserted in holes of the same size that have been drilled in salt deposits, deep underground. There are several reasons given for using salt deposits. First, the very existence of a salt deposit indicates long-term absence of underground water that could slowly dissolve the wastes and move them about. Second, salt deposits are typically found in areas that have been geologically undisturbed (for example, by earthquakes) for millions of years. The assumption is they will continue to be undisturbed. Third, salt tends to flow slowly under pressure to reseal any cracks introduced as the drilling and storage take place. It is widely—but by no means uniformly—believed that all such wastes can thus be kept out of the way until they are no longer dangerous.

Although the dangers of nuclear power will be a source of genuine concern for some time, those associated with present systems are said to be manageable. But a move to develop a new kind of power source poses dangers of a wholly different magnitude. As Table 23.4 indicates, if nuclear power plants are built and operated at the rate of current projections, the known reserves of high-grade uranium could run out around the year 2000. To forestall nuclear fuel depletion just as the world comes to depend on it, intensive research and development efforts are now directed to the development of **breeder reactors.** As these reactors consume a rare isotope of uranium, they convert a much greater amount of the common, unusable form of uranium to an isotope of plutonium—which is a usable nuclear fuel. Indeed, the plutonium could then be isolated and used for conventional reactors as well as for another breeder reactor. Such reactors would produce more nuclear fuel than they consume. If they come into use, the world's uranium reserves could be used to generate nearly a hundred times as much energy and would last a hundred times as long as conventional reactors. But if breeder reactors were to expand the operating period for our uranium reserves, they would also expand the dangers.

Breeder reactors could not be cooled with water. In the design being most actively pursued, liquid metallic sodium would be used as the coolant. It is a highly dangerous, corrosive substance. The problems of containing the molten sodium and of maintaining the reactors in a functional state are horrendous. More than this, although conventional reactors cannot explode like atomic bombs, breeder reactors potentially could do so. Certain mishaps could cause the entire reactor to explode, spreading radioactive wastes equivalent to those produced by 1,000 Hiroshima-sized bombs over many square miles. In any event, the problems associated with the development of breeder reactors are so complex that some former advocates of breeder technology are now gloomy about the prospects of developing functional breeders before the uranium is gone.

Plutonium, in the form produced and used in reactors, is the most toxic substance known. A speck the size of a pollen grain almost certainly will cause fatal lung cancer. It is also the stuff that makes atomic bombs. Making a bomb from uranium requires a major scientific and industrial effort that can be mounted only by a country with substantial resources. But fashioning a bomb from plutonium is, by comparison, relatively simple. A small amount of plutonium in the hands of terrorists could be used for extortion; a somewhat larger amount could be used for mindless devastation. If we enter an era of breeder reactors, we will surely enter an era of plutonium economy in which all aspects of the nuclear power industry—processing plants, transportation systems, reactors, disposal sites—will require utmost security precautions.

A third type of nuclear power source that may become available some day is **fusion power,** in which hydrogen atoms are fused to form helium atoms, with a considerable release of energy. It is a process somewhat analogous to the reactions that cause the heat energy of the sun. But the problems associated with developing fusion power are so great that, without a major breakthrough, we cannot expect fusion power before more than four decades.

In sum, nuclear reactors of the type now in use are probably safer than many of their critics proclaim, but far less safe than their advocates claim. The most negative aspects are that they have a low net energy yield, achieved at a much higher cost than expected. Besides, the supply of naturally occurring fuel will be quickly exhausted. Breeder reactors could get around the problem of fuel depletion, but only if we are willing to gamble on the possibility of not having major accidents, of not encountering sabotage or terrorist activities. Fusion, while ultimately a possibility, is so far into the future as to hold no hope for the energy problems that will be with us this century.

Wind Energy In certain geographic regions, winds are predictable enough and strong enough to be a source of potential energy. Such high prevailing winds sweep through the vast plains of the American Midwest, from the Dakotas to Texas. **Wind energy** is thus seen to be one of the most attractive energy options; not only would it be an unlimited energy source, it would not generate pollution (and it certainly would be harmless in the hands of terrorists). In one proposal, perhaps 300,000 turbine towers could harness the winds of the plains and thereby provide fully half of our present energy needs. It is true that rows of giant turbines erupting like warts from one end of the nation's heartland to the other could be something of an aesthetic insult. On the other hand, they would be non-polluting; they would not even generate an appreciable amount of waste heat. Besides, many windmills presently being introduced in Europe are of a simple design that will disfigure the landscape far less than several other energy alternatives. Turbine towers would take up less land than, say, solar-energy generating plants, and they most assuredly would be far less disruptive of the environment than, say, strip-mining for coal. It is also possible that windmills constructed on top of existing electrical transmission towers could feed their electrical output directly into regional utility lines as a supplemental energy source. The possibility of harnessing the winds is economically and technologically feasible right now. But interest is essentially zero, for reasons that will become apparent below.

Solar Energy Without question, a virtually limitless, free energy source is potentially available to us, as it has always been to all life on earth. **Solar energy** ranks with wind energy as the most abundant, the cleanest, and the safest of all potential energy sources now being considered as alternatives to fossil fuels. Solar power plants would not release heat as entropy to any significant extent, as coal-burning and nuclear power plants do. They would pollute neither the earth nor the atmosphere. We have yet to tap directly into the sunlight energy streaming all around us. It has been estimated that if 39,000 square kilometers (15,000 square miles) of desert were set aside in Arizona and California for solar energy collectors that convert sunlight energy directly into electricity, we could develop a system for providing half our nation's annual energy needs even decades from today.

Until recently, the seeming abundance of fossil fuels lulled us into thinking there was no need to carry our search for energy to other sources, other places. But more than this, in an economic system based on profit motivation, there has not been much of a commitment to finance research that is unlikely to yield a sustained profit. Who is going to design and develop more efficient solar energy collectors and storage devices for home use? Utilities corporations? Oil companies? What would they "gain" if every home, every building had a self-sustaining or at least a supplemental energy source of its own, and therefore required less fossil fuel and electric power? To date, research funds have come mainly from environmentally concerned groups and citizens. Today, owning solar heaters and collectors is a financial luxury. No one is going to pay homeowners to install their own; it will be a cost out of their own pocket. And how many people are ready to make that kind of financial investment in a future that will belong not to them, but to the generations to come? But times are slowly changing. In 1977 the Energy Research and Development Agency elevated solar energy research from obscurity to the status of poor relative of nuclear energy research.

On Humans and Other Endangered Species

You may have noticed, now that you have just about completed this book, that we did not include a token section on vanishing species, along with a list of most endangered ones in ascending order of vulnerability. It is not that the problem is overrated; far from it. The slow process of natural selection is not keeping pace with the accelerating pressures of human activities on the biosphere, and species are disappearing fifty times faster than even a century ago. Many outstanding organizations are working to buy time for the species we threaten most with extinction. But this is not an isolated problem, with an isolated solution. Throughout earth history, the nature of the community of life has changed constantly, with new species arising to take the place of those which have been left behind in time. But always, since the beginning, it has been a *community* of life forms, whatever the kaleidoscopic shifts in its character. *Like all life forms in the past, we are all potentially endangered species, kin to one another in vulnerability because of intricate, often invisible threads of ecological interdependence stretched through the biosphere.* Timber wolf and puffin, humpback whale and golden eagle, ourselves and our children—this is our shared hour on earth. It is not for the sake of one but for the sake of all that we must make our commitment. We must face not just the isolated symptoms but the major sources of stress that have brought about the heightened state of vulnerability that now marks the entire biosphere:

Human overpopulation—by far the most serious—with its consequent stresses on available energy, food, space, and other resources, with its consequent pollution that the environment cannot absorb.

Overconsumption, founded on throwaway goods, planned obsolescence, minimal recycling, and acquisition of goods not essential for a life of dignity and quality.

Ecosystem oversimplification, with the low-diversity crops of human agriculture, with its disruption of functional and self-sustaining webs of life.

Shotgun technology, wielded without regard for biological and environmental consequences.

Lack of leadership, with moves toward the future mired in misplaced priorities, pork-barrel politics, and allegiance to totally unchecked economic growth.

Me-first behavior, with regard only for the individual's present, with nonregard for future generations and for the biosphere.

All these sources of stress on the biosphere can be traced, ultimately, to indifference to the principles of energy flow and materials reuse. It is going to take massive public education and economic redirection before the situation gets better. Even now, increased population in a local region is still seen as a healthy indicator of economic growth. Even now, as Dow Chemical Company fights environmental groups and the California State Resources Department in the courts for its right to build thirteen petrochemical plants in the Sacramento/San Joaquin River Delta—which happens to be California's major water system—construction workers stomp about with posters reading "Hungry? Unemployed? Eat an Environmentalist." In the most dangerous kind of oversimplification, industrial growth—it doesn't matter what kind of industry, or where—is equated with food on the table.

We have, most of us, been so steeped in the philosophy of growth and consumption that it might seem almost silly to say we should make attempts at a transition to a new way of life. But we can and must make these attempts, you and I, as individuals:

We can become conscious of our expectations, and learn to ask where they are leading us. We have, in general, been lulled into thinking that no matter how bad things look today, "technology" will do something to save us tomorrow. In all ways, large and small, even technology can't get us something for nothing. Medical advances applied on a global scale can wipe out many horrible infectious diseases, and alleviate much human suffering. But in doing so, they contribute also to explosive population growth—and which is the "better" choice: death by disease or starvation? And so technology moves on to agriculture to find better ways of feeding the multitudes—and ecosystems are stripped for monocrop agriculture, and left vulnerable to weeds and insect predators. Technology moves to chemical control—DDT, malathion—and the effects of biological magnification extend back to those who don't stop to ask what it is we really are expecting to get. Technology can and must be used wisely to help us out of this twentieth-century dilemma. But we must get into the habit of looking for the hidden costs; we must learn to follow through by asking what applied technology may mean to long-term stability, not short-term magic.

We can make an ecological impact statement for our own way of life. In recent years, one of the best steps our government has taken is to insist that no major new construction or development projects proceed until some attention is given to their ecological impact. Think of how much could be accomplished if we *all* did the same thing in our daily lives. What if we got into the habit of asking ourselves: What will be the ecological impact if I drive a car instead of walk? Buy milk in plastic bottles instead of returnable bottles? Buy a home in a new housing development built on what had been the rich soil of a productive farm? Become a parent once, twice, three times? Take this job, instead of that one? Vote for one candidate for public office instead of another? Speak out (at the risk of possible embarrassment) or keep my mouth shut?

We can avoid tokenism. It is so easy for each of us to take a few small steps toward achieving the goal of a better world of life, and then to reward ourselves for our virtue by lapsing back into the life-style that has brought us to our present predicament. If you feel you really want and deserve some of the luxuries of the industrialized society you live in, try to give up enough other luxuries so that you come out, on balance, contributing to the solution instead of the problem.

We can get involved. We can't assume that just because we develop an ecological perspective for our personal life style that we have done all we can for the future. We can join public interest groups, demanding that public and private institutions alike serve the public interest in a long-term, ecologically sound way. As you can gather by looking over the list of readings and the list of public interest groups at this chapter's end, there are many individuals and groups already working effectively to bring about change.

We can expect an ecological perspective from our elected representatives. Many of the crucial decisions affecting both the short-range and long-range future of the biosphere are going to be made by our elected representatives. At the turbulent close of the 1960s, many students in this country lost faith in the capacity of our system of government to respond to the people. They developed the cynical attitude that a system so corrupt would not—could not—be responsive. Since that time, there has been enough evidence that the system *can* be made responsive *if enough people demand it.* If we wish for other goals, we must work to make our presence felt. *Our system of government has become far more responsive and open than any other system in the recent history of human experience.* If it fails to move in the directions we think best, then more than ever it will be the fault of the people. In the near future, more than the recent past, we will have the kind of government we ask for and deserve. If we feel strongly about issues of public policy, it is up to us to keep our local, state, and national representatives informed of our positions.

In the words of the human ecologist G. Tyler Miller, *the secret of sustained action is to think and work on two levels at once.* We can constantly whittle away at making major changes in our political and economic systems, and in our world view. At the same time, we can do a number of little things each day to help return to dynamic compatibility with the total world of life. Each small act can be used to expand awareness of the need for basic changes in our political, economic, and social systems over the next few decades. They also can help us avoid psychic numbness every time we think about the magnitude of the job to be done. We can begin at the individual level and work in ever widening circles. We can join with others and amplify our commitment. This is the way of change.

Recommended Readings

Audubon (bimonthly). National Audubon Society. 1130 Fifth Avenue, New York, New York 10028. $10/year. Conservationist viewpoint; more than bird-watching. Exceptionally well written, outstanding graphics. Gives a current picture of environmental issues the world over.

BioScience (monthly). American Institute of Biological Sciences. 3900 Wisconsin Avenue NW, Washington, D.C. 20016. $24/year to institutions. Official publication of AIBS; major coverage of biological concerns.

Catalyst for Environmental Quality (quarterly). 274 Madison Avenue, New York, New York 10016. $5/year. Popular treatment of all aspects of environment, including population control. Reviews books and films for environmental education.

Family Planning Perspectives (bimonthly). Planned Parenthood—World Population. Editorial Offices, 666 Fifth Avenue, New York, New York 10019. Free on request. Detailed, wide-ranging, liberal attitude. Useful for persons concerned with the problem of overpopulation.

Mother Earth News (6 issues yearly). Mother Earth News, Inc. P.O. Box 70, Hendersonville, North Carolina 28739. $6/year. Good articles on organic farming, alternative energy systems, and alternative life-styles.

National Wildlife (bimonthly). National Wildlife Federation. 1412 Sixteenth Street NW, Washington, D.C. 20036. $5/year. Good summaries of wildlife issues.

New Scientist (monthly). 128 Long Acre, London, W.C. 2, England. $16/year. Excellent journal on general science.

Population Bulletin (bimonthly). Population Reference Bureau. 1755 Massachusetts Avenue NW, Washington, D.C. 20036. $8/year. Nontechnical articles on age structure, fertility rates, migration, and mortality.

Science (weekly). The American Association for the Advancement of Science. 1515 Massachusetts Avenue NW, Washington, D.C. 20036. $20/year. Outstanding forum for American science. Probably the single best source for keeping up with research—and with what researchers are thinking in terms of applications and consequences of research.

Scientific American (monthly). 415 Madison Avenue, New York, New York 10017. $10/year. Excellent journal; accessible writing style, excellent graphic presentation of scientific research and discoveries.

The Sierra Club Bulletin (monthly). The Sierra Club. 1050 Mills Tower, San Francisco, California 94104. $5/year. Good coverage of a range of environmental issues and citizen action.

Recommended Doings

Common Cause. P.O. Box 220, Washington, D.C. 20004. One of the most important citizens' organizations. More than 100,000 members lobby actively on a range of issues. Try to join this group. It is working for you.

Conservation Foundation. 1717 Massachusetts Avenue NW, Washington, D.C. 20036. Active in conservation.

The Defenders of Wildlife. 2000 N Street NW, Washington, D.C. 20036. Its goal is the preservation of all forms of wildlife.

Environmental Defense Fund, Inc. 162 Old Town Road, East Setauket, New York 11733. A public interest organization of scientists, lawyers, and laymen.

League of Women Voters of the United States. 1730 M Street NW, Washington, D.C. 20036. Outstanding organization with local and state leagues working for political responsibility through informed and active citizens.

National Audubon Society. 1130 Fifth Avenue, New York, New York 10028. Provides a spectrum of ecology education services. Operates forty wildlife sanctuaries throughout the country.

National Parks and Conservation Association. 1701 18th Street NW, Washington, D.C. 20009. Urges acquiring and protecting public park lands. Active in such issues as resource management, pesticide control, and pollution. Led coalition against Everglades jet port in Florida.

National Wildlife Federation. 1412 16th Street NW, Washington, D.C. 20036. Encourages citizen and government action for conservation. Annual conservation directory available at $1.50.

The Nature Conservancy. 1522 K Street NW, Washington, D.C. 20005. Seeks to preserve natural environments. Acquires vulnerable property and holds it for later resale to public agencies.

Population Reference Bureau. 1775 Massachusetts Avenue NW, Washington, D.C. 20036. Gathers data on the effects of the world population explosion.

Scientists Institute for Public Information. 30 East 68th Street, New York, New York 10021. Draws on scientists from all disciplines in public information programs on many social issues. Serves as national coordinating body for local scientific information committees.

Sierra Club. 1050 Mills Tower, San Francisco, California 94104. Devoted to protecting the nation's resources. Provides films, books, exhibits, and lecturers. The Washington office is an effective lobbying group.

Wildlife Society. Suite S-176, 3900 Wisconsin Avenue NW, Washington, D.C. 20016. Major concern is wildlife preservation.

Zero Population Growth. 1346 Connecticut Avenue NW, Washington, D.C. 20036. Main emphasis is on family planning and stabilizing population levels.

Appendix
Brief System of Classification

This appendix, based on Robert Whittaker's scheme (*Science*, 1969, 163: 150–160), reflects current understandings of phylogenetic relationships. It is by no means all-encompassing; it simply includes the kinds of organisms you have read about in the text. The numbers refer to pages on which representative organisms for the group are discussed.

Kingdom Monera Single-celled prokaryotes; autotrophs and heterotrophs lacking true nucleus and other internal, membrane-enclosed organelles

Phylum Schizomycetes. Bacteria	253–254, 255
Phylum Cyanophyta. Blue-green algae	257–258

Kingdom Protista Single-celled eukaryotes; heterotrophs and some autotrophs having true nucleus and other internal, membrane-enclosed organelles

Phylum Pyrrophyta. Dinoflagellates	258, 259–260, 269
Phylum Chrysophyta. Golden algae, diatoms	260, 269
Phylum Euglenophyta. Photosynthetic flagellates	260, 261
Phylum Flagellata. Heterotrophic flagellates	260, 261, 285, 286
Phylum Sarcodina. Amoebas, foraminiferans	72, 132, 261
Phylum Sporozoa. Parasitic sporozoans	262
Phylum Ciliata. Ciliates	130, 262–264

Kingdom Plantae Multicellular eukaryotes; autotrophic forms able to build their own food molecules through photosynthesis

Phylum Rhodophyta. Red algae	268–269
Phylum Phaeophyta. Brown algae	268–270
Phylum Chlorophyta. Green algae	269, 270–271
Phylum Bryophyta. Bryophytes; mosses, liverworts	271, 272
Phylum Tracheophyta. Vascular plants	155–169
Subphylum Spenopsida. Horsetails	271–272, 273
Subphylum Lycopsida. Lycopods	272–273
Subphylum Pteropsida. Ferns	272–273
Subphylum Spermopsida. Seed plants	273–280
Class Cycadae. Cycads	274
Class Ginkgoae. Ginkgoes	274
Class Coniferae. Conifers	266, 274–275, 380
Class Angiospermae. Flowering plants	276–280
Subclass Monocotyledonae. Monocots; grasses, palms, lilies, orchids	155, 156, 160, 276
Subclass Dicotyledonae. Dicots; most common flowering plants not mentioned above	155, 156, 160, 276

Kingdom Fungi Multicellular eukaryotes; generally heterotrophs relying on extracellular digestion and absorption (enzyme secretions break down organic matter in environment, which is absorbed across body wall)

Phylum Eumycophyta. True fungi	280–281, 282
Class Phycomycetes. Algal-like fungi	34
Class Ascomycetes. Sac fungi	280, 282
Class Basidiomycetes. Club fungi	280, 281, 282

Kingdom Animalia Multicellular eukaryotes; heterotrophs that ingest other organisms for food

Subkingdom Parazoa	
Phylum Porifera. Sponges	6, 285–286
Subkingdom Radiata. Radially symmetrical animals	286, 287, 297, 298
Phylum Cnidaria. Coelenterates	284, 286–287
Subkingdom Protostomia. Protostomes	289–290, 296
Phylum Platyhelminthes. Flatworms	133, 134, 288–289
Phylum Nematoda. Nematodes; roundworms	289
Phylum Annelida. Annelids; segmented worms	280–292
Phylum Mollusca. Molluscs	293–295, 296
Phylum Arthropoda. Arthropods	292–293, 294–295
Subkingdom Deuterostomia. Deuterostomes	289–290, 296, 299
Phylum Echinodermata. Echinoderms	296–298
Phylum Chordata. Chordates	296–306
Subphylum Urochordata. Tunicates	298, 299
Subphylum Cephalochordata. Lancelets	299–300
Subphylum Vertebrata. Vertebrates	285–290
Class Agnatha. Jawless fishes	247, 299, 300
Class Chondricthyes. Cartilaginous fishes; sharks, rays	
Class Osteoichthyes. Bony fishes	6–7, 300
Class Amphibia. Amphibians; frogs, toads, salamanders	245, 247, 301–303
Class Reptilia. Reptiles; lizards, snakes, turtles	22, 296, 297, 301
Class Aves. Birds	303, 328, 331, 334
Class Mammalia. Mammals	
Subclass Prototheria. Egg-laying mammals; spiny anteater, duck-billed platypus	303
Subclass Metatheria. Pouched mammals; opossum, kangaroo, koala	303
Subclass Eutheria. Placental mammals	225
Order Insectivora. Insect-eating mammals; shrews, moles, hedgehogs	304
Order Edentata. Toothless mammals; sloths, armadillos	
Order Rodentia. Rodents; mice, rats, squirrels	312, 365, 382
Order Perissodactyla. Odd-toed ungulates; horses, rhinoceros	9, 11, 250
Order Artiodactyla. Even-toed ungulates; deer, cattle, camels, hippopotamus	366, 371–372, 380
Order Proboscidea. Elephants	14, 15
Order Lagomorpha. Rabbits, hares	343
Order Sirenia. Manatees, dugongs, sea cows	
Order Carnivora. Wolves, dogs, cats, bears, seals, weasels	12, 212, 324, 325
Order Cetacea. Whales, porpoises	133, 328, 338, 339
Order Chiroptera. Bats	316
Order Primates. Lemurs, monkeys, apes, humans	304, 305–306

Glossary

A Few Metric Equivalents

Length:	kilometer	0.62 mile
	meter	39.37 inches
	centimeter	0.39 inch
	millimeter	0.039 inch
Mass:	gram	0.035 ounce
	kilogram,	2.205 pounds
Volume:	cubic centimeter	0.061 cubic inch
	liter	1.057 quarts

abortion Induced expulsion of the embryo from the uterus before the third trimester of pregnancy.

abscission (ab-SIH-zhun) In some plants, the dropping of a leaf (or fruit or flower) after hormonal action causes a corky cell layer to form where the leaf stalk joins the stem, which shuts off nutrient and water flow.

acid A substance that releases hydrogen ions (H$^+$) in a water solution, where it has a pH of less than 7.

action potential The rapid, cyclic alternation in the electric gradient across a nerve cell membrane; basis for the nerve impulse that travels along a neuron.

activation energy For each set of reactants, some minimum amount of energy from an outside source that must be applied before the reaction will begin.

active site A three-dimensional groove or pocket in an enzyme in which R groups project outward and form an exact three-dimensional complement of chemical groups on those substances the enzyme acts upon.

active transport Movement of a substance across a plasma membrane, against a concentration gradient, by the active expenditure of ATP. The substance is moved in a direction other than the one in which simple diffusion would take it.

adaptation A structural, physiological, or behavioral trait of an individual that promotes the likelihood of survival and reproduction under prevailing environmental conditions.

adaptive radiation Process whereby several isolated populations of a species accumulate different adaptive traits that allow them to exploit different aspects of the environment in different ways.

addiction State in which the body is physically dependent on some drug; it no longer can maintain normal functioning without the drug.

adenine (AH-deh-neen) A purine; one of the nitrogen-containing bases in nucleic acids; ATP, and ADP (among other things).

adenosine diphosphate (ah-DEN-uh-seen die-FOSS-fate) ADP, the compound formed by hydrolysis of an ATP molecule.

adenosine triphosphate ATP, a compound that is a major source of usable energy in all cells. When an ATP molecule loses a phosphate group to form ADP, energy is released.

ADH Antidiuretic hormone.

adipose tissue (AH-di-pose) Tissue of fat-filled cells acting mainly as the body's energy reserves.

ADP Adenosine diphosphate.

adrenalin (uh-DREN-uh-lin) Animal hormone produced by the adrenal medulla. A potent stimulant for the breakdown of glycogen and lipids; increases rate and force of the heart's pumping action; raises blood pressure through constriction and dilation of blood vessels.

aerobic pathway (air-OH-bik) In a cell, a pathway of energy metabolism that uses oxygen as an electron acceptor.

aldosterone (al-DAW-stir-own) Animal hormone produced by the adrenal cortex; stimulates sodium reabsorption from the kidneys. (Often called a mineralocorticoid because its main effects are on sodium and potassium metabolism.)

alga, plural **algae** (AL-guh, AL-jee) Photosynthetic, autotrophic organisms without complex reproductive organs.

allele (uh-LEEL) An alternative form of a gene for any given trait.

allele frequency For any population, the number of individuals carrying a given allele, divided by the total number of individuals in the population. Also called gene frequency, which is something of a non sequitur.

alpha islet cells Pancreatic cells that make and secrete the hormone glucagon.

alternation of generations The alternation of a diploid, spore-producing generation (sporophyte) with a haploid, gamete-producing generation (gametophyte) in the life cycle of a plant.

alveolus, plural **alveoli** (ahl-VEE-uh-luss) One of many small, thin-walled pouches in the lungs that are sites of gas exchange between the air in the lungs and the bloodstream.

amino acid (uh-MEE-no) A substance containing an amino group (NH$_2$) and an acid group (—COOH); a subunit for protein synthesis. Twenty amino acids commonly occur in living organisms.

amnion (AM-nee-on) In reptiles, birds, and mammals, a membrane that arises from the inner cell mass of a blastocyst. It becomes the fluid-filled sac in which the embryo develops freely.

anaerobic pathway An energy-extracting pathway that does not require oxygen as an electron acceptor; both fermentation and glycolysis are anaerobic pathways (even though glycolysis *can* proceed in the presence of oxygen).

anaphase In mitosis the stage at which the two sister chromatids of each chromosome are separated from each other and moved to opposite poles of the cell.

angiosperm (AN-jee-oh-sperm) A flowering plant.

animal Multicellular organism that ingests other organisms for food.

annual plant A vascular plant that completes its life cycle in one growing season.

anther In flowering plants, the pollen-bearing part of the stamen (the male reproductive organ).

antibody A type of protein molecule, carried in the blood, that is able to combine with a specific antigen and thereby inactivate it.

antidiuretic hormone ADH; animal hormone secreted by the hypothalamus and promoting water reabsorption from the kidneys.

antigen Foreign cell or substance that has penetrated the body and that triggers an immune response.

anus In some invertebrates and all vertebrates, the terminal opening of the gut through which solid residues of digestion are eliminated.

aorta (ay-OR-tah) The main artery of systemic circulation; it leaves the heart and supplies oxygenated blood to all body regions except the lungs.

apical dominance (AY-pickle) In many vascular plants, the restriction of lateral bud growth on a given stem; as long as the growing tip remains intact, lateral branching is inhibited.

asexual reproduction Production of new individuals by any process that does not require the fusion of gametes.

atom The smallest unit of an element that still retains the characteristics of that element.

atomic number A relative number assigned to each kind of element based on the number of protons in one of its atoms.

atomic weight The weight of an atom of any element relative to the weight of the most abundant isotope of carbon (which is set at 12).

ATP Adenosine triphosphate.

autonomic nervous system (auto-NOM-ik) The part of the vertebrate nervous system regulating such internal organs as the heart, stomach, and blood vessels, which usually are not under conscious control.

autosome Any chromosome that has the same physical appearance in both sexes of a species. (Humans normally have twenty-two pairs of autosomes, and one pair of sex chromosomes.)

autotroph "Self-feeder"; an organism capable of building all the complex organic molecules that it needs as its own food source, from simple compounds. Compare **heterotroph.**

auxin Plant hormone regulating growth, especially cell elongation; it is active in certain standard tests, such as the oat coleoptile curvature test.

axon Nerve cell extension that generally carries impulses away from the nerve cell body.

bacillus, plural **bacilli** (BAH-sill-us, BAH-sill-eye) A rodlike form of bacterium.

bacteriophage (bak-TEER-ee-oh-fahj) A category of viruses that infect and destroy certain bacterial cells.

basal body A structure that gives rise to a cilium or flagellum.

base A substance that accepts hydrogen ions (H$^+$); often it gives up hydroxyl ions (OH$^-$) in a water solution, where it has a pH above 7.

beta islet cells Pancreatic cells that make, store, and secrete the hormone insulin.

bilateral symmetry Arrangement whereby the left half of an animal body is approximately equivalent to the right half.

binary fission Asexual reproduction by division of one cell into two daughter cells.

biodegradable Refers to those complex molecules that living organisms can break down into simpler molecules that are not harmful in the environment.

biological clock Internal mechanisms that correlate cyclic changes in behavioral and physiological responses to cyclic changes in the environment. In some animals, the pituitary and pineal glands are known to be involved.

biological magnification Increasing concentration of a relatively nondegradable substance in body tissues, beginning at low trophic levels and moving up through those organisms which are diners and then are dined upon as part of community food webs.

biomass Dry weight of an organism or group of organisms.

biome Broad geographic region distinctly recognizable by its dominant array of primary producers.

biosphere Zone of the planet's surface, encompassing earth, water, and air, in which life can exist.

blastocyst In mammalian development, a modified blastula stage consisting of a hollow ball of surface cells (the trophoblast) with inner cells massed at one end.

blastula In many animal species, an embryonic stage consisting of a hollow, fluid-filled ball of cells, one layer thick.

blood pressure The level of pressure exerted on arteries as they are distended by blood being forced out of the heart. Compare **pulse pressure.**

brain stem An extension of the spinal cord; integrates movements of the sense organs and head at the reflex level.

breed A distinct variety of organisms within a species.

bronchus, plural **bronchi** (BRONG-cuss, BRONG-kee) Passageway leading from the trachea to the lungs.

budding Asexual reproduction in which an extension of the parent body enlarges, then breaks away to form a new individual.

buffer In living cells, a substance that prevents sudden swings in pH by combining with hydrogen ions (H$^+$) when their concentration rises, or by releasing them when the H$^+$ concentration falls.

calorie The amount of heat needed to raise the temperature of 1 gram of water by 1°C.

Calvin–Benson cycle Stage of the dark reactions of photosynthesis, in which carbon-containing compounds are used to form carbohydrates (such as glucose) and to regenerate a sugar phosphate (RuDP) required in carbon dioxide fixation. The first product is a three-carbon compound (PGA).

cambium (KAM-bee-um) In vascular plants, a zone of lateral meristematic tissue.

cancer A malignant tumor; a mass resulting from uncontrollable cell division. Because of a breakdown of controls over cell surface recognition factors, malignant cells lose their identity and can migrate to various body regions, there to multiply uncontrollably into secondary tumors.

capillary One of the smallest blood vessels. Through capillary walls, substances such as oxygen and carbon dioxide are exchanged between the bloodstream and the cells of surrounding body tissues.

carbohydrate Organic compound of carbon, hydrogen, and oxygen; sugar is an example.

carbon dioxide fixation First stage of the dark reactions of photosynthesis; carbon dioxide from the air is combined with organic molecules.

carcinogen (car-SIN-oh-jen) Any agent capable of promoting cancer.

cardiac cycle Two alternating stages in which the heart relaxes (diastole) and contracts (systole).

carnivore Heterotroph that ingests other animals for food.

carpel In flowering plants, one of several pistils fused together into a compound female reproductive structure.

carrying capacity of the environment For any species, the maximum population size that the environment will support and maintain on a continuing basis.

cell The basic *living* unit. There may be organic molecules of life below this level of organization, but such molecules by themselves are nonliving.

cell plate In plant cell division, a partition that forms from vesicles at the equator of the mitotic spindle, between the two newly forming cells.

cellular respiration The Krebs cycle and oxidative phosphorylation: two related pathways in which considerable energy is extracted from sugar molecule fragments left over from glycolysis.

cell wall An extracellular layer of material that surrounds the plasma membrane of some cell types, mainly in plants.

central nervous system The brain and spinal cord.

centriole A cytoplasmic organelle arranged in pairs near the nucleus; gives rise to the microtubule system of eukaryotic cilia and flagella.

centromere (SEN-troh-meer) Localized, differentiated region of a chromosome where two sister chromatids are attached after DNA replication and before mitotic division.

cerebellum Outgrowth of the brain stem; concerned with stimulation and responsiveness of muscles and skin, with coordination and equilibrium.

cerebrum Brain region concerned with integration of sensory inputs; main control center for coordinating all aspects of the nervous system.

cervix Lower region of the uterus that opens to the vagina.

chemical reaction Energetic interactions between atoms or molecules in which existing bonds are broken and/or new bonds are formed.

chemoreceptor A sensory cell that directly or indirectly transforms chemical stimuli into nerve impulses.

chiasma, plural chiasmata (kai-AZ-mah, kai-az-MAH-tah) During meiosis, a region of contact between sister chromatids that have undergone synapsis and that indicates they have exchanged corresponding segments.

chlorinated hydrocarbon Synthetic, chlorine-containing organic compound; often such compounds are not readily degradable in the environment. DDT is an example.

chlorophyll Pigment that acts as an electron donor in photosynthesis.

chloroplast Small, membrane-enclosed organelle in the cytoplasm that houses membranes, pigments, and enzymes for photosynthesis.

chorion (CORE-ee-on) In reptiles, birds, and mammals, the outermost membrane around the developing embryo, with external, vascularized villi that become embedded in the endometrium.

chromatid (CROW-mah-tid) One of the two coiled strands making up a chromosome following replication. A pair of chromatids joined at the centromere represent duplicate sets of genetic information.

chromatin Masses of fibers within the nucleus; contains DNA, RNA, and protein. Deep-staining in slide preparation.

chromosome In a eukaryotic cell, a distinctly organized form of the hereditary material, consisting of a single DNA molecule and its associated proteins (before replication) or of two identical DNA molecules and their associated proteins (after replication).

cilium, plural cilia (SILL-ee-um) Short, hairlike structure composed of a regular array of microtubules. Usually occur together in patterns that promote rapid movement at the cellular level.

cleavage Rapid, successive cell divisions in a zygote; leads to increase in cell number, not to increased size.

climax community An array of species locked together in such an efficient use of materials and energy that it is relatively stable and self-perpetuating. Final step in community succession.

coccus, plural cocci (COCK-us, COCK-eye) Spherical form of bacterium.

coelom (SEE-lum) Between the gut and body wall of more complex animals, a fluid-filled cavity lined with mesoderm.

coenzyme An organic molecule that becomes attached to and assists in the catalytic action of an enzyme. NAD, FAD, and NADP are co-enzymes that act as electron acceptors and donors in metabolic reactions.

cohesion theory of water transport The theory that hydrogen bonds between water molecules are enough to pull water up through a plant body as water molecules are given up for cell growth, photosynthesis, or transpiration.

coitus (COE-eh-tuss) Sexual intercourse.

colon Region of the large intestine in which excess water is removed from undigested food residues.

commensalism Type of symbiotic relationship in which one species benefits but the other species neither benefits nor is harmed.

community An interacting association of organisms in a common environment.

competitive exclusion The idea that no two species requiring precisely the same resource can coexist indefinitely; one eventually displaces the other because it has a competitive edge in some characteristic.

complement A group of plasma proteins that become activated in sequence by the presence of antigen-antibody complexes. The activated proteins work as a system to amplify a local inflammatory response.

compound Chemical substance made of the atoms of two or more different kinds of elements bonded to one another.

concentration gradient State in which the molecules of a substance are concentrated more in one region of a system than in another.

conditioning Form of learning in which a response becomes associated (by means of a reinforcing stimulus) with a new stimulus that was not connected before with the response.

conjugation In some bacteria and protistans, the transfer of DNA segments between two individual cells, the outcome being genetic recombination.

contractile vacuole In some single-celled organisms, a membrane-enclosed chamber that takes up excess water in the cell body, then contracts, expelling the water through a pore to the outside.

control genes Genes that regulate when and to what extent structural genes will be expressed.

copulation Among many animals, the injection of sperm by the male into the female's reproductive tract.

corpus luteum (CORE-puss LOO-tee-um) An endocrine gland that forms in the ovary in the site of a ruptured follicle.

cotyledon (cot-ill-EE-don) A "seed leaf," often containing stored nutrients to be used by a plant embryo during germination.

covalent bond A sharing of one or more electrons between atoms or groups of atoms. When the electrons are shared equally, it is a nonpolar covalent bond. When they are shared unequally, it is a polar covalent bond.

cristae In mitochondria, the inner, deeply folded membranes within which the main energy conversions occur.

crossing over In meiosis, the exchange of corresponding chromatid segments between homologous chromosomes.

cyclic photophosphorylation An ATP-yielding pathway in which electrons excited by sunlight energy move from a photosystem to a transport chain, then back to the photosystem. Probably represents the most ancient photosynthetic pathway.

cytochrome (SIGH-toe-krome) Iron-containing protein molecule that acts as an electron carrier in transport chains of photosynthesis and respiration.

cytokinin (SIGH-toe-KINE-in) Plant hormone that mainly promotes cell division.

cytoplasm Semifluid substance of a cell in which organelles are embedded; the zone exclusive of the cell wall (if any), plasma membrane, and nucleus.

cytosine (SIGH-toe-seen) A pyrimidine; one of the nitrogen-containing bases in nucleic acids.

dark reactions A two-stage photosynthetic process. First, carbon dioxide is combined with a sugar phosphate (RuDP), then the carbon atoms are locked into intermediate compounds. Second, ATP and NADPH$_2$ are used in the conversion of those compounds to food molecules, and the RuDP is replaced.

DDT Dichlorodiphenyltrichloroethane; a chlorinated hydrocarbon developed as a pesticide but subject to biological magnification in food webs.

dehydration synthesis A covalent bonding of two molecules in which one is stripped of a hydrogen atom and the other of an —OH group.

dendrite Branched, usually short extension of a neuron; generally carries nerve impulses toward the nerve cell body.

denitrification A process by which certain bacteria in the soil convert nitrates to gaseous nitrogen (N_2).

deoxyribonucleic acid (dee-ox-ee-rye-bow-new-CLAY-ik) DNA; a nucleic acid built from nucleotide subunits containing the sugar deoxyribose. The double-stranded carrier molecule of genetic information in all cells.

diaphragm (DIE-uh-fram) A sheet of muscle tissue between the thoracic and abdominal cavities that functions in breathing movements. Also, a contraceptive device used to temporarily close off and thus prevent sperm from entering the uterus during sexual intercourse.

dicot (DIE-kot) Short for dicotyledon; a subclass of flowering plants characterized primarily by seeds having embryos with two cotyledons (seed leaves).

differentiation The processes by which cells of identical genetic makeup become structurally and functionally different from one another, according to the genetically controlled developmental program of the species.

diffusion The net movement of molecules from a region of greater concentration to a region where they are less concentrated; occurs because of energetic movements of individual molecules, which tend to become dispersed uniformly through a given system.

diploid (DIP-loyd) A state in which a cell contains two sets of chromosomes. Compare **haploid.**

direct development Developmental program in which the embryo develops directly into the young adult form.

directional selection In a population, an increasing frequency of a heritable trait that corresponds to a specific direction of change in the environment.

disaccharide A carbohydrate made of two sugar units.

disruptive selection A splitting up of a population through selection for markedly variant forms that are in some way more adaptive to environmental conditions than is the existing species form.

distal convoluted tubule In the kidney nephron, the main region in which concentrations of the body's water and salts are regulated.

divergence Process whereby an isolated population of a species accumulates specialized traits that are adaptive to a different environment, and thereby becomes progressively different from the existing species form. The process may lead to a separate line of descent.

DNA Deoxyribonucleic acid.

dominant allele In a diploid cell, the expression of one allele to the extent that it masks expression of its partner.

dormancy Among some plants, a period of suspended activity preceding seed germination or regrowth of buds.

ecology Study of the interactions of organisms with one another and with their physical environment.

ecosystem Sum total of all the interactions linking organisms in a community with one another and with their environment.

ectoderm In an animal embryo, the outermost cell layer destined to give rise to skin and nerve tissue.

effector A muscle (or gland) that responds to nerve impulses by producing movement (or chemical changes) to counterbalance changes in internal and external conditions.

electric charge A property of matter that enables ions, atoms, and molecules to attract or to repel one another.

electron Negatively charged particle that orbits the nucleus of an atom.

electron transport chain An organized sequence of molecules that transfer relatively high-energy electrons from one molecule to the next in line. Energy released during such transfers may be used to phosphorylate ADP present in the cell.

element A substance composed of a single kind of atom.

embryonic development Process by which cells of an embryo become different in position, developmental potential, appearance, composition, and function.

embryonic induction The process by which one group of embryonic cells signals an adjacent group of cells in a way that causes it to differentiate.

embryonic regulation The ability of embryonic cells to modify the developmental program in ways that compensate for missing or extra parts in that program.

endergonic reaction (en-dur-GONE-ik) Reaction in which energy from an outside source must be added before the reaction will proceed.

endocrine gland (EN-doh-krin) Small, ductless gland that helps govern animal body functioning through the production and controlled release of a hormone into the bloodstream.

endoderm In an animal embryo, the innermost cell layer, which differentiates into internal organs such as the liver and stomach.

endometrium (en-doh-MEET-ree-um) Mucous membrane of the uterus, consisting of epithelium and uterine glands.

endoplasmic reticulum, rough Ribosome-studded membrane system on which proteins are assembled and within which they may be temporarily stored.

endoplasmic reticulum, smooth Membrane system on which lipids are assembled and within which they may be temporarily stored.

endoskeleton In chordates, the internal framework of bone and/or cartilage.

endosperm Food storage tissue in a flowering plant seed.

endospore Resistant body that forms within the cell body of some bacteria.

entropy A measure of how much energy in a system has been so randomly dispersed (usually in the form of evenly distributed heat) that it is no longer readily available to do work.

enzyme A type of protein that speeds up the rate of a metabolic reaction by lowering the activation energy for that reaction.

epidermis Outermost tissue layer of a multicellular animal or plant.

epithelium Sheets of cells, one or more layers thick, lining internal and external surfaces of the multicellular animal body.

equilibrium, dynamic A stable condition. The point at which a chemical reaction runs forward as fast as it runs in reverse, so that there is no further net change in the concentrations of reactants or products.

erythrocyte (eh-RITH-row-site) Red blood cell.

estrogen Sex hormone that stimulates development of secondary sex characteristics; initiates cyclic changes in the endometrium in preparation for pregnancy.

estrus (ESS-truss) Among mammals, the cyclic period of a female's sexual receptivity to the male.

estuary (ESS-chew-airy) A region where fresh water from a river or stream mixes with salt water from the sea.

ethylene Gaseous plant hormone that promotes fruit ripening and many other responses in plants.

eukaryote (yoo-CARRY-oht) A cell that has membrane-enclosed organelles, the most notable of which is the nucleus.

eutrophication (yoo-trofe-ih-KAY-shun) Successional process whereby a body of water becomes so enriched with inorganic nutrients that many populations (e.g., algal, plant) grow explosively. Often many aerobic organisms die because of the concurrent depletion of oxygen.

evolution Change in the allele frequencies of a population over time, hence in the diversity and adaptations embodied in that population.

exergonic reaction (ex-ur-GONE-ik) A chemical reaction in which energy is released from the reactants, so that the products contain less chemical potential energy than the reactants.

exoskeleton An external skeleton, as in arthropods.

exponential growth (ex-poe-NEN-shul) In populations, the increasingly accelerated rate of growth due to the increasing number of individuals being added to the reproductive base. In the absence of control factors, all populations experience exponential growth.

extrinsic limiting factor An environmental factor setting some limit on the maximum size of a population.

fat A molecule composed of glycerol and three fatty acid molecules, linked together by dehydration synthesis.

fatty acid Component of oils and fats; long-chain hydrocarbon with an acid group (—COOH) attached.

feedback inhibition A control mechanism whereby an increase in some substance or activity inhibits the very process leading to (or allowing) the increase.

fermentation pathways Those pathways of glycolysis in which pyruvate is converted to alcohol, lactic acid, or a similar end product.

fertilization Fusion of the nuclei of gametes.

fetus (FEE-tuss) Name given to the human embryo after the first trimester of pregnancy.

first principle of energy The amount of energy in the universe is finite, and that amount never changes.

fixed action pattern Stereotyped, innate motor response linked to relatively simple environmental stimuli.

flagellum, plural **flagella** (fluh-JELL-um) Structure involved in rapid cell movement through the environment. Longer and less numerous than cilia; composed of microtubule subunits.

flexion reflex The bending of one or more joints as a simple reflex response of withdrawal from an (often-painful) stimulus.

fluid mosaic membrane structure Current model of cell membrane structure, in which diverse proteins and other molecules are more or less suspended in a fluid lipid bilayer. The lipid bilayer may represent the barrier itself, with proteins functioning in communication and control.

follicle In the ovary, one of the spherical chambers containing a ripening egg.

follicle-stimulating hormone FSH; a pituitary hormone. In females it stimulates follicle maturation; in males it helps promote sperm formation and development.

food web Complex patterns of who eats whom among producers, consumers, and decomposers in a community.

fruit In flowering plants, a mature ovary (or aggregate of ovaries) that functions in the protection and dispersal of seeds.

fungus A multicellular organism that secretes enzymes for breaking down organic material in the outside environment; the resulting small molecules are then absorbed across the fungal body wall.

gamete (GAM-eet) Mature haploid cell that functions as a sexual reproductive cell.

gametophyte (gam-EET-oh-fight) The haploid, gamete-producing stage in a plant life cycle.

ganglion A cluster of nerve cell bodies, usually located in the peripheral nervous system.

gastrula Among some animals, the embryo in a stage of development following the blastula; a two- or three-layered structure enclosing a central cavity that has an opening to the outside.

gene A unit of inheritance; a stretch of DNA coding either for an RNA molecule or for the translation product of an RNA molecule (a polypeptide). The actual expression of a gene may be influenced by interactions with other genes and by conditions in the internal and external environments.

gene flow Movement of alleles into or out of a population as different individuals move into or out of the group.

gene frequency More precisely, allele frequency; the number of individuals in a population that carry a given allele, divided by the total number of individuals in the population.

gene pool The sum total of all genes in a population.

genetic code The universal, biochemical "language" by which nucleotide sequences in DNA call for specific amino acids used in protein synthesis.

genetic counseling Providing prospective parents with information on the likelihood that their children may have specific genetic diseases; based on family histories and test methods.

genetic drift Random fluctuations in the relative allele frequencies of a small breeding population.

genetic equilibrium Theoretical baseline for measuring change in allele frequency in a population; the assumption is that stable allele ratios occur only if all individuals have equal probability of reproducing.

genetic recombination In sexually reproducing organisms, the introduction of new combinations of genes into a chromosome as a result of crossing over during meiosis. In some asexually reproducing organisms, the transfer of short DNA segments from one individual to another through conjugation also leads to genetic recombination.

genotype (JEE-no-type) The genetic constitution of an individual. Compare **phenotype.**

genus, plural **genera** (JEE-nuss, JEN-er-uh) A broad category into which similar yet distinct species may be grouped, based on their implied descent from a fairly recent common ancestor.

geotropism (JEE-oh-TROPE-izm) Growth or movement in relation to the direction of the pull of gravity.

gibberellin (JIB-er-ELL-in) Plant hormone that promotes cell elongation and several other responses in plants, especially stem elongation in intact plants.

gill Body region of thin tissue flaps, richly supplied with blood vessels and specialized for gas exchange in aquatic environments.

gland Secretory cell or organ.

glomerulus (glow-MARE-yoo-luss) Cluster of capillaries in Bowman's capsule of the kidney.

glucagon Animal hormone secreted by alpha islet cells in the pancreas; essential in glucose breakdown and thereby promotes increased blood glucose concentrations.

glycerol (GLISS-er-ohl) Three-carbon molecule with three hydroxyl groups attached; combines with fatty acids to form fat or oil.

glycogen In animals, a starch that is a main food reserve; it can be readily broken down into individual glucose molecules.

glycolysis (gly-CALL-ih-sis) The initial breaking apart of sugar molecules such as glucose, with the release of energy. Glycolysis may proceed under aerobic as well as anaerobic conditions, but it does not require oxygen to do so.

Golgi complex (GOAL-jee) Membrane system in which large carbohydrates are assembled and modified before use or before secretion from the cell.

gonad (GO-nad) Reproductive organ in which gametes are produced.

green revolution Term for attempts to improve crop production in developing countries by creating high-yield crop varieties, and by encouraging use of modern agricultural practices and equipment.

gymnosperm Plant in which seeds are produced on cone scales, without protective tissue around them. Conifers are examples.

habitat The physical environment, with its characteristic array of organisms, in which a species lives and reproduces.

haploid (HAP-loyd) State in which a cell contains only one of each type of chromosome characteristic of its species. Compare **diploid.**

Hardy–Weinberg rule A theoretical baseline for measuring change in allele frequency; the basic assumption is that if all individuals of a population have an equal chance of surviving and reproducing, and if no individuals leave or join the population, the allele frequency should remain constant from generation to generation.

heart Muscular organ that acts as a pump for circulating blood through the body.

hemoglobin (HEE-moe-GLOW-bin) An iron-containing protein that gives red blood cells their color; functions in oxygen transport.

herbivore Plant-eating animal.

heterospory In plants, the production of spores that are differentiated in size.

heterotroph (HET-er-oh-trofe) An organism that depends on substances produced by other organisms for its food source.

heterozygous (HET-er-oh-ZIE-guss) In sexually reproducing organisms, having nonidentical alleles for a given trait.

homeostasis (HOE-me-oh-STAY-sis) For cells and multicellular organisms, maintaining the constancy of internal conditions even when environmental conditions change.

home range Area occupied over long periods by one group or individual animal but not aggressively defended against others of the species. Compare **territory.**

hominid Any primate in the human family; *Homo sapiens* is the only living representative.

homologous chromosome One chromosome of a pair that have equivalent gene sequences. For any gene locus in that linear sequence, the pair may have identical or nonidentical alleles. (Typically the two chromosomes of a homologous pair are derived from two different parents, but exceptions do occur, as in the case of self-fertilizing plants).

homologous structure Body part constructed of the same kinds of materials according to the same basic, functional plan, yet occurring in entirely separate species.

homospory In some plants, the production of only one type of spore, rather than differentiated types.

homozygous (HOE-moe-ZIE-guss) In sexually reproducing organisms, having identical alleles for a given trait.

hormone Chemical substance that is produced by cells in one region of the body and that affects the functioning of specific cells or tissues in other regions.

hybrid Offspring of two parents that differ in one or more genetic traits.

hydrocarbon Substance composed only of hydrogen and carbon atoms.

hydrogen bond Chemical bond in which adjacent molecules share a hydrogen atom.

hydrolysis (high-DRAWL-ih-sis) Breaking of a molecular bond, in which a hydrogen atom derived from a water molecule becomes attached to one product, and the —OH group becomes attached to another product of the reaction.

hydrophilic Having an attraction for water molecules; refers to a polar substance that readily dissolves in water.

hydrophobic Repelled by water molecules; refers to a nonpolar substance that does not readily dissolve in water.

hypha, plural **hyphae** In true fungi, cellular filaments that grow underground into a matlike structure (mycelium).

hypothalamus Region of the brain governing metabolism of the entire body.

immune response Defense response by the body that has a specific target; usually involves antibody formation against a specific type of invading cell or substance.

indirect development In some animals, developmental program moving through a larval stage, followed by metamorphosis into the adult form.

inflammation General body response to tissue injury; involves redness, swelling, and pain.

insulin Animal hormone produced by beta islet cells in the pancreas. It is vital in glucose utilization; also promotes formation and storage of glycogen and fats.

intermediate neuron Type of neuron concentrated in the brain and spinal cord, between sensory neurons leading to and from the central nervous system.

interphase Time interval in which a cell is growing and maintaining itself but is not dividing; also during interphase, the hereditary material is replicated prior to division.

intrauterine device (in-tra-YOO-ter-in) IUD; a device inserted into the uterine lining as a way of preventing pregnancy.

ion, negatively charged Atom or group of atoms that has drawn one or more extra electrons into its domain.

ion, positively charged Atom or group of atoms that has given up one or more of its electrons to another atom or group of atoms.

isogamy Sexual reproductive strategy in which gametes of opposite sexes are identical in appearance.

isotope One of two or more forms of an element having the number of protons characteristic of that element but having a different number of neutrons.

kidney Organ concerned with regulating the volume of water leaving the body as urine, and the site of selective filtration, reabsorption, and elimination of solutes from the blood.

kinetic energy Energy of motion.

Krebs cycle Stage of cellular respiration in which pyruvate fragments are completely broken down into carbon dioxide; molecules reduced in the process can be used in ATP formation.

larva, plural **larvae** Immature form of an animal, which undergoes metamorphosis to the adult form.

larynx Organ of voice production that lies between the pharynx and trachea; made of cartilage and membranes.

lateral meristem Tissue that increases plant root or stem diameter as a result of the direction in which its cells divide.

learning An enduring potential for adapting future responses as a result of past experience.

lethal mutation A permanent, generally fatal change in a gene that is involved in some vital step in structural development.

leucoplast Colorless plastid that stores starch grains and other substances in a plant cell.

leukocyte (LOO-coe-site) A white blood cell.

light reactions First stage of photosynthesis, concerned with trapping light and using it as an energy source for ATP and/or $NADPH_2$ synthesis.

linkage mapping Plotting the position of genes relative to one another on a given chromosome.

lipid Hydrophobic molecule to which at least some hydrophilic atoms or atomic groups are attached; a lipid thus has both water-insoluble and water-soluble parts. Includes fats, waxes, steroids, and phospholipids.

liver Large gland concerned with glucose metabolism, interconversion of substances in the maintenance of blood sugar levels, and the secretion of substances such as bile for the digestion of fat molecules.

luteinizing hormone (LOO-ten-eyes-ing) LH; a pituitary hormone that stimulates development of the corpus luteum and its secretion of progesterone.

lymph node Small, bean-shaped body that functions as a filter and defense station against disease; part of the lymphatic system.

lymphocyte Specialized white blood cell that engulfs foreign particles such as bacterial cells invading the body.

lysosome Membrane-enclosed organelle containing hydrolytic enzymes, which may dispose of malfunctioning and worn-out cell parts and foreign particles.

mantle In molluscs, a body wall surrounding internal parts; secretes substances that form the molluscan shell.

medulla Generally, the innermost portion of an organ. Also, the inner region of the brain stem that governs automatic reflexes of breathing, heartbeat, blood pressure, sucking, and swallowing.

medusa (meh-DOO-sah) Free-swimming, bell-shaped stage in coelenterate life cycles.

meiosis (my-OH-sis). Two-stage nuclear division process in which a diploid number of chromosomes is halved; at the end of meiosis, the chromosome number in each daughter nucleus is haploid.

menopause End of the period of a human female's reproductive potential.

menstrual cycle Recurring physiological changes in the uterine lining, synchronized with the events of the ovarian cycle.

menstruation Periodic sloughing off of the blood-enriched lining of the uterus when pregnancy does not occur.

meristem In complex land plants, a tissue of undifferentiated cells that have the ability to undergo division.

mesoderm In an animal embryo, the middle tissue layer.

mesophyll An inner zone of photosynthetic parenchyma tissue within a leaf.

metabolic pathways Interconnected routes of chemical reactions inside living organisms, with the products of one reaction serving as starting material for the next reaction in the series.

metamorphosis In development, genetically programmed change in form of an organ or structure; also, drastic changes of a larval stage of an animal into the adult form.

metaphase In mitosis, the stage when microtubules form a bipolar structure (spindle apparatus) to which chromosomes become attached and by which they become lined up halfway between the two poles of the cell.

metastasis (met-uh-STAY-sis) In cancer, the ability of malignant cells to migrate throughout the body.

metazoan A multicellular animal.

microbody An organelle in which fats and some amino acids may be converted to other substances the cell requires.

microfilament Thin structural element composed of the protein actin; involved in contractile movements.

microtubule Small, hollow cylinder composed of protein filaments; takes part in many cellular events involving directed movement.

miscarriage Spontaneous expulsion of an embryo from the uterus before the third trimester of pregnancy.

mitochondrion, plural **mitochondria** (MY-toe-KON-dree-on) In eukaryotic cells, a membrane-enclosed organelle in which the main energy-extracting pathways of respiration occur.

mitosis (my-TOE-sis) Process by which the nucleus of a eukaryotic cell divides in two, so that each daughter nucleus ends up with a complete set of hereditary instructions. DNA replication precedes mitosis; actual cell division follows afterward at some point.

mitotic spindle During nuclear division, a microtubular structure that separates sister chromatids of a chromosome from each other.

molecule Two or more atoms linked by chemical bonds.

moneran Single-celled prokaryote; a bacterium or blue-green alga.

monocot Short for monocotyledon; a flowering plant in which only one cotyledon is found in the seed. Compare **dicot.**

monomer Small molecular subunit that may join with others to form a larger molecule, or polymer.

monosaccharide Carbohydrate made only of one sugar unit.

morphogenesis (more-foe-JEN-ih-sis) Process by which groups of similar cells become spatially coordinated, producing structures of genetically programmed shapes and patterns.

morula (MORE-yoo-lah) In some animals, an embryonic stage that contains about sixteen cells clustered into a sphere.

motor neuron A sensory neuron leading away from the central nervous system and carrying impulses to skeletal muscle.

mutation Any permanent change in the nucleotide sequence of a gene.

mutualism Type of symbiotic relationship from which both species benefit.

mycelium, plural **mycelia** In true fungi, a mat-like, often underground structure formed from hyphae.

mycoplasma (MY-coe-PLAZ-mah) One of the smallest organisms known; typically a disease agent lacking a cell wall and living in the moist tissues of animals.

myofibril Filamentous unit made of alternating bands of two proteins (myosin and actin); found in skeletal muscle.

NAD Nicotinamide adenine dinucleotide.

NADP Nicotinamide adenine dinucleotide phosphate.

nanometer A billionth of a meter; standard unit of measure in microscopy.

natural selection Result of selective agents in environment acting on phenotypes in ways that lead to differential reproduction of genotypes. Traits that are most adaptive in a given environment become increasingly represented among individuals of a population, for their bearers (which in some way have a better chance of surviving and reproducing) contribute proportionally more offspring to the next generation.

nephridium, plural **nephridia** In invertebrates such as annelids, an organ for excreting fluids and metabolic wastes.

nephron In the kidney, one of numerous functional units involved in filtration of the blood.

nerve In the peripheral nervous system, a functional unit made of nerve processes arranged in bundles and held together by connective tissue. Such nerves and their branchings transmit signals to and from distinct body regions. (A cutaneous nerve servicing blood vessels, muscles, and glands in the skin is an example.)

nerve impulse A series of rapid, cyclic alternations in the electric gradient across a nerve cell membrane; the action potentials transmitted along a neuron.

nerve net In some invertebrates, a network of cells that transmit nerve impulses but that have no control center for integrating neural activity.

nervous system In vertebrates, the brain, spinal cord, and nerves; the activity of which is integrated for communication among and control over the body parts.

net primary productivity For any community, the potential energy available from photosynthetically produced food, less the amount the autotrophic producers themselves require for growth and reproduction.

neuron The nerve cell of all complex multicellular animals.

neurotransmitter A chemical substance released from an activated neuron and that diffuses across a synapse to an adjacent neuron or muscle cell, thereby changing the membrane potential of the receiving cell. Acts as a chemical bridge by which a nerve impulse is transmitted from neuron to neuron, and from motor neuron to muscle cell.

neutron Subatomic particle of about the same size and mass as a proton but having no electric charge.

niche (nitch) A functional description of a species' role in a community; an expression of the range of all factors that influence whether a species has all the resources it needs, and whether it can carry out all the activities necessary for surviving and reproducing.

nicotinamide adenine dinucleotide NAD; a molecule that functions as an electron acceptor in oxidation–reduction reactions.

nicotinamide adenine dinucleotide phosphate NADP; a molecule that functions as an electron acceptor and carrier in several metabolic pathways, most notably photosynthesis. (The energy-rich molecule $NADPH_2$ is apparently produced when an NADP molecule combines with the two electrons, and the two H^+ ions released form a water molecule at the start of noncyclic photophosphorylation.)

nitrification Process by which certain bacteria in the soil convert ammonium ions (NH_4^+) to nitrite (NO_2^-), which other kinds of bacteria convert to nitrate (NO_3^-).

nitrogen fixation Incorporation of atmospheric nitrogen (N_2) into compounds such as NH_3, which can be cycled through a community.

noncyclic photophosphorylation Photosynthetic pathway in which new electrons derived from water molecules flow through two photosystems and two electron transport chains, the result being the formation of ATP and $NADPH_2$.

non-identical twins Individuals resulting from the fertilization of two different eggs by two different sperm.

notochord Somewhat flexible rod of cartilage; probable forerunner to the chordate endoskeleton. During embryonic development of complex vertebrates, it is replaced by a vertebral column.

nuclear envelope A double membrane surrounding a eukaryotic nucleus.

nucleic acid (new-CLAY-ik) Compound assembled by dehydration synthesis from nucleotides; DNA and RNA are examples.

nucleolus Within the nucleus, a dense mass in which ribosomes are assembled.

nucleoplasm Semifluid substance enclosed by the nuclear envelope.

nucleotide (NEW-klee-oh-tide) Compound containing carbon, hydrogen, oxygen, and nitrogen;

nucleotides are the subunits from which DNA and RNA are assembled.

nucleus In atoms, the central core of one or more positively charged protons and (in all but hydrogen) electrically neutral neutrons. In eukaryotic cells, the membrane-enclosed organelle that houses the chromosomes.

obligate relationship A total dependence of one species on another for survival.

oogamy (OO-AH-gam-ee) Sexual reproductive strategy in which gametes are differentiated in size and in motility. One gamete typically is small and motile (a sperm, for example); the other is larger and nonmotile (an egg).

operon Any set of control genes and structural genes operating as a unit.

organ Body unit composed of different tissues having a common function.

organelle Organized functional structure, usually enclosed in a membrane, within a cell's cytoplasm.

organic compound A carbon-containing substance.

osmosis (OSS-MOE-sis) Movement of water molecules across any differentially permeable membrane in response to a gradient in the concentration of solutes, and/or to a pressure gradient.

ovarian cycle Recurring events synchronized with the menstrual cycle; the ripening of an egg in a fluid-filled cavity inside the ovary, the transformation of the cavity (which ruptures when the egg is expelled) into an endocrine gland, and the concurrent production of ovarian hormones.

ovary In animals, the primary female reproductive organ in which eggs are formed.

oviduct Passageway through which eggs travel from the ovary to the uterus.

ovule In seed-bearing plants, an oval structure composed of the sporangium, female gametophyte, and tissue layers.

oxidation The loss of one or more electrons from an atom or molecule.

oxidation–reduction reaction Process by which an atom or molecule (an electron acceptor) picks up an extra electron from another atom or molecule; the electron acceptor thus becomes reduced and the donor becomes oxidized. In photosynthesis and respiration, the energy released during such electron transfers is used to phosphorylate ADP.

oxidative phosphorylation The use of electron energy being released during cellular respiration to phosphorylate (tack a phosphate group onto) ADP, thereby yielding energy-rich ATP molecules.

pancreas (PAN-kree-us) Organ that secretes digestive enzymes for breaking down proteins,

lipids, and carbohydrates in the duodenum; also secretes the hormones insulin and glucagon.

parallel evolution In two entirely separate lines of descent, the development of similar structures having similar functions as a result of the same kinds of selective pressures acting on individuals in each line.

parasitism An extreme form of symbiosis in which one species benefits at the expense of another.

parasympathetic nervous system The part of the autonomic nervous system that usually dominates muscles and organs during times that permit normal metabolic functioning, in contrast to times of stress, danger, excitement, or heightened awareness. Interacts with the sympathetic system to adjust the body's organs and organ systems to prevailing conditions.

parenchyma (peh-REN-kih-ma) In plants, a tissue of thin-walled cells.

passive transport Movement of a substance across a plasma membrane in response to a concentration gradient.

pathogen Disease-causing organism.

penis Male sex organ by which sperm are ejaculated into the female reproductive tract during sexual intercourse.

pepsin Digestive enzyme that splits the peptide bonds holding amino acids together in proteins.

peripheral nervous system (per-IF-ur-uhl) All the neurons leading to and from the spinal cord and brain. Here, processes of many individual neurons are combined into bundles called nerves.

peristalsis (pare-ih-STALL-sis) Rhythmic waves of muscular contraction that force food through the digestive tract.

phagocytosis "Cell-eating"; a cytoplasmic extension forms a vacuole around a food particle or foreign particle in the extracellular environment, then lysosomes fuse with and release digestive enzymes into the vacuole.

pharynx The throat region; gateway to the digestive tract and to the windpipe (the trachea of the respiratory system).

phenotype (FEE-no-type) The physical form and functional behavior that are expressions of an individual's genotype; arises from interactions between genes, and between genes and the environment.

pheromone (FARE-oh-moan) A chemical secretion by an organism that functions as a behavioral releaser for another organism of the same species.

phloem (FLOW-um) In vascular plants, a tissue for transporting the products of photosynthesis throughout the plant body.

phospholipid Lipid molecule assembled from glycerol, fatty acids, a phosphorus-containing

compound, and (usually) a nitrogen-containing base. Phospholipids are the foundation for cell membranes.

phosphorylation (FOSS-for-ih-LAY-shun) Addition of a phosphate group or groups to a molecule.

photoperiodism Responses to the relative length of light and darkness (day and night).

photoreceptors Light-sensitive sensory cells.

photosynthesis In most autotrophs, the trapping of solar energy and its conversion to chemical energy, which is used in manufacturing food molecules from carbon dioxide and water.

photosystem Functional light-trapping unit in photosynthetic membranes; contains pigment molecules and enzymes.

phototropism Growth in response to the direction of light.

pH scale A scale used in describing the concentration of H^+ in a solution. Concentrations of H^+ and OH^- are related, so that solutions having equal amounts of both ions are neutral; the pH is equal to 7. Those having more H^+ are acidic; the pH is less than 7. Those having more OH^- are basic; the pH is greater than 7.

phytoplankton Community of photosynthetic microorganisms in freshwater or saltwater environments.

pinocytosis "Cell-drinking"; the plasma membrane of a cell dimples inward around particles adhering to it and forms a vesicle around them, which moves into the cytoplasm.

pistil In flowering plants, the female reproductive organ; composed of an ovary, style, and stigma.

placenta (play-SEN-tuh) In the uterus, an organ made of extensions of the chorion and the endometrium. Through this composite of embryonic and maternal tissues and vessels, nutrients reach the embryo and wastes are carried away.

placental development Program in which the developing embryo is retained within the mother's body for a predictable time span; it attaches to the uterine lining and is nourished by the maternal bloodstream.

plant A multicellular organism able to build its own food molecules through photosynthesis.

plasma Liquid component of blood.

plasma membrane Outer membrane of a cell; the differentially permeable boundary layer between the cytoplasm and the external environment.

plastid In plant cells, a storage organelle; some plastids also function in photosynthesis.

platelet Component of blood that functions in clotting; contains thromboplastin, a substance necessary for coagulation of blood to occur.

pollen grain In gymnosperms and in flowering plants, the immature male gametophyte.

pollination Process whereby pollen grains are transported to the female reproductive structure of a plant.

pollutant Any naturally occurring or synthetic substance that accumulates to levels that are harmful to living things.

polymer Molecule formed from small monomer subunits.

polyp Vase-shaped, sedentary stage of coelenterate life cycles.

polypeptide chain A sequence of amino acids linked by dehydration synthesis.

polyribosome A number of ribosomes grouped together during protein synthesis; sometimes attached to inner cell membranes.

polysaccharide A carbohydrate consisting of at least three sugar monomers.

population A group of individuals that in some way is separated or isolated from other members of their species.

population growth The difference between the birth rate and death rate for a population, plus or minus any inward or outward migration.

potential energy Energy in a potentially usable form that is not, for the moment, being used.

predator A free-living organism that captures and feeds on other living organisms as a means of obtaining nutrients.

primary growth In vascular plant roots and stems, the cell division and differentiation into specialized tissues resulting from activity of the apical meristem.

primary structure For proteins, the sequence of amino acids that forms the protein backbone.

producer An autotrophic organism; usually a photosynthesizer that contributes to the net primary productivity of a community.

progesterone (pro-JESS-ter-OWN) Female sex hormone produced by the corpus luteum; helps prepare the uterus for pregnancy.

prokaryote (pro-CARRY-oht) Single-celled organism that has no membrane-enclosed nucleus; a bacterium or blue-green alga.

prophase In mitosis, the stage when chromatin coils up and becomes visible under the microscope as compact chromosome bodies; the nuclear envelope and nucleolus usually break down.

protein Molecule made of one or more chains of amino acids.

protistan Single-celled eukaryote that not only resembles the simpler monerans but that may show similarities to plants, animals, and/or fungi.

pseudopod (SOO-doe-pod) "False foot"; a non-permanent cytoplasmic projection from an amoeboid cell.

pulmonary circulation Pathways of blood flow leading to and from the lungs.

pulse pressure The difference between the maximum pressure exerted on arteries when blood is forced from the heart and the least pressure, which occurs just before blood is forced out. The difference is usually measured in terms of how much it can raise a column of mercury (Hg).

purine Nucleotide base having a double ring structure. Examples are adenine and guanine.

pyrimidine (pih-RIM-ih-deen) Nucleotide base having a single ring structure. Cytosine and thymine are examples.

pyruvate (PIE-roo-vate) A three-carbon compound produced during glycolysis (the initial breakdown of sugar molecules).

radial symmetry General arrangement of body parts in a regular pattern about a central axis, much like spokes of a bike wheel.

receptor One of many nerve cells and organs located throughout the animal body, each of which is specific to a certain kind of stimulus from the internal or external environment.

recessive allele In a diploid cell, an allele that is expressed only in the homozygous state (when both alleles for the trait are identical); in the heterozygous state, a recessive allele is masked by the expression of its dominant partner.

recombinant DNA New combinations of DNA produced in the laboratory by breaking apart and reassembling the DNA molecules from entirely different organisms.

reduction The gaining of one or more electrons by an atom or molecule.

reflex response One of the "simplest" pathways through the nervous system, involving receptor cells, sensory and intermediate neurons, and effectors.

releaser Any relatively simple environmental stimulus that triggers a fixed action pattern.

respiration In cells, the release of energy from food molecules, which is then used in metabolism. In whole animals, breathing; the taking in of oxygen from the environment and the release of carbon dioxide (a waste product of cellular metabolism).

resting membrane potential An actively maintained electric gradient across a nerve cell membrane when the nerve cell is not being stimulated into action.

ribonucleic acid (RYE-bow-new-CLAY-ik) RNA; a category of nucleic acids that are needed to translate the genetic message of DNA into actual protein structure. Built from nucleotide subunits containing the sugar ribose.

ribosome In both prokaryotic and eukaryotic cells, an organelle made of RNA and protein; the site of protein synthesis.

ritualized behavior Formal behavioral display that is an exaggeration of ordinary functional movements, but that is a clear signal laden with social meaning for members of the same species.

RNA Ribonucleic acid.

saprophyte (SAP-row-fight) Heterotrophic organism that feeds on nonliving organic matter.

sarcomere The fundamental unit of contraction in skeletal muscle.

sclerenchyma (skleh-REN-kih-ma) In vascular plants, a supportive or protective tissue of thick-walled cells.

scrotum (SKROE-tum) Pouch of skin containing the testes; located at the base of the penis.

secondary growth In vascular plants, an increase in stem or root diameter, made possible by lateral meristems, which produce additional xylem and phloem after primary tissue formation.

secondary principle of energy Left to itself, any system (the matter in a defined region), along with its surroundings (the rest of the universe), undergoes spontaneous energy conversions into less organized forms. Each time that happens, some energy gets randomly dispersed in a form that is not readily available to do work.

secondary structure For proteins, the twisting of the protein backbone about its own axis, largely as a result of hydrogen bonds that tend to form between certain groups along its length.

seed In gymnosperms and flowering plants, a structure consisting of an embryo surrounded by stored food reserves and contained within a protective coat.

semen The sperm-bearing fluid ejaculated from the penis during male orgasm.

semiconservative replication The manner in which a DNA molecule is reproduced; the formation of a complementary strand on each of the unzipped strands of a DNA double helix, the outcome being two "half-old, half-new" molecules.

sensory neuron A neuron that carries impulses from a receptor cell to the central nervous system; or one that carries impulses from the central nervous system to the body's muscles or glands.

sepal A modified leaf that encloses the developing floral bud; a flower bud scale.

sex chromosomes Those chromosomes that differ either in number or in kind between the male and female of a species. All other chromosomes are called autosomes.

sexual reproduction Formation of a new diploid individual by the fusion of gametes, which are usually derived from two different parent organisms.

sieve element In phloem, an elongated conducting cell.

social behavior Interactions among members of a species that are the outcome of dispersive and cohesive behaviors.

solute (SOL-yoot) A substance dissolved in some solution. A mixture in which molecules of two or more substances are uniformly dispersed. For example, a mixture in which water molecules are uniformly dispersed among and hydrogen-bonded to atoms or molecules of different substances.

solvent A fluid in which one or more substances are dissolved.

somatic nervous system The motor pathways and controls involved in reflex responses; this system potentially is under voluntary control.

speciation Process whereby separate populations of a species become adapted to different environments and, given enough time, accumulate so many differences in form and behavior that their members no longer can interbreed even if brought together again.

species All organisms of a given type that are at least potentially capable of interbreeding in the natural environment. (This definition applies to sexually reproducing organisms; biologists are still trying to formulate a good definition that applies to asexually reproducing organisms as well.)

sphincter Ring of smooth muscle that helps control the flow of substances through the body's passageways; works through alternating contraction and relaxation.

spindle Structure assembled from microtubules; the physical basis for movement of chromosomes during mitosis and meiosis.

spirillum, plural **spirilla** (spih-RILL-um) Spiral form of bacterium.

spleen Organ of the lymphatic system that is a site of lymphocyte production, and of degradation of worn-out or damaged red blood cells. Also stores blood in reserve, which is released when blood pressure drops (for example, during hemorrhaging or when blood vessels open wide in times of stress).

sporangium, plural **sporangia** In plants, a cell produced by the sporophyte generation and that divides to produce the gametophyte. Often enters a resting stage that is resistant to adverse environmental conditions.

sporophyte The diploid, spore-producing stage in plant life cycles.

stabilizing selection In an environment that remains fairly constant, a process whereby there is selection for an existing species form that has proved most adaptive to that environment; at the same time, there is selection against variant forms. Hence the existing species form is maintained.

stamen In flowering plants, the male reproductive organ.

steroid Lipid consisting of multiple carbon ring structures to which various atoms may be attached.

stimulus Any detected change in an organism's external environment or within its body. Every stimulus is a change in some form of energy (heat, sound wave, chemical, light energy).

stoma (more popularly, **stomate**) A small passageway through which carbon dioxide, essential for photosynthesis, enters a plant leaf. Each stomate is bounded by guard cells that act to increase or decrease the size of the opening, thereby regulating the rate of water loss from the leaf.

structural genes Genes that code for the structure of proteins.

substrate Molecule or molecules of a reactant on which an enzyme acts.

succession A sequential replacement of communities until a stable, self-perpetuating array of producers, consumers, and decomposers is reached; this stable array is the climax stage typical of a given region.

symbiosis A form of attachment between individuals of two or more different species, in which at least one party benefits.

sympathetic nervous system The part of the autonomic nervous system that generally dominates internal organs and events during times of stress, danger, and excitement; prepares the body for rapid response to change.

synapse A junction where the axon of one neuron terminates near the cell body or dendrites of another neuron, or where the axon of a motor neuron terminates near a muscle cell membrane.

synapsis In meiosis, the alignment of the two sister chromatids of one chromosome with the two sister chromatids of its homologue.

systemic circulation Pathways of blood flow leading to and from all body parts except the lungs.

telophase Final stage of mitosis, during which the separated chromosomes unwind and become enclosed within newly forming nuclear membrane.

territory For some animal groups, a more-or-less geographically definable area to which behavioral expressions of dominance may become attached.

tertiary structure For proteins, the asymmetrical bending and folding of the protein backbone in space as it assumes its final shape.

testis, plural **testes** Male organ in which male gametes and sex hormones are produced.

testosterone Hormone involved in sperm production, in maintaining normal sexuality, and in growth and maintenance of male secondary sex traits.

thalamus In the brain, a region in which impulses from all major nerves converge before being sent on to appropriate regions in the cerebrum; the crossroads for the brain.

threshold value The minimal intensity of chemical stimulation that is required to activate a neuron.

thymine Nitrogen-containing base in nucleic acids.

tidal ventilation Alternating expansion and compression of lungs, which creates rhythmic changes in the pressure gradient between the lungs and the atmosphere, and thereby helps assure constant gas exchange.

tissue Permanent grouping of cells having a similar function.

trachea, plural **tracheae** A tube for breathing; in land vertebrates, the "windpipe" that carries air from the larynx to bronchi on the way to (and from) the lungs.

tracheid A long, tapered conducting cell in plant xylem.

transcription Assembly of an RNA molecule on some region of a DNA molecule (which has been temporarily unwound for the process to occur). The RNA has a base sequence that is complementary to the DNA region on which it is assembled.

transduction The transfer of genes from one cell to another by a bacteriophage.

transformation In genetics, the transfer of "naked" DNA fragments derived from one cell to a recipient cell, which incorporates the fragments into its own genetic material.

translation Process by which the genetic message in an mRNA molecule dictates the precise order of amino acids assembled into a given type of protein.

transpiration Evaporative water loss from moist mesophyll cells in leaves.

trophic levels For any community, an interlocked array of producer, consumer, and decomposer species through which materials and energy flow.

tumor, benign A mass resulting from breakdown of cellular controls, such that cells grow and divide faster than they otherwise would.

tumor, malignant Cancerous growth. Malignant cells have lost surface recognition factors, hence their identity. They can migrate to various body regions and multiply into secondary tumors.

ureter A duct that transports urine from the kidney to the bladder.

urethra A duct that leads from the bladder to the exterior.

uterus In some mammalian females, a chamber in which the developing embryo is contained and nurtured during pregnancy.

vacuole In plant cells, a membrane-enclosed chamber containing solutions of mineral ions and other materials. In animal cells, a membrane-enclosed chamber active in material transport.

vagina Female organ that receives sperm from the male penis; forms part of the birth canal, and acts as a channel to the exterior for uterine secretions and menstrual flow.

vascular cambium In vascular plants, a lateral meristem whose organization and direction of cell division increases stem or root diameter.

vascular tissue Internal conducting tissue for fluids and nutrients.

vein A vessel that carries blood back to the heart.

vertebrae A series of hard bones that form the backbone in chordates.

vertebrate An animal with a backbone made of bony segments called vertebrae.

vesicle In plant and animal cells, a small, membrane-enclosed sac in which various substances are stored.

vestigial structure Body part that no longer has any apparent role in the functioning of an organism.

virus An infectious particle consisting of nucleic acid encased in protein; incapable of metabolism or reproduction without a host cell, hence is often not considered alive.

vitamin One of a class of organic molecules that an organism requires to help build essential compounds that it cannot produce itself.

vulva Female external genitals.

white matter Myelin sheaths that surround axons of neurons in the spinal cord.

xylem (ZEYE-lem) In vascular plants, a tissue for transporting water and minerals through the plant body.

zygote (ZEYE-goat) A diploid cell resulting from the fusion of male and female gametes.

Index